SIGNAL RESILIENT TO INTERPOLATION:

AN EXPLORATION ON THE APPROXIMATION PROPERTIES OF THE MATHEMATICAL FUNCTIONS

CARLO CIULLA

Signal Resilient to Interpolation:
An Exploration on the Approximation Properties of the Mathematical Functions

ISBN-10: 1-477-56748-8
ISBN-13: 978-1-477-56748-7

CONTENTS

CHAPTER 10. CLASSIC-RESILIENT-CURVATURE HYBRID INTERPOLATION: TRIVARIATE POLYNOMIALS WITH AND WITHOUT EMBEDDING   228

CHAPTER 11. RESILIENT INTERPOLATION: AN INVESTIGATION ON THE DIMENSIONALITY OF THE POLYNOMIALS, THE MATHEMATICS   262

OVERVIEW

The purpose of this overview is to convey the intent of the author as regards what he has written in this book. It is aimed at a readership composed of undergraduate and graduate students. The intent is that of fostering creativity while preparing the students in engineering related disciplines within the domain of signal (image) processing, applied mathematics, biomedical imaging, biomedical engineering, imaging technology, and more generally applied computational engineering.

Resilient interpolation is a concept illustrated earlier in Ciulla (2009). The concept has its theoretical foundations in the intensity-curvature terms of a given signal, said terms being employed to improve the approximation properties of the interpolation functions. The intensity-curvature terms can be calculated for mathematical functions that admit second order differentiability, which would therefore be continuous and admit non-null second order derivatives.

The calculation of the intensity-curvature terms is related to the discrete sample of signal values and is also related to the analog/continuous transformation necessary to the function to become an interpolator. Such calculation is undertaken within the entire spatial extent of the sampling location, and as shown in Ciulla (2009), leads to the measurement of the energy level change determined through the interpolation function which is called: *the Intensity-Curvature Functional*. The idea is therefore to equate the intensity-curvature term calculated with the signal (image) through the given mathematical function in two conditions: (i) the given signal as it has been sampled and (ii) the signal calculated at locations where is unknown because of the limitations of the sampling instrument.

Through mathematical developments that illustrate concepts of algebra and calculus, the equation of the two intensity-curvature terms furnish the instrument apt for deriving a new signal at the desired time or space locations. This new signal is dependent on the given model function and is called *Signal Resilient to Interpolation (SRI)*.

The book includes illustrations of the math processes, logical reasoning, and results (of quantitative and qualitative nature) which show how to generate the *Signal Resilient to Interpolation*. Within the book, the *Signal Resilient to Interpolation* is derived on the basis of quadratic and cubic polynomials in two and three dimensions, embedding and not embedding the pixel (voxel in three dimensions) to be re-sampled.

Once the signal is modelled through an interpolator, it is possible to calculate the second order derivatives, and thus the curvature of the model interpolator which is nonetheless the modelled representation of the curvature of the signal. Geometrically, the curvature of the signal is the tangent to the first order derivative curve of the signal. There are two types of curvature treated in this book and they are: (i) *classic curvature* and (ii) *resilient curvature*. Through the use of the curvature, the book introduces a novel conception of the Pixel Intensity Correction (PIC) and uses this new concept to derive transfer functions obtained deterministically from the model interpolation function.

The transfer function of the model interpolation function is that function necessary to scale the convolution made through the interpolator, it is thus the function which the interpolator needs in order to convolve the signal and to obtain the estimate of the signal at unknown locations. The transfer function employed to process the entire convolution can be of arbitrary choice in the interpolation process, it can also be the Z-transform employed to calculate the B-Spline (or more generally polynomials) coefficients (Unser et al. 1993a, 1993b; Ramani et al. 2010). Moreover, like wise in this book, the transfer function to process the entire polynomial convolution can be obtained deterministically from the polynomial form of the interpolator. Generally, because of the arbitration, bias is introduced in the quantification of the interpolation error and also affects signal reconstruction. It is certainly useful to introduce a novel methodology free from arbitration, deterministic, without introduced bias and which is capable of devising the transfer function directly from the math form of the interpolator.

In this book the PIC derived from the math form of the interpolator replaces the transfer function and it does that through the use of the curvature.

Given that there are two types of curvature (classic and resilient), the PIC is calculated with the two of them, leading to two types of signal reconstruction which are called: *classic-curvature* and *resilient-curvature* interpolation formulae. They are unaffected by bias and they include the transfer function which is necessary in order to convolve (scale) the signal and to release the estimate at the given shift (misplacement).

Neither of the two formulae: *classic-curvature interpolation* and *resilient-curvature interpolation* can overpower the other one in an absolute sense within the full set of discrete samples of which the signal is constituted of. Therefore, a hybrid formula called *classic-resilient-curvature interpolation formula*, which combines the two formulae (classic-curvature and resilient-curvature) on a node-to-node basis (1D), or pixel-to-pixel basis (2D), or voxel-to-voxel basis (3D), makes a further improvement in interpolating the signal at the location (shift, or misplacement, both have the same meaning) where is to be reconstructed.

Validation of the three novel interpolation functions: *classic-curvature*, *resilient-curvature* and *classic-resilient-curvature* formulae, is conducted with synthetic data and real Magnetic Resonance Imaging (MRI) data. The validation process is of a quantitative nature and a qualitative nature. Graphs showing the potentiality of the new interpolation functions in terms of error improvement and also figures of the signal resilient to interpolation accompanied with curvature maps of the original images are reported in the chapters of the book devoted to the presentation of the results.

Chapters 1, 4, 8, 10 and 12 reports on the exploration and the results obtained by building up a special theory for the improvement of the interpolation error, which evolves from the previously reported unifying theory.

Chapters 2, 3, 5, 6, 7, 9 and 11 report on the math developments undertaken to deduce the three new classes of interpolation functions and they give a detailed treatise on the math of the special theory.

The beneficial scope of the book might be considered to be that of: (i) introducing students to applied mathematics through the illustration of basic concepts in algebra and calculus, (ii) devising innovative methods aimed at generating signal estimation and (iii) increasing students' potentiality in creativity. Also, introducing students to the methodology employed to validate the proposed exploration, while at the same time fostering creativity and developing students' skills and performance in computing related disciplines.

A keynote of the book is that the students are given the possibility to expand and to produce knowledge that goes beyond what is currently known and also to realize how to develop research.

REFERENCES

Ciulla, C. (2009). *Improved Signal and Image Interpolation in Biomedical Applications: The Case of Magnetic Resonance Imaging (MRI)* – Medical Information Science Reference - IGI Global Publisher, Hershey, PA, U.S.A.

Ramani, S., Thevenaz, P. & Unser, M. (2010). *Regularized Interpolation for Noisy Images.* IEEE Transactions on Medical Imaging. 29(2), 543-558.

Unser, M., Aldroubi, A., & Eden, M. (1993a). *B-spline signal processing: Part I – theory.* IEEE Transactions on Signal Processing, 41(2), 821-833.

Unser, M., Aldroubi, A., & Eden, M. (1993b). *B-spline signal processing: Part II – efficient design and applications.* IEEE Transactions on Signal Processing, 41(2), 834-848.

CHAPTER 1

THE PHILOSOPHY OF THOUGHT

INTRODUCTION

This book treats of signal interpolation, focusing specifically on a special theory on the improvement of the approximation properties of mathematical functions. It is aimed at a readership of undergraduate and graduate students in colleges and universities. Readers will learn that given an interpolation function (interpolator), we can formulate a special theory which derives the transfer function directly from the math form of the interpolator, avoiding: (i) the bias in the approximation of which the interpolation error is measure and (ii) the arbitration in the choice of the math form of the transfer function. The transfer function is the function obtained from the interpolation function in order to convolve the signal with misplacements and to obtain the estimate of the signal at unknown locations. During this process, two new classes of interpolation functions are derived from the original interpolator, and one more interpolation function is also derived which combines the two of them on a pixel-to-pixel basis.

The tree of knowledge shown in figure 1(a) allows us to move from theory to practice and also gives support to the author in connecting up with the rationale of the developmental effort made in order to produce the math development of the signal resilient to interpolation. The subsequent developments in deriving new classes of interpolation functions outlined in this book connects up with another main finding, which is that of calculating customized transfer functions to any reconstruction techniques which make use of interpolation formulae. The tree of knowledge gives support to the exploration of the approximation properties of the mathematical functions.

The starting point is the scattered knowledge. There are two questions to ask, namely: *(i) what is the scattered knowledge?* And, should a special theory be deduced *(ii) what is the statement which we would expect to make as a summary when researching, debating, and experimenting on the approximate nature of the mathematical functions?* The very basic scattered knowledge consists in the fact that the curvature of the mathematical function provides us with a means to improve the approximation properties of the mathematical (interpolation) functions. The final stage is strictly related to the improvement of the interpolation error. Thus, the statement of the special theory which we would expect to make as a summary is: *'it is possible, through the use of the curvature, to improve the estimation of the value not existing in the original sampled signal, and such improvement benefits the approximation properties of the mathematical functions.*

Some definitions are in order. The pixel intensity correction (PIC) is the 'fraction of pixel intensity to be added to the pixel intensity found at the location of signal reconstruction to obtain the estimate of the true unknown value of the signal'. The total curvature of an interpolation function is defined to be the sum of second order derivatives of its math form respect to the dimensional variables. The classic-curvature signal reconstruction (interpolation) formula is: the sum of the pixel intensity found at the location of signal reconstruction plus the PIC: $I(x) = f(0) + PIC(x)$ in 1D, $I(x, y) = f(0, 0) + PIC(x, y)$ in 2D and $I(x, y, z) = f(0, 0, 0) + PIC(x, y, z)$ in 3D. The resilient-curvature signal reconstruction (interpolation) formula is: the sum of the pixel intensity found at the location of signal reconstruction plus the PIC: $R(x) = f(0) + PIC(x)$ in 1D, $R(x, y) = f(0, 0) + PIC(x, y)$ in 2D and $R(x, y, z) = f(0, 0, 0) + PIC(x, y, z)$ in 3D. What is the difference between classic-curvature and resilient-curvature formulae? The difference consists in the fact that the PIC is calculated from the total curvature of the classic interpolation formula and the resilient interpolation formulae respectively. Therefore the two values of PIC are not the same (as shall be seen in the math developments), because the total curvature of the classic interpolation formula is not the same as the total curvature of the resilient interpolation formula.

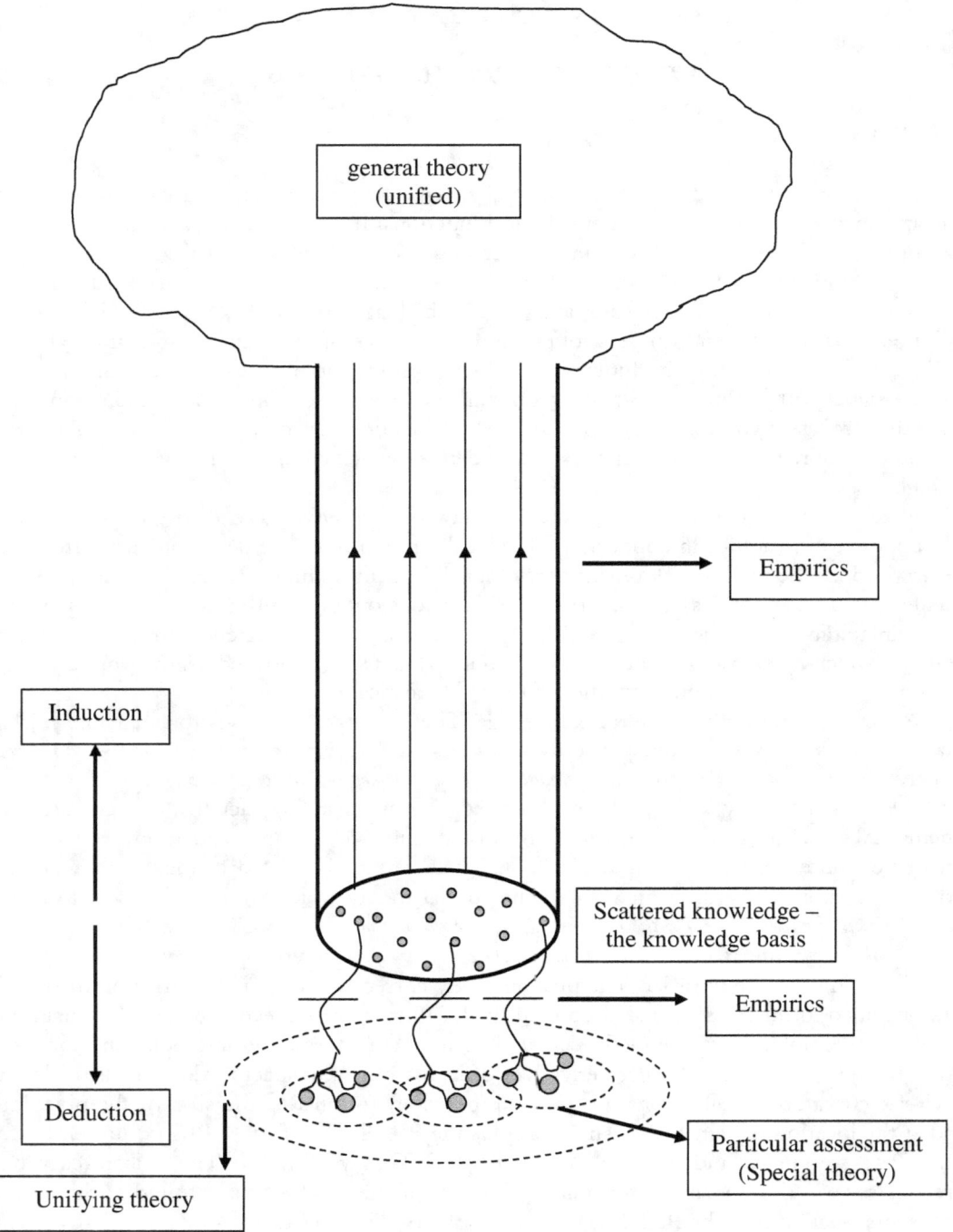

Figure 1(a): The tree of knowledge, from theory to practice.

The classic-resilient-curvature hybrid signal reconstruction (interpolation) formula is defined on a pixel-by-pixel basis depending on which one of the two: I (x) or R(x), in 1D, I (x, y) or R(x, y), in 2D,

I (x, y, z) or R(x, y, z), in 3D, gives the smaller Mean Absolute Error (MAE). The MAE is the measurement chosen in this book to measure the interpolation error.

The transfer function of the interpolation formula is the mathematical function chosen to process the convolution of signal and misplacement. It is usually arbitrary in its math form and it does produce the effect of scaling the convolution such that its output becomes a fraction of pixel intensity to be employed as pixel intensity correction. This way, the pixel intensity changes, so as to produce the estimation which takes place through the signal reconstruction technique.

Thus the knowledge basis triggers the deduction of the pixel intensity correction in order to obtain low interpolation error; this also takes place in relation to the process of scaling the convolution. The knowledge basis also devises the transfer function, which removes the arbitration in the choice and the consequential bias induced in the interpolation error and thus in the signal reconstruction technique.

The capability of the special theory in determining the transfer function from the math form of the interpolator is new and not found in literature in similar available books (De Boor, 1986; Schoenberg, 1987; Kalton, 1988; Bennett, 1988; Szabados, & Vértesi, 1990; Krugljak, 1991; Gohberg, 1992; Agarwal & Wong, 1993; Späth, 1995; Dym et al., 1997; Foias, 1998; Gohberg, 1998; Lorentz, et al., 1984; McLeod, R. J. Y. & Baart, M. L. 1998; Stein, 1999; Bercovici & Foias 2000; Agler & McCarthy 2002; Dym, et al., 2002; Krein, et al., 2002; Peetre, J. & Cwikel, M. 2002; Tartar, L. 2002; McArtney Phillips, 2003; Shi, 2003; Seip, 2004; Alpay & Gohberg 2006; Bolotnikov & Dym 2006; Jetter, et al., 2006; Cwikel, et al. 2007; Popescu, 2006; Astola, 2007; Mastroianni, & Milovanović 2008; Bannore, 2009; Fraser, 2009; Buzzi-Ferraris & Manenti 2010; Goodearl, 2010).

What is, then, the logical path to undertake from the scattered knowledge in order to arrive at the statement which supports the special theory? What is, then, embedded in the quest to research, to debate and to experiment? Answering these two questions would clarify the conceptual framework underlying what one is making an effort to develop in this book. The logical path undertaken in this book has its foundation in removing the arbitration in the choice of the transfer function through the idea which has taken shape and which may be stated thus: *it is possible to formulate the transfer function of the given interpolation formula.* How is this possible? In the quest to research, debate and experiment *this is possible through the novel conceptualization of the pixel intensity correction brought to light in this book.*

BACKGROUND AND RELATED LITERATURE

The Error Measure and the Energy Functions Employed in this Research

The energy functions (intensity-curvature terms) employed in this research are called intensity-curvature term before interpolation (E_0) and intensity-curvature term after interpolation (E_{IN}).

There is a conceptual correlation between interpolation carried out with the classic-curvature formulae or the resilient-curvature formulae or the classic-resilient curvature formulae (hybrid), and the validation paradigm, which through the use of the Mean Absolute Error (MAE), presents the idea of energy preservation in the quest for the minimization of the change of energy level of the signal as measured through the ratio between E_0 and E_{IN}, which is the Intensity-Curvature Functional (Ciulla, 2009).

The idea behind the choice of the intensity-curvature term is that of combining the information content of both signal intensity and second order derivative of the interpolation function.

The rationale of the intensity-curvature term is driven by the need of elucidating the true behavior of the signal intensity in relationship to the neighboring signal intensity values for what regards the local concavity or local convexity of the model interpolator. Clearly, only the numerical value of the signal intensity is not sufficient to clarify the local behavior of the signal when modeled with the interpolation function, thus the information content provided through the curvature (calculated as the sum of all the second order derivatives) completes the true picture on the behavior of the signal in relationships to the neighboring pixel intensities.

The reason for the choice of the intensity-curvature terms is the lack in literature of an energy measure capable to combine signal intensity with curvature (convexity or concavity) and the advantage consists in that the intensity-curvature terms allow to bring in the possibility to add information about the true and unknown (not sampled) signal value which is being estimated through the model interpolation function. The added value is brought in through the curvature as measured through the second order derivatives.

Literature on Error Measures, Energy Functions, Non-Uniform Sampling and the Use of the Derivatives

This section address the review of the signal-image interpolation processing literature as far as regards interpolation error measures, energy functions, non-uniform sampling and the use of the derivatives of the interpolation function. Energy functions, intended to measure the interpolation error, such as the Normalized Mean Square Error (NMSE) have been earlier devised in: Wang et al., (1992), or the Root Mean Square Error (RME) in: Frijns et al., (2000), Fraser, (1989), or on the basis of Least Square Techniques in: Unser & Daubechies (1997), or the Mean Square Error (MSE) found in: Wang (2001).

Within the context of non-uniform re-sampling, error measures, used as energy to minimize in order to reduce the interpolation error, have been constructed on the basis of derivatives of the interpolation function (B-Splines in the specifics, De Boor, (1978)). Within the context of contour detection though B-Spline (Unser, et al. 1993a; Unser, et al. 1993b) snakes an energy function formulation based on the Gaussian-smoothed gradient of the image was presented (Brigger et al., 2000). Emphasis on non-rigid (non-uniform) spacing of the re-sampling points has been given (Amidror, 2002) and also it has been object of study within the context of the linear interpolation function while using the Normalized Mutual Information as energy measure and B-Splines as optimization strategy (Penney et al., 2004). Within the context of the design of univariate interpolation functions, a method was reported that computes the coefficient of the third order (also called degree (Meijering, 2002)) polynomial interpolator on the basis of the first order derivatives (Akima, 1991). Integration of first order derivatives has been also object of study at the aim to estimate the interpolation error (Waldron, 2000).

As far as regards the use of second order derivatives Appledorn, (1996) had devised a novel set of interpolation kernels which made use of the derivatives in order to keep the Fourier Transform of the interpolation kernel as flat as possible such to minimize the change in Fourier properties of the resulting signal after interpolation. Error bounds for linear interpolation which are dependent on the second order derivative of the interpolation function have been reported (Waldron, 1998) and such math expressions can be used to derive more general expressions of error bounds that are dependent on the step size and that also apply to higher order interpolation functions such as the Lagrange. This work describes the interpolation error as the integral of: the polynomial multiplied by its second order derivative; which is the Kowalewski reminder (Kowalewski, 1932), and also creates a connection with Ciulla, (2009) where the multiplicative term of the polynomial multiplied by its second order derivative is calculated before (E_o) and after interpolation (E_{IN}) and is included into the Intensity-Curvature Functional (ΔE), which is minimized through the search for the Sub-pixel Efficacy Region (SRE).

RESILIENT INTERPOLATION: THE CONCEPTUALIZATION

The concept of resilient interpolation is closely linked to that of the equation between the intensity curvature terms before (no interpolation) and after interpolation.

This concept is applicable to any interpolation formula (given its second order differentiability), when convolving signals with misplacements and processing the convolution with a transfer function chosen arbitrarily. Such signal-image processing is thus avoided, giving support to the assertion that states: *'from the total curvature of the signal it is possible to formulate the math of the transfer function'*.

Nevertheless, removing the approximation in its absolute form is unlikely to happen, thus the resilient signal is another way of approximating the signal, reconstructing the signal on the basis of the equation of a non interpolated signal (intensity-curvature term E_0 before interpolation) with an interpolated signal (intensity curvature term E_{IN} after interpolation).

It shall be seen in the developmental chapters, which are mainly concerned with presenting the math of this book, that the starting point and the scattered knowledge recalls the equation between the intensity curvature terms before and after interpolation. Thus, the resilient signal is derived from the aforementioned equation, and this happens partially (but not absolutely) regardless of the degree and/or dimensionality of the interpolator. The classic-curvature signal is also derived as scattered knowledge. From the whole of the scattered knowledge, additional math developments converge to create three new classes of interpolation formulae: classic-curvature, resilient-curvature and classic-resilient-curvature hybrid.

The degree of the interpolation function is the maximum exponent of the formula (Meijering, 2002). As far as regards the math deduction of the transfer function derived from the total curvature of the interpolator, the specificity of the generalization inside the special theory allows one to disregard the degree as long as it is 2 in two spatial dimensions and 3 in three dimensional formulae. Overall, the special theory is fully applicable in three dimensions in space to interpolators of the third degree or above, and is applicable in one, two and three dimensions in space to interpolators of the second degree or above.

For a review on signal-image interpolation techniques and procedures, the reader may refer to previous works (Meijering, 2002; Ciulla, 2009) since such a task is outside the scope of the present book.

FROM SCATTERED KNOWLEDGE THROUGH EMPIRICISM, TO EXPLORE THE POSSIBILITY OF DEDUCING A SPECIAL THEORY

In this section, the author refers to the tree of knowledge and reports on the possibility of deducing a special theory. The branch of the unifying theory called resilient interpolation reported in Ciulla (2009) is in its essence an example of the aforementioned possibility. This book undertakes an exploration with the explicit purpose of deriving a special theory which at its largest extension is the unifying theory. The connection to the aforementioned book is made evident in those sections of the book, published in the year 2009, which treated resilient interpolation and which met with a favorable reception on the part of the readership. Bearing in mind the assertion revealing the purposes of the present book in relation to the previous one, the connection between the two is being made *in that the present book constitutes an exploration which is geared to go further into the specifics of resilient interpolation*, thus making progress in the direction of the development of the special theory.

The starting point of this book is the scattered knowledge provided in order to reinforce the special theory, in relation to the concept that the curvature of the interpolator can be used to improve the approximation properties of signal-image interpolation. Such a notion has an additional conceptual connotation in that *of devising the transfer function (Pixel Intensity Correction) of an interpolator from the total curvature.*

More specifically, the scattered knowledge is constituted of: (i) the curvature of the model interpolation function, (ii) the classic-curvature signal and (iii) the Pixel Intensity Correction (PIC), which allows one to derive, through the total curvature, *the transfer function as customized to the interpolator and thus to remove the arbitration in choosing a transfer function to process the convolution of the interpolation formula.*

It shall be seen in chapters 4, 8, 10 and 12 that one of the main achievements of this exploratory research is that of producing three different classes of interpolators: classic-curvature formulation based, resilient-curvature formulation based, and classic-resilient-curvature hybrid formulation based and this will be shown to happen through the induced process which starts from the knowledge that the curvature can improve the approximation properties of the interpolator and at the same time can derive transfer functions without arbitration. The transfer function is instead systematically derived from the math form of the interpolator.

THE CURVATURE

Normally, calculus and math books provide the following notion about the curvature and the differential of a mathematical function such for instance the book by Stewart (2009).

The curvature is a local property of a mathematical function f(x) (curve) defined to be the Limit of the ratio between the two increments $\Delta\alpha$ and Δs (Lim ($\Delta\alpha$ / Δs) = (dα / ds); with $\Delta s \to 0$). The increment at the numerator is that one of the angle subtended by the direction of the two tangents to the curve at the locations x_m and x_n (which are spatial points located on the arc of the curve, with the arc of the curve beginning at x_m and ending at x_n). The increment at the numerator is thus an angle. The increment at the denominator is the length of the arc in between x_m and x_n.

The differential is also a local property of the mathematical function (curve) defined as the linear increment of the mathematical function y = f(x), versus the increment (Δx = dx) of the independent variable x. The differential of the mathematical function is calculated as the multiplication of: the derivative of the mathematical function and the increment of the independent variable. Therefore, given an arc of the mathematical function in between the locations x_m and x_n (with the arc of the curve beginning at x_m and ending at x_n), the linear increment called differential (dy) of the mathematical function, is measured on the tangent to the curve at the location x_m. Such linear increment is calculated through the formula dy = (∂ f(x) / ∂x) * dx.

The second order differential of the mathematical function relates to the concept of curvature because the curve is the first order derivative of the mathematical function, and still the linear increment of the curve is measured on the tangent to the curve (the first order derivative in such case) at the location x_m where the local property called curvature is defined.

In summary, *the curvature of a mathematical function, which allows not null second order derivative, is an arc tangent (arctg, arctag) of an angle.* Let us call the arc tangent of the angle as atan (θ), where θ is the angle.

The length of an arc has its unit in radians. Thus atan (θ) has its unit in radians and both atan (θ) and 2π share the same unit. Practically, the curvature atan (θ) is admissible in the range [0, $\pm(\pi/2)$ - ε], where $\varepsilon > 0$ when $\theta = (\pi/2)$ - ε and $\varepsilon < 0$ when $\theta = -(\pi/2) - \varepsilon$, is a very small fraction of radians. The reason is two folded. One is that $\theta = 0$ implies that the tangent to the first order derivative is horizontal, thus the second order derivative is zero and this is admissible. The other one is that to admit the tangent to the first order derivative making a ninety degree angle with the horizontal is like admitting a discontinuity point in the math function chosen to model the signal and this is not admissible. Therefore ε needs to be different from zero.

TECHNICAL CHALLENGE

The technical challenge is that the total curvature (sum of second order derivatives of the interpolation function respect to the dimensional variables) has the unit not radians and this means that it takes some processing in order to input the total curvature to the trigonometric function tangent. The unit of the total curvature is $[mm * I]^k$ where mm is millimeters, I is the unit of pixel intensity and k (integer number) is the exponent.

One solution provided through the exploratory research treated in this book is that of calculating the Pixel Intensity Correction (PIC) with the formula $PIC = \Delta x * \tan([mm * I]^k / 2\pi)$. Δx is the increment of the independent variable (either in 1D, 2D or 3D); $[mm * I]^k$ is the unit of the total curvature; 2π is the length of the circle measured in radians (rad).

An alternative provided through this exploratory research is the calculation of the Pixel Intensity Correction with the formula $PIC = \Delta x * \tan([mm * I]^k * 2\pi)$. In such case the input is $[mm * I]^k * 2\pi$, which unit is $[mm * I]^k * (rad)$. It is not exact to input neither of: the ratio $[mm * I]^k / 2\pi$, nor the factor $[mm * I]^k * 2\pi$ into the math function tangent (tan).

Let us call S_{scaled} the segment provided with the transfer function of the model interpolation function (which is that function necessary to scale the convolution made through the interpolator) Let us call S the convolution which unit is $\Sigma_i [mm^i * I]$ where $i = 1...n$ with n the maximum exponent of the polynomial form defining the convolution, and I is the signal intensity.

Another solution has been that one of scaling the convolution S with the following formulae: (i) $S_{scaled} = \cos(2\pi * S)$; (ii) $S_{scaled} = \sin(2\pi * S)$; (iii) $S_{scaled} = (1.0 / (1.0 + e^{-S}))$; (iv) $S_{scaled} = \log(|1.0 / S|)$. In such cases the transfer functions are: the cosine function (cos); the sine function (sin); the function $(1.0 / (1.0 + e^{-S}))$, where e is the exponential; and the function $\log(|1.0 / S|)$ where log is the natural logarithm respectively. The unit of: $[2\pi * S]$, is $(rad * \Sigma_i [mm^i * I])$; the unit of: $[1.0 / (1.0 + e^{-S})]$, is $(e^{- (\Sigma_i [mm^i * I])})^{-1}$; and the unit of: $[\log(|1.0 / S|)]$, is $(\log(|1.0 / \Sigma_i [mm^i * I]|))$. In both of the cases: (i) $S_{scaled} = \cos(2\pi * S)$ and (ii) $S_{scaled} = \sin(2\pi * S)$; is not exact to input the factor $2\pi * \Sigma_i [mm^i * I]$ into the math functions cosine (cos) and sine (sin).

This paragraph provides the theoretical solution to the aforementioned, which has general applicability and validity. The solution is consistency among the conflicting units of the following variables: space (s), time (t) and the angle (χ). The following logical reasoning provides the reader with the solution. Let w be any of the variables: s, t, χ. Let the unit of space be for instance mm (millimeters), let the unit of time be for instance sec (seconds), let the unit of the angle be for instance degrees.

To input any numerical values to the trigonometric functions sine (sin), cosine (cos) or tangent (tan), it requires the numerical value having the unit to be in radians, therefore it requires the numerical value not to be a pure number. Let $w_{scaled} = (w - w_{min}) / (w_{max} - w_{min})$ be the scaling formula capable to remove the unit (either: mm of the variable space, or sec of the variable time, or degree of the variable angle) from the numerical value w_{scaled}, where w is defined into the closed interval $[w_{min}, w_{max}]$. The consistency among the conflicting units of the aforementioned variables, happens because of the ratio $(w - w_{min}) / (w_{max} - w_{min})$, where mm is divided by mm in the case of space, sec is divided by sec in the case of time, and deg is divided by deg in the case of the angle. It follows that the ratios mm/mm, sec/sec, and deg/deg are consistent because all of their values is 1. Any numerical values of w (in the variables s, t and χ), which is inserted into: $(w - w_{min}) / (w_{max} - w_{min})$ becomes a pure number because $w_{scaled} = (w - w_{min}) / (w_{max} - w_{min})$ is a pure number.

Now, to use correctly an input to the trigonometric functions sin (sine), cos (cosine) and tan (tangent) requires that either space or time or degrees are processed as follows. Let w be either s (space), t (time) or χ (the angle). The correct input to the functions sin, cos, and tan is given by: $L = 2\pi * (w - w_{min}) / (w_{max} - w_{min})$ because 2π is multiplied times a pure number and therefore L is radians. Thus, it makes sense to calculate $\sin(L)$, $\cos(L)$ and $\tan(L)$.

SIGNAL RESILIENT TO INTERPOLATION: THE CONCEPT AND THE EXAMPLE

Theory

This summary presents the theoretical basis of a concept recently introduced in signal processing (Ciulla, 2009), its meaning and implications.

Let it be given a signal $s(x)$ where x is defined in the domain of the real numbers. Let $[x_i, x_{i+1}]$ be the spatial resolution where $i = -\infty\ldots0\ldots+\infty$ and $|x_i - x_{i+1}| > 0$. Let the signal be estimated at the inter-node location x_k where $i < k < i+1$. The simplest approximation of $s(x_k)$ is through admitting the denial of the need of the estimation and thus hypothesizing $s(x_k) = s(x_i)$.

When such hypothesis is denied, the estimate of $s(x_k)$ can be made through an interpolation function. Let it thus be given a model function $f(x)$ which is continuous and in its domain admits non null second order derivatives, and where x is defined in the domain real numbers.

Let us define mathematical terms related to $f(x)$, derived on the basis of the math form the model function. They are: (a) the intensity-curvature term without interpolation $E_0(x) = \int [f(0) * (\partial^2 f(x)/\partial x^2)]_{(x = xi)}\, dx$; and (b) the intensity-curvature term with interpolation $E_{IN}(x) = \int [f(x) * (\partial^2 f(x)/\partial x^2)]_{(x = xk)}\, dx$.

It has been empirically demonstrated in Ciulla (2009) that to consider the curvature, which is represented through the second order derivative of the model function, into the interpolation improvement paradigm, yields betterment of the signal estimate. Conceptually, this is equivalent to the assertion stating that the signal information content included into the intensity can be increased adding the second order derivative. Thus, the information content about the signal $s(x)$, which is embedded into $E_0(x)$ and $E_{IN}(x)$, is higher than the information content embedded into $f(x)$ alone and this can be locally true inside the domain of $f(x)$.

The Signal Resilient to Interpolation (SRI) is derived at any intra-node location from the solution of the equation $E_0(x) = E_{IN}(x)$ such to explicit the value of $f(0)$ given by $f(x_i) = f(0)$ and $f(x_k) = f(0) + \xi(\Omega\, x_k)$. Where $f(0)$ is the value of the SRI (thus changing at each node) and $\xi(\Omega\, x)$ (with $x = x_i$ and $x = x_k$) is the convolution of the neighborhood pixel intensities with the polynomial form of the model function. The key question to answer is: *'what is the meaning of the SRI ?'* Where the SRI is given by $\int [f(0)] * [(\partial^2 f(x)/\partial x^2)]_{(x = xi)}\, dx = \int [f(0) + \xi(\Omega\, x_k)] * [(\partial^2 f(x)/\partial x^2)]_{(x = xk)}\, dx$. From such equation is possible to derive $f(0) * \int [(\partial^2 f(x)/\partial x^2)]_{(x = xi)}\, dx = f(0) * \int [(\partial^2 f(x)/\partial x^2)]_{(x = xk)}\, dx + \int [\xi(\Omega\, x_k)] * [(\partial^2 f(x)/\partial x^2)]_{(x = xk)}\, dx$, thus $f(0) * \{ \int [(\partial^2 f(x)/\partial x^2)]_{(x = xi)}\, dx - \int [(\partial^2 f(x)/\partial x^2)]_{(x = xk)}\, dx \} = \int [\xi(\Omega\, x_k)] * [(\partial^2 f(x)/\partial x^2)]_{(x = xk)}\, dx$ yields $f(0) = \{ \int [\xi(\Omega\, x_k)] * [(\partial^2 f(x)/\partial x^2)]_{(x = xk)}\, dx \} / \{ \int [(\partial^2 f(x)/\partial x^2)]_{(x = xi)}\, dx - \int [(\partial^2 f(x)/\partial x^2)]_{(x = xk)}\, dx \}$.

Let us introduce the concept of Intensity Correction (IC) as the multiplication between the misplacement $(x_i - x_k)$ and the angle subtended by the tangent to the first order derivative (geometrically, the second order derivative $[(\partial^2 f(x)/\partial x^2)]_{(x = xi)}$), which takes the form of $IC = (x_i - x_k) * \tan ([(\partial^2 f(x)/\partial x^2)]_{(x = xi)})$ (see figure 1(b)). The value of $[(\partial^2 f(x)/\partial x^2)]_{(x = xi)}$ relates to the tangent of the angle θ angle formed by the tangent to the first order derivative and the horizontal, at the location x_i (which is the geometrical representation of the curvature).

The implications of these theoretical results are several. (a) It is possible to devise interpolation formulas having the math form defined as $f(x_k) = f(0) + IC$, where $f(x_k)$ is the estimate of the signal $s(x_k)$. (b) The IC can be calculated using either the curvature of the math model $f(x)$, or the curvature of the SRI given by $f(0)$. (c) The interpolation paradigm can be further improved on a pixel-by-pixel basis, either using the curvature of $f(x)$ or the curvature of $f(0)$, and in such case the reconstructed signal is called hybrid. (d) The interpolation formula can benefit of a transfer function (employed to scale the convolution of pixel intensity of the neighborhood with the polynomial form) derived from its math form and thus tailored to its math form. In two and three dimensions, the IC is called Pixel Intensity Correction (PIC). Figure 1(b) shows the concept of Intensity Correction.

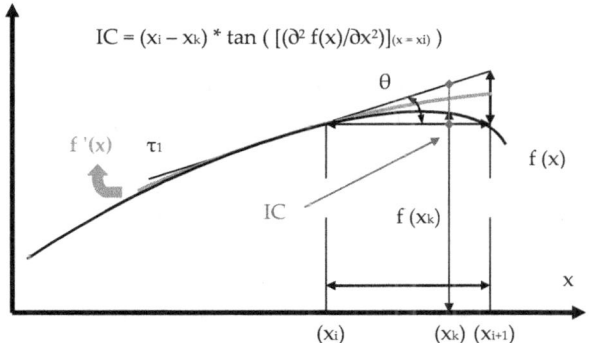

$$IC = (x_i - x_k) * \tan ([(\partial^2 f(x)/\partial x^2)]_{(x = xi)})$$

Figure 1(b): The Intensity Correction (IC). The angle θ is subtended by the tangent τ_1 to f ', where τ_1 is the tangent to the first order derivative f '.

The concept expressed in figure 1(b) is explained here to follow. The IC segment may end at a location placed onto the interpolation surface or may not. That depends on the value of the estimate made of IC. Such uncertainty is the direct consequence of the fact that the true value of the signal to estimate is unknown. What is necessary instead is that the IC segment ends at the location placed on the tangent to the first order derivative curve. This is a direct consequence of the fact that the angle θ (related to the value of the curvature) is a measure of the increment of the model function and such a

measure is made on the tangent to the first order derivative curve (see figure 1(b)). What is congruent is that the IC is the misplacement times the tangent of the value of the curvature (related to the angle θ, which geometrical counterpart is the value of the angle subtended by the tangent to the first order derivative curve and the horizontal). Furthermore, congruency demands that the angle θ is independent from the increment (misplacement $|x_i - x_k|$), and also that θ is defined at any location in the domain of the model function, because so it is the second order derivative of the model function. The angle θ is thus a local property of the model function. Additionally, the angle θ is defined even when the increment is zero, which is the case of $x_i \equiv x_k$. This means that the node to interpolate is coincidental with the location where the signal has to be interpolated, which means basically no interpolation. In such case of pure theoretical nature, the IC ends at the intersection of the tangent to the first order derivative curve and the first order derivative curve, and still θ is measured on the tangent to the first order derivative, and consequentially IC equals zero.

The Mathematical Process to Derive the SRI

The Interpolation Formula

Let $f(\mathbf{x}) = f(\mathbf{0}) + \xi(\Omega \mathbf{x})$ be an interpolation formula with $\mathbf{x} = [a_1, a_2... a_n]$ defined in n dimensional variables and Ω the neighborhood form, and ξ the convolution of the neighborhood pixel intensities with the polynomial form. Ω can have the math form of the type $\Omega = \Sigma_q f(q)$ with $q = 1, 2...p$, where p is the number of pixels' intensity included in the neighborhood form.

The Intensity-Curvature Terms

Let us calculate the two intensity-curvature terms, without and with interpolation. Let $\mathbf{x} = \mathbf{x_i}$ be the pixel to interpolate, $E_0(\mathbf{x}) = \int [f(\mathbf{0}) * \Delta(\mathbf{x})]_{(x = xi)} d\mathbf{x}$ where $\Delta(\mathbf{x})_{(x = xi)} = [\Sigma_v (\partial^2 f(\mathbf{x})/\partial a_v{}^2 + \Sigma_f (\partial^2 f(\mathbf{x})/\partial a_v \partial a_w)]_{(x = xi)}$ with $v = 1, 2...n$; $w = 1, 2...n$; $v \neq w$ and f running from 1 up to the number of second order partial derivatives for any combination of v and w with $v \neq w$. Let $\mathbf{x} = \mathbf{x_k}$ be the inter-node location inside $[x_i, x_{i+1}]$. Now consider that $E_{IN}(x) = \int [f(x) * \Delta(x)]_{(x = xk)} \, dx$ where $\Delta(x)_{(x = xk)} = [\Sigma_v (\partial^2 f(x)/\partial a_v{}^2) + \Sigma_f (\partial^2 f(x)/\partial a_v \, \partial a_w)]_{(x = xk)}$ with v, w and f defined the same way as it is for $E_0(x)$.

The Signal Resilient to Interpolation

The function $f(x) = f(0) + \xi(\Omega \, x)$ is calculated at the pixel location $x = xk$ and thus it becomes $f(xk) = f(0) + \xi(\Omega \, xk)$. It is also true that $f(xi) = f(0)$. The SRI is calculated through the solution in $f(0)$ of the equation $E_{IN}(x) = E_0(x)$.

$E_0(x) = \int [f(0) * \Delta(xi)]_{(x = xi)} \, dx = \int \{ f(0) * [\Sigma_v (\partial^2 f(xi)/\partial a_v{}^2) + \Sigma_f (\partial^2 f(xi)/\partial a_v \, \partial a_w)] \} \, dx = \{ f(0) \} * \int [\Sigma_v (\partial^2 f(xi)/\partial a_v{}^2) + \Sigma_f (\partial^2 f(xi)/\partial a_v \, \partial a_w)] \, dx$

$E_{IN}(x) = \int [f(xk) * \Delta(xk)]_{(x = xk)} \, dx = \int \{ f(xk) * [\Sigma_v (\partial^2 f(xk)/\partial a_v{}^2) + \Sigma_f (\partial^2 f(xk)/\partial a_v \, \partial a_w)] \} \, dx = \int \{ f(0) + \xi(\Omega \, xk) \} * [\Sigma_v (\partial^2 f(xk)/\partial a_v{}^2) + \Sigma_f (\partial^2 f(xk)/\partial a_v \, \partial a_w)] \, dx = \{ f(0) \} * \int [\Sigma_v (\partial^2 f(xk)/\partial a_v{}^2) + \Sigma_f (\partial^2 f(xk)/\partial a_v \, \partial a_w)] \, dx + \int \{ \xi(\Omega \, xk) \} * [\Sigma_v (\partial^2 f(xk)/\partial a_v{}^2) + \Sigma_f (\partial^2 f(xk)/\partial a_v \, \partial a_w)] \, dx.$

Thus $E_{IN}(x) = E_0(x)$ is: $\{ f(0) \} * \int [\Sigma_v (\partial^2 f(xi)/\partial a_v{}^2) + \Sigma_f (\partial^2 f(xi)/\partial a_v \, \partial a_w)] \, dx = \{ f(0) \} * \int [\Sigma_v (\partial^2 f(xk)/\partial a_v{}^2) + \Sigma_f (\partial^2 f(xk)/\partial a_v \, \partial a_w)] \, dx + \int \{ \xi(\Omega \, xk) \} * [\Sigma_v (\partial^2 f(xk)/\partial a_v{}^2) + \Sigma_f (\partial^2 f(xk)/\partial a_v \, \partial a_w)] \, dx$ yields

$f(0) = \{ \int \{ \xi(\Omega \, xk) \} * [\Sigma_v (\partial^2 f(xk)/\partial a_v{}^2) + \Sigma_f (\partial^2 f(xk)/\partial a_v \, \partial a_w)] \, dx \} / \{ \int [\Sigma_v (\partial^2 f(xi)/\partial a_v{}^2) + \Sigma_f (\partial^2 f(xi)/\partial a_v \, \partial a_w)] \, dx - \int [\Sigma_v (\partial^2 f(xk)/\partial a_v{}^2) + \Sigma_f (\partial^2 f(xk)/\partial a_v \, \partial a_w)] \, dx \}$

Example

An example is given in this paragraph for what is relevant to the implications of the theoretical results outlined in the theory section. Let us consider the bivariate interpolation formula $f(x, y) = f(0, 0) + [f(1, 0) - f(0, 0)] (a x^2 + y) + [f(0, 1) - f(0, 0)] (b y^2 - x) + [f(1, 1) + f(0, 0) - f(0, 1) - f(1, 0)] (a x^2 y + b y^2 x)$, with the neighborhood form shown in figure 1(c). Let us posit $\varphi_a = [f(1, 0) - f(0, 0)]$; $\varphi_b = [f(0, 1) -$

$f(0, 0)]$ and $\varphi_{ab} = [f(1, 1) + f(0, 0) - f(0, 1) - f(1, 0)]$. Thus $f(x, y) = f(0, 0) + \varphi_a (a x^2 + y) + \varphi_b (b y^2 - x) + \varphi_{ab} (a x^2 y + b y^2 x)$.

f (0, 1)	f (1, 1)
f (0, 0)	f (1, 0)

The Classic Curvature

Let us calculate the curvature of the interpolation formula. $(\partial f(x, y) / \partial x) = \varphi_a (2a \, x) - \varphi_b + \varphi_{ab} (2a \, xy + b \, y^2)$; $(\partial f(x, y) / \partial y) = \varphi_a + \varphi_b (2b \, y) + \varphi_{ab} (a \, x^2 + 2b \, yx)$.

$(\partial^2 f(x, y) / \partial x^2) = \varphi_a (2a) + \varphi_{ab} (2ay)$; $(\partial^2 f(x, y) / \partial y^2) = \varphi_b (2b) + \varphi_{ab} (2bx)$;

$(\partial^2 f(x, y) / \partial x \partial y) = \varphi_{ab} (2a \, x + 2b \, y)$; $(\partial f(x, y) / \partial y \partial x) = \varphi_{ab} (2a \, x + 2b \, y)$.

Figure 1(c): The neighborhood form of $f(x, y)$.

The classic curvature is: $\Upsilon_C(x, y) = (\partial^2 f(x, y) / \partial x^2) + (\partial^2 f(x, y) / \partial y^2) + (\partial^2 f(x, y) / \partial x \partial y) + (\partial^2 f(x, y) / \partial y \partial x) = \varphi_a (2a) + \varphi_{ab} (2ay) + \varphi_b (2b) + \varphi_{ab} (2bx) + 2 \varphi_{ab} (2a\, x + 2b\, y)$.

The Intensity-Curvature Terms

Let us proceed with the calculation of the two intensiy-curvature terms: $E_0(x, y) = \iint f(0, 0) * [\Upsilon_C(x, y)]_{(0, 0)}\, dxdy = f(0, 0) * \{ \varphi_a (2a) + \varphi_b (2b) \} * xy$; and $E_{IN}(x, y) = \iint f(x, y) * [\Upsilon_C(x, y)]_{(x, y)}\, dxdy$. Where: $\Upsilon_C(x, y) = \{ \varphi_a (2a) + \varphi_{ab} (2a\, y) + \varphi_b (2b) + \varphi_{ab} (2b\, x) + 2 \varphi_{ab} (2a\, x + 2b\, y) \}$ and $f(x, y) = f(0, 0) + \varphi_a (a\, x^2 + y) + \varphi_b (b\, y^2 - x) + \varphi_{ab} (a\, x^2 y + b\, y^2 x)$.

$E_{IN}(x, y) = \iint \{ f(0, 0) + \varphi_a (a\, x^2 + y) + \varphi_b (b\, y^2 - x) + \varphi_{ab} (a\, x^2 y + b\, y^2 x) \} * \{ \varphi_a (2a) + \varphi_{ab} (2ay) + \varphi_b (2b) + \varphi_{ab} (2bx) + 2 \varphi_{ab} (2a\, x + 2b\, y) \}\, dxdy = \iint \{ f(0, 0) * \Upsilon_C(x, y) + \varphi_a (a\, x^2 + y) * \Upsilon_C(x, y) + \varphi_b (b\, y^2 - x) * \Upsilon_C(x, y) + \varphi_{ab} (a\, x^2 y + b\, y^2 x) * \Upsilon_C(x, y) \}\, dxdy$ (1)

Let us call the primitive of $\Upsilon_C(x, y)$ in dxdy as $H(x, y) = \{ \varphi_a (2a\, xy) + \varphi_{ab} (a\, xy^2) + \varphi_b (2b\, xy) + \varphi_{ab} (b\, x^2 y) + 2 \varphi_{ab} (a\, x^2 y + b\, xy^2) \}$. Let us call:

$A(x, y) = \iint \{ f(0, 0) * \Upsilon_C(x, y) \}\, dxdy = \{ f(0, 0) * H(x, y) \}$ (2)

$B(x, y) = \iint \{ \varphi_a (a\, x^2 + y) * \Upsilon_C(x, y) \}\, dxdy = \iint \{ \varphi_a (a\, x^2 + y) * \varphi_a (2a) \} + \{ \varphi_a (a\, x^2 + y) * \varphi_{ab} (2 ay) \} + \{ \varphi_a (a\, x^2 + y) * \varphi_b (2b) \} + \{ \varphi_a (a\, x^2 + y) * \varphi_{ab} (2b\, x) \} + \{ \varphi_a (a\, x^2 + y) * 2 \varphi_{ab} (2a\, x + 2b\, y) \}\, dxdy = \{ \varphi_a (a\, x^3 y/3 + xy^2/2) * \varphi_a (2a) \} + \iint \{ \varphi_a (2a^2 x^2 y + 2a\, y^2) * \varphi_{ab} \} + \{ \varphi_a (a\, x^2 + y) * \varphi_b (2b) \} + \{ \varphi_a (2ab\, x^3 + 2b\, xy) * \varphi_{ab} \} + \{ \varphi_a (2a^2 x^3 + 2a\, xy + 2ab\, x^2 y + 2b\, y^2) * 2 \varphi_{ab} \}\, dxdy =$

$\{ \varphi_a (a\, x^3 y/3 + xy^2/2) * \varphi_a (2a) \} + \{ \varphi_a (a^2 x^3 y^2/3 + a\, xy^3\, 2/3) * \varphi_{ab} \} + \{ \varphi_a (a\, yx^3/3 + xy^2/2) * \varphi_b (2b) \} + \{ \varphi_a (ab\, yx^4/2 + b\, x^2 y^2/2) * \varphi_{ab} \} + \{ \varphi_a (a^2 yx^4/2 + a\, x^2 y^2/2 + ab\, x^3 y^2/3 + b\, xy^3\, 2/3) * 2 \varphi_{ab} \}$ (3)

$C(x, y) = \iint \{ \varphi_b (b\, y^2 - x) * \Upsilon_C(x, y) \}\, dxdy = \iint \{ \varphi_b (b\, y^2 - x) * \varphi_a (2a) \} + \{ \varphi_b (b\, y^2 - x) * \varphi_{ab} (2a\, y) \} + \{ \varphi_b (b\, y^2 - x) * \varphi_b (2b) \} + \{ \varphi_b (b\, y^2 - x) * \varphi_{ab} (2b\, x) \} + \{ \varphi_b (b\, y^2 - x) * 2 \varphi_{ab} (2a\, x + 2b\, y) \}\, dxdy = \{ \varphi_b (b\, xy^3/3 - yx^2/2) * \varphi_a (2a) \} + \iint \{ \varphi_b (2ab\, y^3 - 2a\, xy) * \varphi_{ab} \} + \{ \varphi_b (b\, y^2 - x) * \varphi_b (2b) \} + \{ \varphi_b (2b^2 xy^2 - 2b\, x^2) * \varphi_{ab} \} + \{ \varphi_b (2ab\, xy^2 - 2a\, x^2 + 2b^2 y^3 - 2b\, xy) * 2 \varphi_{ab} \}\, dxdy =$

$\{ \varphi_b (b\, xy^3/3 - yx^2/2) * \varphi_a (2a) \} + \{ \varphi_b (ab\, xy^4/2 - a\, x^2 y^2/2) * \varphi_{ab} \} + \{ \varphi_b (b\, xy^3/3 - x^2 y/2) * \varphi_b (2b) \} + \{ \varphi_b (b^2 x^2 y^3/3 - b\, yx^3\, 2/3) * \varphi_{ab} \} + \{ \varphi_b (ab\, x^2 y^3/3 - a\, yx^3\, 2/3 + b^2 xy^4/2 - b\, x^2 y^2/2) * 2 \varphi_{ab} \}$ (4)

$D(x, y) = \iint \{ \varphi_{ab} (a\, x^2 y + b\, y^2 x) * \Upsilon_C(x, y) \}\, dxdy = \iint \{ \varphi_{ab} (a\, x^2 y + b\, y^2 x) * \varphi_a (2a) \} + \{ \varphi_{ab} (a\, x^2 y + b\, y^2 x) * \varphi_{ab} (2a\, y) \} + \{ \varphi_{ab} (a\, x^2 y + b\, y^2 x) * \varphi_b (2b) \} + \{ \varphi_{ab} (a\, x^2 y + b\, y^2 x) * \varphi_{ab} (2b\, x) \} + \{ \varphi_{ab} (a\, x^2 y + b\, y^2 x) * 2 \varphi_{ab} (2a\, x + 2b\, y) \}\, dxdy = \{ \varphi_{ab} (a\, x^3 y^2/6 + b\, y^3 x^2/6) * \varphi_a (2a) \} + \iint \{ \varphi_{ab} (2a^2 x^2 y^2 + 2ab\, y^3 x) * \varphi_{ab} \} + \{ \varphi_{ab} (a\, x^2 y + b\, y^2 x) * \varphi_b (2b) \} + \{ \varphi_{ab} (2ab\, x^3 y + 2 b^2 y^2 x^2) * \varphi_{ab} \} + \{ \varphi_{ab} (2a^2 x^3 y + 2ab\, y^2 x^2 + 2ab\, x^2 y^2 + 2b^2 y^3 x) * 2 \varphi_{ab} \}\, dxdy =$

$\{ \varphi_{ab} (a\ x^3y^2/6 + b\ y^3x^2/6) * \varphi_a (2a) \} + \{ \varphi_{ab} (a^2\ x^3y^3\ 2/9 + ab\ y^4x^2/4) * \varphi_{ab} \} + \{ \varphi_{ab} (a\ x^3y^2/6 + b\ y^3x^2/6) * \varphi_b (2b) \} + \{ \varphi_{ab} (ab\ x^4y^2/4 + b^2\ y^3x^3\ 2/9) * \varphi_{ab} \} + \{ \varphi_{ab} (a^2\ x^4y^2/4 + ab\ y^3x^3\ 2/9 + ab\ x^3y^3\ 2/9 + b^2\ y^4x^2/4) * 2\ \varphi_{ab} \}$ (5)

$E_{IN}(x, y) = A(x, y) + B(x, y) + C(x, y) + D(x, y) = \{ f(0, 0) * H(x, y) \} + B(x, y) + C(x, y) + D(x, y)$ (6)

The Signal Resilient to Interpolation

The signal resilient to interpolation results from the solution in f(0) of the equation $E_0(x, y) = E_{IN}(x, y)$ which is: $f(0, 0) * \{ \varphi_a (2a) + \varphi_b (2b) \} * xy = \{ f(0, 0) * H(x, y) \} + B(x, y) + C(x, y) + D(x, y)$. Let us posit $F(x, y) = \{ \{ \varphi_a (2a) + \varphi_b (2b) \} * xy - H(x, y) \} = \{ \{ \varphi_a (2a) + \varphi_b (2b) \} * xy - \{ \varphi_a (2a\ xy) + \varphi_{ab} (a\ xy^2) + \varphi_b (2b\ xy) + \varphi_{ab} (b\ x^2y) + 2\ \varphi_{ab} (a\ x^2y + b\ xy^2) \} \}$.

$SRI(x, y) = f(0, 0) = \{ B(x, y) + C(x, y) + D(x, y) \} / \{ F(x, y) \}$ (7)

The Resilient Curvature

To calculate the resilient curvature, necessitates calculation of the second order partial derivatives of B(x, y), C(x, y), D(x, y) and F(x, y).

$(\partial\ B(x, y) / \partial x) = (\partial\ \{ \varphi_a (a\ x^3y/3 + xy^2/2) * \varphi_a (2a) \} + \{ \varphi_a (a^2\ x^3y^2/3 + a\ xy^3\ 2/3) * \varphi_{ab} \} + \{ \varphi_a (a\ yx^3/3 + xy^2/2) * \varphi_b (2b) \} + \{ \varphi_a (ab\ yx^4/2 + b\ x^2y^2/2) * \varphi_{ab} \} + \{ \varphi_a (a^2\ yx^4/2 + a\ x^2y^2/2 + ab\ x^3y^2/3 + b\ xy^3\ 2/3) * 2\ \varphi_{ab} \} / \partial x) = \{ \varphi_a (a\ x^2y + y^2/2) \times \varphi_a (2a) \} + \{ \varphi_a (a^2\ x^2y^2 + a\ y^3\ 2/3) * \varphi_{ab} \} + \{ \varphi_a (a\ yx^2 + y^2/2) * \varphi_b (2b) \} + \{ \varphi_a (2ab\ yx^3 + b\ xy^2) * \varphi_{ab} \} + \{ \varphi_a (2a^2\ yx^3 + a\ xy^2 + ab\ x^2y^2 + b\ y^3\ 2/3) * 2\ \varphi_{ab} \}$ (8)

$(\partial\ B(x, y) / \partial y) = (\partial\ \{ \varphi_a (a\ x^3y/3 + xy^2/2) * \varphi_a (2a) \} + \{ \varphi_a (a^2\ x^3y^2/3 + a\ xy^3\ 2/3) * \varphi_{ab} \} + \{ \varphi_a (a\ yx^3/3 + xy^2/2) * \varphi_b (2b) \} + \{ \varphi_a (ab\ yx^4/2 + b\ x^2y^2/2) * \varphi_{ab} \} + \{ \varphi_a (a^2\ yx^4/2 + a\ x^2y^2/2 + ab\ x^3y^2/3 + b\ xy^3\ 2/3) * 2\ \varphi_{ab} \} / \partial y) = \{ \varphi_a (a\ x^3/3 + xy) * \varphi_a (2a) \} + \{ \varphi_a (a^2\ x^3y\ 2/3 + 2a\ xy^2) * \varphi_{ab} \} + \{ \varphi_a (a\ x^3/3 + xy) * \varphi_b (2b) \} + \{ \varphi_a (ab\ x^4/2 + b\ x^2y) * \varphi_{ab} \} + \{ \varphi_a (a^2\ x^4/2 + a\ x^2y + ab\ x^3y\ 2/3 + 2b\ xy^2) * 2\ \varphi_{ab} \}$ (9)

$(\partial^2\ B(x, y) / \partial x^2) = (\partial\ \{ \varphi_a (a\ x^2y + y^2/2) * \varphi_a (2a) \} + \{ \varphi_a (a^2\ x^2y^2 + a\ y^3\ 2/3) * \varphi_{ab} \} + \{ \varphi_a (a\ yx^2 + y^2/2) * \varphi_b (2b) \} + \{ \varphi_a (2ab\ yx^3 + b\ xy^2) * \varphi_{ab} \} + \{ \varphi_a (2a^2\ yx^3 + a\ xy^2 + ab\ x^2y^2 + b\ y^3\ 2/3) * 2\ \varphi_{ab} \} / \partial x) = \{ \varphi_a (2a\ xy) * \varphi_a (2a) \} + \{ \varphi_a (2a^2\ xy^2) * \varphi_{ab} \} + \{ \varphi_a (2a\ xy) * \varphi_b (2b) \} + \{ \varphi_a (6ab\ yx^2 + by^2) * \varphi_{ab} \} + \{ \varphi_a (6a^2\ yx^2 + a\ y^2 + 2ab\ xy^2) * 2\ \varphi_{ab} \}$ (10)

$(\partial^2\ B(x, y) / \partial y^2) = (\partial\ \{ \varphi_a (a\ x^3/3 + xy) * \varphi_a (2a) \} + \{ \varphi_a (a^2\ x^3y\ 2/3 + 2a\ xy^2) * \varphi_{ab} \} + \{ \varphi_a (a\ x^3/3 + xy) * \varphi_b (2b) \} + \{ \varphi_a (ab\ x^4/2 + b\ x^2y) * \varphi_{ab} \} + \{ \varphi_a (a^2\ x^4/2 + a\ x^2y + ab\ x^3y\ 2/3 + 2b\ xy^2) * 2\ \varphi_{ab} \} / \partial y) = \{ \varphi_a (x) * \varphi_a (2a) \} + \{ \varphi_a (a^2\ x^3\ 2/3 + 4a\ xy) * \varphi_{ab} \} + \{ \varphi_a (x) * \varphi_b (2b) \} + \{ \varphi_a (b\ x^2) * \varphi_{ab} \} + \{ \varphi_a (a\ x^2 + ab\ x^3\ 2/3 + 4b\ xy) * 2\ \varphi_{ab} \}$ (11)

$(\partial^2 B(x, y) / \partial x \partial y) = (\partial \{ \varphi_a (a\, x^2 y + y^2/2) * \varphi_a (2a) \} + \{ \varphi_a (a^2\, x^2 y^2 + a\, y^3\, 2/3) * \varphi_{ab} \} + \{ \varphi_a (a\, yx^2 + y^2/2) * \varphi_b (2b) \} + \{ \varphi_a (2ab\, yx^3 + b\, xy^2) * \varphi_{ab} \} + \{ \varphi_a (2a^2\, yx^3 + a\, xy^2 + ab\, x^2 y^2 + b\, y^3\, 2/3) * 2 \varphi_{ab} \} / \partial y) = \{ \varphi_a (a\, x^2 + y) * \varphi_a (2a) \} + \{ \varphi_a (2a^2\, x^2 y + 2a\, y^2) * \varphi_{ab} \} + \{ \varphi_a (a\, x^2 + y) * \varphi_b (2b) \} + \{ \varphi_a (2ab\, x^3 + 2b\, xy) * \varphi_{ab} \} + \{ \varphi_a (2a^2\, x^3 + 2a\, xy + 2ab\, x^2 y + 2b\, y^2) * 2 \varphi_{ab} \} = \{ \varphi_a (a\, x^2 + y) * \varphi_a (2a) \} + \{ \varphi_a (a\, x^2 + y) * \varphi_{ab} (2a\, y) \} + \{ \varphi_a (a\, x^2 + y) * \varphi_b (2b) \} + \{ \varphi_a (a\, x^2 + y) * \varphi_{ab} (2b\, x) \} + \{ \varphi_a (a\, x^2 + y) * 2 \varphi_{ab} (2a\, x + 2b\, y) \}$ \hfill (12)

$(\partial^2 B(x, y) / \partial y \partial x) = (\partial \{ \varphi_a (a\, x^3/3 + xy) * \varphi_a (2a) \} + \{ \varphi_a (a^2\, x^3 y\, 2/3 + 2a\, xy^2) * \varphi_{ab} \} + \{ \varphi_a (a\, x^3/3 + xy) * \varphi_b (2b) \} + \{ \varphi_a (ab\, x^4/2 + b\, x^2 y) * \varphi_{ab} \} + \{ \varphi_a (a^2\, x^4/2 + a\, x^2 y + ab\, x^3 y\, 2/3 + 2b\, xy^2) * 2 \varphi_{ab} \} / \partial x) = \{ \varphi_a (a\, x^2 + y) * \varphi_a (2a) \} + \{ \varphi_a (2a^2\, x^2 y + 2a\, y^2) * \varphi_{ab} \} + \{ \varphi_a (a\, x^2 + y) * \varphi_b (2b) \} + \{ \varphi_a (2ab\, x^3 + 2b\, xy) * \varphi_{ab} \} + \{ \varphi_a (2a^2\, x^3 + 2a\, xy + 2ab\, x^2 y + 2b\, y^2) * 2 \varphi_{ab} \} = \{ \varphi_a (a\, x^2 + y) * \varphi_a (2a) \} + \{ \varphi_a (a\, x^2 + y) * \varphi_{ab} (2a\, y) \} + \{ \varphi_a (a\, x^2 + y) * \varphi_b (2b) \} + \{ \varphi_a (a\, x^2 + y) * \varphi_{ab} (2b\, x) \} + \{ \varphi_a (a\, x^2 + y) * 2 \varphi_{ab} (2a\, x + 2b\, y) \}$ \hfill (13)

The last two derivative calculations are a reminder that it is true that: $(\partial^2 C(x, y) / \partial x \partial y) = (\partial^2 C(x, y) / \partial y \partial x) = \{ \varphi_b (b\, y^2 - x) * \varphi_a (2a) \} + \{ \varphi_b (b\, y^2 - x) * \varphi_{ab} (2a\, y) \} + \{ \varphi_b (b\, y^2 - x) * \varphi_b (2b) \} + \{ \varphi_b (b\, y^2 - x) * \varphi_{ab} (2b\, x) \} + \{ \varphi_b (b\, y^2 - x) * 2 \varphi_{ab} (2a\, x + 2b\, y) \}$ and $(\partial^2 D(x, y) / \partial x \partial y) = (\partial^2 D(x, y) / \partial y \partial x) = \{ \varphi_{ab} (a\, x^2 y + b\, y^2 x) * \varphi_a (2a) \} + \{ \varphi_{ab} (a\, x^2 y + b\, y^2 x) * \varphi_{ab} (2a\, y) \} + \{ \varphi_{ab} (a\, x^2 y + b\, y^2 x) * \varphi_b (2b) \} + \{ \varphi_{ab} (a\, x^2 y + b\, y^2 x) * \varphi_{ab} (2b\, x) \} + \{ \varphi_{ab} (a\, x^2 y + b\, y^2 x) * 2 \varphi_{ab} (2a\, x + 2b\, y) \}$

$(\partial C(x, y) / \partial x) = (\partial \{ \varphi_b (b\, xy^3/3 - yx^2/2) * \varphi_a (2a) \} + \{ \varphi_b (ab\, xy^4/2 - a\, x^2 y^2/2) * \varphi_{ab} \} + \{ \varphi_b (b\, xy^3/3 - x^2 y/2) * \varphi_b (2b) \} + \{ \varphi_b (b^2\, x^2 y^3/3 - b\, yx^3\, 2/3) * \varphi_{ab} \} + \{ \varphi_b (ab\, x^2 y^3/3 - a\, yx^3\, 2/3 + b^2\, xy^4/2 - b\, x^2 y^2/2) * 2 \varphi_{ab} \} / \partial x) = \{ \varphi_b (b\, y^3/3 - yx) * \varphi_a (2a) \} + \{ \varphi_b (ab\, y^4/2 - a\, xy^2) * \varphi_{ab} \} + \{ \varphi_b (b\, y^3/3 - xy) * \varphi_b (2b) \} + \{ \varphi_b (b^2\, xy^3\, 2/3 - 2b\, yx^2) * \varphi_{ab} \} + \{ \varphi_b (ab\, xy^3\, 2/3 - 2a\, yx^2 + b^2\, y^4/2 - b\, xy^2) * 2 \varphi_{ab} \}$ \hfill (14)

$(\partial C(x, y) / \partial y) = (\partial \{ \varphi_b (b\, xy^3/3 - yx^2/2) * \varphi_a (2a) \} + \{ \varphi_b (ab\, xy^4/2 - a\, x^2 y^2/2) * \varphi_{ab} \} + \{ \varphi_b (b\, xy^3/3 - x^2 y/2) * \varphi_b (2b) \} + \{ \varphi_b (b^2\, x^2 y^3/3 - b\, yx^3\, 2/3) * \varphi_{ab} \} + \{ \varphi_b (ab\, x^2 y^3/3 - a\, yx^3\, 2/3 + b^2\, xy^4/2 - b\, x^2 y^2/2) * 2 \varphi_{ab} \} / \partial y) = \{ \varphi_b (b\, xy^2 - x^2/2) * \varphi_a (2a) \} + \{ \varphi_b (2ab\, xy^3 - a\, x^2 y) * \varphi_{ab} \} + \{ \varphi_b (b\, xy^2 - x^2/2) * \varphi_b (2b) \} + \{ \varphi_b (b^2\, x^2 y^2 - b\, x^3\, 2/3) * \varphi_{ab} \} + \{ \varphi_b (ab\, x^2 y^2 - a\, x^3\, 2/3 + 2b^2\, xy^3 - b\, x^2 y) * 2 \varphi_{ab} \}$ \hfill (15)

$(\partial^2 C(x, y) / \partial x^2) = (\partial \{ \varphi_b (b\, y^3/3 - yx) * \varphi_a (2a) \} + \{ \varphi_b (ab\, y^4/2 - a\, xy^2) * \varphi_{ab} \} + \{ \varphi_b (b\, y^3/3 - xy) * \varphi_b (2b) \} + \{ \varphi_b (b^2\, xy^3\, 2/3 - 2b\, yx^2) * \varphi_{ab} \} + \{ \varphi_b (ab\, xy^3\, 2/3 - 2a\, yx^2 + b^2\, y^4/2 - b\, xy^2) * 2 \varphi_{ab} \} / \partial x) = \{ -\varphi_b (y) * \varphi_a (2a) \} + \{ -\varphi_b (ay^2) * \varphi_{ab} \} + \{ -\varphi_b (y) * \varphi_b (2b) \} + \{ \varphi_b (b^2\, y^3\, 2/3 - 4b\, yx) * \varphi_{ab} \} + \{ \varphi_b (ab\, y^3\, 2/3 - 4a\, yx - b\, y^2) * 2 \varphi_{ab} \}$ \hfill (16)

$(\partial^2 C(x, y) / \partial y^2) = (\partial \{ \varphi_b (b\, xy^2 - x^2/2) * \varphi_a (2a) \} + \{ \varphi_b (2ab\, xy^3 - a\, x^2 y) * \varphi_{ab} \} + \{ \varphi_b (b\, xy^2 - x^2/2) * \varphi_b (2b) \} + \{ \varphi_b (b^2\, x^2 y^2 - b\, x^3\, 2/3) * \varphi_{ab} \} + \{ \varphi_b (ab\, x^2 y^2 - a\, x^3\, 2/3 + 2b^2\, xy^3 - b\, x^2 y) * 2 \varphi_{ab} \} / \partial y) = \{ \varphi_b (2b\, xy) * \varphi_a (2a) \} + \{ \varphi_b (6ab\, xy^2 - a\, x^2) * \varphi_{ab} \} + \{ \varphi_b (2b\, xy) * \varphi_b (2b) \} + \{ \varphi_b (2b^2\, x^2 y) * \varphi_{ab} \} + \{ \varphi_b (2ab\, x^2 y + 6b^2\, xy^2 - b\, x^2) * 2 \varphi_{ab} \}$ \hfill (17)

$(\partial\, D(x, y) / \partial x) = (\partial\, \{\, \varphi_{ab}\, (a\, x^3y^2/6 + b\, y^3x^2/6) * \varphi_a\, (2a)\, \} + \{\, \varphi_{ab}\, (a^2\, x^3y^3\, 2/9 + ab\, y^4x^2/4) * \varphi_{ab}\, \} + \{\, \varphi_{ab}\, (a\, x^3y^2/6 + b\, y^3x^2/6) * \varphi_b\, (2b)\, \} + \{\, \varphi_{ab}\, (ab\, x^4y^2/4 + b^2\, y^3x^3\, 2/9) * \varphi_{ab}\, \} + \{\, \varphi_{ab}\, (a^2\, x^4y^2/4 + ab\, y^3x^3\, 2/9 + ab\, x^3y^3\, 2/9 + b^2\, y^4x^2/4) * 2\, \varphi_{ab}\, \} / \partial x) = \{\, \varphi_{ab}\, (a\, x^2y^2/2 + b\, y^3x/3) * \varphi_a\, (2a)\, \} + \{\, \varphi_{ab}\, (a^2\, x^2y^3\, 2/3 + ab\, y^4x/2) * \varphi_{ab}\, \} + \{\, \varphi_{ab}\, (a\, x^2y^2/2 + b\, y^3x/3) * \varphi_b\, (2b)\, \} + \{\, \varphi_{ab}\, (a^2\, x^3y^2 + b^2\, y^3x^2\, 2/3) * \varphi_{ab}\, \} + \{\, \varphi_{ab}\, (a^2\, x^3y^2 + ab\, y^3x^2\, 2/3 + ab\, x^2y^3\, 2/3 + b^2\, y^4x/2) * 2\, \varphi_{ab}\, \}$ \hfill (18)

$(\partial\, D(x, y) / \partial y) = (\partial\, \{\, \varphi_{ab}\, (a\, x^3y^2/6 + b\, y^3x^2/6) * \varphi_a\, (2a)\, \} + \{\, \varphi_{ab}\, (a^2\, x^3y^3\, 2/9 + ab\, y^4x^2/4) * \varphi_{ab}\, \} + \{\, \varphi_{ab}\, (a\, x^3y^2/6 + b\, y^3x^2/6) * \varphi_b\, (2b)\, \} + \{\, \varphi_{ab}\, (ab\, x^4y^2/4 + b^2\, y^3x^3\, 2/9) * \varphi_{ab}\, \} + \{\, \varphi_{ab}\, (a^2\, x^4y^2/4 + ab\, y^3x^3\, 2/9 + ab\, x^3y^3\, 2/9 + b^2\, y^4x^2/4) * 2\, \varphi_{ab}\, \} / \partial y) = \{\, \varphi_{ab}\, (a\, x^3y/3 + b\, y^2x^2/2) * \varphi_a\, (2a)\, \} + \{\, \varphi_{ab}\, (a^2\, x^3y^2\, 2/3 + ab\, y^3x^2) * \varphi_{ab}\, \} + \{\, \varphi_{ab}\, (a\, x^3y/3 + b\, y^2x^2/2) * \varphi_b\, (2b)\, \} + \{\, \varphi_{ab}\, (ab\, x^4y/2 + b^2\, y^2x^3\, 2/3) * \varphi_{ab}\, \} + \{\, \varphi_{ab}\, (a^2\, x^4y/2 + ab\, y^2x^3\, 2/3 + ab\, x^3y^2\, 2/3 + b^2\, y^3x^2) * 2\, \varphi_{ab}\, \}$ \hfill (20)

$(\partial^2\, D(x, y) / \partial x^2) = (\partial\, \{\, \varphi_{ab}\, (a\, x^2y^2/2 + b\, y^3x/3) * \varphi_a\, (2a)\, \} + \{\, \varphi_{ab}\, (a^2\, x^2y^3\, 2/3 + ab\, y^4x/2) * \varphi_{ab}\, \} + \{\, \varphi_{ab}\, (a\, x^2y^2/2 + b\, y^3x/3) * \varphi_b\, (2b)\, \} + \{\, \varphi_{ab}\, (ab\, x^3y^2 + b^2\, y^3x^2\, 2/3) * \varphi_{ab}\, \} + \{\, \varphi_{ab}\, (a^2\, x^3y^2 + ab\, x^2y^3\, 2/3 + b^2\, y^4x/2) * 2\, \varphi_{ab}\, \} / \partial x) = \{\, \varphi_{ab}\, (a\, xy^2 + b\, y^3/3) * \varphi_a\, (2a)\, \} + \{\, \varphi_{ab}\, (a^2\, xy^3\, 4/3 + ab\, y^4/2) * \varphi_{ab}\, \} + \{\, \varphi_{ab}\, (a\, xy^2 + b\, y^3/3) * \varphi_b\, (2b)\, \} + \{\, \varphi_{ab}\, (3ab\, x^2y^2 + b^2\, y^3x\, 4/3) * \varphi_{ab}\, \} + \{\, \varphi_{ab}\, (3a^2\, x^2y^2 + ab\, y^3x\, 4/3 + ab\, xy^3\, 4/3 + b^2\, y^4/2) * 2\, \varphi_{ab}\, \}$ \hfill (21)

$(\partial^2\, D(x, y) / \partial y^2) = (\partial\, \{\, \varphi_{ab}\, (a\, x^3y/3 + b\, y^2x^2/2) * \varphi_a\, (2a)\, \} + \{\, \varphi_{ab}\, (a^2\, x^3y^2\, 2/3 + ab\, y^3x^2) * \varphi_{ab}\, \} + \{\, \varphi_{ab}\, (a\, x^3y/3 + b\, y^2x^2/2) * \varphi_b\, (2b)\, \} + \{\, \varphi_{ab}\, (ab\, x^4y/2 - b^2\, y^2x^3\, 2/3) * \varphi_{ab}\, \} + \{\, \varphi_{ab}\, (a^2\, x^4y/2 + ab\, y^2x^3\, 2/3 + ab\, x^3y^2\, 2/3 + b^2\, y^3x^2) * 2\, \varphi_{ab}\, \} / \partial y) = \{\, \varphi_{ab}\, (a\, x^3/3 + b\, yx^2) * \varphi_a\, (2a)\, \} + \{\, \varphi_{ab}\, (a^2\, x^3y\, 4/3 + 3ab\, y^2x^2) * \varphi_{ab}\, \} + \{\, \varphi_{ab}\, (a\, x^3/3 + b\, yx^2) * \varphi_b\, (2b)\, \} + \{\, \varphi_{ab}\, (ab\, x^4/2 + b^2\, yx^3\, 4/3) * \varphi_{ab}\, \} + \{\, \varphi_{ab}\, (a^2\, x^4/2 + ab\, yx^3\, 4/3 + ab\, x^3y\, 4/3 + 3b^2\, y^2x^2) * 2\, \varphi_{ab}\, \}$ \hfill (22)

The second order partial derivatives of F(x, y) are hereto calculated.

$(\partial\, F(x, y)/ \partial x) = \{\, \{\, \varphi_a\, (2a) + \varphi_b\, (2b)\, \} * y - \{\, \varphi_a\, (2a\, y) + \varphi_{ab}\, (a\, y^2) + \varphi_b\, (2b\, y) + \varphi_{ab}\, (2b\, xy) + 2\, \varphi_{ab}\, (2a\, xy + b\, y^2)\, \}\, \}$ \hfill (23)

$(\partial\, F(x, y)/ \partial y) = \{\, \{\, \varphi_a\, (2a) + \varphi_b\, (2b)\, \} * x - \{\, \varphi_a\, (2a\, x) + \varphi_{ab}\, (2a\, xy) + \varphi_b\, (2b\, x) + \varphi_{ab}\, (b\, x^2) + 2\, \varphi_{ab}\, (a\, x^2 + 2b\, xy)\, \}\, \}$ \hfill (24)

$(\partial^2\, F(x, y) / \partial x^2) = \{\, -\, \{\, \varphi_{ab}\, (2b\, y) + 2\, \varphi_{ab}\, (2a\, y)\, \}\, \}$ \hfill (25)

$(\partial^2\, F(x, y) / \partial y^2) = \{\, -\, \{\, \varphi_{ab}\, (2a\, x) + 2\, \varphi_{ab}\, (2b\, x)\, \}\, \}$ \hfill (26)

$(\partial^2\, F(x, y) / \partial x \partial y) = \{\, \{\, \varphi_a\, (2a) + \varphi_b\, (2b)\, \} - \{\, \varphi_a\, (2a) + \varphi_{ab}\, (2a\, y) + \varphi_b\, (2b) + \varphi_{ab}\, (2b\, x) + 2\, \varphi_{ab}\, (2a\, x + 2b\, y)\, \}\, \} = \{\, \{\, \varphi_a\, (2a) + \varphi_b\, (2b)\, \} - \Upsilon_C(x, y)\, \}$ \hfill (27)

$(\partial^2\, F(x, y) / \partial y \partial x) = \{\, \{\, \varphi_a\, (2a) + \varphi_b\, (2b)\, \} - \{\, \varphi_a\, (2a) + \varphi_{ab}\, (2a\, y) + \varphi_b\, (2b) + \varphi_{ab}\, (2b\, x) + 2\, \varphi_{ab}\, (2a\, x + 2b\, y)\, \}\, \} = \{\, \{\, \varphi_a\, (2a) + \varphi_b\, (2b)\, \} - \Upsilon_C(x, y)\, \}$ \hfill (28)

Consider the formula $SRI(x, y) = f(0, 0) = \{\, B(x, y) + C(x, y) + D(x, y)\, \} / \{\, F(x, y)\, \}$.

$$(\partial \, SRI(x, y) \, / \, \partial x) = \{ \, \{ \, F(x, y) \, \} \, * \, \{ \, (\partial \, B(x, y) \, / \, \partial x) + (\partial \, C(x, y) \, / \, \partial x) + (\partial \, D(x, y) \, / \, \partial x) \, \} - \{ \, (\partial \, F(x, y) \, / \, \partial x) \, \} \, * \, \{ \, B(x, y) + C(x, y) + D(x, y) \, \} \, \} \, / \, \{ \, F(x, y) \, \}^2 \tag{29}$$

$$(\partial \, SRI(x, y) \, / \, \partial y) = \{ \, \{ \, F(x, y) \, \} \, * \, \{ \, (\partial \, B(x, y) \, / \, \partial y) + (\partial \, C(x, y) \, / \, \partial y) + (\partial \, D(x, y) \, / \, \partial y) \, \} - \{ \, (\partial \, F(x, y) \, / \, \partial y) \, \} \, * \, \{ \, B(x, y) + C(x, y) + D(x, y) \, \} \, \} \, / \, \{ \, F(x, y) \, \}^2 \tag{30}$$

$$(\partial^2 \, SRI(x, y) \, / \, \partial x^2) = \{ \, \{ \, \{ \, \{ \, (\partial \, F(x, y) \, / \, \partial x) \, \} \, * \, \{ \, (\partial \, B(x, y) \, / \, \partial x) + (\partial \, C(x, y) \, / \, \partial x) + (\partial \, D(x, y) \, / \, \partial x) \, \} - \{ \, (\partial^2 \, F(x, y) \, / \, \partial x^2) \, \} \, * \, \{ \, B(x, y) + C(x, y) + D(x, y) \, \} + \{ \, F(x, y) \, \} \, * \, \{ \, (\partial^2 \, B(x, y) \, / \, \partial x^2) + (\partial^2 \, C(x, y) \, / \, \partial x^2) + (\partial^2 \, D(x, y) \, / \, \partial x^2) \, \} - \{ \, (\partial \, F(x, y) \, / \, \partial x) \, \} \, * \, \{ \, (\partial \, B(x, y) \, / \, \partial x) + (\partial \, C(x, y) \, / \, \partial x) + (\partial \, D(x, y) \, / \, \partial x) \, \} \, \} \, * \, \{ \, F(x, y) \, \}^2 \, \} - \{ \, \{ \, \{ \, F(x, y) \, \} \, * \, \{ \, (\partial \, B(x, y) \, / \, \partial x) + (\partial \, C(x, y) \, / \, \partial x) + (\partial \, D(x, y) \, / \, \partial x) \, \} - \{ \, (\partial \, F(x, y) \, / \, \partial x) \, \} \, * \, \{ \, B(x, y) + C(x, y) + D(x, y) \, \} \, \} \, * \, \{ \, 2* \, F(x, y) \, * \, (\partial \, F(x, y) \, / \, \partial x) \, \} \, \} \, \} \, / \, \{ \, F(x, y) \, \}^4 \tag{31}$$

$$(\partial^2 \, SRI(x, y) \, / \, \partial y^2) = \{ \, \{ \, \{ \, \{ \, (\partial \, F(x, y) \, / \, \partial y) \, \} \, * \, \{ \, (\partial \, B(x, y) \, / \, \partial y) + (\partial \, C(x, y) \, / \, \partial y) + (\partial \, D(x, y) \, / \, \partial y) \, \} - \{ \, (\partial^2 \, F(x, y) \, / \, \partial y^2) \, \} \, * \, \{ \, B(x, y) + C(x, y) + D(x, y) \, \} + \{ \, F(x, y) \, \} \, * \, \{ \, (\partial^2 \, B(x, y) \, / \, \partial y^2) + (\partial^2 \, C(x, y) \, / \, \partial y^2) + (\partial^2 \, D(x, y) \, / \, \partial y^2) \, \} - \{ \, (\partial \, F(x, y) \, / \, \partial y) \, \} \, * \, \{ \, (\partial \, B(x, y) \, / \, \partial y) + (\partial \, C(x, y) \, / \, \partial y) + (\partial \, D(x, y) \, / \, \partial y) \, \} \, \} \, * \, \{ \, F(x, y) \, \}^2 \, \} - \{ \, \{ \, \{ \, F(x, y) \, \} \, * \, \{ \, (\partial \, B(x, y) \, / \, \partial y) + (\partial \, C(x, y) \, / \, \partial y) + (\partial \, D(x, y) \, / \, \partial y) \, \} - \{ \, (\partial \, F(x, y) \, / \, \partial y) \, \} \, * \, \{ \, B(x, y) + C(x, y) + D(x, y) \, \} \, \} \, * \, \{ \, 2* \, F(x, y) \, * \, (\partial \, F(x, y) \, / \, \partial y) \, \} \, \} \, \} \, / \, \{ \, F(x, y) \, \}^4 \tag{32}$$

$$(\partial^2 \, SRI(x, y) \, / \, \partial x \partial y) = \{ \, \{ \, \{ \, \{ \, (\partial \, F(x, y) \, / \, \partial y) \, \} \, * \, \{ \, (\partial \, B(x, y) \, / \, \partial x) + (\partial \, C(x, y) \, / \, \partial x) + (\partial \, D(x, y) \, / \, \partial x) \, \} - \{ \, (\partial^2 \, F(x, y) \, / \, \partial x \partial y) \, \} \, * \, \{ \, B(x, y) + C(x, y) + D(x, y) \, \} + \{ \, F(x, y) \, \} \, * \, \{ \, (\partial^2 \, B(x, y) \, / \, \partial x \partial y) + (\partial^2 \, C(x, y) \, / \, \partial x \partial y) + (\partial^2 \, D(x, y) \, / \, \partial x \partial y) \, \} - \{ \, (\partial \, F(x, y) \, / \, \partial x) \, \} \, * \, \{ \, (\partial \, B(x, y) \, / \, \partial y) + (\partial \, C(x, y) \, / \, \partial y) + (\partial \, D(x, y) \, / \, \partial y) \, \} \, \} \, * \, \{ \, F(x, y) \, \}^2 \, \} - \{ \, \{ \, \{ \, F(x, y) \, \} \, * \, \{ \, (\partial \, B(x, y) \, / \, \partial x) + (\partial \, C(x, y) \, / \, \partial x) + (\partial \, D(x, y) \, / \, \partial x) \, \} - \{ \, (\partial \, F(x, y) \, / \, \partial x) \, \} \, * \, \{ \, B(x, y) + C(x, y) + D(x, y) \, \} \, \} \, * \, \{ \, 2 \, * \, F(x, y) \, * \, (\partial \, F(x, y) \, / \, \partial y) \, \} \, \} \, \} \, / \, \{ \, F(x, y) \, \}^4 \tag{33}$$

$$(\partial^2 \, SRI(x, y) \, / \, \partial y \partial x) = \{ \, \{ \, \{ \, \{ \, (\partial \, F(x, y) \, / \, \partial x) \, \} \, * \, \{ \, (\partial \, B(x, y) \, / \, \partial y) + (\partial \, C(x, y) \, / \, \partial y) + (\partial \, D(x, y) \, / \, \partial y) \, \} - \{ \, (\partial^2 \, F(x, y) \, / \, \partial y \partial x) \, \} \, * \, \{ \, B(x, y) + C(x, y) + D(x, y) \, \} + \{ \, F(x, y) \, \} \, * \, \{ \, (\partial^2 \, B(x, y) \, / \, \partial y \partial x) + (\partial^2 \, C(x, y) \, / \, \partial y \partial x) + (\partial^2 \, D(x, y) \, / \, \partial y \partial x) \, \} - \{ \, (\partial \, F(x, y) \, / \, \partial y) \, \} \, * \, \{ \, (\partial \, B(x, y) \, / \, \partial x) + (\partial \, C(x, y) \, / \, \partial x) + (\partial \, D(x, y) \, / \, \partial x) \, \} \, \} \, * \, \{ \, F(x, y) \, \}^2 \, \} - \{ \, \{ \, \{ \, F(x, y) \, \} \, * \, \{ \, (\partial \, B(x, y) \, / \, \partial y) + (\partial \, C(x, y) \, / \, \partial y) + (\partial \, D(x, y) \, / \, \partial y) \, \} - \{ \, (\partial \, F(x, y) \, / \, \partial y) \, \} \, * \, \{ \, B(x, y) + C(x, y) + D(x, y) \, \} \, \} \, * \, \{ \, 2 \, * \, F(x, y) \, * \, (\partial \, F(x, y) \, / \, \partial x) \, \} \, \} \, \} \, / \, \{ \, F(x, y) \, \}^4 \tag{34}$$

Finally the resilient curvature is: $Y_R(x, y) = (\partial^2 \, SRI(x, y) \, / \, \partial x^2) + (\partial^2 \, SRI(x, y) \, / \, \partial y^2) + (\partial^2 \, SRI(x, y) \, / \, \partial x \partial y) + (\partial^2 \, SRI(x, y) \, / \, \partial y \partial x)$.

Let us call the misplacement $m(x, y) = \sqrt{x^2 + y^2}$.

The PIC Classic and Resilient

The classic pixel intensity correction is: $PIC_C(x, y) = m(x, y) \, * \, \tan(Y_C(x, y))$. The resilient pixel intensity correction is: $PIC_R(x, y) = m(x, y) \, * \, \tan(Y_R(x, y))$. Both $PIC_C(x, y)$ and $PIC_F(x, y)$ are transfer functions employed to scale the convolution of pixel intensity of the neighborhood with the polynomial form, and they are derived from the math form of the interpolation formula, they are thus tailored to the interpolation formula. Using the classic curvature, the estimate of the signal is: $S(x, y) = s(x, y) + PIC_C(x, y)$. Whereas using the resilient curvature, the estimate of the signal is: $S(x, y) = s(x, y) + PIC_R(x, y)$.

ASPECTS OF THE TECHNOLOGICAL EFFORT

The technological effort of this book has been devoted to the creation of application programs in Visual C++/OpenGL and Matlab®. What follows are some detailed specifications of three of the four Visual C++/OpenGL programs which aided the exploration.

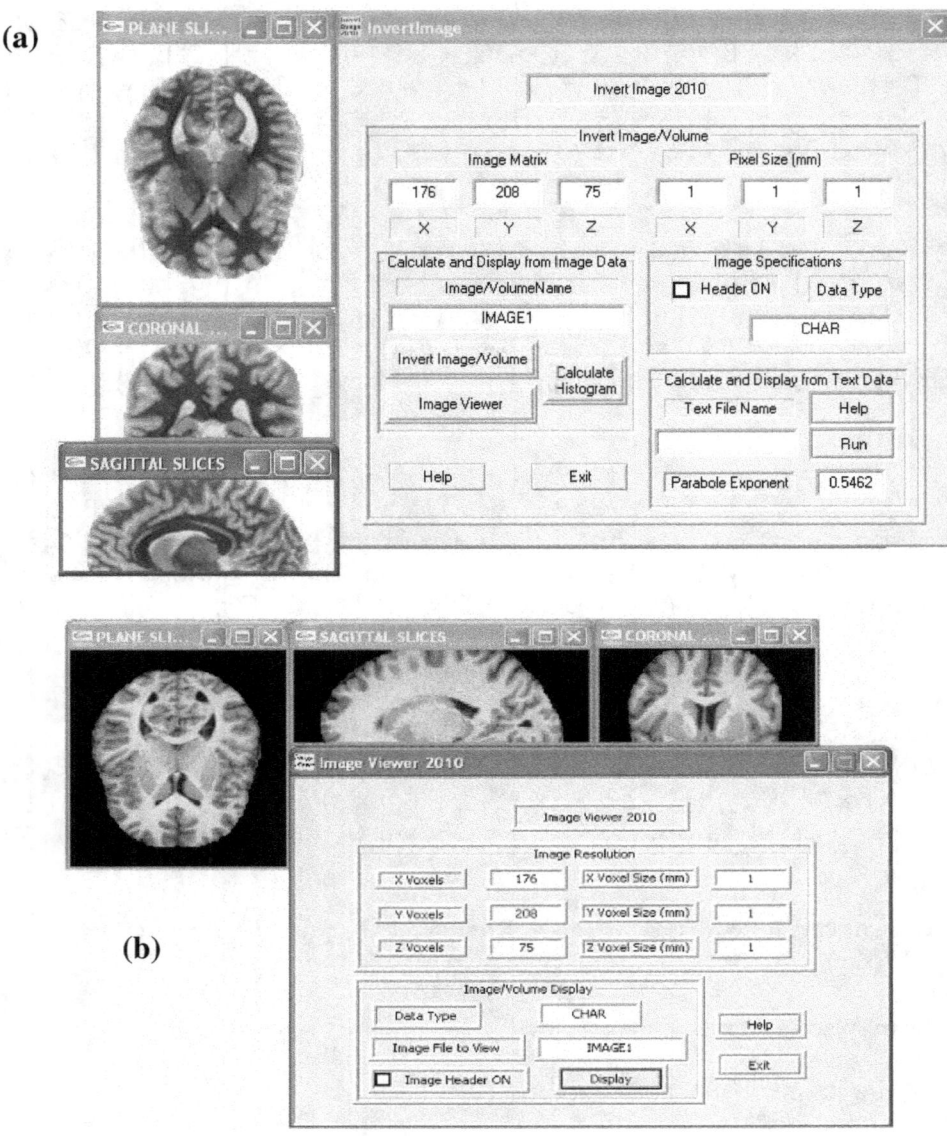

Figure 2: InvertImage 2010 Graphical User Interface (GUI) (a). This program has multiple functionalities: (i) image processing techniques in inverting an image/volume; (ii) the routine to visualize data in 2D or 3D in three composite 2D views: ImageViewer 2010 (b). © Carlo Ciulla. The Magnetic Resonance Imaging (MRI) can be found in the OASIS database: www.oasis-brains.org (Buckner et al., 2004; Fotenos et al., 2005; Marcus et al., 2007; Morris, 1993; Rubin et al., 1998; Zhang et al., 2001) and is courtesy of Dr. Daniel Marcus.

InvertImage 2010 is an application which calculates the inverse of an image/volume. Images and volumes can be of two types: (i) with a header, and (ii) without a header. Regardless of whether the image/volume has the header, the program writes one text header of the resulting inverse image/volume and this is done in a way that is transparent to the user. The user is asked to work with one single image resolution (2D or 3D) at the same time. Should the user want to process a different image, he is asked to exit the program and to run it again. The resulting inverse does have a binary header in it. The result consists of pixel data in DOUBLE format (64 bits).

The image header information is in the following order: X, Y, and Z matrix sizes; X, Y and Z pixel size [mm]; DataType; StdLabel. The data types available are: CHAR, SHORT, INT, FLOAT and DOUBLE. The StdLabel is either Y or N to indicate whether the image data is in standardized form or not. Input to the program can be either an image from the application named ImageTool 2010 or any relevant Matlab® application with conforming header information, or raw images (without the header) containing pixel data in the following formats: CHAR, SHORT, INT, FLOAT and DOUBLE.

When processing a raw image the user specifies the format in the edit box: 'Image Specifications' Also, the user makes sure that the check box: 'Header ON' is properly checked or not and also makes sure that the image/volume name is located in the edit box just above the push button 'Invert Image/Volume'.

The Button: 'Calculate Histogram' will save the text file containing the histogram values of the image/volume. The local folder contains a template text file which is overwritten, and is the interface to Microsoft Excel®, which is employed so as to make sure that the user can compile the resulting graph. The panel: 'Calculate and Display from Text Data' requests the user not to type the extension: '.txt' of the text file name in the edit box. The text data must be in scientific format %12.20f, which is: FLOAT with 20 digits after the decimal point. The user is asked to check the header has been clicked on.

The user is asked to place the name of the text file containing data in the edit box located at the left of the button: 'Run'. Then the user chooses a value between 0 and 1 to assign to the parabolic function. The program calculates and saves the histogram and also displays the text data in either 2D or 3D. Visual data display is aided through OpenGL.

The result consists of pixel data in DOUBLE format (64 bits). Histogram data is also saved, but this takes place, however, in text format. Help to use the application is provided through a log file, obtained when one pushes the 'Help' button, and saved in the same directory where the program runs, which is where the user is asked to place the image data. When viewing data through OpenGL graphics, the user is asked to run ImageViewer 2010 (see figure 2.b).

Through ImageViewer 2010, 3D data is displayed with 2D windows, giving plane, sagittal and coronal sections. Moving the mouse arrow allows one to give control to the windows. 'U' and 'D' keys are used to see plane sections, 'L' and 'R' keys are used to see sagittal sections, 'F' and 'B' keys are used to see coronal sections. Once one window is reshaped, the user is asked to reshape the other two windows too. Also, when viewing 2D data, a single 2D plane window is on display.

Image Tool 2010 creates, reads and displays 2D and 3D image data. Also, the application calculates difference images in either 2D or 3D.

(i) Creation of image data. Chaotic image data can be created employing randomization with the push button named: 'Cube and Slice'. Image data conforming to the header format can also be imported into this application. The program writes the image header, in a way which is transparent to the user. The image header information is in the following order: X, Y, and Z matrix sizes; X, Y and Z pixel sizes [mm]; DataType; StdLabel. The data types available are: CHAR, SHORT, INT, FLOAT and DOUBLE. The StdLabel is either Y or N to indicate whether the image data is in standardized form or not.

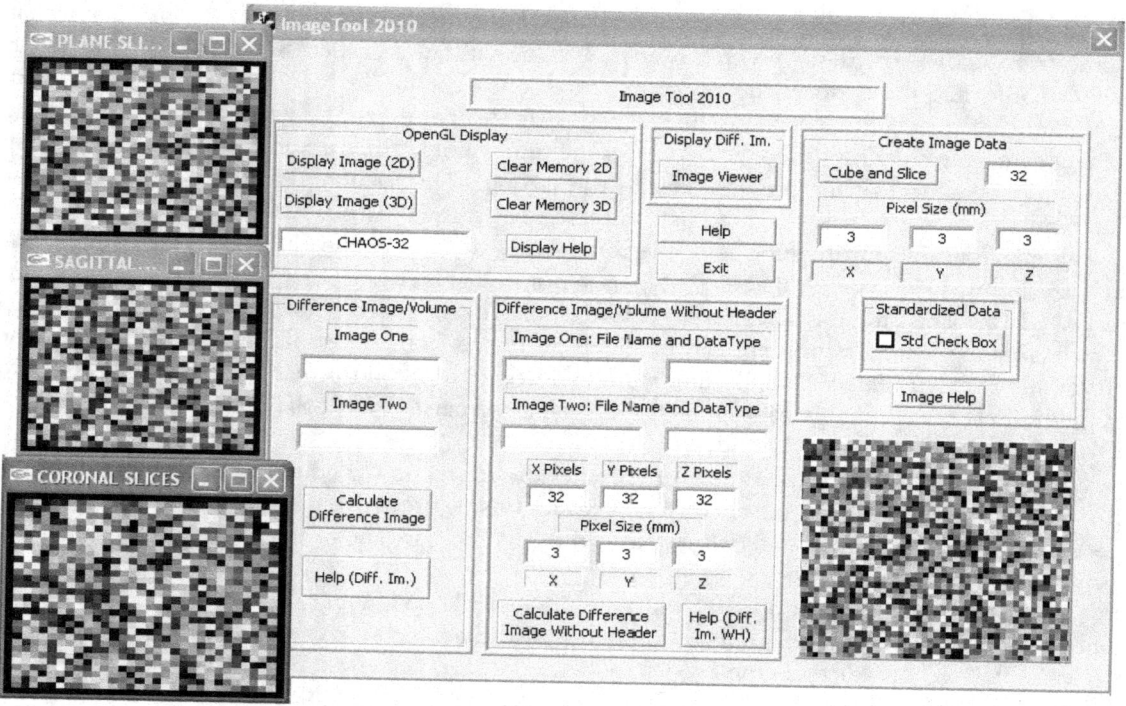

Figure 3: ImageTool 2010. Visual C++/OpenGL application employed to create the artificial data in 2D and 3D and to calculate difference images with or without header information. © Carlo Ciulla.

Should the user want to create standardized image data, he needs to keep the 'Std Check Box' checked before pushing the button named 'Cube and Slice'. Should the user want to import image data with header information conforming to this application it is not necessary to keep the 'Std Check Box' checked.

(ii) OpenGL display. To display an image either in 2D or 3D: insert in the relevant edit box the file name of the image being displayed: either 2D or 3D. Multiple sessions are allowed. The user makes sure that memory is cleared after display, either 2D or 3D. Specifically he clears the memory 2D after 2D display and clears the memory 3D after 3D display. The program displays the image data and windows for 2D plane, sagittal and coronal sections will be visualized; the user moves the mouse arrow to give control to the windows and hits 'U' and 'D' keys to see plane sections, hits 'L' and 'R' keys to see sagittal sections and hits 'F' and 'B' keys to see coronal sections. When viewing an image volume, once one window is reshaped, the user reshapes the other two windows also.

(iii) Difference image/volume. To calculate a difference image between images/volumes which do have binary header information either in 2D or 3D, the user inserts in the relevant edit boxes the two file names of the images and makes sure that the two images are consistent: (a) in their dimensionality, both of them either 2D or 3D; (b) in their matrix size, both of them having the same number of pixels; (c) in their pixel size, both of them having the same pixel sizes in X, Y and Z; (d) in their data type, both of them having the same data type. Then the user simply pushes the button named 'Calculate Difference Image'. The result consists of the difference image/volume, and the header text file. To visualize the resulting difference image/volume, the user employs ImageViewer 2010 after reading the header text file.

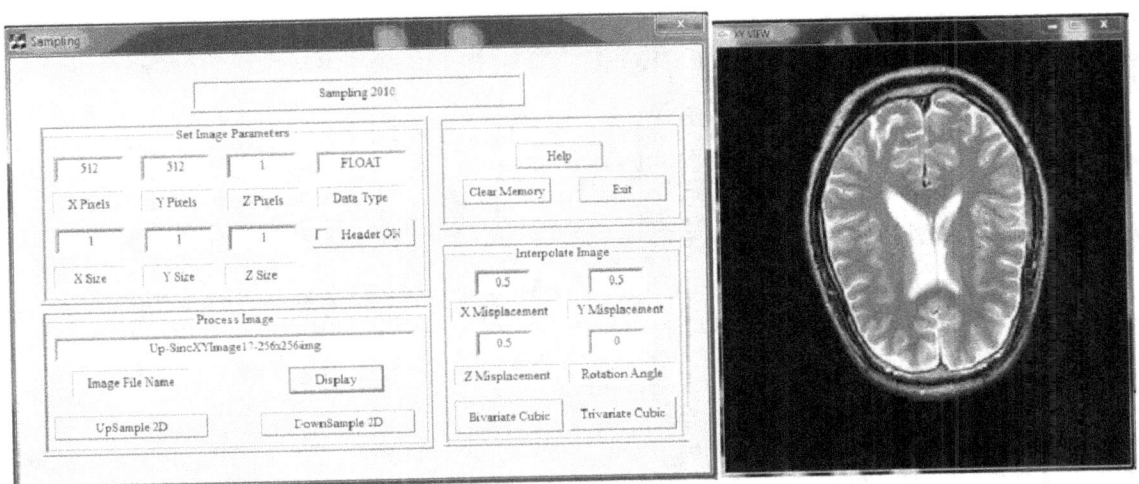

Figure 4: Sampling 2010, the Visual C++/OpenGL application employed to calculate the curvature images and the resilient signals seen in figures 10 and 13 through 19. © Carlo Ciulla. The MRI plane view is courtesy of Casa di Cure Triolo-Zancla, Palermo - Italy.

Sampling 2010: the user is asked to set the image/volume parameters; if the image/volume has a header then the user checks the corresponding box; in this case the image/volume parameters input from the GUI will be compared with those stored in the file. The 'Image File Name' edit box is used to input the name of the image/volume to be processed.

The 'Clear Memory' button is used to clear the OpenGL memory allocated for graphics at each display after the OpenGL window is closed.

The program offers the following functionalities. (i) Up-sampling: the 2D image is doubled in size. This feature can be used iteratively to reach aspect ratios greater than 200%. Each time the up-sample is of 200%. Up-sampling is done through bivariate linear and Sinc interpolation functions.

(ii) Down-sampling: the 2D image is halved in size. This feature can be used iteratively to reach aspect ratios smaller than 50%, however it is not recommended because of the effect of the interpolation. Each time the down-sample is of 50%. Down-sampling is done through the Sinc interpolation function.

(iii) Display: this button is used after processing and/or interpolation to view an image/volume using OpenGL. The image/volume is seen in the one plane window and when a volume is viewed the use of the keys 'U' and 'D' slides the volume up and down.

(iv) Image Interpolation: the user is requested to set the X, Y, Z misplacements and the rotation angle. Pushing the button 'Bivariate Cubic' interfaces with Matlab® and runs the program called: 'LGR2D2010RCCurvature'. At the end of processing, several images are saved in the local directory and can be displayed. File names can be retrieved from the local directory and placed in the edit box for display. Should the user wish to interpolate an image volume, he is then asked to use the button named: "Trivariate Cubic". On each occasion a log file is saved where the image file names can be retrieved and placed in the edit box for display. Each time any image is saved in the format called FLOAT (32 bits Real).

ON THE ASSUMPTION THAT THE TRUE VALUE OF THE SIGNAL WHICH IS BEING RECONSTRUCTED MOST CLOSELY MATCHES THE ONE SIGNAL VALUE AT THE GRID POINT

This section addresses the assumption made to undertake the experimentations relevant to the new classes of interpolators built in order to perform signal reconstruction. The assumption is that the true value of the signal which is being reconstructed most closely matches the original signal at the grid point.

From this assumption it may be derived that the Mean Absolute Error is calculated as $MAE = abs(C(s_1) - C(s_2)) = \varepsilon$, where the content of s_1, herein called $C(s_1)$, differs from the content of s_2, herein called $C(s_2)$, of the attenuation ε, which is not equal to zero ($\varepsilon \neq 0$).

The concept needs support from the definition of virtual shift-rotation. A signal is virtually shifted-rotated when the sampling grid is not moved and the signal is simply re-calculated at the location say (x, y), in 2D and this is done through an interpolation formula. The signals we are working with in this book are shifted-rotated and re-calculated, but the grid of the original samples does not change.

The argument as stated earlier might be that one should expect $\varepsilon = 0$ when the neighborhood admits constant pixel intensity, thus flat curvature (zero), not otherwise. Indeed when the signal is virtually shifted-rotated, the question to answer is why would it be that the attenuation ε is expected to be to zero? Why so, even if the pixels' intensity is not constant in the neighborhood, thus the curvature is not zero?

It is surely possible to admit that ε equal to zero implies that the signals should be the closest possible match to the original signal at the grid points. To expect this to be so is like expecting to have moved the signal (in space for instance) away from its original position, and that its value has its new sample as similar as possible to the original value. Thus, signal reconstruction in such cases is like simulated signal shift (rotation), and therefore virtual shift-rotation, done in such a way as to attempt to preserve the value of the samples. How rational is this?

Consider a digitized object (say a 3D object) having a given location in space at a given time, and a new location at a new time. The object is to be reconstructed at the new location. Would it be expected that the object would change the value of the digital samples?

The object would be different if the samples change their values. *Thus to simulate the virtual shift-rotation has the rationale behind the signal reconstruction of signals which move from one location to another, being a space or a temporal location.*

Then why not simply move the object and avoid re-sampling through interpolation? Interpolation has been invented (Newton & Huygens, 1934) mostly to find the value of a signal at intra-node (1D), intra-pixel (2D) and intra-voxel (3D) locations. *When the object is to be reconstructed at different locations, which means we are not sampling unknown values but simply reconstructing what is known already in its digitized samples, then interpolation needs to take care of the need of not changing the value of the original samples; in other words, re-sampling without affecting the values of the samples (see section QUESTIONS AND ANSWERS: Q & A. 4).*

These works demonstrate that makes sense to interpolate the signal and to measure the interpolation error through the MAE given the assumption that interpolation is used to reconstruct the signal at a different spatial or temporal location. Within the aforementioned context, the use of the Mean Absolute Error (MAE) relates to the search for minimal energy change and so the use of the MAE relates to the minimal change in curvature as shown in *Q & A. 4*. The energy change of the signal is quantifiable through the change in curvature because the Intensity-Curvature Functional, which is a measure of the *energy change of the signal* (Ciulla, 2009) embeds the geometrical meaning of the curvature through the second order derivatives.

Under the assumption that interpolation is used to reconstruct the signal at a different spatial location is the signal reconstruction technique called up-sampling.

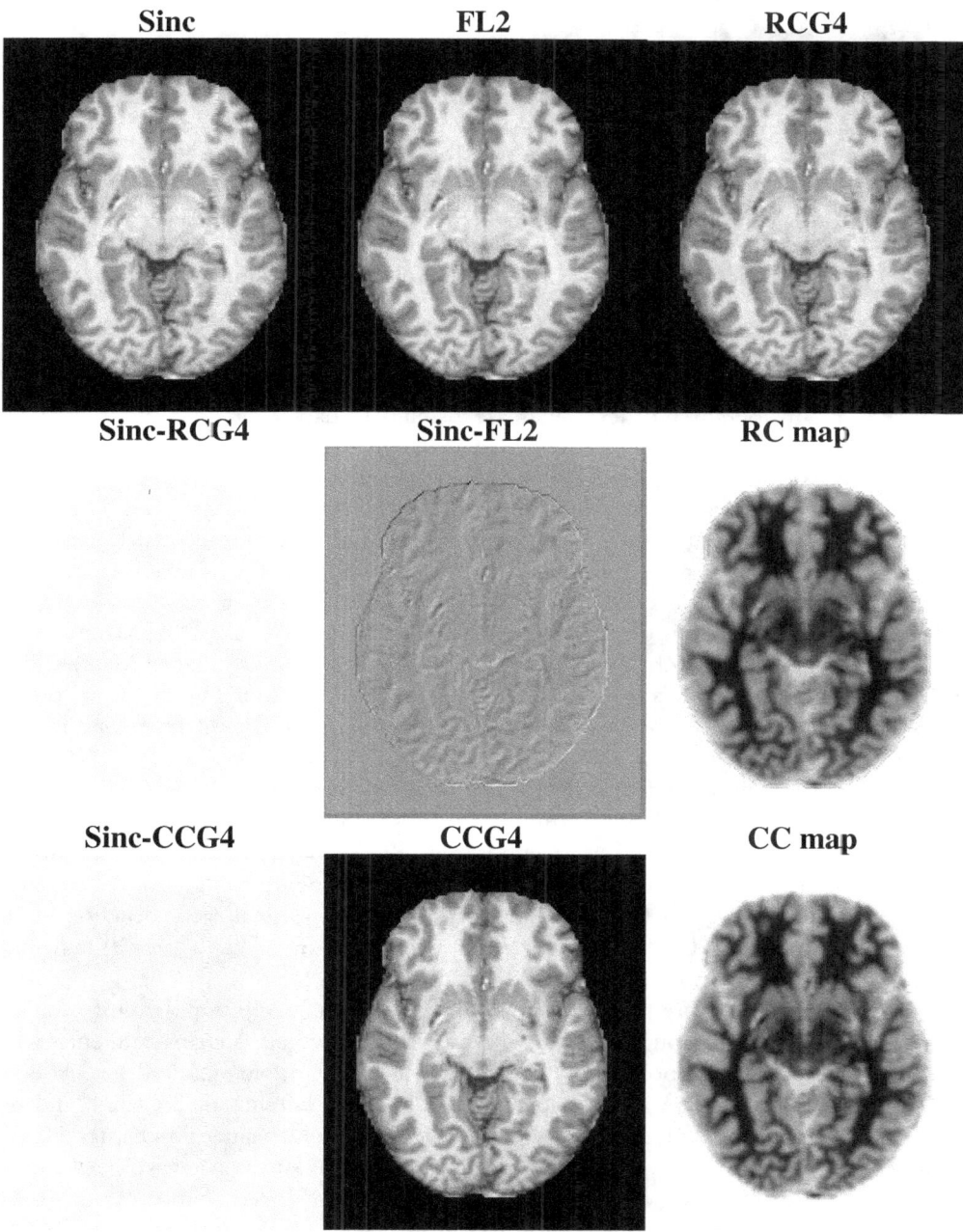

Figure 5: Experimentation with a Magnetic Resonance Imaging (MRI) plane slice. Top row: Sinc, bivariate linear (FL2), and resilient-curvature (RCG4) signal reconstructions. Second row from left to right: difference images (residuals) Sinc-RCG4, Sinc-FL2, and the curvature map of the RCG4 (RC map). In the third row from left to right we give: the residual Sinc-CCG4, classic curvature (CCG4) signal reconstruction, and the curvature map of the CCG4 (CC map).

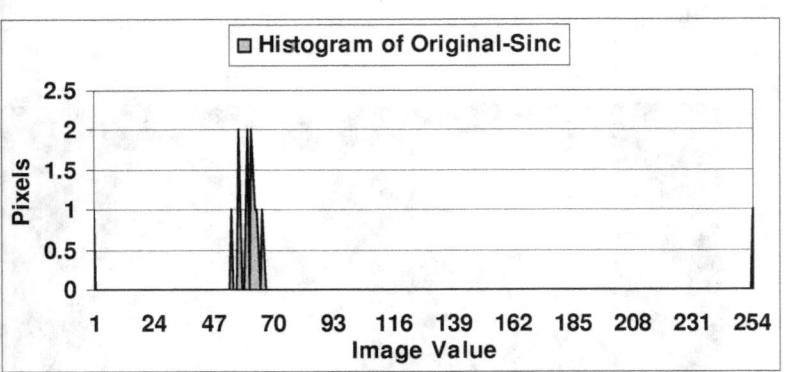

Figure 6: Residual Original-Sinc (left), and the histogram of the residual image (right), where the pixels' intensity has been scaled in the range [0, 255]. The value 3.583 E+04 is not shown at location 61 on the abscissa of the histogram in order to increase visibility of data.

Given this premise, let us have a look at an experiment involving Sinc, bivariate linear (FL2) interpolation functions, and the two new interpolation formulae called in this book classic-curvature and resilient-curvature interpolation formulae in their math form called G4 of which details are given in chapter 11. The images seen in the top row of figure 5 are Magnetic Resonance Imaging (MRI) slices with a matrix resolution of 176 x 208 pixels. The original image (see figure 7) can be found in the OASIS database: www.oasis-brains.org (Buckner et al., 2004; Fotenos et al., 2005; Marcus et al., 2007; Morris, 1993; Rubin et al., 1998; Zhang et al., 2001) and is courtesy of Dr. Daniel Marcus. The image is virtually shifted in figure 5 of (x, y) = (0.0055, 0.0055) with the four interpolation formulae, resampling is performed, the energy level changes and the result is different from one interpolation formula to another.

Figure 5 shows, in the top row from left to right: the Magnetic Resonance Images (MRI) resulting from Sinc, FL2 and RCG4 interpolation formulae. In the second row from the top from left to right the difference images (residuals) of the subtractions: Sinc-RCG4, Sinc-FL2, and the curvature map of the RCG4 (RC map) are shown. In the third row from the top from left to right, the difference image (residual) of the subtraction Sinc-CCG4, the image resulting from signal reconstruction with CCG4 and the curvature map of the CCG4 (CC map) are shown. The value of *the_a_const* (the parameter of the G4 function) was set equal to -4.554.

The gold standard to compare to has been assumed to be the Sinc interpolation formula. Deviations from the results offered through the Sinc function are interpreted as changes in energy level and the energy level of the Sinc interpolated image is, nevertheless, the gold standard too. The two new signal reconstruction formulae are the classic-curvature formula, herein called CCG4, and the resilient-curvature formula, herein called RCG4, the math of which is explained in chapter 11. Also, the curvature maps are shown to reinforce the significance of the curvature in devising an interpolation formula and more generally those which have a transfer function derived from the math form of the interpolator. The two new signal reconstruction formulae: classic-curvature, herein called CCG4, and resilient-curvature, herein called RCG4, are those with a curvature closest to that of the Sinc reconstructed signal (see residuals in the left-most column of figure 5, second and third rows from the top). Figure 6 at the left, shows the residual of the subtraction (Original-Sinc), which is between the original image and Sinc reconstructed image, and at the right, shows the histogram of the residual image, where every pixel's intensity has been scaled in the range [0, 255].

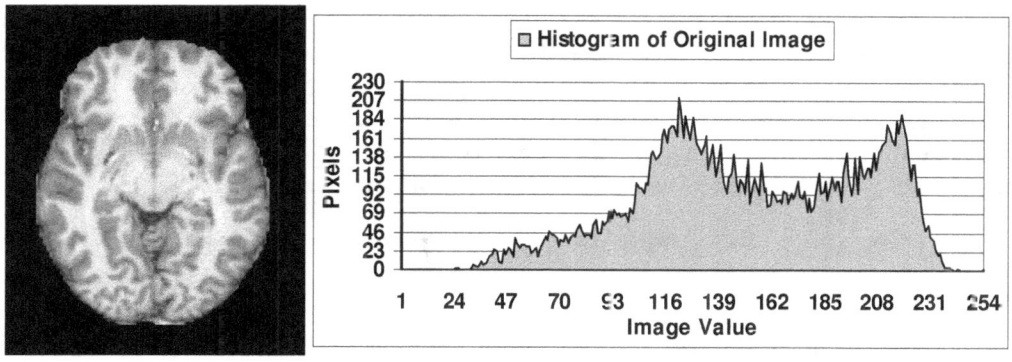

Figure 7: From left to right the original image and its histogram calculated with Invert Image 2010 is shown. The value 1.730 E+04 has been removed from location 0 in the abscissa of the histogram so as to increase visibility. This large value takes into account the large number of pixels outside the brain with intensity value equal to zero.

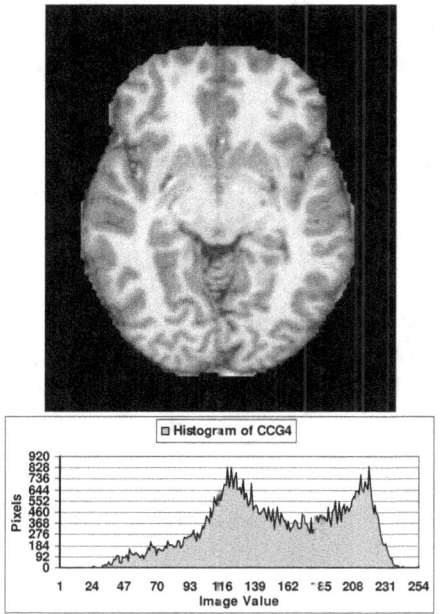

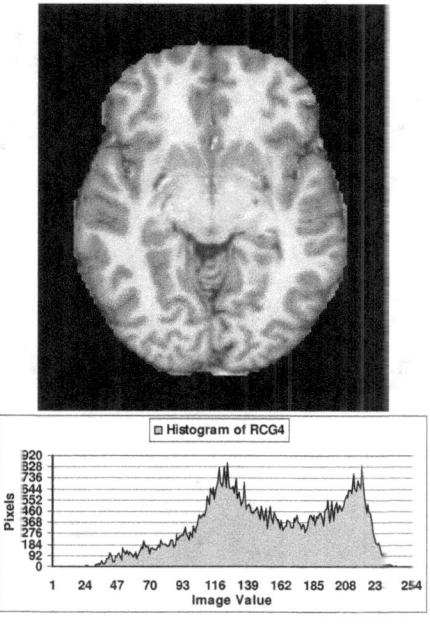

Figure 8: From top to bottom the original image up-sampled with the classic-curvature signal reconstruction technique and its histogram is shown. The value 7.071 E+04 has been removed from location 0 in the abscissa of the histogram so as to increase visibility. This large value takes into account the large number of pixels outside the brain with intensity value equal to zero.

Figure 9: From top to bottom the original image up-sampled with the resilient-curvature signal reconstruction technique and its histogram is shown. The value 7.071 E+04 has been removed from location 0 in the abscissa of the histogram so as to increase visibility.

In another experiment which we present here the initial matrix of 176 x 208 was brought up to 352 x 416, re-sampling at the center of each pixel, thus the misplacement was (x, y) = (0.5, 0.5). In figure 7 the original image is displayed along with the its histogram, whereas in figures 8 and 9 the up-sampled images, obtained with the classic-curvature signal reconstruction formula CCG4 and the resilient-curvature signal reconstruction formula RCG4 respectively, are shown along with their histograms. Noticeably the histograms in figures 7, 8 and 9 are similar; they change in the number of pixels but not change much in their profile and this suggests that the two signal reconstruction techniques are effective in changing the matrix size but not the pixel intensity map.

CURVATURE IMAGES AND THEIR ADAPTIVE BEHAVIOR TO THE SAMPLED SIGNAL

In this paragraph an introduction to the meaning of the curvature images (curvature maps) is presented. Indeed the curvature image conveys the meaning of the second order derivative image of the sampled signal, obtained on the basis of the math form of the model fitted to the data.

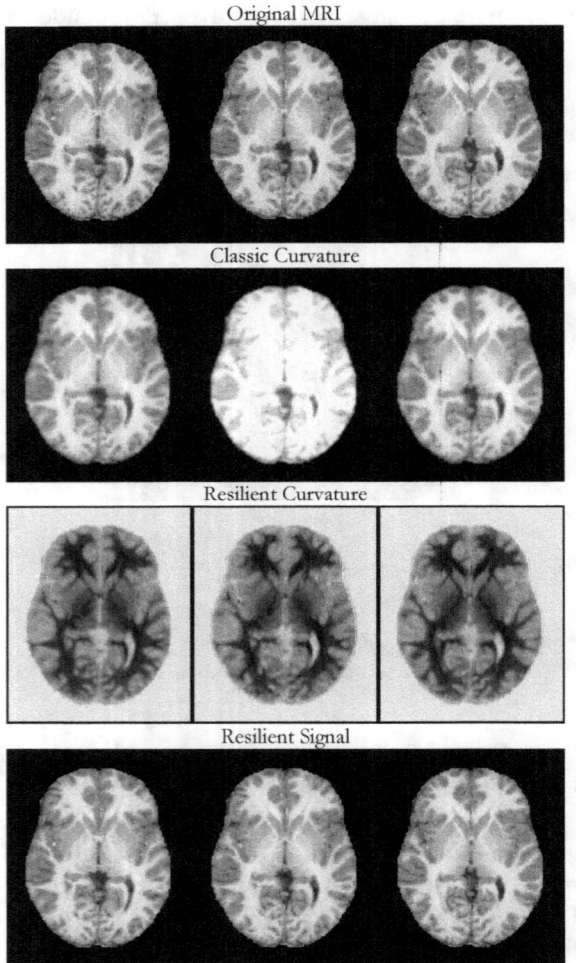

Original MRI

Classic Curvature

Resilient Curvature

Resilient Signal

Figure 10: The top row shows three plane views of an MRI (the images come from the OASIS database: www.oasis-brains.org and they are courtesy of Dr. Daniel Markus), the second row from the top shows the classic curvature images and the third row from the top shows the resilient curvature images. The fourth row from the top shows the signal resilient to interpolation from which the resilient curvature images were derived. The classic curvature images were derived from the classic form of signal through the model fitted to the data which is that of the trivariate cubic Lagrange interpolation formula.

In the set of experiments presented in this paragraph the model is the three-dimensional cubic Lagrange formulae in their variants: classic-curvature interpolation formula and resilient-curvature interpolation formula, both with embedding. The math of these interpolators is explained in chapter 2. Naturally, to calculate the derivatives of the image, the model interpolation function is fitted to the data and the derivatives are obtained from the math form of the interpolator. Figure 10 shows, in the top row, three MRI plane views in their original form; the signals were sampled at the center of the pixel $(x, y, z) = (0.0967, 0.0999, 0.1031)$ with the trivariate cubic Lagrange interpolation formula with embedding in its classic and resilient forms respectively.

The focus of this presentation is on Magnetic Resonance Imaging (MRI) signals seen in figures 11, 12 and 20. In figure 10 and in figures 13 through 19 the curvature images are presented along with the resilient signal. Calculation of the sum of second order derivatives (the total curvature) of the resilient

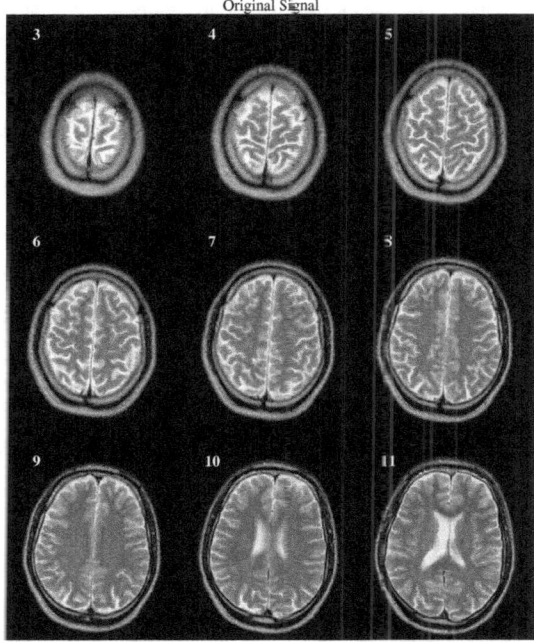

Figure 11: MRI slices 3 through 11.

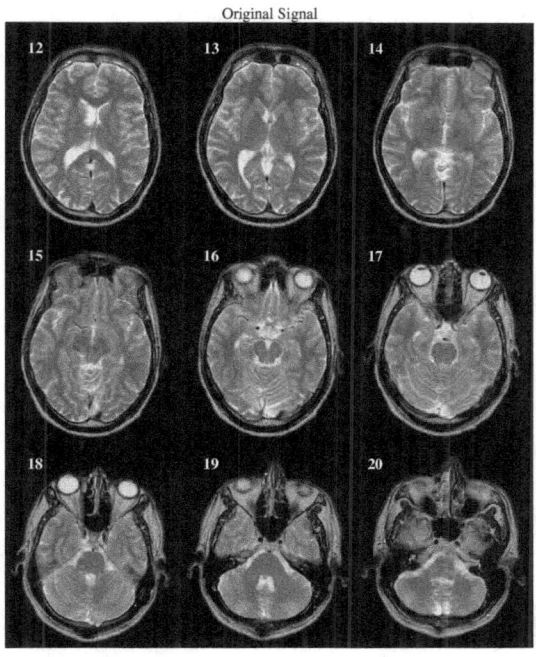

Figure 12: MRI slices 12 through 20.

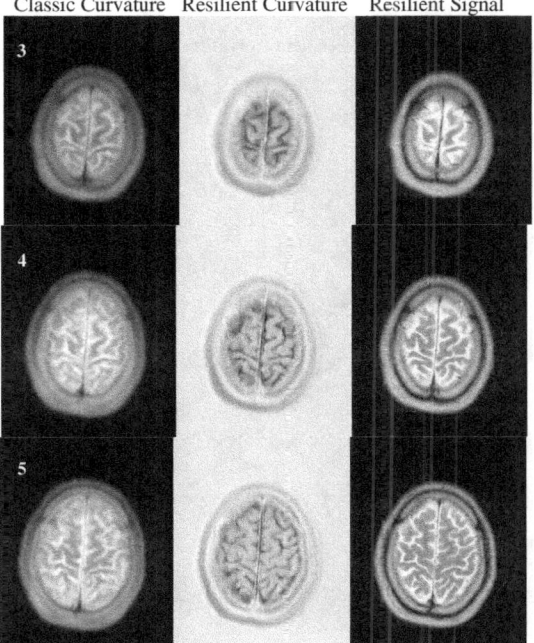

Figure 13: Classic curvature, resilient curvature and resilient signal images: slices 3 through 5.

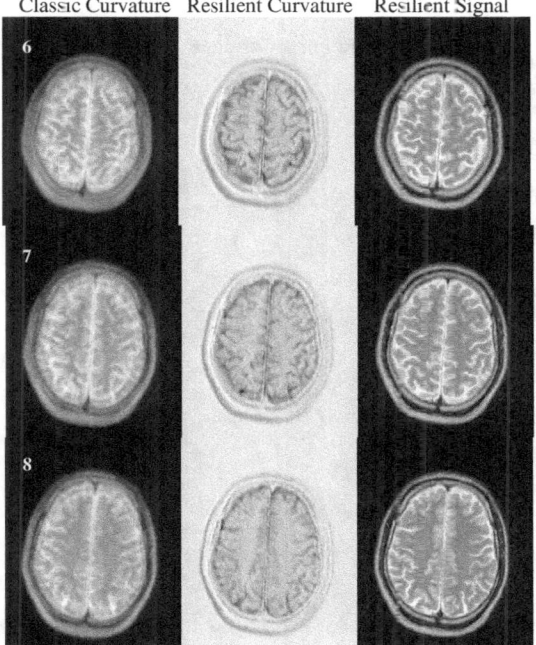

Figure 14: Classic curvature, resilient curvature and resilient signal images: slices 6 through 8.

Classic Curvature   Resilient Curvature   Resilient Signal

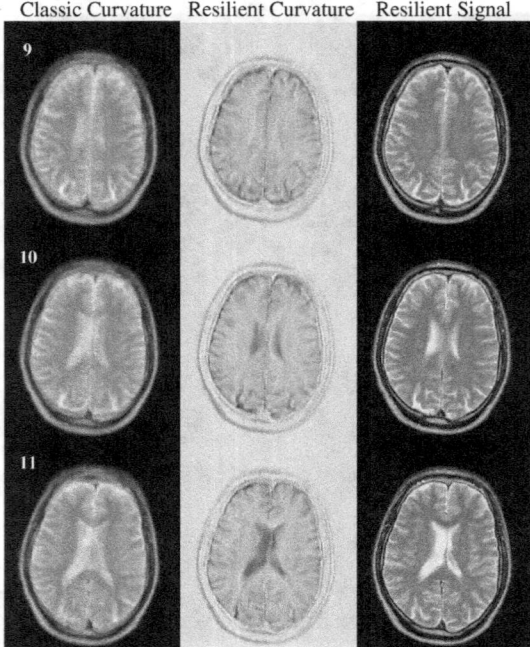

Figure 15: Classic curvature, resilient curvature and resilient signal images: slices 9 through 11.

Classic Curvature   Resilient Curvature   Resilient Signal

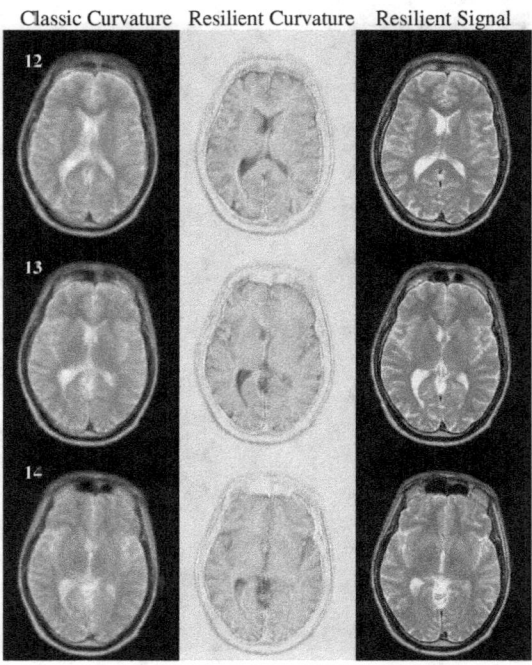

Figure 16: Classic curvature, resilient curvature and resilient signal images: slices 12 through 14.

Classic Curvature   Resilient Curvature   Resilient Signal

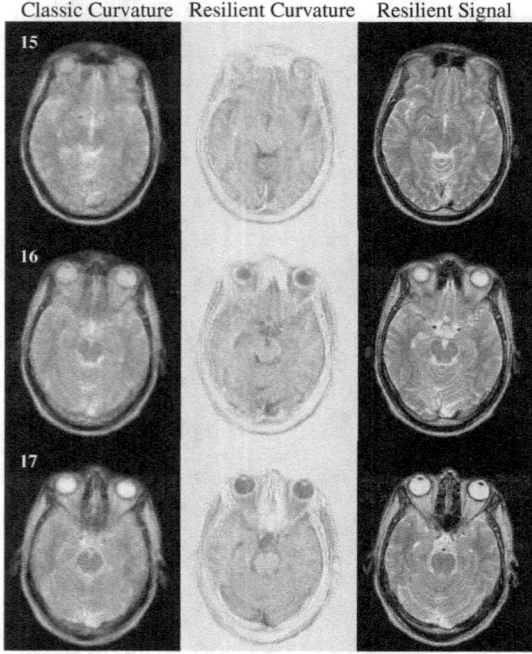

Figure 17: Classic curvature, resilient curvature and resilient signal images: slices 15 through 17.

Classic Curvature   Resilient Curvature   Resilient Signal

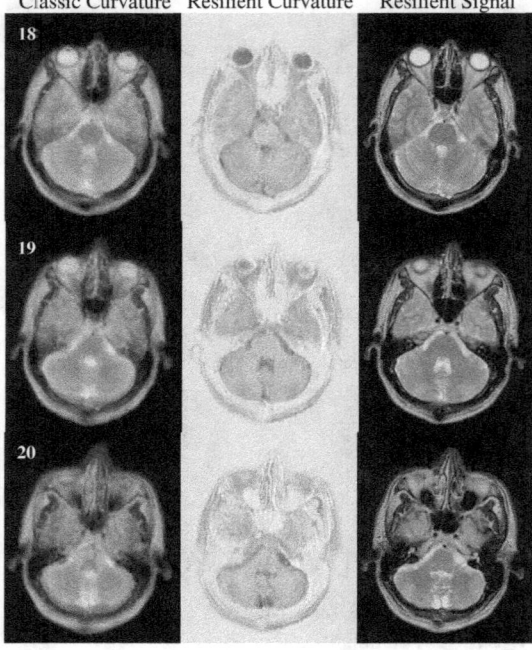

Figure 18: Classic curvature, resilient curvature and resilient signal images: slices 18 through 20.

Classic Curvature   Resilient Curvature   Resilient Signal

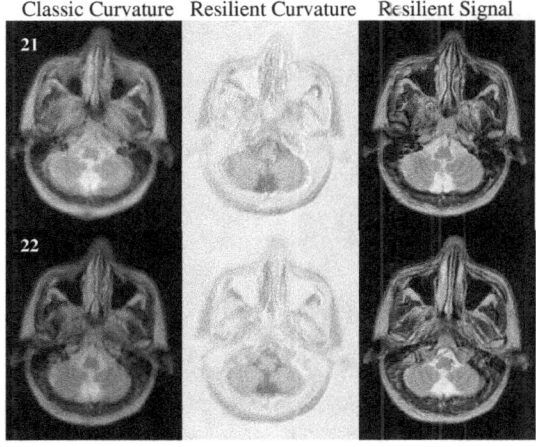

Original Signal

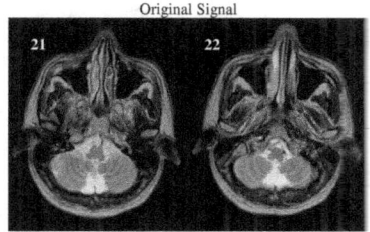

Figure 20: MRI slices 21 and 22.

Figure 19: Classic curvature, resilient curvature and resilient signal images: slices 21 through 22.

signal yields the resilient curvature map. Similarly the classic curvature map is derived from the classic signal through the calculation of the total curvature. The model fitted to the data is the interpolator used to sample the signal. The interpolator was the trivariate cubic Lagrange interpolation formula with embedding, which is the basis for calculating classic curvature, resilient curvature and the resilient signal shown in the figures.

From these two signals: classic and resilient, it is possible to calculate the second order derivatives and thus the maps of the total curvature.

SUMMARY

Interpolators do have the capability of estimating signals at time and/or space locations where they are unknown. Their mathematical form is established in such a way as to convolve signal (pixels' intensity) with misplacement (which can either be measured in space or time). This makes what is called the signal intensity correction. In this book the signal intensity correction is referred to as pixel intensity correction (PIC) for the sake of convenience since tests are conducted with images (2D) and image volumes (3D). The signal intensity correction needs to be scaled, or as one might say filtered, or as one might say adjusted to the values of the current signal and in order to do so, the transfer function, whose choice is as relevant as much as it may be arbitrary, can introduce bias and arbitration. Through the use of the total curvature of the classic signal or the resilient signal, this book demonstrates that the pixel intensity correction can have a new definition, and at the same time can serve as the transfer function that is derived deterministically from the math form of the interpolator, thus removing bias and arbitration.

REFERENCES

Akima, H. (1991). *A Method of Univariate Interpolation that Has the Accuracy of a Third-Degree Polynomial.* ACM Transactions on Mathematical Software, 17(3), 341-366.

Agarwal, R. P. & Wong, P. J. Y. (1993). *Error inequalities in polynomial interpolation and their applications. Mathematics and its applications.* Dordrecht, The Netherlands: Kluwer Academic Publishers.

Agler, J. & McCarthy J. E. (2002). *Pick interpolation and Hilbert function spaces.* Amer Mathematical Society.

Alpay, D. & Gohberg, I. (2006). *Interpolation, Schur functions, and moment problems.* Birkhauser.

Amidror, I. (2002). *Scattered Data Interpolation Methods for Electronic Imaging Systems: A Survey.* J Electr Imag, 11(2), 157-176.

Appledorn, C. G. (1996). *A New Approach to the Interpolation of Sampled Data*. IEEE Transactions on Medical Imaging, 15(3), 369-376.

Astola, J. (2007). *Advances In Signal Transforms, Theory and Applications*. Hindawi Publishing Corporation.

Bannore, V. (2009). *Iterative-Interpolation Super-Resolution Image Reconstruction, A Computationally Efficient Technique*. Springer Verlag.

Bennett, C. & Sharpley, R. C. (1988). *Interpolation of operators*. Academic Press.

Bercovici, H. & Foiaş, C. (2000). *Operator theory and interpolation*. Springer.

Bolotnikov, V. & Dym, H. (2006). *On boundary interpolation for matrix valued Schur functions*. Amer Mathematical Society.

Brigger, P. Hoeg, J. & Unser, M. (2000). *B-Spline Snakes: A Flexible Tool for parametric Contour Detection*. IEEE Transactions on Image Processing, 9(9), 1484-1496.

Buckner, R. L., Head, D., Parker, J., Fotenos, A. F., Marcus, D., Morris, J. C. & Snyder, A. Z. (2004). *A unified approach for morphometric and functional data analysis in young, old, and demented adults using automated atlas-based head size normalization: Reliability and validation against manual measurement of total intracranial volume*. Neuroimage, 23(2), 724-738.

Buzzi-Ferraris, G. & Manenti F. (2010). *Interpolation and Regression Models for the Chemical Engineer, Solving Numerical Problems*. Wiley-VCH.

Ciulla, C. (2009). *Improved Signal and Image Interpolation in Biomedical Applications: The Case of Magnetic Resonance Imaging (MRI)* – Medical Information Science Reference - IGI Global Publisher, Hershey, PA, U.S.A.

Cwikel, M., De Carli, L. & Milman M. (2007). *Interpolation theory and application*. American Mathematical Society.

De Boor, C. (1978). *A practical guide to splines. Applied mathematical sciences*. Springer-Verlag.

De Boor, C. (1986). *Approximation theory*. Amer Mathematical Society.

Deng, X. & Denney, Jr, T. S. (2004). *On Optimizing Knot Positions for Multi-dimensional B-Spline Models*. Proc. IS & T SPIE.

Dym, H. Alpay, D. Gohberg, I. & Vinnikov, V. (2002). *Interpolation theory, systems theory and related topics*. Birkhäuser.

Dym, H. Fritzsche, B. & KatsnelsonBernd V. (1997). *Topics in interpolation theory*. Birkhauser.

Foiaş, C. (1998). *Metric constrained interpolation, commutant lifting, and systems*. Birkhauser.

Fotenos, A. F., Snyder, A. Z., Girton, L. E., Morris, J. C. & Buckner, R. L. (2005). *Normative estimates of cross-sectional and longitudinal brain volume decline in aging and AD*. Neurology, 64, 1032-1039.

Fraser, D. (1989). *Interpolation by the FFT Revisited – An Experimental Investigation*. IEEE Transactions on Acoustics, Speech, and Signal Processing, 37(5), 665-675.

Fraser, D. C. (2009). *Newton's Interpolation Formulas*. Bibliolife.

Frijns, J. H. M., de Snoo, S. L. & Schoonhoven, R. (2000). *Improving the Accuracy of the Boundary Element Method by the Use of Second-Order Interpolation Functions*. IEEE Transactions on Biomedical Engineering, 47(10), 1336-1346.

Gohberg, I. (1992). *Time-variant systems and interpolation*. Princeton Architectural Press.

Gohberg, I. (1998). *Topics in operator theory and interpolation*. Birkhauser.

Goodearl, K. R. (2010). *Partially Ordered Abelian Groups with Interpolation*. AMS Bookstore.

Jetter, K., Buhmann, M. & HaussmannRobert W. (2006). *Topics in multivariate approximation and interpolation*. Elsevier Science.

Kalton, N. J. (1988). *Nonlinear commutators in interpolation theory*. Amer Mathematical Society.

Kowalewski, G. (1932). *Interpolation und genäherte Quadratur*. Teubner, Berlin.

Krein, S. G., Petunin, J. I. & Semenov, E.M. (2002). *Interpolation of linear operators*. Amer Mathematical Society.

Krugljak, N. Ya. (1991). *Interpolation functors and interpolation spaces*. North Holland.

Lorentz, G. G. Jetter, K. & Riemenschneider, S. D. (1984). *Birkhoff interpolation*. Cambridge Univ Press.

McArtney Phillips, G. (2003). *Interpolation and approximation by polynomials*. Springer Verlag.

McLeod, R. J. Y. & Baart, M. L. (1998) *Geometry and interpolation of curves and surfaces*. Cambridge Univ Press.

Marcus, D. S., Wang, T. H., Parker, J., Csernansky, J. G., Morris, J. C. & Buckner, R. L. (2007). *Open Access Series of Imaging Studies (OASIS): Cross-sectional MRI data in young, middle aged, nondemented, and demented older adults*. Journal of Cognitive Neuroscience, 19(9), 1498-1507.

Meijering, E. (2002). *A chronology of interpolation: From ancient astronomy to modern signal and image processing*. Proceedings of the IEEE, 90(3), 319-342.

Morris, J. C. (1993). *The clinical dementia rating (CDR): current version and scoring rules*. Neurology, 43(11), 2412b-2414b.

Newton I. & Huygens, C. (1934). The motion of the moon's nodes (lemma 5). In R. M. Aynard Hutchins (Ed.), *Mathematical Principles of Natural Philosophy* (pp. 338-339). William Benton.

Peetre, J. & Cwikel, M. (2002). *Function spaces, interpolation theory, and related topics*. Walter De Gruyter Inc.

Penney, G. P. Schnabel, J. A. Rueckert, D. Viergever, M. A. & Niessen, W. J. (2004). *Registration- based interpolation*. IEEE Transactions on Medical Imaging, 23(7), 922-926.

Popescu, G. (2006). *Entropy and multivariable interpolation*. Amer Mathematical Society.

Rubin, E. H., Storandt, M., Miller, J. P., Kinscherf, D. A., Grant, E. A., Morris, J. C. & Berg, L. A. (1998). *A prospective study of cognitive function and onset of dementia in cognitively healthy elders*. Archives of Neurology, 55(3), 395-401.

Schoenberg, I. J. (1987). *Cardinal Spline Interpolation*. Society for Industrial Mathematics.

Seip, K. (2004). *Interpolation and sampling in spaces of analytic functions*. Amer Mathematical Society.

Shi, Y. G. (2003). *Theory of Birkhoff interpolation*. Nova Science Pub Inc.

Späth, H. (1995). *Two dimensional spline interpolation algorithms*. A K Peters, Ltd.

Stein, M. L. (1999) *Interpolation of spatial data, some theory for kriging*. Springer Verlag.

Stewart, J. (2009). *Calculus Imperial*. Thomson Brooks/Cole.

Szabados, J. P. & Vértesi, P. (1990) *Interpolation of functions*. World Scientific Pub. Co. Inc.

Tartar, L. (2002). *An introduction to Sobolev spaces and interpolation spaces*. Springer Verlag.

Unser, M., Aldroubi, A., & Eden, M. (1993a). *B-spline signal processing: Part I – theory*. IEEE Transactions on Signal Processing, 41(2), 821-833.

Unser, M., Aldroubi, A., & Eden, M. (1993b). *B-spline signal processing: Part II – efficient design and applications*. IEEE Transactions on Signal Processing, 41(2), 834-848.

Unser, M. & Daubechies, I. (1997). *On the Approximation of Power of Convolution-Based Least Squares Versus Interpolation*. IEEE Transactions on Signal Processing, 45(7), 1697-1711.

Waldron, S. (1998). *The Error in Linear Interpolation at the Vertices of a Simplex*. SIAM J Numer Anal, 35, 1191-1200.

Waldron, S. (2000). *Minimally Supported Error Representations and Approximation by the Constants*. Numer Math, 85, 469-484.

Wang, J. (2001). *Optimal Design for Linear Interpolation of Curves*. Statistics in Medicine, 20, 2467-2477.

Wang, Z. W., Soltis, J. J. & Miller, W. C. (1992). *Improved Approach to Interpolation Using the FFT*. Eletronic Letters, 28(25), 2320-2321.

Zhang, Y., Brady, M., & Smith, S. (2001). *Segmentation of brain MR images through a hidden Markov random field model and the expectation maximization algorithm*. IEEE Transactions on Medical Imaging, 20(1), 45-57.

CHAPTER 2

SIGNAL RESILIENT TO INTERPOLATION WITH EMBEDDING: THE MATHEMATICAL FORMULATION

INTRODUCTION

This chapter shows how to calculate the signal resilient to interpolation (or so called resilient signal) of an interpolation function. The only requirement to the interpolation function is that of second order differentiability, which means that the second order derivatives of the function exist, are continuous and not null (Meijering, 2002).

This ensures the existence of the tangent to the first order derivative of the function, which will be come useful in subsequent chapters when the total curvature will be calculated. The first step in the calculation of the signal resilient to interpolation is that of the computation of the second order partial derivatives.

For instance in a bivariate function $h_3(x, y)$, the derivatives to calculate are: $(\partial^2 (h_3(x, y)) / \partial x^2)$, $(\partial^2 (h_3(x, y)) / \partial x \partial y)$, $(\partial^2 (h_3(x, y)) / \partial y \partial x)$, $(\partial^2 (h_3(x, y)) / \partial y^2)$ so as to consider the mono-variate x and y and the bivariates xy and yx.

The second step is to calculate the intensity curvature terms (Ciulla, 2009) before and after interpolation: $E_o(x, y)$ and $E_{IN}(x, y)$ respectively.

The third step is that of solving the equation $E_o(x, y) = E_{IN}(x, y)$ in $f(0, 0)$, which is the pixel to re-sample. The reader shall see that the value of $f(0, 0)$ is a convolution of the pixels' intensity and misplacement.

Why then is it necessary to calculate the signal resilient to interpolation? At this stage of the exploratory research, $f(0, 0)$ is the new signal expected to provide interpolation error improvement over the classic model signal $h_3(x, y)$.

In chapter 4 the reader shall see that through the proof of concept, $f(0, 0)$, obtained from the equation $E_o(x, y) = E_{IN}(x, y)$, interpolation error improvement is able to furnished, however with too high a value of the MAE because no transfer functions has been used to scale its convolution.

Thus, at that stage of the exploratory research $f(0, 0)$ is initially that fraction of pixel intensity to be added to the original value so as to obtain the estimate at the location (x, y) of the unknown value of the signal through the model function $h_3(x, y)$ and it will become the means to calculate the new conception of Pixel Intensity Correction (PIC) introduced in this book.

PART I – BIVARIATE QUADRATIC AND CUBIC FUNCTIONS WITH EMBEDDING

BIVARIATE QUADRATIC AND CUBIC B-SPLINES

Let us consider the quadratic $h_3(x, y)$ and cubic $h_4(x, y)$ bivariate B-Splines as per equations (1) and (2) with $((3 \times 3)-1)$ and $((3 \times 3)-1)$ neighbouring nodes:

$h_3(x, y) = f (0, 0) + [f (1/2, 1/2) + f (-1/2, -1/2) + f (2/3, 2/3) + f (-2/3, -2/3)] * [- 2a (x +$

$y)^2 + 1/2 (a+1)] + [f (1/2, 1/2) + f (-1/2, -1/2) + f (2/3, 2/3) + f (-2/3, -2/3) + f (-1, -1) +$

$f (1, 1) + f (3/2, 3/2) + f (-3/2, -3/2)] * [a (x + y)^2 - (2a + 1/2) (x + y) + 3/4 (a+1)]$ (1)

$h_4(x, y) = f (0, 0) + [f (1/2, 1/2) + f (-1/2, -1/2) + f (-1, -1) + f (1, 1)] * [1/2 (x + y)^3 - (x$

$+ y)^2 + 2/3] + [f (1/2, 1/2) + f (-1/2, -1/2) + f (2/3, 2/3) + f (-2/3, -2/3) + f (-1, -1) +$

$f (1, 1) + f (3/2, 3/2) + f (-3/2, -3/2)] * [-1/6 (x + y)^3 + (x + y)^2 - 2 (x + y) + 4/3]$ (2)

An illustration of the neighborhood of f (0, 0) is given in figure 1.

f (-2/3, -2/3)	f (-1, -1)	f (3/2, 3/2)
f (-1/2, -1/2)	f (0, 0)	f (1/2, 1/2)
f (-3/2, -3/2)	f (1, 1)	f (2/3, 2/3)

Figure 1: The organization of the neighborhood of f(0, 0) for the B-Splines $h_3(x, y)$ and $h_4(x, y)$.

Let us posit:

$\alpha = [f (1/2, 1/2) + f (-1/2, -1/2) - f (2/3, 2/3) + f (-2/3, -2/3)]$

$\alpha_2 = [\, f(1/2, 1/2) + f(-1/2, -1/2) + f(2/3, 2/3) + f(-2/3, -2/3) + f(-1, -1) + f(1, 1) +$

$f(3/2, 3/2) + f(-3/2, -3/2)\,]$

$\alpha_3 = [\, f(1/2, 1/2) + f(-1/2, -1/2) + f(-1, -1) + f(1, 1)\,]$

Thus equations (1) and (2) can be written as:

$h_3(x, y) = f(0, 0) + \alpha_1 * [-2a(x + y)^2 + 1/2(a+1)] + \alpha_2 * [a(x + y)^2 - (2a + 1/2)(x + y)$

$+ 3/4(a+1)]$
$\hspace{11cm}(3)$

$h_4(x, y) = f(0, 0) + \alpha_3 * [1/2(x + y)^3 - (x + y)^2 + 2/3] + \alpha_2 * [-1/6(x + y)^3 + (x + y)^2 - 2$

$(x + y) + 4/3]$
$\hspace{11cm}(4)$

CALCULATION OF THE PARTIAL SECOND ORDER DERIVATIVES: BIVARIATE QUADRATIC B-SPLINE

First Order Partial Derivative with Respect to the x Variable

$(\partial(h_3(x, y))/\partial x) = \partial\{f(0, 0) + \alpha_1 * [-2a(x + y)^2 + 1/2(a+1)] + \alpha_2 * [a(x + y)^2 - (2a$

$+ 1/2)(x + y) + 3/4(a+1)]\}/\partial x = \{\alpha_1 * [-2a\,\partial(x + y)^2/\partial x] + \alpha_2 * [a\,\partial(x + y)^2/\partial x -$

$(2a + 1/2)\,\partial(x + y)/\partial x]\} = \{\alpha_1 * [-2a * 2(x + y)] + \alpha_2 * [a * 2(x + y) - (2a + 1/2)]\}$

Therefore:

$(\partial(h_3(x, y))/\partial x) = \{\alpha_1 * [-2a * 2(x + y)] + \alpha_2 * [a * 2(x + y) - (2a + 1/2)]\}$
$\hspace{6cm}(5)$

First Order Partial Derivative with Respect to the y Variable

$(\partial(h_3(x, y))/\partial y) = \partial\{f(0, 0) + \alpha_1 * [-2a(x + y)^2 + 1/2(a+1)] + \alpha_2 * [a(x + y)^2 - (2a$

$+ 1/2)(x + y) + 3/4(a+1)]\}/\partial y = \{\alpha_1 * [-2a\,\partial(x + y)^2/\partial y] + \alpha_2 * [a\,\partial(x + y)^2/\partial y - (2a$

$+ 1/2)\,\partial(x + y)/\partial y]\} = \{\alpha_1 * [-2a * 2(x + y)] + \alpha_2 * [a * 2(x + y) - (2a + 1/2)]\}$

Therefore:

$(\partial(h_3(x, y))/\partial y) = \{\alpha_1 * [-2a * 2(x + y)] + \alpha_2 * [a * 2(x + y) - (2a + 1/2)]\}$
$\hspace{6cm}(6)$

and it follows that:

$(\partial(h_3(x, y))/\partial x) = (\partial(h_3(x, y))/\partial y).$

Second Order Partial Derivative with Respect to the x Variable

$(\partial^2 (h_3(x, y)) / \partial x^2) = \{\partial \{\partial (h_3(x, y)) / \partial x\} / \partial x\} = \{\partial \{ \alpha_1 * [- 2a * 2 (x + y)] + \alpha_2 * [a *$

$2 (x + y) - (2a + 1/2)] \} / \partial x\} = - 4a \, \alpha_1 + 2a \, \alpha_2$ (7)

Second Order Partial Derivative with Respect to the y Variable

$(\partial^2 (h_3(x, y)) / \partial y^2) = \{\partial \{\partial (h_3(x, y)) / \partial y\} / \partial y\} = \{\partial \{ \alpha_1 * [- 2a * 2 (x + y)] + \alpha_2 * [a *$

$2 (x + y) - (2a + 1/2)] \} / \partial y\} = - 4a \, \alpha_1 + 2a \, \alpha_2$ (8)

Therefore:

$(\partial^2 (h_3(x, y)) / \partial x^2) = (\partial^2 (h_3(x, y)) / \partial y^2) = - 4a \, \alpha_1 + 2a \, \alpha_2$

Second Order Partial Derivatives with Respect to the x and y Variables

$(\partial^2 (h_3(x, y)) / \partial x \partial y) = \{\partial (\partial (h_3(x, y)) / \partial x) / \partial y\} = \{\partial \{ \alpha_1 * [- 2a * 2 (x + y)] + \alpha_2 * [a$

$* 2 (x + y) - (2a + 1/2)] \} / \partial y\} = - 4a \, \alpha_1 + 2a \, \alpha_2$ (9)

$(\partial^2 (h_3(x, y)) / \partial y \partial x) = \{\partial (\partial (h_3(x, y)) / \partial y) / \partial x\} = \{\partial \{ \alpha_1 * [- 2a * 2 (x + y)] + \alpha_2 * [a$

$* 2 (x + y) - (2a + 1/2)] \} / \partial x\} = - 4a \, \alpha_1 + 2a \, \alpha_2$ (10)

Therefore:

$(\partial^2 (h_3(x, y)) / \partial x \partial y) = (\partial^2 (h_3(x, y)) / \partial y \partial x) = - 4a \, \alpha_1 + 2a \, \alpha_2$

CALCULATION OF THE INTENSITY-CURVATURE TERMS: BIVARIATE QUADRATIC B-SPLINE

Let the intensity-curvature term before interpolation be defined as:

$E_\circ = E_\circ (x, y) = \int_0^x \int_0^y f (0, 0) * \{ (\partial^2 (h_3(x, y)) / \partial x^2) + (\partial^2 (h_3(x, y)) / \partial x \partial y) + (\partial^2 (h_3(x, y))$

$/ \partial y \partial x) + (\partial^2 (h_3(x, y)) / \partial y^2) \} (0, 0) \, dx \, dy$ (11)

Where it follows from equations (7), (8), (9) and (10) that:

$(\partial^2 (h_3(x, y)) / \partial x^2) (0, 0) = - 4a \, \alpha_1 + 2a \, \alpha_2$

$(\partial^2 (h_3(x, y)) / \partial x \partial y) (0, 0) = - 4a \, \alpha_1 + 2a \, \alpha_2$

$(\partial^2 (h_3(x, y)) / \partial y \partial x) (0, 0) = - 4a \, \alpha_1 + 2a \, \alpha_2$

$(\partial^2 (h_3(x, y)) / \partial y^2) (0, 0) = -4a \; \alpha_1 + 2a \; \alpha_2$

Therefore:

$$E_o (x, y) = \int_0^x \int_0^y 4 * f (0, 0) \{ -4a \; \alpha_1 + 2a \; \alpha_2 \} \, dx \, dy = 4xy * f (0, 0) \{ -4a \; \alpha_1 + 2a \; \alpha_2 \} \tag{12}$$

Let the intensity-curvature term after interpolation be defined as:

$$E_{IN} = E_{IN} (x, y) = \int_0^x \int_0^y h_3(x, y) * \{ (\partial^2 (h_3(x, y)) / \partial x^2) + (\partial^2 (h_3(x, y)) / \partial x \partial y) + (\partial^2 (h_3(x, y))$$

$$) / \partial y \partial x) + (\partial^2 (h_3(x, y)) / \partial y^2) \} \, dx \, dy \tag{13}$$

Where it follows from equations (7), (8), (9) and (10) that:

$(\partial^2 (h_3(x, y)) / \partial x^2) = -4a \; \alpha_1 + 2a \; \alpha_2$

$(\partial^2 (h_3(x, y)) / \partial x \partial y) = -4a \; \alpha_1 + 2a \; \alpha_2$

$(\partial^2 (h_3(x, y)) / \partial y \partial x) = -4a \; \alpha_1 + 2a \; \alpha_2$

$(\partial^2 (h_3(x, y)) / \partial y^2) = -4a \; \alpha_1 + 2a \; \alpha_2$

Therefore:

$$E_{IN} (x, y) = \int_0^x \int_0^y 4 * h_3(x, y) \{ -4a \; \alpha_1 + 2a \; \alpha_2 \} \, dx \, dy =$$

$$\int_0^x \int_0^y 4 * \{ f (0, 0) + \alpha_1 * [- 2a (x + y)^2 + 1/2 (a+1)] + \alpha_2 * [a (x + y)^2 - (2a + 1/2) (x +$$

$$y) + 3/4 (a+1)] \} * \{ -4a \; \alpha_1 + 2a \; \alpha_2 \} \, dx \, dy \tag{14}$$

Given that: $(x + y)^2 = (x^2 + 2xy + y^2)$, equation (14) is written as:

$$E_{IN} (x, y) = \int_0^x \int_0^y 4 * \{ f (0, 0) + \alpha_1 * [- 2a (x^2 + 2xy + y^2) + 1/2 (a+1)] + \alpha_2 * [a (x^2 + 2xy +$$

$$y^2) - (2a + 1/2) (x + y) + 3/4 (a+1)] \} * \{ -4a \; \alpha_1 + 2a \; \alpha_2 \} \, dx \, dy =$$

$$4 * \{ -4a \; \alpha_1 + 2a \; \alpha_2 \} * \{ xy f (0, 0) + \int_0^x \int_0^y \alpha_1 * [- 2a (x^2 + 2xy + y^2) + 1/2 (a+1)] + \alpha_2 * [a$$

$$(x^2 + 2xy + y^2) - (2a + 1/2) (x + y) + 3/4 (a+1)] \, dx \, dy \} =$$

$4 * \{ - 4a \; \alpha_1 + 2a \; \alpha_2 \} * \{ xy \; f(0, 0) + \int_0^x \alpha_1 * [- 2a (x^2y + xy^2 + y^3/3) + 1/2 \; y \; (a+1)] + \alpha_2 *$

$[a (x^2y + xy^2 + y^3/3) - (2a + 1/2) (xy + y^2/2) + 3/4 \; y \; (a+1)] \; dx \} =$

$4 * \{ - 4a \; \alpha_1 + 2a \; \alpha_2 \} * \{ xy \; f(0, 0) + \alpha_1 * [- 2a (x^3y/3 + x^2y^2/2 + xy^3/3) + 1/2 \; xy \; (a+1)]$

$- \alpha_2 * [a (x^3y/3 + x^2y^2/2 + xy^3/3) - (2a + 1/2) (x^2y/2 + xy^2/2) + 3/4 \; xy \; (a+1)] \}$ (15)

EQUATING THE TWO INTENSITY-CURVATURE TERMS BEFORE AND AFTER INTERPOLATION: BIVARIATE QUADRATIC B-SPLINE

Equating $E_o(x, y)$ as given through equation (12) to $E_{IN}(x, y)$ as given through equation (15), it can be written:

$E_o(x, y) = E_{IN}(x, y) = 4xy * f(0, 0) \{ - 4a \; \alpha_1 + 2a \; \alpha_2 \} = 4 * \{ - 4a \; \alpha_1 + 2a \; \alpha_2 \} * \{ xy$

$f(0, 0) + \alpha_1 * [- 2a (x^3y/3 + x^2y^2/2 + xy^3/3) + 1/2 \; xy \; (a+1)] + \alpha_2 * [a (x^3y/3 + x^2y^2/2 +$

$xy^3/3) - (2a + 1/2) (x^2y/2 + xy^2/2) + 3/4 \; xy \; (a+1)] \}$ (16)

which implies:

$xy * f(0, 0) = \{ xy \; f(0, 0) + \alpha_1 * [- 2a (x^3y/3 + x^2y^2/2 + xy^3/3) + 1/2 \; xy \; (a+1)] + \alpha_2 * [a$

$(x^3y/3 + x^2y^2/2 + xy^3/3) - (2a + 1/2) (x^2y/2 + xy^2/2) + 3/4 \; xy \; (a+1)] \}$

and yields:

$\{ \alpha_1 * [- 2a (x^3y/3 + x^2y^2/2 + xy^3/3) + 1/2 \; xy \; (a+1)] + \alpha_2 * [a (x^3y/3 + x^2y^2/2 + xy^3/3) -$

$(2a + 1/2) (x^2y/2 + xy^2/2) + 3/4 \; xy \; (a+1)] \} = 0$ (17)

Recall that:

$\alpha_1 = [f(1/2, 1/2) + f(-1/2, -1/2) + f(2/3, 2/3) + f(-2/3, -2/3)]$

$\alpha_2 = [f(1/2, 1/2) + f(-1/2, -1/2) + f(2/3, 2/3) + f(-2/3, -2/3) + f(-1, -1) + f(1, 1) +$

$f(3/2, 3/2) + f(-3/2, -3/2)]$

Let us posit:

$\alpha_1 = [f(1/2, 1/2) + \beta_1]$

$\alpha_2 = [\, f(1/2, 1/2) + \beta_2 \,]$

where:

$\beta_1 = f(-1/2, -1/2) + f(2/3, 2/3) + f(-2/3, -2/3)$

$\beta_2 = f(-1/2, -1/2) + f(2/3, 2/3) + f(-2/3, -2/3) + f(-1, -1) + f(1, 1) +$

$f(3/2, 3/2) + f(-3/2, -3/2)$

Equation (17) can be written as:

$\{\, [\, f(1/2, 1/2) + \beta_1 \,] * [-\, 2a\,(x^3y/3 + x^2y^2/2 + xy^3/3) + 1/2\,xy\,(a+1)\,] + [\, f(1/2, 1/2) + \beta_2 \,]$

$* [\, a\,(x^3y/3 + x^2y^2/2 + xy^3/3) - (2a + 1/2)\,(x^2y/2 + xy^2/2) + 3/4\,xy\,(a+1)\,]\,\} = 0$

$\{\, [\,\beta_1\,] * [-\,2a\,(x^3y/3 + x^2y^2/2 + xy^3/3) + 1/2\,xy\,(a+1)\,] + [\,\beta_2\,] * [\, a\,(x^3y/3 + x^2y^2/2 +$

$xy^3/3) - (2a + 1/2)\,(x^2y/2 + xy^2/2) + 3/4\,xy\,(a+1)\,]\,\} = \{\, -\,f(1/2, 1/2) * [-\,2a\,(x^3y/3 +$

$x^2y^2/2 + xy^3/3) + 1/2\,xy\,(a+1)\,] - f(1/2, 1/2) * [\, a\,(x^3y/3 + x^2y^2/2 + xy^3/3) - (2a + 1/2)$

$(x^2y/2 + xy^2/2) + 3/4\,xy\,(a+1)\,]\,\}$

Therefore:

$\{\, [\,\beta_1\,] * [-\,2a\,(x^3y/3 + x^2y^2/2 + xy^3/3) + 1/2\,xy\,(a+1)\,] + [\,\beta_2\,] * [\, a\,(x^3y/3 + x^2y^2/2 +$

$xy^3/3) - (2a + 1/2)\,(x^2y/2 + xy^2/2) + 3/4\,xy\,(a+1)\,]\,\} = \{\, -\,f(1/2, 1/2)\,\} * \{\, [-\,2a\,(x^3y/3 +$

$x^2y^2/2 + xy^3/3) + 1/2\,xy\,(a+1)\,] + [\, a\,(x^3y/3 + x^2y^2/2 + xy^3/3) - (2a + 1/2)\,(x^2y/2 + xy^2/2)$

$+ 3/4\,xy\,(a+1)\,]\,\}$ (18)

Let us posit:

$\gamma_1 = [-\,2a\,(x^3y/3 + x^2y^2/2 + xy^3/3) + 1/2\,xy\,(a+1)\,]$

$\gamma_2 = [\, a\,(x^3y/3 + x^2y^2/2 + xy^3/3) - (2a + 1/2)\,(x^2y/2 + xy^2/2) + 3/4\,xy\,(a+1)\,]$

Thus equation (18) is written as:

$\{\, \beta_1 * \gamma_1 + \beta_2 * \gamma_2 \,\} = \{\, -\,f(1/2, 1/2)\,\} * \{\, \gamma_1 + \gamma_2 \,\}$ (19)

And hereto follow equation (19) can be solved in $f(1/2, 1/2)$.

$f(1/2, 1/2) = -\,\{\, \beta_1 * \gamma_1 + \beta_2 * \gamma_2 \,\} / \{\, \gamma_1 + \gamma_2 \,\}$ (20)

which is the value of the signal resilient to the bivariate B-Spline interpolation function $h_4(x, y)$ given in equation (1).

CALCULATION OF THE PARTIAL SECOND ORDER DERIVATIVES: BIVARIATE CUBIC B-SPLINE

First Order Partial Derivative with Respect to the x Variable

$(\partial\,(\,h_4(x, y)\,)\,/\partial x) = \partial\,\{\,f\,(0, 0) + \alpha_3 * [\,1/2\,(x + y)^3 - (x + y)^2 + 2/3\,] + \alpha_2 * [\,-1/6\,(x + y)^3$

$+ (x + y)^2 - 2\,(x + y) + 4/3\,]\}/\partial x = \alpha_3 * [\,1/2\,\partial\,(x + y)^3/\partial x - \partial\,(x + y)^2/\partial x\,] + \alpha_2 * [\,-1/6$

$\partial\,(x + y)^3\,/\partial x + \partial\,(x + y)^2/\partial x - 2\,\partial\,(x + y)\,/\partial x\,] = \alpha_3 * [\,3/2\,(x + y)^2 - 2\,(x + y)\,] + \alpha_2 * [\,-$

$3/6\,(x + y)^2 + 2\,(x + y) - 2\,]$

Therefore:

$(\partial\,(\,h_4(x, y)\,)\,/\partial x) = \alpha_3 * [\,3/2\,(x + y)^2 - 2\,(x + y)\,] + \alpha_2 * [\,-3/6\,(x + y)^2 + 2\,(x + y) - 2\,]$ \hfill (21)

First Order Partial Derivative with Respect to the y Variable

$(\partial\,(\,h_3(x, y)\,)\,/\partial y) = \partial\,\{\,f\,(0, 0) + \alpha_3 * [\,1/2\,(x + y)^3 - (x + y)^2 + 2/3\,] + \alpha_2 * [\,-1/6\,(x + y)^3$

$+ (x + y)^2 - 2\,(x + y) + 4/3\,]\}/\partial y = \alpha_3 * [\,1/2\,\partial\,(x + y)^3/\partial y - \partial\,(x + y)^2/\partial y\,] + \alpha_2 * [\,-1/6$

$\partial\,(x + y)^3\,/\partial y + \partial\,(x + y)^2/\partial y - 2\,\partial\,(x + y)\,/\partial y\,] = \alpha_3 * [\,3/2\,(x + y)^2 - 2\,(x + y)\,] + \alpha_2 * [\,-$

$3/6\,(x + y)^2 + 2\,(x + y) - 2\,]$

Therefore:

$(\partial\,(\,h_4(x, y)\,)\,/\partial y) = \alpha_3 * [\,3/2\,(x + y)^2 - 2\,(x + y)\,] + \alpha_2 * [\,-3/6\,(x + y)^2 + 2\,(x + y) - 2\,]$ \hfill (22)

Second Order Partial Derivative with Respect to the x Variable

$(\partial^2\,(\,h_4(x, y)\,)\,/\partial x^2) = \{\partial\,\{\partial\,(\,h_4(x, y)\,)\,/\partial x\}\,/\partial x\} = \{\partial\,\{\,\alpha_3 * [\,3/2\,(x + y)^2 - 2\,(x + y)\,] +$

$\alpha_2 * [\,-3/6\,(x + y)^2 + 2\,(x - y) - 2\,]\,\}\,/\partial x\} = \alpha_3 * [\,3\,(x + y) - 2\,] + \alpha_2 * [\,-(x + y) + 2\,]$

Therefore:

$(\partial^2\,(\,h_4(x, y)\,)\,/\partial x^2) = \alpha_3 * [\,3\,(x + y) - 2\,] + \alpha_2 * [\,-(x + y) + 2\,]$ \hfill (23)

Second Order Partial Derivative with Respect to the y Variable

$(\partial^2\,(\,h_4(x, y)\,)\,/\partial y^2) = \{\partial\,\{\partial\,(\,h_4(x, y)\,)\,/\partial y\}\,/\partial y\} = \{\partial\,\{\,\alpha_3 * [\,3/2\,(x + y)^2 - 2\,(x + y)\,] +$

$\alpha_2 * [-3/6 (x + y)^2 + 2 (x + y) - 2] \} /\partial y\} = \alpha_3 * [3 (x + y) - 2] + \alpha_2 * [- (x + y) + 2]$

Therefore:

$$(\partial^2 (h_4(x, y)) /\partial x^2) = (\partial^2 (h_4(x, y)) /\partial y^2) \tag{24}$$

Second Order Partial Derivatives with Respect to the x and y Variables

$(\partial^2 (h_4(x, y)) /\partial x \partial y) = \{\partial (\partial (h_4(x, y)) /\partial x) /\partial y\} = \{\partial \{ \alpha_3 * [3/2 (x + y)^2 - 2 (x + y)] +$

$\alpha_2 * [-3/6 (x + y)^2 + 2 (x + y) - 2] \} /\partial y\} = \alpha_3 * [3 (x + y) - 2] + \alpha_2 * [- (x + y) + 2]$

$(\partial^2 (h_4(x, y)) /\partial y \partial x) = \{\partial (\partial (h_4(x, y)) /\partial y) /\partial x\} = \{\partial \{ \alpha_3 * [3/2 (x + y)^2 - 2 (x + y)] +$

$\alpha_2 * [-3/6 (x + y)^2 + 2 (x + y) - 2] \} /\partial x\} = \alpha_3 * [3 (x + y) - 2] + \alpha_2 * [- (x + y) + 2]$

Therefore:

$$(\partial^2 (h_4(x, y)) /\partial x \partial y) = (\partial^2 (h_4(x, y)) /\partial y \partial x) = \alpha_3 * [3 (x + y) - 2] + \alpha_2 * [- (x + y) + 2] \tag{25}$$

CALCULATION OF THE INTENSITY-CURVATURE TERMS: BIVARIATE CUBIC B-SPLINE

Let the intensity-curvature term before interpolation be defined as:

$E_o = E_o (x, y) = \int_0^x \int_0^y f (0, 0) * \{ (\partial^2 (h_4(x, y)) /\partial x^2) + (\partial^2 (h_4(x, y)) /\partial x \partial y) + (\partial^2 (h_4(x, y))$

$$/\partial y \partial x) + (\partial^2 (h_4(x, y)) /\partial y^2) \} (0, 0) \, dx \, dy \tag{26}$$

It follows from equations (23), (24) and (25) that:

$(\partial^2 (h_4(x, y)) /\partial x^2) (0, 0) = - 2 \alpha_3 + 2 \alpha_2$

$(\partial^2 (h_4(x, y)) /\partial y^2) (0, 0) = - 2 \alpha_3 + 2 \alpha_2$

$(\partial^2 (h_4(x, y)) /\partial x \partial y) (0, 0) = - 2 \alpha_3 + 2 \alpha_2$

$(\partial^2 (h_4(x, y)) /\partial y \partial x) (0, 0) = - 2 \alpha_3 + 2 \alpha_2$

Therefore:

$$E_o (x, y) = \int_0^x \int_0^y 4 * f (0, 0) \{ - 2 \alpha_3 + 2 \alpha_2 \} \, dx \, dy = 4xy * f (0, 0) \{ - 2 \alpha_3 + 2 \alpha_2 \} \tag{27}$$

Let the intensity-curvature term after interpolation be defined as:

$E_{IN} = E_{IN} (x, y) = \int_0^x \int_0^y h_4(x, y) * \{ (\partial^2 (h_4(x, y)) /\partial x^2) + (\partial^2 (h_4(x, y)) /\partial x \partial y) +$

$(\partial^2 (h_4(x, y)) /\partial y\partial x) + (\partial^2 (h_4(x, y)) /\partial y^2) \} \, dx \, dy$ (28)

From equations (23), (24) and (25), it follows that:

$(\partial^2 (h_4(x, y)) /\partial x^2) = \alpha_3 * [3 (x + y) - 2] + \alpha_2 * [- (x + y) + 2]$

$(\partial^2 (h_4(x, y)) /\partial y^2) = \alpha_3 * [3 (x + y) - 2] + \alpha_2 * [- (x + y) + 2]$

$(\partial^2 (h_4(x, y)) /\partial x\partial y) = \alpha_3 * [3 (x + y) - 2] + \alpha_2 * [- (x + y) + 2]$

$(\partial^2 (h_4(x, y)) /\partial y\partial x) = \alpha_3 * [3 (x + y) - 2] + \alpha_2 * [- (x + y) + 2]$

Therefore:

$E_{IN} (x, y) = \int_0^x \int_0^y 4 * h_4(x, y) \{ \alpha_3 * [3 (x + y) - 2] + \alpha_2 * [- (x + y) + 2] \} \, dx \, dy$ (29)

From equation (4) it follows that:

$E_{IN} (x, y) = \int_0^x \int_0^y 4 * \{ f (0, 0) + \alpha_3 * [1/2 (x + y)^3 - (x + y)^2 + 2/3] + \alpha_2 * [-1/6 (x + y)^3 +$

$(x + y)^2 - 2 (x + y) + 4/3] \} * \{ \alpha_3 * [3 (x + y) - 2] + \alpha_2 * [- (x + y) + 2] \} \, dx \, dy =$

$\int_0^x \int_0^y 4 * \{ f (0, 0) \, \alpha_3 * [3 (x + y) - 2] + f (0, 0) \, \alpha_2 * [- (x + y) + 2] + \alpha_3 * [1/2 (x + y)^3 - (x + y)^2 + 2/3] * \alpha_3 * [3 (x + y) - 2] + \alpha_3 * [1/2 (x + y)^3 - (x + y)^2 + 2/3] * \alpha_2 * [- (x + y) + 2] + \alpha_2 * [-1/6 (x + y)^3 + (x + y)^2 - 2 (x + y) + 4/3] * \alpha_3 * [3 (x + y) - 2] + \alpha_2 * [-1/6 (x + y)^3 + (x + y)^2 - 2 (x + y) + 4/3] * \alpha_2 * [- (x + y) + 2] \} \, dx \, dy =$

$\int_0^x \int_0^y 4 * \{ f (0, 0) \, \alpha_3 * [3 (x + y) - 2] + f (0, 0) \, \alpha_2 * [- (x + y) + 2] + \alpha_3^2 * [3/2 (x + y)^4 - 3(x + y)^3 + 2(x + y) - (x + y)^3 + 2 (x + y)^2 - 4/3] + \alpha_3 \, \alpha_2 * [-1/2 (x + y)^4 + (x + y)^3 - 2/3 (x + y) + (x + y)^3 - 2 (x + y)^2 + 4/3] + \alpha_3 \, \alpha_2 * [-3/6 (x + y)^4 + 3 (x + y)^3 - 6 (x + y)^2 + 4 (x + y) + 2/6 (x + y)^3 - 2 (x + y)^2 + 4 (x + y) - 8/3] + \alpha_2^2 * [1/6 (x + y)^4 - (x + y)^3 + 2 (x + y)^2 - 4/3 (x + y) - 2/6 (x + y)^3 + 2 (x + y)^2 - 4 (x + y) + 8/3] \} \, dx \, dy =$

$\int_0^x \int_0^y 4 * \{ f (0, 0) \, \alpha_3 * [3 (x + y) - 2] + f (0, 0) \, \alpha_2 * [- (x + y) + 2] + \alpha_3^2 * [3/2 (x + y)^4 - 4(x + y)^3 + 2(x + y) + 2 (x + y)^2 - 4/3] + \alpha_3 \, \alpha_2 * [-1/2 (x + y)^4 + 2 (x + y)^3 - 2/3 (x$

$+ y) - 2 (x + y)^2 + 4/3] + \alpha_3 \alpha_2 * [-3/6 (x + y)^4 + 4/3 (x + y)^3 - 8 (x + y)^2 + 8 (x +$

$y) - 8/3] + \alpha_2^2 * [1/6 (x + y)^4 - 4/3(x + y)^3 + 4 (x + y)^2 - 16/3 (x + y) + 8/3] \} \, dx \, dy =$

$\int_0^x \int_0^y 4 * \{ f (0, 0) \, \alpha_3 * [3 (x + y) - 2] + f (0, 0) \, \alpha_2 * [- (x + y) + 2] + [3/2 \, \alpha_3^2 - 1/2 \, \alpha_3 \, \alpha_2 -$

$3/6 \, \alpha_3 \, \alpha_2 + 1/6 \, \alpha_2^2] (x + y)^4 + [-4 \, \alpha_3^2 + 2 \, \alpha_3 \, \alpha_2 + 4/3 \, \alpha_3 \, \alpha_2 - 4/3 \, \alpha_2^2] (x + y)^3 + [2 \, \alpha_3^2 - 2$

$\alpha_3 \, \alpha_2 - 8 \, \alpha_3 \, \alpha_2 + 4 \, \alpha_2^2] (x + y)^2 + [2 \, \alpha_3^2 - 2/3 \, \alpha_3 \, \alpha_2 + 8 \, \alpha_3 \, \alpha_2 - 16/3 \, \alpha_2^2] (x + y) + [- 4/3$

$\alpha_3^2 + 4/3 \, \alpha_3 \, \alpha_2 - 8/3 \, \alpha_3 \, \alpha_2 + 8/3 \, \alpha_2^2] \} \, dx \, dy$ 　　　　　　(30)

It can be written that:

$(x + y)^2 = (x^2 + 2xy + y^2)$ 　　　　　　(30.a)

$(x + y)^3 = (x^2 + 2xy + y^2) (x + y) = (x^3 + 2x^2y + xy^2 + x^2y + 2xy^2 + y^3) =$

$(x^3 + 3x^2y + 3xy^2 + y^3)$ 　　　　　　(30.b)

$(x + y)^4 = (x + y)^3 (x + y) = (x^3 + 3x^2y + 3xy^2 + y^3) (x + y) =$

$(x^4 + 3x^3y + 3x^2y^2 + xy^3 + x^3y + 3x^2y^2 + 3xy^3 + y^4)$ 　　　　　　(30.c)

$\int_0^x \int_0^y (x + y) \, dx \, dy = \int_0^x (xy + y^2/2) \, dx = (x^2y/2 + xy^2/2)$ 　　　　　　(30.d)

$\int_0^x \int_0^y (x + y)^2 \, dx \, dy = \int_0^x \int_0^y (x^2 + 2xy + y^2) \, dx \, dy = \int_0^x (x^2y + xy^2 + y^3/3) \, dx =$
$(x^3y/3 + x^2y^2/2 + xy^3/3)$ 　　　　　　(30.e)

$\int_0^x \int_0^y (x + y)^3 \, dx \, dy = \int_0^x \int_0^y (x^3 + 3x^2y + 3xy^2 + y^3) \, dx \, dy = \int_0^x (x^3y + 3/2 \, x^2y^2 + xy^3 + y^4/4) \, dx =$

$(x^4y/4 + x^3y^2/2 + x^2y^3/2 + xy^4/4)$ 　　　　　　(30.f)

$\int_0^x \int_0^y (x + y)^4 \, dx \, dy = \int_0^x \int_0^y (x^4 + 3x^3y + 3x^2y^2 + xy^3 + x^3y + 3x^2y^2 + 3xy^3 + y^4) \, dx \, dy =$
$\int_0^x (x^4y + 3/2 \, x^3y^2 + x^2y^3 + xy^4/4 + x^3y^2/2 + x^2y^3 + 3/4 \, xy^4 + y^5/5) \, dx =$

$(x^5y/5 + 3/8\ x^4y^2 + x^3y^3/3 + x^2y^4/8 + x^4y^2/8 + x^3y^3/3 + 3/8\ x^2y^4 + xy^5/5)$ (30.g)

Equation (30) can be written as:

$$\int_0^x \int_0^y 4 * \{ f(0,0)\ \alpha_3 * [3(x+y)-2] + f(0,0)\ \alpha_2 * [-(x+y)+2] \}\,dx\,dy +$$

$$\int_0^x \int_0^y 4 * \{ [3/2\ \alpha_3^2 - 1/2\ \alpha_3\ \alpha_2 - 3/6\ \alpha_3\ \alpha_2 + 1/6\ \alpha_2^2](x+y)^4 + [-4\ \alpha_3^2 + 2\ \alpha_3\ \alpha_2 + 4/3\ \alpha_3\ \alpha_2 -$$

$$-4/3\ \alpha_2^2](x+y)^3 + [2\ \alpha_3^2 - 2\ \alpha_3\ \alpha_2 - 8\ \alpha_3\ \alpha_2 + 4\ \alpha_2^2](x+y)^2 + [2\ \alpha_3^2 - 2/3\ \alpha_3\ \alpha_2 + 8\ \alpha_3\ \alpha_2$$

$$- 16/3\ \alpha_2^2](x+y) + [-4/3\ \alpha_3^2 + 4/3\ \alpha_3\ \alpha_2 - 8/3\ \alpha_3\ \alpha_2 + 8/3\ \alpha_2^2]\}\,dx\,dy$$ (31)

Also solving:

$$\int_0^x \int_0^y 4 * \{ f(0,0)\ \alpha_3 * [3(x+y)-2] + f(0,0)\ \alpha_2 * [-(x+y)+2] \}\,dx\,dy =$$

$$\int_0^x 4 * \{ f(0,0)\ \alpha_3 * [3(xy + y^2/2) - 2y] + f(0,0)\ \alpha_2 * [-(xy + y^2/2) + 2y] \}\,dx =$$

$$4 * \{ f(0,0)\ \alpha_3 * [3(x^2y/2 + xy^2/2) - 2xy] + f(0,0)\ \alpha_2 * [-(x^2y/2 + xy^2/2) + 2xy] \}$$

Equation (31) furnishes:

$$E_{IN}(x,y) = 4 * \{ f(0,0)\ \alpha_3 * [3(x^2y/2 + xy^2/2) - 2xy] + f(0,0)\ \alpha_2 * [-(x^2y/2 + xy^2/2) +$$

$$2xy]\} + 4 * \{ [3/2\ \alpha_3^2 - 1/2\ \alpha_3\ \alpha_2 - 3/6\ \alpha_3\ \alpha_2 + 1/6\ \alpha_2^2][x^5y/5 + 3/8\ x^4y^2 + x^3y^3/3 +$$

$$x^2y^4/8 + x^4y^2/8 + x^3y^3/3 + 3/8\ x^2y^4 + xy^5/5] + [-4\ \alpha_3^2 + 2\ \alpha_3\ \alpha_2 + 4/3\ \alpha_3\ \alpha_2 - 4/3\ \alpha_2^2]$$

$$[x^4y/4 + x^3y^2/2 + x^2y^3/2 - xy^4/4] + [2\ \alpha_3^2 - 2\ \alpha_3\ \alpha_2 - 8\ \alpha_3\ \alpha_2 + 4\ \alpha_2^2][x^3y/3 + x^2y^2/2 +$$

$$xy^3/3] + [2\ \alpha_3^2 - 2/3\ \alpha_3\ \alpha_2 + 8\ \alpha_3\ \alpha_2 - 16/3\ \alpha_2^2][x^2y/2 + xy^2/2] + [-4/3\ \alpha_3^2 + 4/3\ \alpha_3\ \alpha_2 -$$

$$8/3\ \alpha_3\ \alpha_2 + 8/3\ \alpha_2^2][xy]\}$$ (32)

EQUATING THE TWO INTENSITY-CURVATURE TERMS BEFORE AND AFTER INTERPOLATION: BIVARIATE CUBIC B-SPLINE

The signal resilient to interpolation as reconstructed through the bivariate cubic B-Spline of equation (2) is obtained (see equation 34)) solving the equation $E_{IN}(x,y) = E_o(x,y)$ resulting from (32) and (27).

$$E_o(x,y) = 4xy * f(0,0)\{ -2\ \alpha_3 + 2\ \alpha_2 \} = E_{IN}(x,y) = 4 * \{ f(0,0)\ \alpha_3 * [3(x^2y/2 +$$

$xy^2/2) - 2xy] + f(0,0) \alpha_2 * [- (x^2y/2 + xy^2/2) + 2xy] \} + 4 * \{ [3/2 \alpha_3^2 - 1/2 \alpha_3 \alpha_2 - 3/6$

$\alpha_3 \alpha_2 + 1/6 \alpha_2^2] [x^5y/5 + 3/8 x^4y^2 + x^3y^3/3 + x^2y^4/8 + x^4y^2/8 + x^3y^3/3 + 3/8 x^2y^4 + xy^5/5]$

$+ [-4 \alpha_3^2 + 2 \alpha_3 \alpha_2 + 4/3 \alpha_3 \alpha_2 - 4/3 \alpha_2^2] [x^4y/4 + x^3y^2/2 + x^2y^3/2 + xy^4/4] + [2 \alpha_3^2 - 2 \alpha_3$

$\alpha_2 - 8 \alpha_3 \alpha_2 + 4 \alpha_2^2] [x^3y/3 + x^2y^2/2 + xy^3/3] + [2 \alpha_3^2 - 2/3 \alpha_3 \alpha_2 + 8 \alpha_3 \alpha_2 - 16/3 \alpha_2^2]$

$[x^2y/2 + xy^2/2] + [-4/3 \alpha_3^2 + 4/3 \alpha_3 \alpha_2 - 8/3 \alpha_3 \alpha_2 + 8/3 \alpha_2^2] [xy] \}$ (33)

Let us posit:

$\Lambda = 4 * \{ [3/2 \alpha_3^2 - 1/2 \alpha_3 \alpha_2 - 3/6 \alpha_3 \alpha_2 + 1/6 \alpha_2^2] [x^5y/5 + 3/8 x^4y^2 + x^3y^3/3 + x^2y^4/8 +$

$x^4y^2/8 + x^3y^3/3 + 3/8 x^2y^4 + xy^5/5] + [-4 \alpha_3^2 + 2 \alpha_3 \alpha_2 + 4/3 \alpha_3 \alpha_2 - 4/3 \alpha_2^2] [x^4y/4 +$

$x^3y^2/2 + x^2y^3/2 + xy^4/4] + [2 \alpha_3^2 - 2 \alpha_3 \alpha_2 - 8 \alpha_3 \alpha_2 + 4 \alpha_2^2] [x^3y/3 + x^2y^2/2 + xy^3/3] +$

$[2 \alpha_3^2 - 2/3 \alpha_3 \alpha_2 + 8 \alpha_3 \alpha_2 - 16/3 \alpha_2^2] [x^2y/2 + xy^2/2] + [-4/3 \alpha_3^2 + 4/3 \alpha_3 \alpha_2 - 8/3 \alpha_3 \alpha_2$

$+ 8/3 \alpha_2^2] [xy] \}$

From equation (33) it follows that:

$4xy * f(0,0) \{ -2 \alpha_3 + 2 \alpha_2 \} - 4 * \{ f(0,0) \alpha_3 * [3 (x^2y/2 + xy^2/2) - 2xy] + f(0,0) \alpha_2 *$

$[- (x^2y/2 + xy^2/2) + 2xy] \} = \Lambda$

$f(0,0) * \{ 4xy * [-2 \alpha_3 + 2 \alpha_2] - 4\alpha_3 * [3 (x^2y/2 + xy^2/2) - 2xy] - 4\alpha_2 * [- (x^2y/2 +$

$xy^2/2) + 2xy] \} = \Lambda$

Finally:

$f(0,0) = \Lambda / \{ 4xy * [-2 \alpha_3 + 2 \alpha_2] - 4\alpha_3 * [3 (x^2y/2 + xy^2/2) - 2xy] - 4\alpha_2 * [- (x^2y/2 +$

$xy^2/2) + 2xy] \}$ (34)

BIVARIATE CUBIC LAGRANGE INTERPOLATION FUNCTION

This function has ((3 x 3)-1) pixels' neighborhood: f(-1/2), f(1/2), f (-1), f (1), f(-3/2), f(3/2), f (-2/3), f (2/3) centered at f (0, 0) (see figure 1).

$LGR_3(x, y) = f(0, 0) + [f(1/2, 1/2) + f(-1/2, -1/2) + f(-1, -1) + f(1, 1)] *$

$[(1/2) (x + y)^3 - (x + y)^2 - 1/2 (x + y) + 1] + [f(1/2, 1/2) + f(-1/2, -1/2) + f(2/3, 2/3) +$

f (-2/3, -2/3) + f (-1, -1) + f (1, 1) + f (3/2, 3/2) + f (-3/2, -3/2)] * [-(1/6) (x + y)^3 + (x +

y)^2 - (11/6) (x + y) + 1] (35)

Let us posit:

α_2 = [f (1/2, 1/2) + f (-1/2, -1/2) + f (2/3, 2/3) + f (-2/3, -2/3) + f (-1, -1) + f (1, 1) +

f (3/2, 3/2) + f (-3/2, -3/2)]

α_3 = [f (1/2, 1/2) + f (-1/2, -1/2) + f (-1, -1) + f (1, 1)]

Equation (35) is re-written as:

LGR$_3$(x, y) = f (0, 0) + α_3 * [(1/2) (x + y)^3 − (x + y)^2 − 1/2 (x + y) + 1] + α_2 * [-(1/6) (x

+ y)^3 + (x + y)^2 - (11/6) (x + y) + 1] (36)

CALCULATION OF THE PARTIAL SECOND ORDER DERIVATIVES: BIVARIATE CUBIC LAGRANGE

First Order Partial Derivative with Respect to the x Variable

(∂ (LGR$_3$(x, y)) /∂x) = ∂ { f (0, 0) + α_3 * [(1/2) (x + y)^3 − (x + y)^2 − 1/2 (x + y) + 1] + α_2

* [-(1/6) (x + y)^3 + (x + y)^2 - (11/6) (x + y) + 1] }/∂x = { α_3 * [(1/2) ∂ (x + y)^3 /∂x − ∂ (x

+ y)^2 /∂x − 1/2 ∂ (x + y) /∂x] + α_2 * [-(1/6) ∂ (x + y)^3 /∂x + ∂ (x + y)^2 /∂x - (11/6) ∂ (x + y)

/∂x] } = { α_3 * [(3/2) (x + y)^2 − 2 (x + y) − 1/2] + α_2 * [-(3/6) (x + y)^2 + 2 (x + y) - (11/6)] } (37)

First Order Partial Derivative with Respect to the y Variable

(∂ (LGR$_3$(x, y)) /∂y) = ∂ { f (0, 0) + α_3 * [(1/2) (x + y)^3 − (x + y)^2 − 1/2 (x + y) + 1] + α_2

* [-(1/6) (x + y)^3 + (x + y)^2 - (11/6) (x + y) + 1] }/∂y = { α_3 * [(1/2) ∂ (x + y)^3 /∂y − ∂ (x

+ y)^2 /∂y − 1/2 ∂ (x + y) /∂y] + α_2 * [-(1/6) ∂ (x + y)^3 /∂y + ∂ (x + y)^2 /∂y - (11/6) ∂ (x + y)

/∂y] } = { α_3 * [(3/2) (x + y)^2 − 2 (x + y) − 1/2] + α_2 * [-(3/6) (x + y)^2 + 2 (x + y) - (11/6)] } (38)

Thus:

(∂ (LGR$_3$(x, y)) /∂x) = (∂ (LGR$_3$(x, y)) /∂y)

Second Order Partial Derivative with Respect to the x Variable

(∂^2 (LGR$_3$(x, y)) /∂x^2) = {∂ {∂ (LGR$_3$(x, y)) /∂x} /∂x} = {∂ { α_3 * [(3/2) (x + y)^2 − 2 (x

$+ y) - 1/2] + \alpha_2 * [-(3/6) (x + y)^2 + 2 (x + y) - (11/6) \} /\partial x\} =$

$\{ \alpha_3 * [3 (x + y) - 2] + \alpha_2 * [- (x + y) + 2] \}$ (39)

Second Order Partial Derivative with Respect to the y Variable

$(\partial^2 (LGR_3(x, y)) /\partial y^2) = \{\partial \{\partial (LGR_3(x, y)) /\partial y\} /\partial y\} = \{\partial \{ \alpha_3 * [(3/2) (x + y)^2 - 2 (x$

$+ y) - 1/2] + \alpha_2 * [-(3/6) (x + y)^2 + 2 (x + y) - (11/6) \} /\partial y\} =$

$\{ \alpha_3 * [3 (x + y) - 2] + \alpha_2 * [- (x + y) + 2] \}$ (40)

Thus:

$(\partial^2 (LGR_3(x, y)) /\partial x^2) = (\partial^2 (LGR_3(x, y)) /\partial y^2)$

Second Order Partial Derivatives with Respect to the x and y Variables

$(\partial^2 (LGR_3(x, y)) /\partial x \partial y) = \{\partial (\partial (LGR_3(x, y)) /\partial x) /\partial y\} = \{\partial \{ \alpha_3 * [(3/2) (x + y)^2 - 2 (x$

$+ y) - 1/2] + \alpha_2 * [-(3/6) (x + y)^2 + 2 (x + y) - (11/6)] \} /\partial y\} =$

$\{ \alpha_3 * [3 (x + y) - 2] + \alpha_2 * [- (x + y) + 2] \}$ (41)

$(\partial^2 (LGR_3(x, y)) /\partial y \partial x) = \{\partial (\partial (LGR_3(x, y)) /\partial y) /\partial x\} = \{\partial \{ \alpha_3 * [(3/2) (x + y)^2 - 2 (x$

$+ y) - 1/2] + \alpha_2 * [-(3/6) (x + y)^2 + 2 (x + y) - (11/6)] \} /\partial x\} =$

$\{ \alpha_3 * [3 (x + y) - 2] + \alpha_2 * [- (x + y) + 2] \}$ (42)

Thus:

$(\partial^2 (LGR_3(x, y)) /\partial x \partial y) = (\partial^2 (LGR_3(x, y)) /\partial y \partial x)$

CALCULATION OF THE INTENSITY-CURVATURE TERMS: BIVARIATE CUBIC LAGRANGE

The intensity-curvature term before interpolation is:

$E_o = E_o (x, y) = \int_0^x \int_0^y f (0, 0) * \{ (\partial^2 (LGR_3(x, y)) /\partial x^2) + (\partial^2 (LGR_3(x, y)) /\partial x \partial y) + (\partial^2$

$(LGR_3(x, y)) /\partial y \partial x) + (\partial^2 (LGR_3(x, y)) /\partial y^2) \} (0, 0) \, dx \, dy$ (43)

It is true that:

$(\partial^2 (LGR_3(x, y)) /\partial x^2) (0, 0) = (\partial^2 (LGR_3(x, y)) /\partial y^2) (0, 0) =$

$(\partial^2 \, (LGR_3(x, y)) \, /\partial x \partial y) \, (0, 0) = (\partial^2 \, (LGR_3(x, y)) \, /\partial y \partial x) \, (0, 0) =$

$$\{ \alpha_3 * [\, 3 \, (x + y) - 2 \,] + \alpha_2 * [\, - (x + y) + 2 \,] \} \, (0, 0) = - 2 \, \alpha_3 + 2 \, \alpha_2 \tag{44}$$

Worth noting that the second order partial derivatives of the bivariate cubic Lagrange function, calculated at the location $(0, 0)$ are the same as those of the bivariate cubic B-Spline $h_4(x, y)$ and this can be seen observing what follows from equations (23), (24) and (25).

Therefore:

$$E_o = E_o \, (x, y) = \int_0^x \int_0^y 4 * f \, (0, 0) \, \{ - 2 \, \alpha_3 + 2 \, \alpha_2 \} \, dx \, dy = 4xy * f \, (0, 0) \, \{ - 2 \, \alpha_3 + 2 \, \alpha_2 \} \tag{45}$$

The intensity-curvature term after interpolation is:

$$E_{IN} = E_{IN} \, (x, y) = \int_0^x \int_0^y LGR_3(x, y) \times \{ (\partial^2 \, (LGR_3(x, y)) \, /\partial x^2) + (\partial^2 \, (LGR_3(x, y)) \, /\partial x \partial y) +$$

$$(\partial^2 \, (LGR_3(x, y)) \, /\partial y \partial x) + (\partial^2 \, (LGR_3(x, y)) \, /\partial y^2) \} \, dx \, dy \tag{46}$$

On the basis of equations (36) and (44), $E_{IN} \, (x, y)$ is written as:

$$E_{IN} \, (x, y) = \int_0^x \int_0^y 4 * \{ f \, (0, 0) + \alpha_3 \times [\, (1/2) \, (x + y)^3 - (x + y)^2 - 1/2 \, (x + y) + 1 \,] + \alpha_2 *$$

$$[-(1/6) \, (x + y)^3 + (x + y)^2 - (11/6) \, (x + y) + 1 \,] \} * \{ \alpha_3 * [\, 3 \, (x + y) - 2 \,] + \alpha_2 * [\, - (x + y)$$

$$+ 2 \,] \} \, dx \, dy =$$

$$\int_0^x \int_0^y 4 * \{ f \, (0, 0) \, \alpha_3 * [\, 3 \, (x + y) - 2 \,] + f \, (0, 0) \, \alpha_2 * [\, - (x + y) + 2 \,] + \alpha_3^2 * [\, (1/2) \, (x + y)^3$$

$$- (x + y)^2 - 1/2 \, (x + y) + 1 \,] [\, 3 \, (x - y) - 2 \,] + \alpha_3 \alpha_2 * [\, (1/2) \, (x + y)^3 - (x + y)^2 - 1/2 \, (x +$$

$$y) + 1 \,] [\, - (x + y) + 2 \,] + \alpha_2 \alpha_3 * [-(1/6) \, (x + y)^3 + (x + y)^2 - (11/6) \, (x + y) + 1 \,] [\, 3 \, (x +$$

$$y) - 2 \,] + \alpha_2^2 [-(1/6) \, (x + y)^3 + (x + y)^2 - (11/6) \, (x + y) + 1 \,] [\, - (x + y) + 2 \,] \} \, dx \, dy =$$

$$\int_0^x \int_0^y 4 * \{ f \, (0, 0) \, \alpha_3 * [\, 3 \, (x + y) - 2 \,] + f \, (0, 0) \, \alpha_2 * [\, - (x + y) + 2 \,] + \alpha_3^2 * [\, (3/2) \, (x + y)^4$$

$$- 3 \, (x + y)^3 - 3/2 \, (x + y)^2 + 3 \, (x + y) - (x + y)^3 + 2 \, (x + y)^2 + (x + y) - 2 \,] + \alpha_3 \alpha_2 * [-(1/2)$$

$$(x + y)^4 + (x + y)^3 + 1/2 \, (x + y)^2 - (x + y) + (x + y)^3 - 2 \, (x + y)^2 - (x + y) + 2 \,] + \alpha_2 \alpha_3 * [-$$

$$(3/6) \, (x + y)^4 + 3 \, (x + y)^3 - (33/6) \, (x + y)^2 + 3 \, (x + y) + (2/6) \, (x + y)^3 - 2 \, (x + y)^2 + (22/6)$$

$(x + y) - 2] + \alpha_2^2 [(1/6) (x + y)^4 - (x + y)^3 + (11/6) (x + y)^2 - (x + y) - (2/6) (x + y)^3 + 2 (x + y)^2 - (22/6) (x + y) + 2] \} \, dx \, dy =$

$$\int_0^x \int_0^y 4 * \{ f(0, 0) \, \alpha_3 * [3 (x + y) - 2] + f(0, 0) \, \alpha_2 * [- (x + y) + 2] + \alpha_3^2 * [(3/2) (x + y)^4$$

$- 4 (x + y)^3 + 1/2 (x + y)^2 + 4 (x + y) - 2] + \alpha_3 \alpha_2 * [-(1/2) (x + y)^4 + 2 (x + y)^3 - 3/2 (x +$

$y)^2 - 2 (x + y) + 2] + \alpha_2 \alpha_3 * [-(3/6) (x + y)^4 + (10/3) (x + y)^3 - (45/6) (x + y)^2 + (40/6) (x$

$+ y) - 2] + \alpha_2^2 [(1/6) (x + y)^4 + (23/6) (x + y)^2 - (8/6) (x + y)^3 - (28/6) (x + y) + 2] \} \, dx$

$dy =$

$$\int_0^x \int_0^y 4 * \{ f(0, 0) \, \alpha_3 * [3 (x + y) - 2] + f(0, 0) \, \alpha_2 * [- (x + y) + 2] + [(3/2) \alpha_3^2 -(1/2) \alpha_3$$

$\alpha_2 -(3/6) \alpha_2 \alpha_3 + (1/6) \alpha_2^2] (x + y)^4 + [- 4 \alpha_3^2 + 2 \alpha_3 \alpha_2 + (10/3) \alpha_2 \alpha_3 - (8/6) \alpha_2^2] (x + y)^3 +$

$[1/2 \alpha_3^2 - 3/2 \alpha_3 \alpha_2 - (45/6) \alpha_2 \alpha_3 + (23/6) \alpha_2^2] (x + y)^2 + [4 \alpha_3^2 - 2 \alpha_3 \alpha_2 + (40/6) \alpha_2 \alpha_3 -$

$(28/6) \alpha_2^2] (x + y) + [- 2 \alpha_3^2 + 2 \alpha_2^2] \} \, dx \, dy$ (47)

Equation (47) is solved through the knowledge provided from equations (30.d), (30.e), (30.f) and (30.g).

$$E_{IN} (x, y) = \int_0^x \int_0^y 4 * \{ f(0, 0) \, \alpha_3 * [3 (x + y) - 2] + f(0, 0) \, \alpha_2 * [- (x + y) + 2] \} \, dx \, dy +$$

$4 * \{ [(3/2) \alpha_3^2 -(1/2) \alpha_3\alpha_2 -(3/6) \alpha_2 \alpha_3 + (1/6) \alpha_2^2] (x^5y/5 + 3/8 \, x^4y^2 + x^3y^3/3 + x^2y^4/8 +$

$x^4y^2/8 + x^3y^3/3 + 3/8 \, x^2y^4 + xy^5/5) + [- 4 \alpha_3^2 + 2 \alpha_3 \alpha_2 + (10/3) \alpha_2 \alpha_3 - (8/6) \alpha_2^2] (x^4y/4 +$

$x^3y^2/2 + x^2y^3/2 + xy^4/4) + [1/2 \alpha_3^2 - 3/2 \alpha_3 \alpha_2 - (45/6) \alpha_2 \alpha_3 + (23/6) \alpha_2^2] (x^3y/3 + x^2y^2/2 +$

$xy^3/3) + [4 \alpha_3^2 - 2 \alpha_3 \alpha_2 + (40/6) \alpha_2 \alpha_3 - (28/6) \alpha_2^2] (x^2y/2 + xy^2/2) +$

$[- 2 \alpha_3^2 + 2 \alpha_2^2] \, (xy) \}$ (48)

Therefore:

$E_{IN} (x, y) = 4 * \{ f(0, 0) \, \alpha_3 * [3 (x^2y/2 + xy^2/2) - 2xy] + f(0, 0) \, \alpha_2 * [- (x^2y/2 + xy^2/2)$

+ 2xy] } + 4* { [(3/2) α_3^2 -(1/2) $\alpha_3\alpha_2$ -(3/6) $\alpha_2 \alpha_3$ + (1/6) α_2^2] ($x^5y/5$ + 3/8 x^4y^2 + $x^3y^3/3$ +

$x^2y^4/8$ + $x^4y^2/8$ + $x^3y^3/3$ + 3/8 x^2y^4 + $xy^5/5$) + [− 4 α_3^2 + 2 $\alpha_3\alpha_2$ + (10/3) $\alpha_2 \alpha_3$ − (8/6) α_2^2]

($x^4y/4$ + $x^3y^2/2$ + $x^2y^3/2$ + $xy^4/4$) + [1/2 α_3^2 - 3/2 $\alpha_3 \alpha_2$ - (45/6) $\alpha_2 \alpha_3$ + (23/6) α_2^2] ($x^3y/3$ +

$x^2y^2/2$ + $xy^3/3$) + [4 α_3^2 − 2 $\alpha_3 \alpha_2$ + (40/6) $\alpha_2 \alpha_3$ − (28/6) α_2^2] ($x^2y/2$ + $xy^2/2$) + [- 2 α_3^2 + 2

α_2^2] (xy) }   (49)

EQUATING THE TWO INTENSITY-CURVATURE TERMS BEFORE AND AFTER INTERPOLATION: BIVARIATE CUBIC LAGRANGE

Equating E_o (x, y) as per equation (45) with E_{IN} (x, y) as per equation (49) yields:

4xy * f (0, 0) { − 2 α_3 + 2 α_2 } = 4 * { f (0, 0) α_3 * [3 ($x^2y/2$ + $xy^2/2$) − 2xy] + f (0, 0) α_2

* [- ($x^2y/2$ + $xy^2/2$) + 2xy] } + Λ_{LRG} (50)

Where it is posited that:

Λ_{LRG} = 4* { [(3/2) α_3^2 -(1/2) $\alpha_3\alpha_2$ -(3/6) $\alpha_2 \alpha_3$ + (1/6) α_2^2] ($x^5y/5$ + 3/8 x^4y^2 + $x^3y^3/3$ +

$x^2y^4/8$ + $x^4y^2/8$ + $x^3y^3/3$ + 3/8 x^2y^4 + $xy^5/5$) + [− 4 α_3^2 + 2 $\alpha_3 \alpha_2$ + (10/3) $\alpha_2 \alpha_3$ − (8/6) α_2^2]

($x^4y/4$ + $x^3y^2/2$ + $x^2y^3/2$ + $xy^4/4$) + [1/2 α_3^2 - 3/2 $\alpha_3 \alpha_2$ - (45/6) $\alpha_2 \alpha_3$ + (23/6) α_2^2] ($x^3y/3$ +

$x^2y^2/2$ + $xy^3/3$) + [4 α_3^2 − 2 $\alpha_3 \alpha_2$ − (40/6) $\alpha_2 \alpha_3$ − (28/6) α_2^2] ($x^2y/2$ + $xy^2/2$) + [- 2 α_3^2 + 2

α_2^2] (xy) }

Equation (50) furnishes:

f (0, 0) { 4xy * [− 2 α_3 + 2 α_2] - 4 * α_3 [3 ($x^2y/2$ + $xy^2/2$) − 2xy] - 4 * α_2 [- ($x^2y/2$ +

$xy^2/2$) + 2xy] } = Λ_{LRG}

Finally the signal resilient to the bivariate cubic Lagrange interpolation function $LGR_3(x, y)$ of equation (36) is given by:

f (0, 0) = Λ_{LRG} / { 4xy * [− 2 α_3 + 2 α_2] - 4 * α_3 [3 ($x^2y/2$ + $xy^2/2$) − 2xy] - 4 * α_2 [-

($x^2y/2$ + $xy^2/2$) + 2xy] }   (51)

PART II - TRIVARIATE QUADRATIC B-SPLINE FUNCTION
WITH EMBEDDING

TRIVARIATE QUADRATIC B-SPLINE

Let us consider the quadratic $h_3(x, y, z)$ trivariate B-Spline as per equation (1) with $((3 \times 3 \times 3) - 1)$ neighbouring nodes:

$h_3(x, y, z) = f(0, 0, 0) + [f(-1, 0, 1) + f(-1, 0, 0) + f(1, 0, 0) + f(1, 0, -1) + f(-1, -1, 1)$

$+ f(-1, -1, 0) + f(1, -1, 0) + f(1, -1, -1) + f(-1, 1, 1) + f(-1, 1, 0) + f(1, 1, 0) + f(1, 1, -$

$1)] * [-2a(x + y + z)^2 + 1/2(a+1)] + [f(-1, 0, 1) + f(0, 0, 1) + f(1, 0, 1) + f(-1, 0, 0) +$

$f(1, 0, 0) + f(-1, 0, -1) + f(0, 0, -1) + f(1, 0, -1) + f(-1, -1, 1) + f(0, -1, 1) + f(1, -1, 1)$

$+ f(-1, -1, 0) + f(0, -1, 0) + f(1, -1, 0) + f(-1, -1, -1) + f(0, -1, -1) + f(1, -1, -1) + f(-1,$

$1, 1) + f(0, 1, 1) + f(1, 1, 1) + f(-1, 1, 0) + f(0, 1, 0) + f(1, 1, 0) + f(-1, 1, -1) + f(0, 1,$

$-1) + f(1, 1, -1)] * [a(x + y + z)^2 - (2a + 1/2)(x + y + z) + 3/4(a+1)]$ (1)

The illustration of the neighborhood of $f(0, 0, 0)$ is given in figures 2, 3 and 4.

Let us posit:

$\omega_1 = [f(-1, 0, 1) + f(-1, 0, 0) + f(1, 0, 0) + f(1, 0, -1) + f(-1, -1, 1) + f(-1, -1, 0) + f(1,$

$-1, 0) + f(1, -1, -1) + f(-1, 1, 1) + f(-1, 1, 0) + f(1, 1, 0) + f(1, 1, -1)]$ (2)

$\omega_2 = [f(-1, 0, 1) + f(0, 0, 1) + f(1, 0, 1) + f(-1, 0, 0) + f(1, 0, 0) + f(-1, 0, -1) + f(0, 0,$

$-1) + f(1, 0, -1) + f(-1, -1, 1) + f(0, -1, 1) + f(1, -1, 1) + f(-1, -1, 0) + f(0, -1, 0) + f(1,$

$-1, 0) + f(-1, -1, -1) + f(0, -1, -1) + f(1, -1, -1) + f(-1, 1, 1) + f(0, 1, 1) + f(1, 1, 1) +$

$f(-1, 1, 0) + f(0, 1, 0) + f(1, 1, 0) + f(-1, 1, -1) + f(0, 1, -1) + f(1, 1, -1)]$ (3)

So that equation (1) is written as:

$h_3(x, y, z) = f(0, 0, 0) + \omega_1 * [-2a(x + y + z)^2 + 1/2(a+1)] + \omega_2 * [a(x + y + z)^2 - (2a +$

$1/2)(x + y + z) + 3/4(a+1)]$ (4)

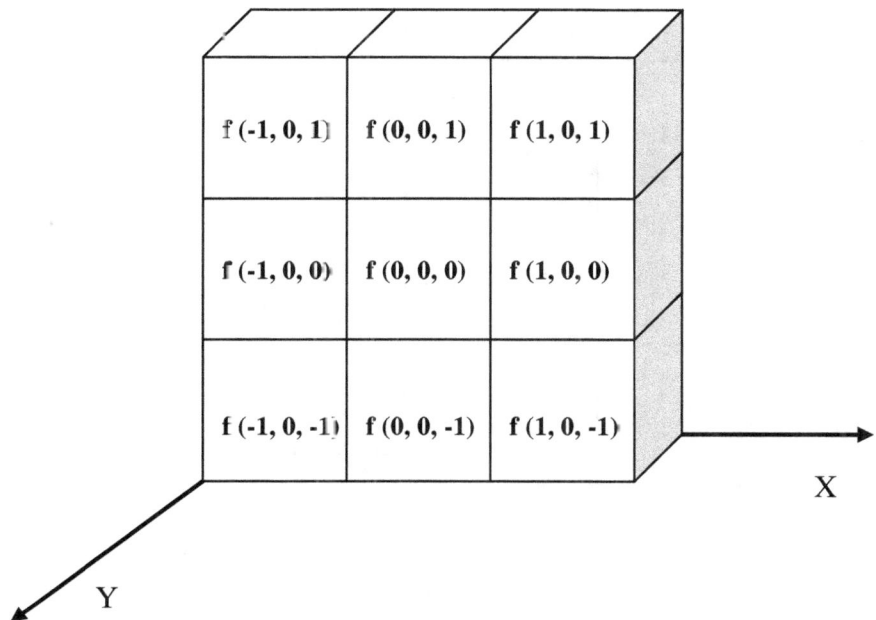

Figure 2: Central 3 x 3 Sub-neighborhood of f (0, 0, 0).

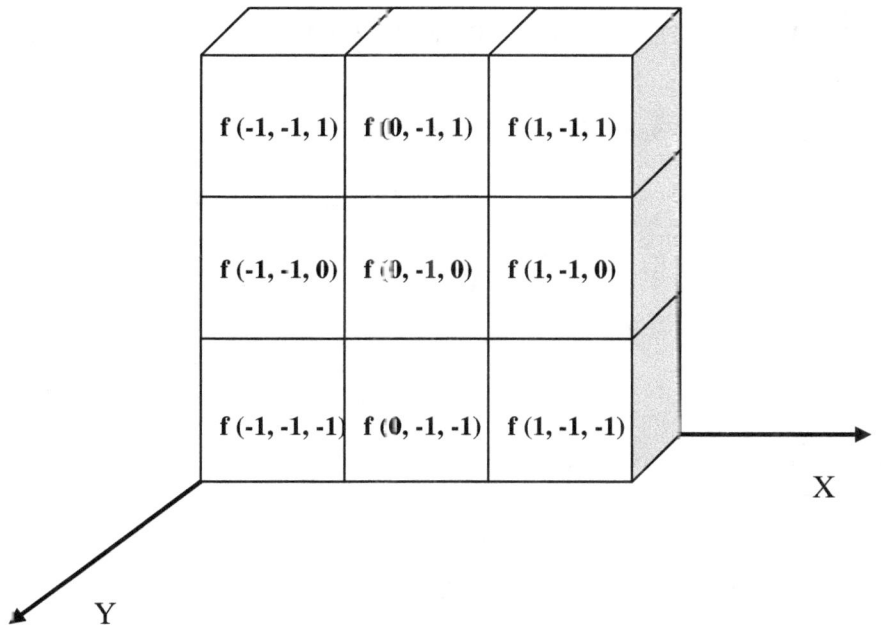

Figure 3: Outer-Right 3 x 3 Sub-neighborhood of f (0, 0, 0).

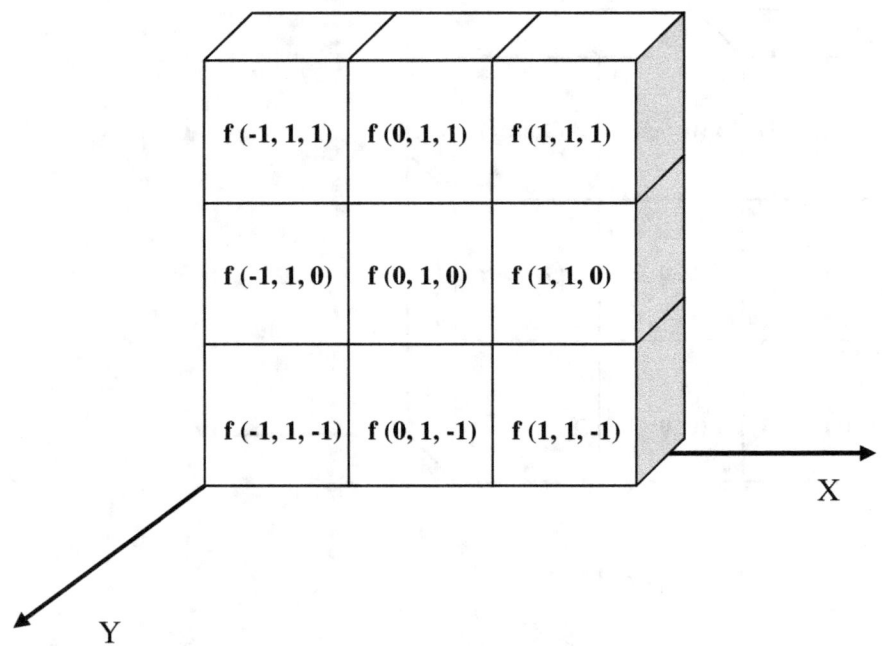

Figure 4: Outer-Left 3 x 3 Sub-neighborhood of f (0, 0, 0).

CALCULATION OF THE PARTIAL SECOND ORDER DERIVATIVES: TRIVARIATE QUADRATIC B-SPLINE

First Order Partial Derivative with Respect to the x Variable

$(\partial \, (\, h_3(x, y, z) \,) \,) \, / \partial x = \partial \, \{ \, f (0, 0, 0) + \omega_1 * [- 2a \, (x + y + z)^2 + 1/2 \, (a{+}1) \,] + \omega_2 * [\, a \, (x + y$

$+ z)^2 - (2a + 1/2) \, (x + y + z) + 3/4 \, (a{+}1) \,] \, \} \, / \partial x = \{ \, \omega_1 * [- 2a \, \partial \, (x + y + z)^2 / \partial x \,] + \omega_2 *$

$[\, a \, \partial \, (x + y + z)^2 / \partial x - (2a + 1/2) \, \partial \, (x + y + z) / \partial x \,] \, \} =$

$\{ \, \omega_1 * [- 4a \, (x + y + z)] + \omega_2 * [\, 2a \, (x + y + z) - (2a + 1/2) \,] \, \}$ 　　　　　(5)

First Order Partial Derivative with Respect to the y Variable

$(\partial \, (\, h_3(x, y, z) \,) \,) \, / \partial y = \partial \, \{ \, f (0, 0, 0) + \omega_1 * [- 2a \, (x + y + z)^2 + 1/2 \, (a{+}1) \,] + \omega_2 * [\, a \, (x + y$

$+ z)^2 - (2a + 1/2) \, (x + y + z) + 3/4 \, (a{+}1) \,] \, \} \, / \partial y = \{ \, \omega_1 * [- 2a \, \partial \, (x + y + z)^2 / \partial y \,] + \omega_2 *$

$[\, a \, \partial \, (x + y + z)^2 / \partial y - (2a + 1/2) \, \partial \, (x + y + z) / \partial y \,] \, \} =$

$\{ \, \omega_1 * [- 4a \, (x + y + z)] + \omega_2 * [\, 2a \, (x + y + z) - (2a + 1/2) \,] \, \}$ 　　　　　(6)

First Order Partial Derivative with Respect to the z Variable

$(\partial\,(\,h_3(x, y, z)\,)\,/\partial z) = \hat{c}\,\{\,f\,(0, 0, 0) + \omega_1 * [- 2a\,(x + y + z)^2 + 1/2\,(a+1)\,] + \omega_2 * [\,a\,(x + y$

$+ z)^2 - (2a + 1/2)\,(x + y + z) + 3/4\,(a+1)\,]\,\} \,/\partial z = \{\,\omega_1 * [- 2a\,\partial\,(x + y + z)^2 /\partial z\,] + \omega_2 *$

$[\,a\,\partial\,(x + y + z)^2 /\partial z - (2a + 1/2)\,\partial\,(x + y + z)\,/\partial z\,]\,\} =$

$\{\,\omega_1 * [- 4a\,(x + y + z)] + \omega_2 * [\,2a\,(x + y + z) - (2a + 1/2)\,]\,\}$ (7)

Therefore:

$(\partial\,(\,h_3(x, y, z)\,)\,/\partial x) = (\partial\,(\,h_3(x, y, z)\,)\,/\partial y) = (\partial\,(\,h_3(x, y, z)\,)\,/\partial z) =$

$\{\,\omega_1 * [- 4a\,(x + y + z)] - \omega_2 * [\,2a\,(x + y + z) - (2a + 1/2)\,]\,\}$ (8)

Second Order Partial Derivative with Respect to the x Variable

$(\partial^2\,(\,h_3(x, y, z)\,)\,/\partial x^2) = \{\partial\,\{\partial\,(\,h_3(x, y, z)\,)\,/\partial x\}\,/\partial x\} = \{\partial\,\{\,\omega_1 * [- 4a\,(x + y + z)] + \omega_2 *$

$[\,2a\,(x + y + z) - (2a + 1/2)\,]\,\}/\partial x\} = \{\,- 4a\,\omega_1 + 2a\,\omega_2\,\}$ (9)

Second Order Partial Derivative with Respect to the y Variable

$(\partial^2\,(\,h_3(x, y, z)\,)\,/\partial y^2) = \{\partial\,\{\partial\,(\,h_3(x, y, z)\,)\,/\partial y\}\,/\partial y\} = \{\partial\,\{\,\omega_1 * [- 4a\,(x + y + z)] + \omega_2 *$

$[\,2a\,(x + y + z) - (2a + 1/2)\,]\,\}/\partial y\} = \{\,- 4a\,\omega_1 + 2a\,\omega_2\,\}$ (10)

Second Order Partial Derivative with Respect to the z Variable

$(\partial^2\,(\,h_3(x, y, z)\,)\,/\partial z^2) = \{\partial\,\{\partial\,(\,h_3(x, y, z)\,)\,/\partial z\}\,/\partial z\} = \{\partial\,\{\,\omega_1 * [- 4a\,(x + y + z)] + \omega_2 *$

$[\,2a\,(x + y + z) - (2a + 1/2)\,]\,\}/\partial z\} = \{\,- 4a\,\omega_1 + 2a\,\omega_2\,\}$ (11)

Second Order Partial Derivatives with Respect to the x and y Variables

$(\partial^2\,(\,h_3(x, y, z)\,)\,/\partial x\partial y) = \{\partial\,(\partial\,(\,h_3(x, y, z)\,)\,/\partial x)\,/\partial y\} = \{\partial\,\{\,\omega_1 * [- 4a\,(x + y + z)] + \omega_2$

$* [\,2a\,(x + y + z) - (2a + 1/2)\,]\,\}\,/\partial y\} = \{\,- 4a\,\omega_1 + 2a\,\omega_2\,\}$ (12)

$(\partial^2\,(\,h_3(x, y, z)\,)\,/\partial y\partial x) = \{\partial\,(\partial\,(\,h_3(x, y, z)\,)\,/\partial y)\,/\partial x\} = \{\partial\,\{\,\omega_1 * [- 4a\,(x + y + z)] + \omega_2$

$* [\,2a\,(x + y + z) - (2a + 1/2)\,]\,\}\,/\partial x\} = \{\,- 4a\,\omega_1 + 2a\,\omega_2\,\}$ (13)

Therefore:

$(\partial^2\,(\,h_3(x, y, z)\,)\,/\partial x\partial y) = (\partial^2\,(\,h_3(x, y, z)\,)\,/\partial y\partial x) = \{\,- 4a\,\omega_1 + 2a\,\omega_2\,\}$ (14)

Second Order Partial Derivatives with Respect to the x and z Variables

$(\partial^2\,(h_3(x,\,y,\,z))\,/\partial x\partial z) = \{\partial\,(\partial\,(h_3(x,\,y,\,z))\,/\partial x)\,/\partial z\} = \{\partial\,\{\,\omega_1\,*\,[\,-\,4a\,(x\,+\,y\,+\,z)\,]\,+\,\omega_2$

$*\,[\,2a\,(x\,+\,y\,+\,z)\,-\,(2a\,+\,1/2)\,]\,\}\,/\partial z\} = \{\,-\,4a\,\omega_1\,+\,2a\,\omega_2\,\}$ 　　　　　(15)

$(\partial^2\,(h_3(x,\,y,\,z))\,/\partial z\partial x) = \{\partial\,(\partial\,(h_3(x,\,y,\,z))\,/\partial z)\,/\partial x\} = \{\partial\,\{\,\omega_1\,*\,[\,-\,4a\,(x\,+\,y\,+\,z)\,]\,+\,\omega_2$

$*\,[\,2a\,(x\,+\,y\,+\,z)\,-\,(2a\,+\,1/2)\,]\,\}\,/\partial x\} = \{\,-\,4a\,\omega_1\,+\,2a\,\omega_2\,\}$ 　　　　　(16)

Second Order Partial Derivatives with Respect to the y and z Variables

$(\partial^2\,(h_3(x,\,y,\,z))\,/\partial y\partial z) = \{\partial\,(\partial\,(h_3(x,\,y,\,z))\,/\partial y)\,/\partial z\} = \{\partial\,\{\,\omega_1\,*\,[\,-\,4a\,(x\,+\,y\,+\,z)\,]\,+\,\omega_2$

$*\,[\,2a\,(x\,+\,y\,+\,z)\,-\,(2a\,+\,1/2)\,]\,\}\ \,/\partial z\} = \{\,-\,4a\,\omega_1\,+\,2a\,\omega_2\,\}$ 　　　　　(17)

$(\partial^2\,(h_3(x,\,y,\,z))\,/\partial z\partial y) = \{\partial\,(\partial\,(h_3(x,\,y,\,z))\,/\partial z)\,/\partial y\} = \{\partial\,\{\,\omega_1\,*\,[\,-\,4a\,(x\,+\,y\,+\,z)\,]\,+\,\omega_2$

$*\,[\,2a\,(x\,+\,y\,+\,z)\,-\,(2a\,+\,1/2)\,]\,\}\ \,/\partial y\} = \{\,-\,4a\,\omega_1\,+\,2a\,\omega_2\,\}$ 　　　　　(18)

Therefore:

$(\partial^2\,(h_3(x,\,y,\,z))\,/\partial x^2)\ = (\partial^2\,(h_3(x,\,y,\,z))\,/\partial y^2) = (\partial^2\,(h_3(x,\,y,\,z))\,/\partial z^2) =$

$(\partial^2\,(h_3(x,\,y,\,z))\,/\partial x\partial y) = (\partial^2\,(h_3(x,\,y,\,z))\,/\partial y\partial x) = (\partial^2\,(h_3(x,\,y,\,z))\,/\partial x\partial z) =$

$(\partial^2\,(h_3(x,\,y,\,z))\,/\partial z\partial x) = (\partial^2\,(h_3(x,\,y,\,z))\,/\partial y\partial z) = (\partial^2\,(h_3(x,\,y,\,z))\,/\partial z\partial y) =$

$\{\,-\,4a\,\omega_1\,+\,2a\,\omega_2\,\}$ 　　　　　(19)

CALCULATION OF THE INTENSITY-CURVATURE TERMS: TRIVARIATE QUADRATIC B-SPLINE

Let the intensity-curvature term before interpolation be defined as:

$$E_o = E_o\,(x,\,y,\,z) = \int_0^x\int_0^y\int_0^z f\,(0,\,0,\,0)\,*\,\{\,(\partial^2\,(h_3(x,\,y,\,z))\,/\partial x^2) + (\partial^2\,(h_3(x,\,y,\,z))\,/\partial y^2) +$$

$(\partial^2\,(h_3(x,\,y,\,z))\,/\partial z^2) + (\partial^2\,(h_3(x,\,y,\,z))\,/\partial x\partial y) + (\partial^2\,(h_3(x,\,y,\,z))\,/\partial y\partial x) + (\partial^2\,(h_3(x,\,y,$

$z))\,/\partial x\partial z) + (\partial^2\,(h_3(x,\,y,\,z))\,/\partial z\partial x) + (\partial^2\,(h_3(x,\,y,\,z))\,/\partial y\partial z) + (\partial^2\,(h_3(x,\,y,\,z))\,/\partial z\partial y)\}$

$(0,\,0,\,0)\,dx\,dy\,dz$ 　　　　　(20)

On the basis of equation (19), it can be written that:

$$E_o = E_o(x, y, z) = \int_0^x \int_0^y \int_0^z 9 * f(0, 0, 0) * \{ -4a\,\omega_1 + 2a\,\omega_2 \}\ dx\ dy\ dz =$$

$$9\,xyz * f(0, 0, 0) * \{ -4a\,\omega_1 + 2a\,\omega_2 \} \tag{21}$$

Let the intensity-curvature term after interpolation be defined as:

$$E_{IN} = E_{IN}(x, y, z) = \int_0^x \int_0^y \int_0^z h_3(x, y, z) * \{ (\partial^2 (h_3(x, y, z))\,/\partial x^2) + (\partial^2 (h_3(x, y, z))\,/\partial y^2) +$$

$(\partial^2 (h_3(x, y, z))\,/\partial z^2) + (\partial^2 (h_3(x, y, z))\,/\partial x \partial y) + (\partial^2 (h_3(x, y, z))\,/\partial y \partial x) + (\partial^2 (h_3(x, y,$

$z))\,/\partial x \partial z) + (\partial^2 (h_3(x, y, z))\,/\partial z \partial x) + (\partial^2 (h_3(x, y, z))\,/\partial y \partial z) + (\partial^2 (h_3(x, y, z))\,/\partial z \partial y) \}$

$$dx\ dy\ dz \tag{22}$$

Consider the following calculations:

$$(x + y + z)^2 = (x + y + z) * (x + y + z) = x^2 + xy + xz + xy + y^2 + yz + xz + yz + z^2 =$$

$$x^2 + y^2 + z^2 + 2\,xy + 2\,xz + 2\,yz \tag{23}$$

$$\int_0^x \int_0^y \int_0^z (x + y + z)^2\ dx\ dy\ dz = \int_0^x \int_0^y \int_0^z (x^2 + y^2 + z^2 + 2\,xy + 2\,xz + 2\,yz)\ dx\ dy\ dz =$$

$$\int_0^x \int_0^y (x^2 z + y^2 z + z^3/3 + 2\,xyz + xz^2 + yz^2)\ dx\ dy =$$

$$\int_0^x (x^2 yz + y^3 z/3 + z^3 y/3 + xy^2 z + xyz^2 + y^2 z^2/2)\ dx =$$

$$(x^3 yz/3 + xy^3 z/3 + xz^3 y/3 + x^2 y^2 z/2 + x^2 yz^2/2 + xy^2 z^2/2) \tag{24}$$

$$\int_0^x \int_0^y \int_0^z (x + y + z)\ dx\ dy\ dz = \int_0^x \int_0^y (xz + yz + z^2/2)\ dx\ dy = \int_0^x (xyz + y^2 z/2 + yz^2/2)\ dx =$$

$$(x^2 yz/2 + xy^2 z/2 + xyz^2/2) \tag{25}$$

Let us proceed to the calculation of equation (22):

$$E_{IN} = E_{IN}(x, y, z) = \int_0^x \int_0^y \int_0^z 9 * \{ f(0, 0, 0) + \omega_1 * [-2a(x + y + z)^2 + 1/2\,(a+1)] + \omega_2 * [a$$

$(x + y + z)^2 - (2a + 1/2)(x + y + z) + 3/4\,(a+1)] \} * \{ -4a\,\omega_1 + 2a\,\omega_2 \}\ dx\ dy\ dz =$

$9 \text{ xyz} * f(0, 0, 0) * \{ -4a \, \omega_1 + 2a \, \omega_2 \} + \int_0^x \int_0^y \int_0^z 9 * \{ \omega_1 * [- 2a (x + y + z)^2 + 1/2 (a+1)] +$

$\omega_2 * [a (x + y + z)^2 - (2a + 1/2) (x + y + z) + 3/4 (a+1)] \} * \{ -4a \, \omega_1 + 2a \, \omega_2 \} \, dx \, dy \, dz =$

$9 \text{ xyz} * f(0, 0, 0) * \{ -4a \, \omega_1 + 2a \, \omega_2 \} + 9 * \{ -4a \, \omega_1 + 2a \, \omega_2 \} * \{ \omega_1 * [- 2a (x^3yz/3 +$

$xy^3z/3 + xz^3y/3 + x^2y^2z/2 + x^2yz^2/2 + xy^2z^2/2) + 1/2 (a+1) \text{ xyz}] + \omega_2 * [a (x^3yz/3 +$

$xy^3z/3 + xz^3y/3 + x^2y^2z/2 + x^2yz^2/2 + xy^2z^2/2) - (2a + 1/2) (x^2yz/2 + xy^2z/2 + xyz^2/2) + 3/4$

$(a+1) \text{ xyz}] \}$ (26)

EQUATING THE TWO INTENSITY-CURVATURE TERMS BEFORE AND AFTER INTERPOLATION: TRIVARIATE QUADRATIC B-SPLINE

Let us proceed to solve the following equation which is derived posing in equality the two intensity-curvature terms E_o (equation (21)) and E_{IN} (equation (26)).

$E_o(x, y, z) = 9 \text{ xyz} * f(0, 0, 0) * \{ -4a \, \omega_1 + 2a \, \omega_2 \} = E_{IN}(x, y, z) = 9 \text{ xyz} * f(0, 0, 0) *$

$\{ -4a \, \omega_1 + 2a \, \omega_2 \} + 9 * \{ -4a \, \omega_1 + 2a \, \omega_2 \} * \{ \omega_1 * [- 2a (x^3yz/3 + xy^3z/3 + xz^3y/3 +$

$x^2y^2z/2 + x^2yz^2/2 + xy^2z^2/2) + 1/2 (a+1) \text{ xyz}] + \omega_2 * [a (x^3yz/3 + xy^3z/3 + xz^3y/3 +$

$x^2y^2z/2 + x^2yz^2/2 + xy^2z^2/2) - (2a + 1/2) (x^2yz/2 + xy^2z/2 + xyz^2/2) + 3/4 (a+1) \text{ xyz}] \}$

(27)

Equation (27) yields:

$9 * \{ -4a \, \omega_1 + 2a \, \omega_2 \} * \{ \omega_1 * [- 2a (x^3yz/3 + xy^3z/3 + xz^3y/3 + x^2y^2z/2 + x^2yz^2/2 +$

$xy^2z^2/2) + 1/2 (a+1) \text{ xyz}] + \omega_2 * [a (x^3yz/3 + xy^3z/3 + xz^3y/3 + x^2y^2z/2 + x^2yz^2/2 +$

$xy^2z^2/2) - (2a + 1/2) (x^2yz/2 + xy^2z/2 + xyz^2/2) + 3/4 (a+1) \text{ xyz}] \} = 0$

$\{ \omega_1 * [- 2a (x^3yz/3 + xy^3z/3 + xz^3y/3 + x^2y^2z/2 + x^2yz^2/2 + xy^2z^2/2) + 1/2 (a+1) \text{ xyz}] + \omega_2$

$* [a (x^3yz/3 + xy^3z/3 + xz^3y/3 + x^2y^2z/2 + x^2yz^2/2 + xy^2z^2/2) - (2a + 1/2) (x^2yz/2 + xy^2z/2$

$+ xyz^2/2) + 3/4 (a+1) \text{ xyz}] \} = 0$ (28)

On the basis of equations (2) and (3), let us posit:

$\omega_1 = [\varphi_1 + f(1, 0, 0)]$ (29)

$$\omega_2 = [\, \varphi_2 + f\,(1, 0, 0)\,] \tag{30}$$

The voxel $f\,(1, 0, 0)$ is visible in figure 2 in the Central 3x3 Sub-neighborhood of $f(0, 0, 0)$. Also it is true that:

$$\varphi_1 = [\, f\,(-1, 0, 1) + f\,(-1, 0, 0) + f\,(1, 0, -1) + f\,(-1, -1, 1) + f\,(-1, -1, 0) + f\,(1, -1, 0) + f\,(1,$$

$$-1, -1) + f\,(-1, 1, 1) + f\,(-1, 1, 0) + f\,(1, 1, 0) + f\,(1, 1, -1)\,] \tag{31}$$

$$\varphi_2 = [\, f\,(-1, 0, 1) + f\,(0, 0, 1) + f\,(1, 0, 1) + f\,(-1, 0, 0) + f\,(-1, 0, -1) + f\,(0, 0, -1) + f\,(1, 0,$$

$$-1) + f\,(-1, -1, 1) + f\,(0, -1, 1) + f\,(1, -1, 1) + f\,(-1, -1, 0) + f\,(0, -1, 0) + f\,(1, -1, 0) + f\,(-1,$$

$$-1, -1) + f\,(0, -1, -1) + f\,(1, -1, -1) + f\,(-1, 1, 1) + f\,(0, 1, 1) + f\,(1, 1, 1) + f\,(-1, 1, 0) + f\,(0,$$

$$1, 0) + f\,(1, 1, 0) + f\,(-1, 1, -1) + f\,(0, 1, -1) + f\,(1, 1, -1)\,] \tag{32}$$

Therefore equation (28) can be written as:

$$\{\, [\, \varphi_1 + f\,(1, 0, 0)\,]\, *\, [- 2a\,(x^3yz/3 + xy^3z/3 + xz^3y/3 + x^2y^2z/2 + x^2yz^2/2 + xy^2z^2/2) + 1/2$$

$$(a+1)\, xyz\,] + [\, \varphi_2 + f\,(1, 0, 0)\,]\, *\, [\, a\,(x^3yz/3 + xy^3z/3 + xz^3y/3 + x^2y^2z/2 + x^2yz^2/2 +$$

$$xy^2z^2/2) - (2a + 1/2)\,(x^2yz/2 + xy^2z/2 + xyz^2/2) + 3/4\,(a+1)\, xyz\,]\,\} = 0 \tag{33}$$

$$- f\,(1, 0, 0)\, *\, \{\, \Xi_1 + \Xi_2\,\} = \{\, \varphi_1 * \Xi_1 + \varphi_2 * \Xi_2\,\} \tag{34}$$

Where it is posited that:

$$\Xi_1 = [- 2a\,(x^3yz/3 + xy^3z/3 + xz^3y/3 + x^2y^2z/2 + x^2yz^2/2 + xy^2z^2/2) + 1/2\,(a+1)\, xyz\,]$$

$$\Xi_2 = [\, a\,(x^3yz/3 + xy^3z/3 + xz^3y/3 + x^2y^2z/2 + x^2yz^2/2 + xy^2z^2/2) - (2a + 1/2)\,(x^2yz/2 +$$

$$xy^2z/2 + xyz^2/2) + 3/4\,(a+1)\, xyz\,]$$

And equation (35) hereto follow furnishes the value of the signal resilient to the trivariate quadratic B-Spline interpolation function of equation (4).

$$f\,(1, 0, 0) = - \{\, \varphi_1 * \Xi_1 + \varphi_2 * \Xi_2\,\}\, /\, \{\, \Xi_1 + \Xi_2\,\} \tag{35}$$

PART III - TRIVARIATE CUBIC B-SPLINE FUNCTION
WITH EMBEDDING

TRIVARIATE CUBIC B-SPLINE

Let us consider the trivariate cubic B-spline of the form:

$$h_4(x, y, z) = f(0, 0, 0) + \omega_1 * [\, 1/2\, (x + y + z)^3 - (x + y + z)^2 + 2/3\,] + \omega_2 * [\,-1/6\, (x + y +$$

$$z)^3 + (x + y + z)^2 - 2\, (x + y + z) + 4/3\,] \tag{36}$$

where the values of the positions ω_1 and ω_2 hold true as per equations (2) and (3).

CALCULATION OF THE PARTIAL SECOND ORDER DERIVATIVES: TRIVARIATE CUBIC B-SPLINE

First Order Partial Derivative with Respect to the x Variable

$$(\partial\, (h_4(x, y, z))\, /\partial x) = \partial\, \{\, f(0, 0, 0) + \omega_1 * [\, 1/2\, (x + y + z)^3 - (x + y + z)^2 + 2/3\,] + \omega_2 *$$

$$[\,-1/6\, (x + y + z)^3 + (x + y + z)^2 - 2\, (x + y + z) + 4/3\,]\, /\partial x\, \} = \{\, \omega_1 * [\, 1/2\, \partial\, (x + y + z)^3$$

$$/\partial x - \partial\, (x + y + z)^2\, /\partial x\,] + \omega_2 * [\,-1/6\, \partial\, (x + y + z)^3\, /\partial x + \partial\, (x + y + z)^2\, /\partial x - 2\, \partial\, (x + y +$$

$$z)\, /\partial x\,]\, \} = \{\, \omega_1 * [\, 3/2\, (x + y + z)^2 - 2\, (x + y + z)\,] + \omega_2 * [\,-3/6\, \partial\, (x + y + z)^2 + 2\, (x +$$

$$y + z) - 2\,]\, \} \tag{37}$$

First Order Partial Derivative with Respect to the y Variable

$$(\partial\, (h_4(x, y, z))\, /\partial y) = \partial\, \{\, f(0, 0, 0) + \omega_1 * [\, 1/2\, (x + y + z)^3 - (x + y + z)^2 + 2/3\,] + \omega_2 *$$

$$[\,-1/6\, (x + y + z)^3 + (x + y + z)^2 - 2\, (x + y + z) + 4/3\,]\, /\partial y\, \} = \{\, \omega_1 * [\, 1/2\, \partial\, (x + y + z)^3$$

$$/\partial y - \partial\, (x + y + z)^2\, /\partial y\,] + \omega_2 * [\,-1/6\, \partial\, (x + y + z)^3\, /\partial y + \partial\, (x + y + z)^2\, /\partial y - 2\, \partial\, (x + y +$$

$$z)\, /\partial y\,]\, \} = \{\, \omega_1 * [\, 3/2\, (x + y + z)^2 - 2\, (x + y + z)\,] + \omega_2 * [\,-3/6\, (x + y + z)^2 + 2\, (x +$$

$$y + z) - 2\,]\, \} \tag{38}$$

First Order Partial Derivative with Respect to the z Variable

$$(\partial\, (h_4(x, y, z))\, /\partial z) = \partial\, \{\, f(0, 0, 0) + \omega_1 * [\, 1/2\, (x + y + z)^3 - (x + y + z)^2 + 2/3\,] + \omega_2 *$$

$$[\,-1/6\, (x + y + z)^3 + (x + y + z)^2 - 2\, (x + y + z) + 4/3\,]\, /\partial z\, \} = \{\, \omega_1 * [\, 1/2\, \partial\, (x + y + z)^3$$

$$/\partial z - \partial\, (x + y + z)^2\, /\partial z\,] + \omega_2 * [\,-1/6\, \partial\, (x + y + z)^3\, /\partial z + \partial\, (x + y + z)^2\, /\partial z - 2\, \partial\, (x + y +$$

$z) /\partial z] \} = \{ \omega_1 * [3/2 \ (x + y + z)^2 - 2 (x + y + z)] + \omega_2 * [-3/6 (x + y + z)^2 + 2 (x + y + z) - 2] \}$ (39)

Therefore:

$\partial (h_4(x, y, z)) /\partial x = (\partial (h_4(x, y, z)) /\partial y = (\partial (h_4(x, y, z)) /\partial z) = \{ \omega_1 * [3/2 (x + y + z)^2 - 2 (x + y + z)] + \omega_2 * [-3/6 (x + y + z)^2 + 2 (x + y + z) - 2] \}$ (40)

Second Order Partial Derivatives

$(\partial^2 (h_4(x, y, z)) /\partial x^2) = \{\partial \{\partial (h_4(x, y, z)) /\partial x\} /\partial x\} = \{\partial \{ \omega_1 * [3/2 (x + y + z)^2 - 2 (x - y + z)] + \omega_2 * [-3/6 (x + y + z)^2 + 2 (x + y + z) - 2] \} /\partial x\} = \{ \omega_1 * [3 (x + y + z) - 2] - \omega_2 * [- (x + y + z) + 2] \}$ (41)

Because of equation (40) it is true that:

$(\partial^2 (h_4(x, y, z)) /\partial x^2) = \{\partial \{\partial (h_4(x, y, z)) /\partial x\} /\partial x\} =$

$(\partial^2 (h_4(x, y, z)) /\partial y^2) = \{\partial \{\partial (h_4(x, y, z)) /\partial y\} /\partial y\} =$

$(\partial^2 (h_4(x, y, z)) /\partial z^2) = \{\partial \{\partial (h_4(x, y, z)) /\partial z\} /\partial z\} =$

$\{ \omega_1 * [3 (x + y + z) - 2] + \omega_2 * [- (x + y + z) + 2] \}$ (42)

Also it is true that:

$(\partial^2 (h_4(x, y, z)) /\partial x \partial y) = \{\partial (\partial (h_4(x, y, z)) /\partial x) /\partial y\} = \{\partial \{ \omega_1 * [3/2 (x + y + z)^2 - 2 (x + y + z)] + \omega_2 * [-3/6 (x + y + z)^2 + 2 (x + y + z) - 2] \} /\partial y\} = \{ \omega_1 * [3 (x + y + z) - 2] + \omega_2 * [- (x + y + z) + 2] \}$ (43)

$(\partial^2 (h_4(x, y, z)) /\partial x \partial z) = \{\partial (\partial (h_4(x, y, z)) /\partial x) /\partial z\} = \{\partial \{ \omega_1 * [3/2 (x + y + z)^2 - 2 (x + y + z)] + \omega_2 * [-3/6 (x + y + z)^2 + 2 (x + y + z) - 2] \} /\partial z\} = \{ \omega_1 * [3 (x + y + z) - 2] + \omega_2 * [- (x + y + z) + 2] \}$ (44)

$(\partial^2 (h_4(x, y, z)) /\partial y \partial z) = \{\partial (\partial (h_4(x, y, z)) /\partial y) /\partial z\} = \{\partial \{ \omega_1 * [3/2 (x + y + z)^2 - 2 (x + y + z)] + \omega_2 * [-3/6 (x + y + z)^2 + 2 (x + y + z) - 2] \} /\partial z\} = \{ \omega_1 * [3 (x + y + z) - 2] + \omega_2 * [- (x + y + z) + 2] \}$ (45)

$(\partial^2 (h_4(x, y, z)) /\partial x \partial y) = \{\partial (\partial (h_4(x, y, z)) /\partial y) /\partial x\} =$

$(\partial^2 (h_4(x, y, z)) /\partial x \partial z) = \{\partial (\partial (h_4(x, y, z)) /\partial z) /\partial x\} =$

$(\partial^2 (h_4(x, y, z)) /\partial y \partial z) = \{\partial (\partial (h_4(x, y, z)) /\partial z) /\partial y\} =$

$\{ \omega_1 * [3 (x + y + z) - 2] + \omega_2 * [- (x + y + z) + 2] \}$ (46)

CALCULATION OF THE INTENSITY-CURVATURE TERMS: TRIVARIATE CUBIC B-SPLINE

The intensity-curvature term before interpolation is defined as:

$E_o = E_o (x, y, z) = \int_0^x \int_0^y \int_0^z f (0, 0, 0) * \{ (\partial^2 (h_4(x, y, z)) /\partial x^2) + (\partial^2 (h_4(x, y, z)) /\partial y^2) +$

$(\partial^2 (h_4(x, y, z)) /\partial z^2) + (\partial^2 (h_4(x, y, z)) /\partial x \partial y) + (\partial^2 (h_4(x, y, z)) /\partial y \partial x) + (\partial^2 (h_4(x, y,$

$z)) /\partial x \partial z) + (\partial^2 (h_4(x, y, z)) /\partial z \partial x) + (\partial^2 (h_4(x, y, z)) /\partial y \partial z) + (\partial^2 (h_4(x, y, z)) /\partial z \partial y)\}$

$(0, 0, 0) \, dx \, dy \, dz$ (47)

On the basis of equations (42) and (46), it can be written that:

$(\partial^2 (h_4(x, y, z)) /\partial x^2) (0, 0, 0) = (\partial^2 (h_4(x, y, z)) /\partial y^2) (0, 0, 0) =$

$(\partial^2 (h_4(x, y, z)) /\partial z^2) (0, 0, 0) = \{ - 2 \omega_1 + 2 \omega_2 \}$ (48.a)

$(\partial^2 (h_4(x, y, z)) /\partial x \partial y) (0, 0, 0) = (\partial^2 (h_4(x, y, z)) /\partial x \partial z) (0, 0, 0) =$

$(\partial^2 (h_4(x, y, z)) /\partial y \partial z) (0, 0, 0) = \{ - 2 \omega_1 + 2 \omega_2 \}$ (48.b)

Therefore:

$E_o = E_o (x, y, z) = \int_0^x \int_0^y \int_0^z 9 * f (0, 0, 0) * \{ - 2 \omega_1 + 2 \omega_2 \} \, dx \, dy \, dz =$

$9 \, xyz * f (0, 0, 0) * \{ - 2 \omega_1 + 2 \omega_2 \}$ (49)

Let the intensity-curvature term after interpolation is defined as:

$E_{IN} = E_{IN} (x, y, z) = \int_0^x \int_0^y \int_0^z h_4(x, y, z) * \{ (\partial^2 (h_4(x, y, z)) /\partial x^2) + (\partial^2 (h_4(x, y, z)) /\partial y^2) +$

$(\partial^2 (h_4(x, y, z)) /\partial z^2) + (\partial^2 (h_4(x, y, z)) /\partial x \partial y) + (\partial^2 (h_4(x, y, z)) /\partial y \partial x) + (\partial^2 (h_4(x, y,$

z)) $/\partial x\partial z$) + (∂^2 (h$_4$(x, y, z)) $/\partial z\partial x$) + (∂^2 (h$_4$(x, y, z)) $/\partial y\partial z$) + (∂^2 (h$_4$(x, y, z)) $/\partial z\partial y$)}

dx dy dz   (50)

On the basis of equations (36), (42) and (46), it can be written that:

$E_{IN} = E_{IN}$ (x, y, z) = $\int\limits_0^x\int\limits_0^y\int\limits_0^z$ 9 * { f (0, 0, 0) + ω_1 * [1/2 (x + y + z)^3 - (x + y + z)^2 + 2/3] + ω_2

* [-1/6 (x + y + z)^3 + (x + y + z)^2 - 2 (x + y + z) + 4/3] } * { ω_1 * [3 (x + y + z) − 2] + ω_2

* [- (x + y + z) + 2] } dx dy dz   (51)

Equation (51) is solved with the following procedure:

$E_{IN} = E_{IN}$ (x, y, z) = 9 * $\int\limits_0^x\int\limits_0^y\int\limits_0^z$ f (0, 0, 0) * { ω_1 * [3 (x + y + z) − 2] + ω_2 *

[- (x + y + z) + 2] } + ω_1^2 * [1/2 (x + y + z)^3 - (x + y + z)^2 + 2/3] [3 (x + y + z) − 2] −

ω_1 ω_2 * [1/2 (x + y + z)^3 - (x + y + z)^2 + 2/3] [- (x + y + z) + 2] + ω_2 ω_1 *

[-1/6 (x + y + z)^3 + (x + y + z)^2 - 2 (x + y + z) + 4/3] [3 (x + y + z) − 2] + ω_2^2

[-1/6 (x + y + z)^3 + (x + y + z)^2 - 2 (x + y + z) + 4/3] [- (x + y + z) + 2] dx dy dz =

9 * $\int\limits_0^x\int\limits_0^y\int\limits_0^z$ f (0, 0, 0) * { ω_1 * [3 (x + y + z) − 2] + ω_2 * [- (x + y + z) + 2] } + ω_1^2 *

3/2 (x + y + z)^4 − 3 (x + y + z)^3 + 2 (x + y + z) - (x + y + z)^3 + 2 (x + y + z)^2 - 4/3] +

ω_1 ω_2 * [-1/2 (x + y + z)^4 + (x + y + z)^3 - 2/3 (x + y + z) + (x + y + z)^3 − 2 (x + y + z)^2 +

4/3] + ω_2 ω_1 * [-3/6 (x + y + z)^4 + 3 (x + y + z)^3 - 6 (x + y + z)^2 + 4 (x + y + z) + 2/6 (x +

y + z)^3 − 2 (x + y + z)^2 + 4 (x + y + z) - 8/3] + ω_2^2 [1/6 (x + y + z)^4 − (x + y + z)^3 + 2 (x +

y + z)^2 - 4/3 (x + y + z) -2/6 (x + y + z)^3 + 2 (x + y + z)^2 - 4 (x + y + z) + 8/3] dx dy dz =

9 * $\int\limits_0^x\int\limits_0^y\int\limits_0^z$ f (0, 0, 0) * { ω_1 * [3 (x + y + z) − 2] + ω_2 * [- (x + y + z) + 2] } + ω_1^2 *

[3/2 (x + y + z)^4 − 4 (x + y + z)^3 + 2 (x + y + z) + 2 (x + y + z)^2 - 4/3] +

ω_1 ω_2 * [-1/2 (x + y + z)^4 + 2 (x + y + z)^3 - 2/3 (x + y + z) − 2 (x + y + z)^2 +

$4/3] + \omega_2\,\omega_1 * [-3/6\,(x + y + z)^4 + 4/3\,(x + y + z)^3 - 8\,(x + y + z)^2 + 8\,(x + y + z) - 8/3]$

$+ \omega_2{}^2\,[1/6\,(x + y + z)^4 - 4/3\,(x + y + z)^3 + 4\,(x + y + z)^2 - 16/3\,(x + y + z) + 8/3]\ dx\ dy$

$dz =$

$9 * \int\limits_0^x\int\limits_0^y\int\limits_0^z f(0, 0, 0) * \{\omega_1 * [3\,(x + y + z) - 2] + \omega_2 * [-(x + y + z) + 2]\} + [3/2\,\omega_1{}^2 - 1/2$

$\omega_1\,\omega_2 - 3/6\,\omega_2\,\omega_1 + 1/6\,\omega_2{}^2]\,(x + y + z)^4 + [-4\,\omega_1{}^2 + 2\,\omega_1\,\omega_2 + 4/3\,\omega_2\,\omega_1 - 4/3\,\omega_2{}^2]\,(x + y$

$+ z)^3 + [2\,\omega_1{}^2 - 2\,\omega_1\,\omega_2 - 8\,\omega_2\,\omega_1 + 4\,\omega_2{}^2]\,(x + y + z)^2 + [2\,\omega_1{}^2 - 2/3\,\omega_1\,\omega_2 + 8\,\omega_2\,\omega_1 -$

$16/3\,\omega_2{}^2]\,(x + y + z) + [-4/3\,\omega_1{}^2 + 4/3\,\omega_1\,\omega_2 - 8/3\,\omega_2\,\omega_1 + 8/3\,\omega_2{}^2]\ dx\ dy\ dz$ 　　　(52)

Now consider that on the basis of equation (23), it is true that:

$(x + y + z)^4 = (x + y + z)^2\,(x + y + z)^2 =$

$(x^2 + y^2 + z^2 + 2\,xy + 2\,xz + 2\,yz)\,(x^2 + y^2 + z^2 + 2\,xy + 2\,xz + 2\,yz) =$

$(x^4 + x^2y^2 + x^2z^2 + 2\,x^3y + 2\,x^3z + 2\,x^2yz + y^4 + x^2y^2 + y^2z^2 + 2\,xy^3 + 2\,xzy^2 + 2\,y^3z +$

$x^2z^2 + z^2y^2 + z^4 + 2\,xyz^2 + 2\,xz^3 + 2\,yz^3 + 2\,x^3y + 2\,xy^3 + 2\,xyz^2 + 4\,x^2y^2 + 4\,x^2yz + 4$

$xy^2z + 2\,x^3z + 2\,xy^2z + 2\,xz^3 + 4\,x^2yz + 4x^2z^2 + 4\,xyz^2 + 2\,x^2yz + 2\,y^3z + 2\,yz^3 + 4\,xy^2z +$

$4\,xyz^2 + 4\,y^2z^2)$ 　　　(53)

$(x + y + z)^3 = (x + y + z)^2\,(x + y + z) = (x^2 + y^2 + z^2 + 2\,xy + 2\,xz + 2\,yz)\,(x + y + z) =$

$(x^3 + xy^2 + xz^2 + 2\,x^2y + 2\,x^2z + 2\,xyz + x^2y + y^3 + yz^2 + 2\,xy^2 + 2\,xyz + 2\,y^2z + x^2z +$

$y^2z + z^3 + 2\,xyz + 2\,xz^2 + 2\,yz^2)$ 　　　(54)

Therefore, on the basis of equation (53) and (54) let us posit:

$\Pi^{(3)} = \int\limits_0^x\int\limits_0^y\int\limits_0^z (x + y + z)^3\ dx\ dy\ dz = \int\limits_0^x\int\limits_0^y\int\limits_0^z (x^3 + xy^2 + xz^2 + 2\,x^2y + 2\,x^2z + 2\,xyz + x^2y + y^3 +$

$yz^2 + 2xy^2 + 2\,xyz + 2\,y^2z + x^2z + y^2z + z^3 + 2\,xyz + 2\,xz^2 + 2\,yz^2)\ dx\ dy\ dz =$

$\int\limits_0^x\int\limits_0^y (x^3z + xy^2z + xz^3/3 + 2\,x^2yz + x^2z^2 + xyz^2 + x^2yz + y^3z + yz^3/3 + 2xy^2z + xyz^2 + y^2z^2$

62

$+ x^2z^2/2 + y^2z^2/2 + z^4/4 + xyz^2 + 2/3\ xz^3 + 2/3\ yz^3)\ dx\ dy =$

$$\int\limits_0^x (x^3yz + xy^3z/3 + xyz^3/3 + x^2y^2z + x^2yz^2 + xy^2z^2/2 + x^2y^2z/2 + y^4z/4 + y^2z^3/6 + 2/3\ xy^3z$$

$+ xy^2z^2/2 + y^3z^2/3 + x^2yz^2/2 + y^3z^2/6 + yz^4/4 + xy^2z^2/2 + 2/3\ xyz^3 + 1/3\ y^2z^3)\ dx =$

$(x^4yz/4 + x^2y^3z/6 + x^2yz^3/6 + x^3y^2z/3 + 1/3\ x^3yz^2 + x^2y^2z^2/4 + x^3y^2z/6 + xy^4z/4 + xy^2z^3/6 +$

$1/3\ x^2y^3z + x^2y^2z^2/4 + xy^3z^2/3 + x^3yz^2/6 + xy^3z^2/6 + xyz^4/4 + x^2y^2z^2/4 + 1/3\ x^2yz^3 + 1/3$

$xy^2z^3)$ (55)

$$\Pi^{(4)} = \int\limits_0^x\int\limits_0^y\int\limits_0^z (x + y + z)^4\ dx\ dy\ dz = \int\limits_0^x\int\limits_0^y\int\limits_0^z (x^4 + x^2y^2 + x^2z^2 + 2\ x^3y + 2\ x^3z + 2\ x^2yz + y^4 +$$

$x^2y^2 + y^2z^2 + 2\ xy^3 + 2\ xzy^2 + 2\ y^3z + x^2z^2 + z^2y^2 + z^4 + 2\ xyz^2 + 2\ xz^3 + 2\ yz^3 + 2\ x^3y + 2$

$xy^3 + 2\ xyz^2 + 4\ x^2y^2 + 4\ x^2yz + 4\ xy^2z + 2\ x^3z + 2\ xy^2z + 2\ xz^3 + 4\ x^2yz + 4x^2z^2 + 4\ xyz^2$

$+ 2\ x^2yz + 2\ y^3z + 2\ yz^3 + 4\ xy^2z + 4\ xyz^2 + 4\ y^2z^2)\ dx\ dy\ dz =$

$$\int\limits_0^x\int\limits_0^y (x^4z + x^2y^2z + x^2z^3/3 + 2\ x^3yz + x^3z^2 + x^2yz^2 + y^4z + x^2y^2z + y^2z^3/3 + 2\ xy^3z + xz^2y^2$$

$+ y^3z^2 + x^2z^3/3 + z^3y^2/3 + z^5/5 + 2/3\ xyz^3 + 1/2\ xz^4 + 1/2\ yz^4 + 2\ x^3yz + 2\ xy^3z + 2/3\ xyz^3$

$+ 4\ x^2y^2z + 2\ x^2yz^2 + 2\ xy^2z^2 + x^3z^2 + xy^2z^2 + 1/2\ xz^4 + 2\ x^2yz^2 + 4/3\ x^2z^3 + 4/3\ xyz^3 +$

$x^2yz^2 + y^3z^2 + 1/2\ yz^4 + 2\ xy^2z^2 + 4/3\ xyz^3 + 4/3\ y^2z^3)\ dx\ dy =$

$$\int\limits_0^x (x^4zy + x^2y^3z/3 + x^2yz^3/3 + x^3y^2z + x^3z^2y + x^2y^2z^2/2 + y^5z/5 + x^2y^3z/3 + y^3z^3/9 + 1/2$$

$xy^4z + xz^2y^3/3 + y^4z^2/4 + x^2yz^3/3 + z^3y^3/9 + z^5y/5 + 1/3\ xy^2z^3 + 1/2\ xyz^4 + 1/4\ y^2z^4 + x^3y^2z$

$+ 1/2\ xy^4z + 1/3\ xy^2z^3 + 4/3\ x^2y^3z + x^2y^2z^2 + 2/3\ xy^3z^2 + x^3z^2y + xy^3z^2/3 + 1/2\ xyz^4 +$

$x^2y^2z^2 + 4/3\ x^2yz^3 + 4/6\ xy^2z^3 + x^2y^2z^2/2 + y^4z^2/4 + 1/4\ y^2z^4 + 2/3\ xy^3z^2 + 4/6\ xy^2z^3 + 4/9$

$y^3z^3)\ dx$

$= (x^5zy/5 + x^3y^3z/9 + x^3yz^3/9 + x^4y^2z/4 + x^4z^2y/4 + x^3y^2z^2/6 + xy^5z/5 + x^3y^3z/9 + xy^3z^3/9 +$

$x^2y^4z + x^2z^2y^3/6 + xy^4z^2/4 + x^3yz^3/9 + xz^3y^3/9 + xz^5y/5 + 1/6\ x^2y^2z^3 + 1/4\ x^2yz^4 + 1/4$

$xy^2z^4 + x^4y^2z/4 + 1/4\ x^2y^4z + 1/6\ x^2y^2z^3 + 4/9\ x^3y^3z + x^3y^2z^2/3 + 1/3\ x^2y^3z^2 + x^4z^2y/4 +$

$x^2y^3z^2/6 + x^2yz^4 + x^3y^2z^2/3 + 4/9\ x^3yz^3 + 4/12\ x^2y^2z^3 + x^3y^2z^2/6 + xy^4z^2/4 + 1/4\ xy^2z^4 +$

$1/3\ x^2y^3z^2 + 4/12\ x^2y^2z^3 + 4/9\ xy^3z^3)$ (56)

Also, from equations (24) and (25) let us posit:

$\Pi^{(2)} = (x^3yz/3 + xy^3z/3 + xz^3y/3 + x^2y^2z/2 + x^2yz^2/2 + xy^2z^2/2)$ (57)

$\Pi^{(1)} = (x^2yz/2 + xy^2z/2 + xyz^2/2)$ (58)

Now equation (52) can be solved considering the knowledge provided in equations (55), (56), (57) and (58).

$E_{IN}(x, y, z) = 9 * \{ f(0, 0, 0) * \{ \omega_1 * [3\ \Pi^{(1)} - 2\ xyz] + \omega_2 * [-\Pi^{(1)} + 2\ xyz] \} + [3/2$

$\omega_1^2 - 1/2\ \omega_1\omega_2 - 3/6\ \omega_2\ \omega_1 + 1/6\ \omega_2^2]\ \Pi^{(4)} + [-4\ \omega_1^2 + 2\ \omega_1\ \omega_2 + 4/3\ \omega_2\ \omega_1 - 4/3\ \omega_2^2]\ \Pi^{(3)} +$

$[2\ \omega_1^2 - 2\ \omega_1\ \omega_2 - 8\ \omega_2\ \omega_1 + 4\ \omega_2^2]\ \Pi^{(2)} + [2\ \omega_1^2 - 2/3\ \omega_1\ \omega_2 + 8\ \omega_2\ \omega_1 - 16/3\ \omega_2^2]\ \Pi^{(1)} +$

$[-4/3\ \omega_1^2 + 4/3\ \omega_1\ \omega_2 - 8/3\ \omega_2\ \omega_1 + 8/3\ \omega_2^2]\ xyz \}$ (59)

EQUATING THE TWO INTENSITY-CURVATURE TERMS BEFORE AND AFTER INTERPOLATION: TRIVARIATE CUBIC B-SPLINE

Let us proceed to equate the two intensity-curvature terms (before and after interpolation) in order to derive the signal resilient to the cubic trivariate B-Spline interpolation function of equation (36). The two intensity curvature terms $E_o(x, y, z)$ and $E_{IN}(x, y, z)$ are given in equations (49) and (59).

$E_o(x, y, z) = 9\ xyz * f(0, 0, 0) * \{ -2\ \omega_1 + 2\ \omega_2 \} = E_{IN}(x, y, z) = 9 * \{ f(0, 0, 0) * \{ \omega_1$

$* [3\ \Pi^{(1)} - 2\ xyz] + \omega_2 * [-\Pi^{(1)} + 2\ xyz] \} + [3/2\ \omega_1^2 - 1/2\ \omega_1\omega_2 - 3/6\ \omega_2\ \omega_1 + 1/6\ \omega_2^2]$

$\Pi^{(4)} + [-4\ \omega_1^2 + 2\ \omega_1\ \omega_2 + 4/3\ \omega_2\ \omega_1 - 4/3\ \omega_2^2]\ \Pi^{(3)} + [2\ \omega_1^2 - 2\ \omega_1\ \omega_2 - 8\ \omega_2\ \omega_1 + 4\ \omega_2^2]$

$\Pi^{(2)} + [2\ \omega_1^2 - 2/3\ \omega_1\ \omega_2 + 8\ \omega_2\ \omega_1 - 16/3\ \omega_2^2]\ \Pi^{(1)} + [-4/3\ \omega_1^2 + 4/3\ \omega_1\ \omega_2 - 8/3\ \omega_2\ \omega_1 +$

$8/3\ \omega_2^2]\ xyz \}$ (60)

Equation (60) can be re-written as:

$xyz * f(0, 0, 0) * \{ -2\ \omega_1 + 2\ \omega_2 \} - \{ f(0, 0, 0) * \{ \omega_1 * [3\ \Pi^{(1)} - 2\ xyz] + \omega_2 * [-\Pi^{(1)}$

$+ 2\ xyz] \} \} = \{ [3/2\ \omega_1^2 - 1/2\ \omega_1\omega_2 - 3/6\ \omega_2\ \omega_1 + 1/6\ \omega_2^2]\ \Pi^{(4)} + [-4\ \omega_1^2 + 2\ \omega_1\ \omega_2 + 4/3\ \omega_2$

$\omega_1 - 4/3\ \omega_2{}^2]\ \Pi^{(3)} + [\ 2\ \omega_1{}^2 - 2\ \omega_1\ \omega_2 - 8\ \omega_2\ \omega_1 + 4\ \omega_2{}^2]\ \Pi^{(2)} + [2\ \omega_1{}^2 - 2/3\ \omega_1\ \omega_2 + 8\ \omega_2\ \omega_1$

$- 16/3\ \omega_2{}^2]\ \Pi^{(1)} + [-\ 4/3\ \omega_1{}^2 + 4/3\ \omega_1\ \omega_2 - 8/3\ \omega_2\ \omega_1 + 8/3\ \omega_2{}^2]\ xyz\ \}$ (61)

Let us posit:

$\Lambda_\gamma = \{\ [3/2\ \omega_1{}^2 - 1/2\ \omega_1\omega_2 - 3/6\ \omega_2\ \omega_1 + 1/6\ \omega_2{}^2]\ \Pi^{(4)} + [-\ 4\ \omega_1{}^2 + 2\ \omega_1\ \omega_2 + 4/3\ \omega_2\ \omega_1 - 4/3$

$\omega_2{}^2]\ \Pi^{(3)} + [\ 2\ \omega_1{}^2 - 2\ \omega_1\ \omega_2 - 8\ \omega_2\ \omega_1 + 4\ \omega_2{}^2]\ \Pi^{(2)} + [2\ \omega_1{}^2 - 2/3\ \omega_1\ \omega_2 + 8\ \omega_2\ \omega_1 - 16/3$

$\omega_2{}^2]\ \Pi^{(1)} + [-\ 4/3\ \omega_1{}^2 + 4/3\ \omega_1\ \omega_2 - 8/3\ \omega_2\ \omega_1 + 8/3\ \omega_2{}^2]\ xyz\ \}$

Equation (61) becomes:

$xyz * f\ (0, 0, 0) * \{\ -2\ \omega_1 + 2\ \omega_2\ \} - \{\ f\ (0, 0, 0) * \{\ \omega_1 * [\ 3\ \Pi^{(1)} - 2\ xyz\ ] + \omega_2 * [\ -\ \Pi^{(1)}$

$+ 2\ xyz\ ]\ \}\ \} = \Lambda_\gamma$ (62)

Ans it is solved hereto:

$f\ (0, 0, 0) * \{\ xyz * [\ -2\ \omega_1 + 2\ \omega_2\ ] - \{\ \omega_1 * [\ 3\ \Pi^{(1)} - 2\ xyz\ ] + \omega_2 * [\ -\ \Pi^{(1)} + 2\ xyz\ ]\ \}\ \}$

$= \Lambda_\gamma$

$f\ (0, 0, 0) = \Lambda_\gamma\ /\ \{\ xyz * [\ -2\ \omega_1 + 2\ \omega_2\ ] - \{\ \omega_1 * [\ 3\ \Pi^{(1)} - 2\ xyz\ ] + \omega_2 * [\ -\ \Pi^{(1)} + 2$

$xyz\ ]\ \}\ \}$ (63)

Where f (0, 0, 0) of equation (63) is the signal resilient to the cubic trivariate B-Spline interpolation function.

PART IV - TRIVARIATE CUBIC LAGRANGE FUNCTION
WITH EMBEDDING

TRIVARIATE LAGRANGE INTERPOLATION FUNCTION

The formula hereto reported is the trivariate Lagrange interpolation function having third degree order:

$$LGR_3(x, y, z) = f(0, 0, 0) + \omega_1 * [(1/2) (x + y + z)^3 - (x + y + z)^2 - 1/2 (x + y + z) + 1] +$$

$$\omega_2 * [-(1/6) (x + y + z)^3 + (x + y + z)^2 - (11/6) (x + y + z) + 1] \tag{64}$$

This function has the same $((3 \times 3 \times 3) - 1)$ pixels' neighborhood of $f(0, 0, 0)$ as the one adopted for the quadratic and cubic B-Splines and this is illustrated in figures 2, 3 and 4. The values of ω_1 and ω_2 are those of equations (2) and (3).

CALCULATION OF THE PARTIAL SECOND ORDER DERIVATIVES: TRIVARIATE CUBIC LAGRANGE

First Order and Second Order Partial Derivatives of the Trivariate Cubic Lagrange

For convenience, let us calculate the first and second order partial derivative of the trivariate Lagrange function with respect to the x variable, also the second order partial derivative with respect to x and y variables. Then, through the knowledge provided while studying the partial second order derivatives of the trivariate cubic B-spline, let us demonstrate the assertions given hereto follow in equation (65).

$$(\partial^2 (LGR_3(x, y, z)) / \partial x^2) = (\partial^2 (LGR_3(x, y, z)) / \partial y^2) = (\partial^2 (LGR_3(x, y, z)) / \partial z^2) =$$

$$(\partial^2 (LGR_3(x, y, z)) / \partial x \partial y) = (\partial^2 (LGR_3(x, y, z)) / \partial y \partial x) =$$

$$(\partial^2 (LGR_3(x, y, z)) / \partial x \partial z) = (\partial^2 (LGR_3(x, y, z)) / \partial z \partial x) =$$

$$(\partial^2 (LGR_3(x, y, z)) / \partial y \partial z) = (\partial^2 (LGR_3(x, y, z)) / \partial z \partial y) \tag{65}$$

$$(\partial (LGR_3(x, y, z)) / \partial x) = \partial \{ f(0, 0, 0) + \omega_1 * [(1/2) (x + y + z)^3 - (x + y + z)^2 - 1/2 (x +$$

$$y + z) + 1] + \omega_2 * [-(1/6) (x + y + z)^3 + (x + y + z)^2 - (11/6) (x + y + z) + 1] \} / \partial x =$$

$$\{ \omega_1 * [(1/2) \partial (x + y + z)^3 / \partial x - \partial (x + y + z)^2 / \partial x - 1/2 \partial (x + y + z) / \partial x] + \omega_2 * [-(1/6)$$

$$\partial (x + y + z)^3 / \partial x + \partial (x + y + z)^2 / \partial x - (11/6) \partial (x + y + z) / \partial x] \} =$$

$$\{ \omega_1 * [(3/2) (x + y + z)^2 - 2 (x + y + z) - 1/2] + \omega_2 * [-(3/6) (x + y + z)^2 + 2 (x + y + z)$$

$$- (11/6)] \} \tag{66}$$

$$(\partial^2 (LGR_3(x, y, z)) / \partial x^2) = (\partial (\partial (LGR_3(x, y, z)) / \partial x) / \partial x) = \partial \{ \omega_1 * [(3/2) (x + y + z)^2$$

$$- 2\,(x + y + z) - 1/2\,] + \omega_2 * [\,-(3/6)\,(x + y + z)^2 + 2\,(x + y + z) - (11/6)\,]\,\}/\partial x =$$

$$\{\,\omega_1 * [\,3\,(x + y + z) - 2\,] + \omega_2 * [\,- (x + y + z) + 2\,]\,\} \tag{67}$$

$$(\partial^2\,(LGR_3(x, y, z)\,)\,)/\partial x \partial y) = \partial\,\{\,\omega_1 * [\,(3/2)\,(x + y + z)^2 - 2\,(x + y + z) - 1/2\,] + \omega_2 * [\,-$$

$$(3/6)\,(x + y + z)^2 + 2\,(x + y + z) - (11/6)\,]\,\}\,/\partial y =$$

$$\{\,\omega_1 * [\,3\,(x + y + z) - 2\,] + \omega_2 * [\,- (x + y + z) + 2\,]\,\} \tag{68}$$

Equations (67) and (68) provide the same exact result of equations (41) and (43), which were derived for the trivariate cubic B-Spline interpolation function of equation (36). Therefore, because of the assertions provided through equations (42) and (46), equation (65) is true.

CALCULATION OF THE INTENSITY-CURVATURE TERMS: TRIVARIATE CUBIC LAGRANGE

The intensity-curvature term E_o of the trivariate cubic Lagrange function is:

$$E_o = E_o\,(x, y, z) = \int_0^x \int_0^y \int_0^z f\,(0, 0, 0) * \{\,(\partial^2\,(LGR_3\,(x, y, z))/\partial x^2) + (\partial^2\,(LGR_3\,(x, y, z))/\partial y^2)$$

$$+ (\partial^2\,(LGR_3\,(x, y, z))/\partial z^2) + (\partial^2\,(LGR_3\,(x, y, z)\,)/\partial x \partial y) + (\partial^2\,(LGR_3\,(x, y, z)\,)/\partial y \partial x) +$$

$$(\partial^2\,(LGR_3\,(x, y, z)\,)/\partial x \partial z) + (\partial^2\,(LGR_3\,(x, y, z)\,)/\partial z \partial x) + (\partial^2\,(LGR_3\,(x, y, z)\,)/\partial y \partial z) +$$

$$(\partial^2\,(LGR_3\,(x, y, z)\,)/\partial z \partial y)\,\}\,(0, 0, 0)\,dx\,dy\,dz \tag{69}$$

On the basis of equation (55), it can be written that:

$$(\partial^2\,(LGR_3(x, y, z)\,)\,)/\partial x^2)\,(0, 0, 0) = (\partial^2\,(LGR_3(x, y, z)\,)/\partial y^2)\,(0, 0, 0) =$$

$$(\partial^2\,(LGR_3(x, y, z)\,)/\partial z^2)\,(0, 0, 0) = (\partial^2\,(LGR_3(x, y, z)\,)/\partial x \partial y)\,(0, 0, 0) =$$

$$(\partial^2\,(LGR_3(x, y, z)\,)/\partial y \partial x)\,(0, 0, 0) = (\partial^2\,(LGR_3(x, y, z)\,)/\partial x \partial z)\,(0, 0, 0) =$$

$$(\partial^2\,(LGR_3(x, y, z)\,)/\partial z \partial x)\,(0, 0, 0) = (\partial^2\,(LGR_3(x, y, z)\,)/\partial y \partial z)\,(0, 0, 0) =$$

$$(\partial^2\,(LGR_3(x, y, z)\,)/\partial z \partial y)\,(0, 0, 0) =$$

$$\{\,\omega_1 * [\,3\,(x + y + z) - 2\,] + \omega_2 * [\,- (x + y + z) + 2\,]\,\}\,(0, 0, 0) = \{\,-2\,\omega_1 + 2\,\omega_2\,\} \tag{70}$$

It follows that:

$$E_o\,(x, y, z) = \int_0^x \int_0^y \int_0^z 9 * f\,(0, 0, 0) * \{\,-2\,\omega_1 + 2\,\omega_2\,\} =$$

$$= 9\,xyz * f\,(0, 0, 0) * \{\,-2\,\omega_1 + 2\,\omega_2\,\} \tag{71}$$

SIGNAL RESILIENT TO INTERPOLATION

The intensity-curvature term E_{IN} of the trivariate cubic Lagrange function is:

$$E_{IN} = E_{IN}(x, y, z) = \int\limits_0^x\int\limits_0^y\int\limits_0^z LGR_3(x, y, z) * \{ (\partial^2 (LGR_3(x, y, z)) / \partial x^2) + (\partial^2 (LGR_3(x, y, z))$$

$$/ \partial y^2) + (\partial^2 (LGR_3(x, y, z)) / \partial z^2) + (\partial^2 (LGR_3(x, y, z)) / \partial x \partial y) + (\partial^2 (LGR_3(x, y, z)) / \partial y \partial x)$$

$$+ (\partial^2 (LGR_3(x, y, z)) / \partial x \partial z) + (\partial^2 (LGR_3(x, y, z)) / \partial z \partial x) + (\partial^2 (LGR_3(x, y, z)) / \partial y \partial z) +$$

$$(\partial^2 (LGR_3(x, y, z)) / \partial z \partial y)\} \, dx \, dy \, dz \tag{72}$$

And because of equations (64) and (70):

$$E_{IN} = E_{IN}(x, y, z) = \int\limits_0^x\int\limits_0^y\int\limits_0^z 9 * \{ f(0, 0, 0) + \omega_1 * [(1/2) (x + y + z)^3 - (x + y + z)^2 - 1/2 (x$$

$$+ y + z) + 1] + \omega_2 * [-(1/6) (x + y + z)^3 + (x + y + z)^2 - (11/6) (x + y + z) + 1]\} *$$

$$\{ \omega_1 * [3 (x + y + z) - 2] + \omega_2 * [- (x + y + z) + 2]\} \, dx \, dy \, dz \tag{73}$$

Let us solve equation (73):

$$E_{IN}(x, y, z) = \int\limits_0^x\int\limits_0^y\int\limits_0^z \{ 9 * f(0, 0, 0) * \{ \omega_1 * [3 (x + y + z) - 2] + \omega_2 * [- (x + y + z) +$$

$$2]\} + 9 * \{ \omega_1^2 * [(1/2) (x + y + z)^3 - (x + y + z)^2 - 1/2 (x + y + z) + 1] * [3 (x + y + z)$$

$$- 2] + \omega_1 \omega_2 * [-(1/6) (x + y + z)^3 + (x + y + z)^2 - (11/6) (x + y + z) + 1]\} * [3 (x + y +$$

$$z) - 2] + \omega_1 \omega_2 * [(1/2) (x + y + z)^3 - (x + y + z)^2 - 1/2 (x + y + z) + 1] * [- (x + y + z)$$

$$+ 2] + \omega_2^2 * [-(1/6) (x + y + z)^3 + (x + y + z)^2 - (11/6) (x + y + z) + 1] * [- (x + y + z)$$

$$+ 2]\}\} \, dx \, dy \, dz =$$

$$\int\limits_0^x\int\limits_0^y\int\limits_0^z \{ 9 * f(0, 0, 0) * \{ \omega_1 * [3 (x + y + z) - 2] + \omega_2 * [- (x + y + z) + 2]\} + 9 * \{$$

$$\omega_1^2 * [(3/2) (x + y + z)^4 - 3 (x + y + z)^3 - 3/2 (x + y + z)^2 + 3 (x + y + z)] + \omega_1^2 * [- (x$$

$$+ y + z)^3 + 2 (x + y + z)^2 + (x + y + z) - 2] + \omega_1 \omega_2 * [-(1/2) (x + y + z)^4 + 3 (x + y + z)^3 -$$

$$(33/6) (x + y + z)^2 + 3 (x + y + z)] + \omega_1 \omega_2 * [(1/3) (x + y + z)^3 - 2 (x + y + z)^2 + (22/6)$$

$$(x + y + z) - 2] + \omega_1 \omega_2 * [-(1/2) (x + y + z)^4 + (x + y + z)^3 + 1/2 (x + y + z)^2 - (x + y +$$

z)] + $\omega_1\omega_2 * [(x + y + z)^3 - 2 (x + y + z)^2 - (x + y + z) + 2] + \omega_2^2 * [(1/6) (x + y + z)^4 -$

$(x + y + z)^3 + (11/6) (x + y + z)^2 - (x + y + z)] + \omega_2^2 * [-(2/6) (x + y + z)^3 + 2 (x + y + z)^2$

$- (22/6) (x + y + z) + 2] \} \} =$

$$\int_0^x \int_0^y \int_0^z \{ 9 * f (0, 0, 0) * \{ \omega_1 * [3 (x + y + z) - 2] + \omega_2 * [- (x + y + z) + 2] \} + 9 * \{$$

$[(3/2) \omega_1^2 - (1/2) \omega_1 \omega_2 - (1/2) \omega_1 \omega_2 + (1/6) \omega_2^2] (x + y + z)^4 + [- 3 \omega_1^2 - \omega_1^2 + 3$

$\omega_1 \omega_2 + (1/3) \omega_1 \omega_2 + 2 \omega_1\omega_2 - \omega_2^2 - (2/6) \omega_2^2] (x + y + z)^3 + [-3/2 \omega_1^2 + 2 \omega_1^2 - (33/6) \omega_2$

$\omega_2 - 2 \omega_1 \omega_2 + 1/2 \omega_1\omega_2 - 2 \omega_1\omega_2 + (11/6) \omega_2^2 + 2 \omega_2^2] (x + y + z)^2 + [3 \omega_1^2 + \omega_1^2 + 3 \omega_1$

$\omega_2 + (22/6) \omega_1 \omega_2 - \omega_1\omega_2 - \omega_1\omega_2 - \omega_2^2 - (22/6) \omega_2^2] (x + y + z) + [- 2 \omega_1^2 - 2 \omega_1 \omega_2 + 2$

$\omega_1\omega_2 + 2 \omega_2^2] \} \} =$

$$\int_0^x \int_0^y \int_0^z \{ 9 * f (0, 0, 0) * \{ \omega_1 * [3 (x + y + z) - 2] + \omega_2 * [- (x + y + z) + 2] \} + 9 * \{$$

$[(3/2) \omega_1^2 - \omega_1 \omega_2 + (1/6) \omega_2^2] (x + y + z)^4 + [- 4 \omega_1^2 + (16/3) \omega_1 \omega_2 - (8/6) \omega_2^2] (x + y +$

$z)^3 + [1/2 \omega_1^2 - (54/6) \omega_1 \omega_2 + (23/6) \omega_2^2] (x + y + z)^2 + [4 \omega_1^2 + (28/6) \omega_1 \omega_2 - (28/6)$

$\omega_2^2] (x + y + z) + [- 2 \omega_1^2 + 2 \omega_2^2] \} \}$ (74)

And equation (74) can be solved considering the knowledge provided in equations (55), (56), (57) and (58) which furnishes the values of $\Pi^{(4)}$ $\Pi^{(3)}$ $\Pi^{(2)}$ and $\Pi^{(1)}$. Therefore:

$E_{IN} (x, y, z) = \{ 9 * f (0, 0, 0) * \{ \omega_1 * [3 \Pi^{(1)} - 2 xyz] + \omega_2 * [- \Pi^{(1)} + 2 xyz] \} + 9 *$

$\{ [(3/2) \omega_1^2 - \omega_1 \omega_2 + (1/6) \omega_2^2] \Pi^{(4)} + [- 4 \omega_1^2 + (16/3) \omega_1 \omega_2 - (8/6) \omega_2^2] \Pi^{(3)} +$

$[1/2 \omega_1^2 - (54/6) \omega_1 \omega_2 + (23/6) \omega_2^2] \Pi^{(2)} + [4 \omega_1^2 + (28/6) \omega_1 \omega_2 - (28/6) \omega_2^2] \Pi^{(1)} +$

$[- 2 \omega_1^2 + 2 \omega_2^2] xyz \} \}$ (75)

EQUATING THE TWO INTENSITY-CURVATURE TERMS BEFORE AND AFTER INTERPOLATION: TRIVARIATE CUBIC LAGRANGE

Let us posit: $E_o (x, y, z) = E_{IN} (x, y, z)$ from equations (71) and (75):

$9 xyz * f (0, 0, 0) * \{ - 2 \omega_1 + 2 \omega_2 \} = \{ 9 * f (0, 0, 0) * \{ \omega_1 * [3 \Pi^{(1)} - 2 xyz] + \omega_2 *$

$[- \Pi^{(1)} + 2 \, xyz \,] \} + 9 * \{ \,[\,(3/2) \, \omega_1{}^2 - \omega_1 \omega_2 + (1/6) \, \omega_2{}^2] \, \Pi^{(4)} + [- 4 \, \omega_1{}^2 + (16/3) \, \omega_1 \omega_2 -$

$(8/6) \, \omega_2{}^2] \, \Pi^{(3)} + [\, 1/2 \, \omega_1{}^2 - (54/6) \, \omega_1 \omega_2 + (23/6) \, \omega_2{}^2 \,] \, \Pi^{(2)} + [\, 4 \, \omega_1{}^2 + (28/6) \, \omega_1 \omega_2 -$

$(28/6) \, \omega_2{}^2] \, \Pi^{(1)} + [\, - 2 \, \omega_1{}^2 + 2 \, \omega_2{}^2] \, xyz \,\} \, \}$
(76)

Let us posit:

$\Lambda_L = 9 * \{ \,[\,(3/2) \, \omega_1{}^2 - \omega_1 \omega_2 + (1/6) \, \omega_2{}^2] \, \Pi^{(4)} + [- 4 \, \omega_1{}^2 + (16/3) \, \omega_1 \omega_2 - (8/6) \, \omega_2{}^2] \, \Pi^{(3)}$

$+ [\, 1/2 \, \omega_1{}^2 - (54/6) \, \omega_1 \omega_2 + (23/6) \, \omega_2{}^2 \,] \, \Pi^{(2)} + [\, 4 \, \omega_1{}^2 + (28/6) \, \omega_1 \omega_2 - (28/6) \, \omega_2{}^2] \, \Pi^{(1)} +$

$[\, - 2 \, \omega_1{}^2 + 2 \, \omega_2{}^2] \, xyz \,\}$
(77)

So that equation (76) is written as:

$9 \, xyz * f \,(0, 0, 0) * \{ - 2 \, \omega_1 + 2 \, \omega_2 \} = \{ \, 9 * f \,(0, 0, 0) * \{ \, \omega_1 * [\, 3 \, \Pi^{(1)} - 2 \, xyz \,] + \omega_2 *$

$[\, - \Pi^{(1)} + 2 \, xyz \,] \} \, \} + \Lambda_L$

And solved hereto:

$9 \, xyz * f \,(0, 0, 0) * \{ - 2 \, \omega_1 + 2 \, \omega_2 \} - \{ \, 9 * f \,(0, 0, 0) * \{ \, \omega_1 * [\, 3 \, \Pi^{(1)} - 2 \, xyz \,] + \omega_2 *$

$[\, - \Pi^{(1)} + 2 \, xyz \,] \} \, \} = \Lambda_L$

$f \,(0, 0, 0) * \{ \, 9 \, xyz * \{ - 2 \, \omega_1 + 2 \, \omega_2 \} - 9 * \{ \, \omega_1 * [\, 3 \, \Pi^{(1)} - 2 \, xyz \,] + \omega_2 * [\, - \Pi^{(1)} + 2$

$xyz \,] \, \} \, \} = \Lambda_L$

$f \,(0, 0, 0) = \Lambda_L \, / \, \{ \, 9 \, xyz * \{ - 2 \, \omega_1 + 2 \, \omega_2 \} - 9 * \{ \, \omega_1 * [\, 3 \, \Pi^{(1)} - 2 \, xyz \,] + \omega_2 * [\, - \Pi^{(1)} +$

$2 \, xyz \,] \, \} \, \}$
(78)

Where equation (78) gives the value of the signal resilient to the trivariate cubic Lagrange interpolation function.

SUMMARY

The signal resilient to interpolation is calculated in this chapter with embedding. That means considering the model interpolation function, as for instance, the quadratic B-Spline:

$h_3(x, y) = f \,(0, 0) + \alpha_1 * [- 2a \,(x + y)^2 + 1/2 \,(a+1) \,] +$

$\alpha_2 * [\, a \,(x + y)^2 - (2a + 1/2) \,(x + y) + 3/4 \,(a+1) \,]$
(79)

Where α_1 and α_2 are sums of pixel intensity, $f(0, 0)$ is the value of the original signal and the value of $f(0, 0)$ known as the signal resilient to interpolation is analogous to the term $h_3(x, y)$, except that it is calculated from the equation $E_o(x, y) = E_{IN}(x, y)$.

To embed signifies to consider $f(0, 0)$ in (79) as the original signal without embedding equation (79) becomes:

$$h_3(x, y) = \alpha_1 * [- 2a(x + y)^2 + 1/2(a+1)] +$$

$$\alpha_2 * [a(x + y)^2 - (2a + 1/2)(x + y) + 3/4(a+1)] \tag{80}$$

And the signal resilient to interpolation $f(0, 0)$ is still calculated from the equation $E_o(x, y) = E_{IN}(x, y)$, therefore the difference between with embedding ('With E') and without embedding ('WE') stands in considering and discarding the value of the original signal $f(0, 0)$ in equation (79) when solving the equation $E_o(x, y) = E_{IN}(x, y)$. The value of $f(0, 0)$ obtained from the equation $E_o(x, y) = E_{IN}(x, y)$ is the signal resilient to interpolation and in chapter 4 it will also become the means to obtain the new conceptualization and math formulation of the Pixel Intensity Correction (PIC). To anticipate concepts that will be object of study in subsequent chapters one can say here that equations (79) and (80) for the specific $h_3(x)$ interpolator are the classic signals which are different from the signals resilient to interpolation with embedding calculated in this chapter. The calculation in this chapter is comprehensive of bivariate and trivariate polynomial functions: B-Splines (quadratic parametric and cubic), and Lagrange (cubic).

REFERENCES

Ciulla, C. (2009). *Improved Signal and Image Interpolation in Biomedical Applications: The Case of Magnetic Resonance Imaging (MRI)* – Medical Information Science Reference - IGI Global Publisher, Hershey, PA, U.S.A.

Meijering, E. (2002). *A chronology of interpolation: From ancient astronomy to modern signal and image processing.* Proceedings of the IEEE, 90(3), 319-342.

CHAPTER 3

SIGNAL RESILIENT TO INTERPOLATION WITHOUT EMBEDDING: THE MATHEMATICAL FORMULATION

INTRODUCTION

The mathematics of this chapter are devoted to the development of the signal resilient to interpolation without embedding. Why would there be the need to calculate the signal in such two experimental conditions, namely with and without embedding? The answer is connected to what will be seen in subsequent chapters to be the instrument to determine the transfer function customized to the specific math form of the interpolator. This instrument is the Pixel Intensity Correction (PIC) and is derived through the total curvature in the case of both classic and resilient signals.

Now the point is to see if the curvature of the signal has to be calculated as that of the signal embedding the original signal or that one of the fraction of pixel intensity to add to the original signal in order to obtained re-sampling.

Conceptually it is the same because the fraction of pixel intensity is nothing more than the Pixel Intensity Correction (PIC), which is a prolongation of the original signal (additive or subtractive, therefore a positive or a negative segment). Thus, the original signal and the fraction of pixel intensity (PIC) to add to the original signal are on the same line, which intersect the model signal: either classic or resilient. Such concepts can be understood visually when looking at the figure 11 in chapter 4.

Still, if there is not conceptual difference, why calculate the curvature in both cases of a signal with embedding and a signal without embedding? Because the resilient signal and the classic signal are different in their math form with or without embedding the pixel to re-sample (there are thus four math models), which provides one with an expectation of different results, which are as many times as the math models.

As a matter of fact the total curvature employed in the calculation of the signals with and without embedding has a different math form and it shall be seen that quantitative and qualitative results are different. So, even in the absence of a conceptual difference, there is a practical difference given from the different mathematics. Is that admissible?

Yes it is, because calculating the curvature with embedding and without embedding is done considering the signals as the sum of the original signal plus the PIC (see equations (24.a) and (24.b) in chapter 4, bivariate case), which in either case with of without embedding, is only a fraction of the pixel intensity to add to the original signal (see figure 11 in chapter 4). The conceptual difference is made congruent through the methodological approach.

PART I – BIVARIATE QUADRATIC AND CUBIC FUNCTIONS WITHOUT EMBEDDING

BIVARIATE QUADRATIC AND CUBIC B-SPLINES

Let us consider the quadratic $h_3(x, y)$ and cubic $h_4(x, y)$ bivariate B-Splines as per equations (1) and (2) with $((3 \times 3) - 1)$ neighbouring nodes:

$h_3(x, y) = [f (1/2, 1/2) - f (-1/2, -1/2) + f (2/3, 2/3) + f (-2/3, -2/3)] * [- 2a (x + y)^2 + \frac{1}{2}$

$(a+1)] + [f (1/2, 1/2) + f (-1/2, -1/2) + f (2/3, 2/3) + f (-2/3, -2/3) + f (-1, -1) + f (1, 1) +$

$f (3/2, 3/2) + f (-3/2, -3/2)] * [a (x + y)^2 - (2a + 1/2) (x + y) + 3/4 (a+1)]$ (1)

$h_4(x, y) = [f (1/2, 1/2) + f (-1/2, -1/2) + f (-1, -1) + f (1, 1)] * [1/2 (x + y)^3 - (x + y)^2 +$

$2/3] + [f (1/2, 1/2) + f (-1/2, -1/2) + f (2/3, 2/3) + f (-2/3, -2/3) + f (-1, -1) + f (1, 1) +$

$f (3/2, 3/2) + f (-3/2, -3/2)] * [-1/6 (x + y)^3 + (x + y)^2 - 2 (x + y) + 4/3]$ (2)

Let us posit:

$\alpha_1 = [f (1/2, 1/2) + f (-1/2, -1/2) + f (2/3, 2/3) + f (-2/3, -2/3)]$

$\alpha_2 = [f (1/2, 1/2) + f (-1/2, -1/2) + f (2/3, 2/3) + f (-2/3, -2/3) + f (-1, -1) + f (1, 1) +$

$f (3/2, 3/2) + f (-3/2, -3/2)]$

$\alpha_3 = [f (1/2, 1/2) + f (-1/2, -1/2) + f (-1, -1) + f (1, 1)]$

Thus equations (1) and (2) can be written as:

$h_3(x, y) = \alpha_1 * [- 2a (x + y)^2 + 1/2 (a+1)] +$

$\alpha_2 * [a (x + y)^2 - (2a + 1/2) (x + y) + 3/4 (a+1)]$ (3)

$h_4(x, y) = \alpha_3 * [1/2 (x + y)^3 - (x + y)^2 + 2/3] +$

$\alpha_2 * [-1/6 (x + y)^3 + (x + y)^2 - 2 (x + y) + 4/3]$ (4)

CALCULATION OF THE PARTIAL SECOND ORDER DERIVATIVES: BIVARIATE QUADRATIC B-SPLINE

First Order Partial Derivative with Respect to the x Variable

$(\partial (h_3(x, y)) / \partial x) = \partial \{ \alpha_1 * [- 2a (x + y)^2 + 1/2 (a+1)] + \alpha_2 * [a (x + y)^2 - (2a + 1/2)$

$(x + y) + 3/4 (a+1)] } /\partial x = { \alpha_1 * [- 2a \partial (x + y)^2 /\partial x] + \alpha_2 * [a \partial (x + y)^2 /\partial x - (2a + 1/2) \partial (x + y) /\partial x] } = { \alpha_1 * [- 2a * 2 (x + y)] + \alpha_2 * [a * 2 (x + y) - (2a + 1/2)] }$

Therefore:

$(\partial (h_3(x, y)) /\partial x) = { \alpha_1 * [- 2a * 2 (x + y)] + \alpha_2 * [a * 2 (x + y) - (2a + 1/2)] }$ 　　　　　(5)

First Order Partial Derivative with Respect to the y Variable

$(\partial (h_3(x, y)) /\partial y) = \partial { \alpha_1 * [- 2a (x + y)^2 + 1/2 (a+1)] + \alpha_2 * [a (x + y)^2 - (2a + 1/2)$

$(x + y) + 3/4 (a+1)] } /\partial y = { \alpha_1 * [- 2a \partial (x + y)^2 /\partial y] + \alpha_2 * [a \partial (x + y)^2 /\partial y - (2a + 1/2)$

$\partial (x + y) /\partial y] } = { \alpha_1 * [- 2a * 2 (x + y)] + \alpha_2 * [a * 2 (x + y) - (2a + 1/2)] }$

Therefore:

$(\partial (h_3(x, y)) /\partial y) = { \alpha_1 * [- 2a * 2 (x + y)] + \alpha_2 * [a * 2 (x + y) - (2a + 1/2)] }$ 　　　　　(6)

and it follows that:

$(\partial (h_3(x, y)) /\partial x) = (\partial (h_3(x, y)) /\partial y).$

Second Order Partial Derivative with Respect to the x Variable

$(\partial^2 (h_3(x, y)) /\partial x^2) = {\partial {\partial (h_3(x, y)) /\partial x} /\partial x} = {\partial { \alpha_1 * [- 2a * 2 (x + y)] + \alpha_2 * [a *$

$2 (x + y) - (2a + 1/2)] } /\partial x} = - 4a \alpha_1 + 2a \alpha_2$ 　　　　　(7)

Second Order Partial Derivative with Respect to the y Variable

$(\partial^2 (h_3(x, y)) /\partial y^2) = {\partial {\partial (h_3(x, y)) /\partial y} /\partial y} = {\partial { \alpha_1 * [- 2a * 2 (x + y)] + \alpha_2 * [a *$

$2 (x + y) - (2a + 1/2)] } /\partial y} = - 4a \alpha_1 + 2a \alpha_2$ 　　　　　(8)

Therefore:

$(\partial^2 (h_3(x, y)) /\partial x^2) = (\partial^2 (h_3(x, y)) /\partial y^2) = - 4a \alpha_1 + 2a \alpha_2$

Second Order Partial Derivatives with Respect to the x and y Variables

$(\partial^2 (h_3(x, y)) /\partial x \partial y) = {\partial (\partial (h_3(x, y)) /\partial x) /\partial y} = {\partial { \alpha_1 * [- 2a * 2 (x + y)] + \alpha_2 * [a$

$* 2 (x + y) - (2a + 1/2)] } /\partial y} = - 4a \alpha_1 + 2a \alpha_2$ 　　　　　(9)

$(\partial^2 (h_3(x, y)) /\partial y \partial x) = {\partial (\partial (h_3(x, y)) /\partial y) /\partial x} = {\partial { \alpha_1 * [- 2a * 2 (x + y)] + \alpha_2 * [a$

$$* \ 2 \ (x + y) - (2a + 1/2) \ ] \ \} \ /\partial x\} = -4a \ \alpha_1 + 2a \ \alpha_2 \tag{10}$$

Therefore:

$$(\partial^2 \ (h_3(x, y)) \ /\partial x \partial y) = (\partial^2 \ (h_3(x, y)) \ /\partial y \partial x) = -4a \ \alpha_1 + 2a \ \alpha_2$$

CALCULATION OF THE INTENSITY-CURVATURE TERMS: BIVARIATE QUADRATIC B-SPLINE

Let the intensity-curvature term before interpolation be defined as:

$$E_o = E_o \ (x, y) = \int_0^x \int_0^y f \ (0, 0) * \{ \ (\partial^2 \ (h_3(x, y)) \ /\partial x^2) + (\partial^2 \ (h_3(x, y)) \ /\partial x \partial y) + (\partial^2 \ (h_3(x, y))$$

$$/\partial y \partial x) + (\partial^2 \ (h_3(x, y)) \ /\partial y^2) \ \} \ (0, 0) \ dx \ dy \tag{11}$$

Where it follows from equations (7), (8), (9) and (10) that:

$$(\partial^2 \ (h_3(x, y)) \ /\partial x^2) \ (0, 0) = -4a \ \alpha_1 + 2a \ \alpha_2$$

$$(\partial^2 \ (h_3(x, y)) \ /\partial x \partial y) \ (0, 0) = -4a \ \alpha_1 + 2a \ \alpha_2$$

$$(\partial^2 \ (h_3(x, y)) \ /\partial y \partial x) \ (0, 0) = -4a \ \alpha_1 + 2a \ \alpha_2$$

$$(\partial^2 \ (h_3(x, y)) \ /\partial y^2) \ (0, 0) = -4a \ \alpha_1 + 2a \ \alpha_2$$

Therefore:

$$E_o \ (x, y) = \int_0^x \int_0^y 4 * f \ (0, 0) \ \{ \ -4a \ \alpha_1 + 2a \ \alpha_2 \ \} dx \ dy = \ 4xy * f \ (0, 0) \ \{ \ -4a \ \alpha_1 + 2a \ \alpha_2 \} \tag{12}$$

Let the intensity-curvature term after interpolation be defined as:

$$E_{IN} = E_{IN} \ (x, y) = \int_0^x \int_0^y h_3(x, y) * \{ \ (\partial^2 \ (h_3(x, y)) \ /\partial x^2) + (\partial^2 \ (h_3(x, y)) \ /\partial x \partial y) + (\partial^2 \ (h_3(x, y)$$

$$) \ /\partial y \partial x) + (\partial^2 \ (h_3(x, y)) \ /\partial y^2) \ \} \ dx \ dy \tag{13}$$

Where it follows from equations (7), (8), (9) and (10) that:

$$(\partial^2 \ (h_3(x, y)) \ /\partial x^2) = -4a \ \alpha_1 + 2a \ \alpha_2$$

$$(\partial^2 \ (h_3(x, y)) \ /\partial x \partial y) = -4a \ \alpha_1 + 2a \ \alpha_2$$

$$(\partial^2 \ (h_3(x, y)) \ /\partial y \partial x) = -4a \ \alpha_1 + 2a \ \alpha_2$$

$$(\partial^2 \ (h_3(x, y)) \ /\partial y^2) = -4a \ \alpha_1 + 2a \ \alpha_2$$

Therefore:

$$E_{IN}(x, y) = \int_0^x \int_0^y 4 * h_3(x, y) \{ -4a\,\alpha_1 + 2a\,\alpha_2 \}\, dx\, dy =$$

$$\int_0^x \int_0^y 4 * \{ \alpha_1 * [-2a\,(x+y)^2 + 1/2\,(a+1)] + \alpha_2 * [a\,(x+y)^2 - (2a+1/2)(x+y) + 3/4\,(a+1)$$

$$]\}* \{ -4a\,\alpha_1 + 2a\,\alpha_2 \}\, dx\, dy \tag{14}$$

Given that: $(x + y)^2 = (x^2 + 2xy + y^2)$, equation (14) is written as:

$$E_{IN}(x, y) = \int_0^x \int_0^y 4 * \{ \alpha_1 * [-2a\,(x^2 + 2xy + y^2) + 1/2\,(a+1)] + \alpha_2 * [a\,(x^2 + 2xy + y^2) -$$

$$(2a + 1/2)(x + y) + 3/4\,(a+1)]\} * \{ -4a\,\alpha_1 + 2a\,\alpha_2 \}\, dx\, dy =$$

$$4 * \{ -4a\,\alpha_1 + 2a\,\alpha_2 \} * \{ \int_0^x \int_0^y \alpha_1 * [-2a\,(x^2 + 2xy + y^2) + 1/2\,(a+1)] + \alpha_2 * [a\,(x^2 + 2xy +$$

$$y^2) - (2a + 1/2)(x + y) + 3/4\,(a+1)]\, dx\, dy \} =$$

$$4 * \{ -4a\,\alpha_1 + 2a\,\alpha_2 \} * \{ \int_0^x \alpha_1 * [-2a\,(x^2 y + xy^2 + y^3/3) + 1/2\, y\,(a+1)] + \alpha_2 *$$

$$[a\,(x^2 y + xy^2 + y^3/3) - (2a + 1/2)(xy + y^2/2) + 3/4\, y\,(a+1)]\, dx \} =$$

$$4 * \{ -4a\,\alpha_1 + 2a\,\alpha_2 \} * \{ \alpha_1 * [-2a\,(x^3 y/3 + x^2 y^2/2 + xy^3/3) + 1/2\, xy\,(a+1)]$$

$$+ \alpha_2 * [a\,(x^3 y/3 + x^2 y^2/2 + xy^3/3) - (2a + 1/2)(x^2 y/2 + xy^2/2) + 3/4\, xy\,(a+1)]\} \tag{15}$$

EQUATING THE TWO INTENSITY-CURVATURE TERMS BEFORE AND AFTER INTERPOLATION: BIVARIATE QUADRATIC B-SPLINE

Equating $E_o(x, y)$ as given through equation (12) to $E_{IN}(x, y)$ as given through equation (15), it can be written:

$$E_o(x, y) = E_{IN}(x, y) = 4xy * f(0, 0)\{ -4a\,\alpha_1 + 2a\,\alpha_2 \} = 4 * \{ -4a\,\alpha_1 + 2a\,\alpha_2 \} * \{ \alpha_1 * [-$$

$$2a\,(x^3 y/3 + x^2 y^2/2 + xy^3/3) + 1/2\, xy\,(a+1)] + \alpha_2 * [a\,(x^3 y/3 + x^2 y^2/2 + xy^3/3) - (2a + 1/2)$$

$$(x^2 y/2 + xy^2/2) + 3/4\, xy\,(a+1)]\} \tag{16}$$

which implies:

$xy * f(0, 0) = \{ \alpha_1 * [- 2a (x^3y/3 - x^2y^2/2 + xy^3/3) + 1/2 \, xy \, (a+1)] + \alpha_2 * [a (x^3y/3 +$

$x^2y^2/2 + xy^3/3) - (2a + 1/2) (x^2y/2 + xy^2/2) + 3/4 \, xy \, (a+1)] \}$

and yields:

$f(0, 0) = \{ \alpha_1 * [- 2a (x^3y/3 + x^2y^2/2 + xy^3/3) + 1/2 \, xy \, (a+1)] + \alpha_2 * [a (x^3y/3 +$

$x^2y^2/2 + xy^3/3) - (2a + 1/2) (x^2y/2 + xy^2/2) + 3/4 \, xy \, (a+1)] \} / \{ xy \}$

Therefore:

$f(0, 0) = \{ \alpha_1 * [- 2a (x^2/3 + xy/2 + y^2/3) + \tfrac{1}{2} (a+1)] + \alpha_2 * [a (x^2/3 +$

$xy/2 + y^2/3) - (2a + 1/2) (x/2 + y/2) + 3/4 (a+1)] \}$ \hfill (17)

which is the value of the signal f (0, 0) resilient to the bivariate B-Spline interpolation function $h_3(x, y)$ given in equation (1).

Recall that:

$\alpha_1 = [f(1/2, 1/2) + f(-1/2, -1/2) + f(2/3, 2/3) + f(-2/3, -2/3)]$ \hfill (18)

$\alpha_2 = [f(1/2, 1/2) + f(-1/2, -1/2) + f(2/3, 2/3) + f(-2/3, -2/3) + f(-1, -1) + f(1, 1) +$

$f(3/2, 3/2) + f(-3/2, -3/2)]$ \hfill (19)

CALCULATION OF THE PARTIAL SECOND ORDER DERIVATIVES: BIVARIATE CUBIC B-SPLINE

First Order Partial Derivative with Respect to the x Variable

$(\partial (h_4(x, y)) /\partial x) = \partial \{ \alpha_3 * [1/2 (x + y)^3 - (x + y)^2 + 2/3] + \alpha_2 * [-1/6 (x + y)^3$

$+ (x + y)^2 - 2 (x + y) + 4/3]\}/\partial x = \alpha_3 * [1/2 \, \partial (x + y)^3/\partial x - \partial (x + y)^2/\partial x] + \alpha_2 * [-1/6$

$\partial (x + y)^3 /\partial x + \partial (x + y)^2/\partial x - 2 \partial (x + y) /\partial x] = \alpha_3 * [3/2 (x + y)^2 - 2 (x + y)] + \alpha_2 * [-$

$3/6 (x + y)^2 + 2 (x + y) - 2]$ \hfill (20)

Therefore:

$(\partial (h_4(x, y)) /\partial x) = \alpha_3 * [3/2 (x + y)^2 - 2 (x + y)] + \alpha_2 * [-3/6 (x + y)^2 + 2 (x + y) - 2]$ \hfill (21)

First Order Partial Derivative with Respect to the y Variable

$(\partial (h_4(x, y)) /\partial y) = \partial \{ \alpha_3 * [1/2 (x + y)^3 - (x + y)^2 + 2/3] + \alpha_2 * [-1/6 (x + y)^3$

77

$+ (x + y)^2 - 2 (x + y) + 4/3]\}/\partial y = \alpha_3 * [1/2 \, \partial (x + y)^3/\partial y - \partial (x + y)^2/\partial y] + \alpha_2 * [-1/6$

$\partial (x + y)^3 /\partial y + \partial (x + y)^2/\partial y - 2 \, \partial (x + y) /\partial y] = \alpha_3 * [3/2 (x + y)^2 - 2 (x + y)] + \alpha_2 * [-$

$3/6 (x + y)^2 + 2 (x + y) - 2]$

Therefore:

$$(\partial (h_4(x, y)) /\partial y) = \alpha_3 * [3/2 (x + y)^2 - 2 (x + y)] + \alpha_2 * [-3/6 (x + y)^2 + 2 (x + y) - 2] \tag{22}$$

Second Order Partial Derivative with Respect to the x Variable

$(\partial^2 (h_4(x, y)) /\partial x^2) = \{\partial \{\partial (h_4(x, y)) /\partial x\} /\partial x\} = \{\partial \{ \alpha_3 * [3/2 (x + y)^2 - 2 (x + y)] +$

$\alpha_2 * [-3/6 (x + y)^2 + 2 (x + y) - 2] \} /\partial x\} = \alpha_3 * [3 (x + y) - 2] + \alpha_2 * [- (x + y) + 2]$

Therefore:

$$(\partial^2 (h_4(x, y)) /\partial x^2) = \alpha_3 * [3 (x + y) - 2] + \alpha_2 * [- (x + y) + 2] \tag{23}$$

Second Order Partial Derivative with Respect to the y Variable

$(\partial^2 (h_4(x, y)) /\partial y^2) = \{\partial \{\partial (h_4(x, y)) /\partial y\} /\partial y\} = \{\partial \{ \alpha_3 * [3/2 (x + y)^2 - 2 (x + y)] +$

$\alpha_2 * [-3/6 (x + y)^2 + 2 (x + y) - 2] \} /\partial y\} = \alpha_3 * [3 (x + y) - 2] + \alpha_2 * [- (x + y) + 2]$

Therefore:

$$(\partial^2 (h_4(x, y)) /\partial x^2) = (\partial^2 (h_4(x, y)) /\partial y^2) \tag{24}$$

Second Order Partial Derivatives with Respect to the x and y Variables

$(\partial^2 (h_4(x, y)) /\partial x \partial y) = \{\partial (\partial (h_4(x, y)) /\partial x) /\partial y\} = \{\partial \{ \alpha_3 * [3/2 (x + y)^2 - 2 (x + y)] +$

$\alpha_2 * [-3/6 (x + y)^2 + 2 (x + y) - 2] \} /\partial y\} = \alpha_3 * [3 (x + y) - 2] + \alpha_2 * [- (x + y) + 2]$

$(\partial^2 (h_4(x, y)) /\partial y \partial x) = \{\partial (\partial (h_4(x, y)) /\partial y) /\partial x\} = \{\partial \{ \alpha_3 * [3/2 (x + y)^2 - 2 (x + y)] +$

$\alpha_2 * [-3/6 (x + y)^2 + 2 (x + y) - 2] \} /\partial x\} = \alpha_3 * [3 (x + y) - 2] + \alpha_2 * [- (x + y) + 2]$

Therefore:

$$(\partial^2 (h_4(x, y)) /\partial x \partial y) = (\partial^2 (h_4(x, y)) /\partial y \partial x) = \alpha_3 * [3 (x + y) - 2] + \alpha_2 * [- (x + y) + 2] \tag{25}$$

CALCULATION OF THE INTENSITY-CURVATURE TERMS: BIVARIATE CUBIC B-SPLINE

Let the intensity-curvature term before interpolation be defined as:

$$E_o = E_o(x, y) = \int_0^x \int_0^y f(0, 0) * \{ (\partial^2 (h_4(x, y)) / \partial x^2) + (\partial^2 (h_4(x, y)) / \partial x \partial y) + (\partial^2 (h_4(x, y))$$

$$/ \partial y \partial x) + (\partial^2 (h_4(x, y)) / \partial y^2) \} (0, 0) \, dx \, dy \tag{26}$$

It follows from equations (23), (24) and (25) that:

$$(\partial^2 (h_4(x, y)) / \partial x^2)(0, 0) = -2 \alpha_3 + 2 \alpha_2$$

$$(\partial^2 (h_4(x, y)) / \partial y^2)(0, 0) = -2 \alpha_3 - 2 \alpha_2$$

$$(\partial^2 (h_4(x, y)) / \partial x \partial y)(0, 0) = -2 \alpha_3 + 2 \alpha_2$$

$$(\partial^2 (h_4(x, y)) / \partial y \partial x)(0, 0) = -2 \alpha_3 + 2 \alpha_2$$

Therefore:

$$E_o(x, y) = \int_0^x \int_0^y 4 * f(0, 0) \{ -2 \alpha_3 + 2 \alpha_2 \} \, dx \, dy = 4xy * f(0, 0) \{ -2 \alpha_3 + 2 \alpha_2 \} \tag{27}$$

Let the intensity-curvature term after interpolation be defined as:

$$E_{IN} = E_{IN}(x, y) = \int_0^x \int_0^y h_4(x, y) * \{ (\partial^2 (h_4(x, y)) / \partial x^2) + (\partial^2 (h_4(x, y)) / \partial x \partial y) +$$

$$(\partial^2 (h_4(x, y)) / \partial y \partial x) + (\partial^2 (h_4(x, y)) / \partial y^2) \} \, dx \, dy \tag{28}$$

From equations (23), (24) and (25), it follows that:

$$(\partial^2 (h_4(x, y)) / \partial x^2) = \alpha_3 * [3(x + y) - 2] + \alpha_2 * [-(x + y) + 2]$$

$$(\partial^2 (h_4(x, y)) / \partial y^2) = \alpha_3 * [3(x + y) - 2] + \alpha_2 * [-(x + y) + 2]$$

$$(\partial^2 (h_4(x, y)) / \partial x \partial y) = \alpha_3 * [3(x + y) - 2] + \alpha_2 * [-(x + y) + 2]$$

$$(\partial^2 (h_4(x, y)) / \partial y \partial x) = \alpha_3 * [3(x + y) - 2] + \alpha_2 * [-(x + y) + 2]$$

Therefore:

$$E_{IN}(x, y) = \int_0^x \int_0^y 4 * h_4(x, y) \{ \alpha_3 * [3(x + y) - 2] + \alpha_2 * [-(x + y) + 2] \} \, dx \, dy \tag{29}$$

From equation (4) it follows that:

$$E_{IN}(x, y) = \int_0^x \int_0^y 4 * \{ \alpha_3 * [1/2(x + y)^3 - (x + y)^2 + 2/3] + \alpha_2 * [-1/6(x + y)^3 +$$

$(x + y)^2 - 2 (x + y) + 4/3] \} * \{ \alpha_3 * [3 (x + y) - 2] + \alpha_2 * [- (x + y) + 2] \}$ dx dy =

$\int_0^x \int_0^y 4 * \{ \alpha_3 * [1/2 (x + y)^3 - (x + y)^2 + 2/3] * \alpha_3 * [3 (x + y) - 2] + \alpha_3 * [1/2 (x + y)^3 - (x$

$+ y)^2 + 2/3] * \alpha_2 * [- (x + y) + 2] + \alpha_2 * [-1/6 (x + y)^3 + (x + y)^2 - 2 (x + y) + 4/3] * \alpha_3 *$

$[3 (x + y) - 2] + \alpha_2 * [-1/6 (x + y)^3 + (x + y)^2 - 2 (x + y) + 4/3] * \alpha_2 * [- (x + y) + 2] \}$

dx dy =

$\int_0^x \int_0^y 4 * \{ \alpha_3{}^2 * [3/2 (x + y)^4 - 3(x + y)^3 + 2(x + y) - (x + y)^3 + 2 (x + y)^2 - 4/3] + \alpha_3 \alpha_2 *$

$[-1/2 (x + y)^4 + (x + y)^3 - 2/3 (x + y) + (x + y)^3 - 2 (x + y)^2 + 4/3] + \alpha_3 \alpha_2 * [-3/6 (x + y)^4$

$+ 3 (x + y)^3 - 6 (x + y)^2 + 4 (x + y) + 2/6 (x + y)^3 - 2 (x + y)^2 + 4 (x + y) - 8/3] + \alpha_2{}^2 * [$

$1/6 (x + y)^4 - (x + y)^3 + 2 (x + y)^2 - 4/3 (x + y) - 2/6 (x + y)^3 + 2 (x + y)^2 - 4 (x + y) + 8/3]$

$\}$ dx dy =

$\int_0^x \int_0^y 4 * \{ \alpha_3{}^2 * [3/2 (x + y)^4 - 4(x + y)^3 + 2(x + y) + 2 (x + y)^2 - 4/3] + \alpha_3 \alpha_2 * [-1/2 (x +$

$y)^4 + 2 (x + y)^3 - 2/3 (x + y) - 2 (x + y)^2 + 4/3] + \alpha_3 \alpha_2 * [-3/6 (x + y)^4 + 4/3 (x + y)^3 - 8 (x$

$+ y)^2 + 8 (x + y) - 8/3] + \alpha_2{}^2 * [1/6 (x + y)^4 - 4/3(x + y)^3 + 4 (x + y)^2 - 16/3 (x + y) + 8/3$

$] \}$ dx dy =

$\int_0^x \int_0^y 4 * \{ [3/2 \alpha_3{}^2 - 1/2 \alpha_3 \alpha_2 - 3/6 \alpha_3 \alpha_2 + 1/6 \alpha_2{}^2] (x + y)^4 + [-4 \alpha_3{}^2 + 2 \alpha_3 \alpha_2 + 4/3 \alpha_3 \alpha_2 -$

$4/3 \alpha_2{}^2] (x + y)^3 + [2 \alpha_3{}^2 - 2 \alpha_3 \alpha_2 - 8 \alpha_3 \alpha_2 + 4 \alpha_2{}^2] (x + y)^2 + [2 \alpha_3{}^2 - 2/3 \alpha_3 \alpha_2 + 8 \alpha_3 \alpha_2$

$- 16/3 \alpha_2{}^2] (x + y) + [- 4/3 \alpha_3{}^2 + 4/3 \alpha_3 \alpha_2 - 8/3 \alpha_3 \alpha_2 + 8/3 \alpha_2{}^2] \}$ dx dy $\qquad$ (30)

It can be written that:

$(x + y)^2 = (x^2 + 2xy + y^2)$ $\qquad$ (30.a)

$(x + y)^3 = (x^2 + 2xy + y^2) (x + y) = (x^3 + 2x^2y + xy^2 + x^2y + 2xy^2 + y^3) =$

$(x^3 + 3x^2y + 3xy^2 + y^3)$ (30.b)

$(x + y)^4 = (x + y)^3 (x + y) = (x^3 + 3x^2y + 3xy^2 + y^3) (x + y) =$

$(x^4 + 3x^3y + 3x^2y^2 + xy^3 + x^3y + 3x^2y^2 + 3xy^3 + y^4)$ (30.c)

$$\int_0^x \int_0^y (x + y)\, dx\, dy = \int_0^x (xy + y^2/2)\, dx = (x^2y/2 + xy^2/2)$$ (30.d)

$$\int_0^x \int_0^y (x + y)^2\, dx\, dy = \int_0^x \int_0^y (x^2 + 2xy + y^2)\, dx\, dy = \int_0^x (x^2y + xy^2 + y^3/3)\, dx =$$

$(x^3y/3 + x^2y^2/2 + xy^3/3)$ (30.e)

$$\int_0^x \int_0^y (x + y)^3\, dx\, dy = \int_0^x \int_0^y (x^3 + 3x^2y + 3xy^2 + y^3)\, dx\, dy = \int_0^x (x^3y + 3/2\, x^2y^2 + xy^3 + y^4/4)\, dx =$$

$(x^4y/4 + x^3y^2/2 + x^2y^3/2 + xy^4/4)$ (30.f)

$$\int_0^x \int_0^y (x + y)^4\, dx\, dy = \int_0^x \int_0^y (x^4 + 3x^3y + 3x^2y^2 + xy^3 + x^3y + 3x^2y^2 + 3xy^3 + y^4)\, dx\, dy =$$

$$\int_0^x (x^4y + 3/2\, x^3y^2 + x^2y^3 + xy^4/4 + x^3y^2/2 + x^2y^3 + 3/4\, xy^4 + y^5/5)\, dx =$$

$(x^5y/5 + 3/8\, x^4y^2 + x^3y^3/3 + x^2y^4/8 + x^4y^2/8 + x^3y^3/3 + 3/8\, x^2y^4 + xy^5/5)$ (30.g)

Equation (30) herein recalled as equation (31):

$$E_{IN}(x, y) = \int_0^x \int_0^y 4 * \{ [3/2\, \alpha_3^2 - 1/2\, \alpha_3\, \alpha_2 - 3/6\, \alpha_3\, \alpha_2 + 1/6\, \alpha_2^2] (x + y)^4 + [-4\, \alpha_3^2 + 2\, \alpha_3\, \alpha_2 +$$

$$4/3\, \alpha_3\, \alpha_2 - 4/3\, \alpha_2^2] (x + y)^3 + [2\, \alpha_3^2 - 2\, \alpha_3\, \alpha_2 - 8\, \alpha_3\, \alpha_2 + 4\, \alpha_2^2] (x + y)^2 + [2\, \alpha_3^2 - 2/3\, \alpha_3$$

$$\alpha_2 + 8\, \alpha_3\, \alpha_2 - 16/3\, \alpha_2^2] (x + y) + [- 4/3\, \alpha_3^2 + 4/3\, \alpha_3\, \alpha_2 - 8/3\, \alpha_3\, \alpha_2 + 8/3\, \alpha_2^2] \}\, dx\, dy$$ (31)

Can be solved into equation (32):

$$E_{IN}(x, y) = 4 * \{ [3/2\, \alpha_3^2 - 1/2\, \alpha_3\, \alpha_2 - 3/6\, \alpha_3\, \alpha_2 + 1/6\, \alpha_2^2] [x^5y/5 + 3/8\, x^4y^2 + x^3y^3/3 +$$

$$x^2y^4/8 + x^4y^2/8 + x^3y^3/3 + 3/8\, x^2y^4 + xy^5/5] + [-4\, \alpha_3^2 + 2\, \alpha_3\, \alpha_2 + 4/3\, \alpha_3\, \alpha_2 - 4/3\, \alpha_2^2]$$

$[x^4y/4 + x^3y^2/2 + x^2y^3/2 + xy^4/4] + [2 \alpha_3{}^2 - 2 \alpha_3 \alpha_2 - 8 \alpha_3 \alpha_2 + 4 \alpha_2{}^2] [x^3y/3 + x^2y^2/2 +$

$xy^3/3] + [2 \alpha_3{}^2 - 2/3 \alpha_3 \alpha_2 + 8 \alpha_3 \alpha_2 - 16/3 \alpha_2{}^2] [x^2y/2 + xy^2/2] + [- 4/3 \alpha_3{}^2 + 4/3 \alpha_3 \alpha_2 -$

$8/3 \alpha_3 \alpha_2 + 8/3 \alpha_2{}^2] [xy] \}$ (32)

EQUATING THE TWO INTENSITY-CURVATURE TERMS BEFORE AND AFTER INTERPOLATION: BIVARIATE CUBIC B-SPLINE

The signal resilient to interpolation as reconstructed through the bivariate cubic B-Spline of equation (2) is obtained (see equation 34)) solving the equation $E_{IN}(x, y) = E_o(x, y)$ resulting from (32) and (27).

$E_o(x, y) = 4xy * f(0, 0) \{ - 2 \alpha_3 + 2 \alpha_2 \} = E_{IN}(x, y) = 4 * \{ [3/2 \alpha_3{}^2 - 1/2 \alpha_3 \alpha_2 - 3/6$

$\alpha_3 \alpha_2 + 1/6 \alpha_2{}^2] [x^5y/5 + 3/8 x^4y^2 + x^3y^3/3 + x^2y^4/8 + x^4y^2/8 + x^3y^3/3 + 3/8 x^2y^4 + xy^5/5]$

$+ [-4 \alpha_3{}^2 + 2 \alpha_3 \alpha_2 + 4/3 \alpha_3 \alpha_2 - 4/3 \alpha_2{}^2] [x^4y/4 + x^3y^2/2 + x^2y^3/2 + xy^4/4] + [2 \alpha_3{}^2 - 2 \alpha_3$

$\alpha_2 - 8 \alpha_3 \alpha_2 + 4 \alpha_2{}^2] [x^3y/3 + x^2y^2/2 + xy^3/3] + [2 \alpha_3{}^2 - 2/3 \alpha_3 \alpha_2 + 8 \alpha_3 \alpha_2 - 16/3 \alpha_2{}^2]$

$[x^2y/2 + xy^2/2] + [- 4/3 \alpha_3{}^2 + 4/3 \alpha_3 \alpha_2 - 8/3 \alpha_3 \alpha_2 + 8/3 \alpha_2{}^2] [xy] \}$ (33)

Let us posit:

$\Lambda = 4 * \{ [3/2 \alpha_3{}^2 - 1/2 \alpha_3 \alpha_2 - 3/6 \alpha_3 \alpha_2 + 1/6 \alpha_2{}^2] [x^5y/5 + 3/8 x^4y^2 + x^3y^3/3 + x^2y^4/8 +$

$x^4y^2/8 + x^3y^3/3 + 3/8 x^2y^4 + xy^5/5] + [-4 \alpha_3{}^2 + 2 \alpha_3 \alpha_2 + 4/3 \alpha_3 \alpha_2 - 4/3 \alpha_2{}^2] [x^4y/4 +$

$x^3y^2/2 + x^2y^3/2 + xy^4/4] + [2 \alpha_3{}^2 - 2 \alpha_3 \alpha_2 - 8 \alpha_3 \alpha_2 + 4 \alpha_2{}^2] [x^3y/3 + x^2y^2/2 + xy^3/3] +$

$[2 \alpha_3{}^2 - 2/3 \alpha_3 \alpha_2 + 8 \alpha_3 \alpha_2 - 16/3 \alpha_2{}^2] [x^2y/2 + xy^2/2] + [- 4/3 \alpha_3{}^2 + 4/3 \alpha_3 \alpha_2 - 8/3 \alpha_3 \alpha_2$

$+ 8/3 \alpha_2{}^2] [xy] \}$

From equation (33) it follows that:

$4xy * f(0, 0) \{ - 2 \alpha_3 + 2 \alpha_2 \} = E_{IN}(x, y) = \Lambda$

Finally:

$f(0, 0) = \Lambda / \{ 4xy * [- 2 \alpha_3 + 2 \alpha_2] \}$ (34)

BIVARIATE CUBIC LAGRANGE INTERPOLATION FUNCTION

This function has $((3 \times 3) - 1)$ pixels' neighborhood: $f(-1/2)$, $f(1/2)$, $f(-1)$, $f(1)$, $f(-3/2)$, $f(3/2)$, $f(-2/3)$, $f(2/3)$ (see figure 1) centered at $f(0, 0)$, which is not included.

$LGR_3(x, y) = [f(1/2, 1/2) + f(-1/2, -1/2) + f(-1, -1) + f(1, 1)] * [(1/2)(x + y)^3 - (x + y)^2 - 1/2(x + y) + 1] + [f(1/2, 1/2) + f(-1/2, -1/2) + f(2/3, 2/3) + f(-2/3, -2/3) + f(-1, -1) + f(1, 1) + f(3/2, 3/2) - f(-3/2, -3/2)] * [-(1/6)(x + y)^3 + (x + y)^2 - (11/6)(x + y) + 1]$ (35)

Let us posit:

$\alpha_2 = [f(1/2, 1/2) + f(-1/2, -1/2) + f(2/3, 2/3) + f(-2/3, -2/3) + f(-1, -1) + f(1, 1) + f(3/2, 3/2) + f(-3/2, -3/2)]$

$\alpha_3 = [f(1/2, 1/2) + f(-1/2, -1/2) + f(-1, -1) + f(1, 1)]$

Equation (35) is re-written as:

$LGR_3(x, y) = \alpha_3 * [(1/2)(x + y)^3 - (x + y)^2 - 1/2(x + y) + 1] + \alpha_2 * [-(1/6)(x + y)^3 + (x + y)^2 - (11/6)(x + y) + 1]$ (36)

CALCULATION OF THE PARTIAL SECOND ORDER DERIVATIVES: BIVARIATE CUBIC LAGRANGE

First Order Partial Derivative with Respect to the x Variable

$(\partial (LGR_3(x, y)))/\partial x = \partial \{ \alpha_3 * [(1/2)(x + y)^3 - (x + y)^2 - 1/2(x + y) + 1] + \alpha_2 * [-(1/6)(x + y)^3 + (x + y)^2 - (11/6)(x + y) + 1] \}/\partial x = \{ \alpha_3 * [(1/2)\partial (x + y)^3/\partial x - \partial (x + y)^2/\partial x - 1/2 \partial (x + y)/\partial x] + \alpha_2 * [-(1/6)\partial (x + y)^3/\partial x + \partial (x + y)^2/\partial x - (11/6)\partial (x + y)/\partial x] \} = \{ \alpha_3 * [(3/2)(x + y)^2 - 2(x + y) - 1/2] + \alpha_2 * [-(3/6)(x + y)^2 + 2(x + y) - (11/6)] \}$ (37)

First Order Partial Derivative with Respect to the y Variable

$(\partial (LGR_3(x, y)))/\partial y = \partial \{ \alpha_3 * [(1/2)(x + y)^3 - (x + y)^2 - 1/2(x + y) + 1] + \alpha_2 * [-(1/6)(x + y)^3 + (x + y)^2 - (11/6)(x + y) + 1] \}/\partial y = \{ \alpha_3 * [(1/2)\partial (x + y)^3/\partial y - \partial (x + y)^2/\partial y - 1/2 \partial (x + y)/\partial y] + \alpha_2 * [-(1/6)\partial (x + y)^3/\partial y + \partial (x + y)^2/\partial y - (11/6)\partial (x + y)/\partial y] \} = \{ \alpha_3 * [(3/2)(x + y)^2 - 2(x + y) - 1/2] + \alpha_2 * [-(3/6)(x + y)^2 + 2(x + y) - (11/6)] \}$ (38)

Thus:
$(\partial (LGR_3(x, y)))/\partial x = (\partial (LGR_3(x, y)))/\partial y$

Second Order Partial Derivative with Respect to the x Variable

$(\partial^2 (LGR_3(x, y)) / \partial x^2) = \{\partial \{\partial (LGR_3(x, y)) / \partial x\} / \partial x\} = \{\partial \{ \alpha_3 * [(3/2) (x + y)^2 - 2 (x$

$+ y) - 1/2] + \alpha_2 * [-(3/6) (x + y)^2 + 2 (x + y) - (11/6) \} / \partial x\} =$

$\{ \alpha_3 * [3 (x + y) - 2] + \alpha_2 * [- (x + y) + 2] \}$ (39)

Second Order Partial Derivative with Respect to the y Variable

$(\partial^2 (LGR_3(x, y)) / \partial y^2) = \{\partial \{\partial (LGR_3(x, y)) / \partial y\} / \partial y\} = \{\partial \{ \alpha_3 * [(3/2) (x + y)^2 - 2 (x$

$+ y) - 1/2] + \alpha_2 * [-(3/6) (x + y)^2 + 2 (x + y) - (11/6) \} / \partial y\} =$

$\{ \alpha_3 * [3 (x + y) - 2] + \alpha_2 * [- (x + y) + 2] \}$ (40)

Thus:

$(\partial^2 (LGR_3(x, y)) / \partial x^2) = (\partial^2 (LGR_3(x, y)) / \partial y^2)$

Second Order Partial Derivatives with Respect to the x and y Variables

$(\partial^2 (LGR_3(x, y)) / \partial x \partial y) = \{\partial (\partial (LGR_3(x, y)) / \partial x) / \partial y\} = \{\partial \{ \alpha_3 * [(3/2) (x + y)^2 - 2 (x$

$+ y) - 1/2] + \alpha_2 * [-(3/6) (x + y)^2 + 2 (x + y) - (11/6)] \} / \partial y\} =$

$\{ \alpha_3 * [3 (x + y) - 2] + \alpha_2 * [- (x + y) + 2] \}$ (41)

$(\partial^2 (LGR_3(x, y)) / \partial y \partial x) = \{\partial (\partial (LGR_3(x, y)) / \partial y) / \partial x\} = \{\partial \{ \alpha_3 * [(3/2) (x + y)^2 - 2 (x$

$+ y) - 1/2] + \alpha_2 * [-(3/6) (x + y)^2 + 2 (x + y) - (11/6)] \} / \partial x\} =$

$\{ \alpha_3 * [3 (x + y) - 2] + \alpha_2 * [- (x + y) + 2] \}$ (42)

Thus:

$(\partial^2 (LGR_3(x, y)) / \partial x \partial y) = (\partial^2 (LGR_3(x, y)) / \partial y \partial x)$

CALCULATION OF THE INTENSITY-CURVATURE TERMS: BIVARIATE CUBIC LAGRANGE

The intensity-curvature term before interpolation is:

$E_o = E_o (x, y) = \int_0^x \int_0^y f (0, 0) * \{ (\partial^2 (LGR_3(x, y)) / \partial x^2) + (\partial^2 (LGR_3(x, y)) / \partial x \partial y) + (\partial^2$

$(LGR_3(x, y)) / \partial y \partial x) + (\partial^2 (LGR_3(x, y)) / \partial y^2) \} (0, 0) \, dx \, dy$ (43)

It is true that:

$$(\partial^2 (LGR_3(x, y)) / \partial x^2) (0, 0) = (\partial^2 (LGR_3(x, y)) / \partial y^2) (0, 0) =$$

$$(\partial^2 (LGR_3(x, y)) / \partial x \partial y) (0, 0) = (\partial^2 (LGR_3(x, y)) / \partial y \partial x) (0, 0) =$$

$$\{ \alpha_3 * [3 (x + y) - 2] + \alpha_2 * [-(x + y) + 2] \} (0, 0) = -2 \alpha_3 + 2 \alpha_2 \qquad (44)$$

Worth of note that the second order partial derivatives of the bivariate cubic Lagrange function, calculated at the location $(0, 0)$ are the same as those of the bivariate cubic B-Spline $h_4(x, y)$ and this can be seen observing what follows from equations (23), (24) and (25).

Therefore:

$$E_o = E_o (x, y) = \int_0^x \int_0^y 4 * f (0, 0) \{ -2 \alpha_3 + 2 \alpha_2 \} \, dx \, dy = 4xy * f (0, 0) \{ -2 \alpha_3 + 2 \alpha_2 \} \qquad (45)$$

The intensity-curvature term after interpolation is:

$$E_{IN} = E_{IN} (x, y) = \int_0^x \int_0^y LGR_3(x, y) * \{ (\partial^2 (LGR_3(x, y)) / \partial x^2) + (\partial^2 (LGR_3(x, y)) / \partial x \partial y) +$$

$$(\partial^2 (LGR_3(x, y)) / \partial y \partial x) + (\partial^2 (LGR_3(x, y)) / \partial y^2) \} \, dx \, dy \qquad (46)$$

On the basis of equations (36) and (44), $E_{IN} (x, y)$ is written as:

$$E_{IN} (x, y) = \int_0^x \int_0^y 4 * \{ \alpha_3 * [(1/2) (x + y)^3 - (x + y)^2 - 1/2 (x + y) + 1] +$$

$$\alpha_2 * [-(1/6) (x + y)^3 + (x + y)^2 - (11/6) (x + y) + 1] \} * \{ \alpha_3 * [3 (x + y) - 2] + \alpha_2 * [-(x$$

$$- y) + 2] \} dx \, dy =$$

$$\int_0^x \int_0^y 4 * \{ \alpha_3^2 * [(1/2) (x + y)^3 - (x + y)^2 - 1/2 (x + y) + 1] [3 (x + y) - 2] + \alpha_3 \alpha_2 * [$$

$$(1/2) (x + y)^3 - (x + y)^2 - 1/2 (x + y) + 1] [-(x + y) + 2] + \alpha_2 \alpha_3 * [-(1/6) (x + y)^3 + (x +$$

$$y)^2 - (11/6) (x + y) + 1] [3 (x + y) - 2] + \alpha_2^2 [-(1/6) (x + y)^3 + (x + y)^2 - (11/6) (x + y) +$$

$$1] [-(x + y) + 2] \} dx \, dy =$$

$$\int_0^x \int_0^y 4 * \{ \alpha_3^2 * [(3/2) (x + y)^4 - 3 (x + y)^3 - 3/2 (x + y)^2 + 3 (x + y) - (x + y)^3 + 2 (x + y)^2$$

$$+ (x + y) - 2] + \alpha_3 \alpha_2 * [-(1/2) (x - y)^4 + (x + y)^3 + 1/2 (x + y)^2 - (x + y) + (x + y)^3 - 2 (x$$

$+ y)^2 - (x + y) + 2] + \alpha_2 \alpha_3 * [-(3/6) (x + y)^4 + 3 (x + y)^3 - (33/6) (x + y)^2 + 3 (x + y) +$

$(2/6) (x + y)^3 - 2 (x + y)^2 + (22/6) (x + y) - 2] + \alpha_2^2 [(1/6) (x + y)^4 - (x + y)^3 + (11/6)$

$(x + y)^2 - (x + y) - (2/6) (x + y)^3 + 2 (x + y)^2 - (22/6) (x + y) + 2] \} dx\, dy =$

$\int\limits_0^x \int\limits_0^y 4 * \{ \alpha_3^2 * [(3/2) (x + y)^4 - 4 (x + y)^3 + 1/2 (x + y)^2 + 4 (x + y) - 2] + \alpha_3 \alpha_2 * [-(1/2)$

$(x + y)^4 + 2 (x + y)^3 - 3/2 (x + y)^2 - 2 (x + y) + 2] + \alpha_2 \alpha_3 * [-(3/6) (x + y)^4 + (10/3) (x +$

$y)^3 - (45/6) (x + y)^2 + (40/6) (x + y) - 2] + \alpha_2^2 [(1/6) (x + y)^4 + (23/6) (x + y)^2 - (8/6) (x +$

$y)^3 - (28/6) (x + y) + 2] \} dx\, dy =$

$\int\limits_0^x \int\limits_0^y 4 * \{ [(3/2) \alpha_3^2 - (1/2) \alpha_3 \alpha_2 - (3/6) \alpha_2 \alpha_3 + (1/6) \alpha_2^2] (x + y)^4 + [- 4 \alpha_3^2 + 2 \alpha_3 \alpha_2 +$

$(10/3) \alpha_2 \alpha_3 - (8/6) \alpha_2^2] (x + y)^3 + [1/2 \alpha_3^2 - 3/2 \alpha_3 \alpha_2 - (45/6) \alpha_2 \alpha_3 + (23/6) \alpha_2^2] (x + y)^2 +$

$[4 \alpha_3^2 - 2 \alpha_3 \alpha_2 + (40/6) \alpha_2 \alpha_3 - (28/6) \alpha_2^2] (x + y) + [- 2 \alpha_3^2 + 2 \alpha_2^2] \} dx\, dy \qquad (47)$

Equation (47) is solved into equation (48) through the knowledge provided from equations (30.d), (30.e), (30.f) and (30.g).

$E_{IN} (x, y) = 4* \{ [(3/2) \alpha_3^2 - (1/2) \alpha_3 \alpha_2 - (3/6) \alpha_2 \alpha_3 + (1/6) \alpha_2^2] (x^5 y/5 + 3/8\, x^4 y^2 + x^3 y^3/3 +$

$x^2 y^4/8 + x^4 y^2/8 + x^3 y^3/3 + 3/8\, x^2 y^4 + xy^5/5) + [- 4 \alpha_3^2 + 2 \alpha_3 \alpha_2 + (10/3) \alpha_2 \alpha_3 - (8/6) \alpha_2^2]$

$(x^4 y/4 + x^3 y^2/2 + x^2 y^3/2 + xy^4/4) + [1/2 \alpha_3^2 - 3/2 \alpha_3 \alpha_2 - (45/6) \alpha_2 \alpha_3 + (23/6) \alpha_2^2] (x^3 y/3 +$

$x^2 y^2/2 + xy^3/3) + [4 \alpha_3^2 - 2 \alpha_3 \alpha_2 + (40/6) \alpha_2 \alpha_3 - (28/6) \alpha_2^2] (x^2 y/2 + xy^2/2) + [- 2 \alpha_3^2 + 2$

$\alpha_2^2] (xy) \} \qquad (48)$

EQUATING THE TWO INTENSITY-CURVATURE TERMS BEFORE AND AFTER INTERPOLATION: BIVARIATE CUBIC LAGRANGE

Equating $E_o (x, y)$ as per equation (45) with $E_{IN} (x, y)$ as per equation (48) yields:

$4xy * f (0, 0) \{ - 2 \alpha_3 + 2 \alpha_2 \} = E_{IN} (x, y) = \Lambda_{LRG} \qquad (49)$

Where it is posited that:

$\Lambda_{LRG} = 4* \{ [(3/2) \alpha_3^2 - (1/2) \alpha_3 \alpha_2 - (3/6) \alpha_2 \alpha_3 + (1/6) \alpha_2^2] (x^5 y/5 + 3/8\, x^4 y^2 + x^3 y^3/3 +$

$x^2y^4/8 + x^4y^2/8 + x^3y^3/3 + 3/8 \, x^2y^4 + xy^5/5) + [- 4 \, \alpha_3{}^2 + 2 \, \alpha_3 \alpha_2 + (10/3) \, \alpha_2 \alpha_3 - (8/6) \, \alpha_2{}^2]$

$(x^4y/4 + x^3y^2/2 + x^2y^3/2 + xy^4/4) + [1/2 \, \alpha_3{}^2 - 3/2 \, \alpha_3 \alpha_2 - (-5/6) \, \alpha_2 \alpha_3 + (23/6) \, \alpha_2{}^2] \, (x^3y/3 - $

$x^2y^2/2 + xy^3/3) + [4 \, \alpha_3{}^2 - 2 \, \alpha_3 \alpha_2 + (40/6) \, \alpha_2 \alpha_3 - (28/6) \, \alpha_2{}^2] \, (x^2y/2 + xy^2/2) + [- 2 \, \alpha_3{}^2 + 2$

$\alpha_2{}^2] \, (xy) \}$ \hfill (50)

Finally the signal resilient to the bivariate cubic Lagrange interpolation function $LGF_3(x, y)$ of equation (36) is:

$f (0, 0) = \Lambda_{LRG} / \{ 4xy * [- 2 \, \alpha_3 + 2 \, \alpha_2] \}$ \hfill (51)

PART II - TRIVARIATE QUADRATIC B-SPLINE FUNCTION WITHOUT EMBEDDING

TRIVARIATE QUADRATIC B-SPLINE

Let us consider the quadratic $h_3(x, y, z)$ trivariate B-Spline as per equation (1) with $((3 \times 3 \times 3) - 1)$ neighbouring nodes:

$h_3(x, y, z) = [f (-1, 0, 1) + f (-1, 0, 0) + f (1, 0, 0) + f (1, 0, -1) + f (-1, -1, 1) + f (-1, -1, 0)$

$+ f (1, -1, 0) + f (1, -1, -1) + f (-1, 1, 1) + f (-1, 1, 0) + f (1, 1, 0) + f (1, 1, -1)] * [- 2a (x +$

$y + z)^2 + 1/2 (a+1)] + [f (-1, 0, 1) + f (0, 0, 1) + f (1, 0, 1) + f (-1, 0, 0) + f (1, 0, 0) + f (-$

$1, 0, -1) + f (0, 0, -1) + f (1, 0, -1) + f (-1, -1, 1) + f (0, -1, 1) + f (1, -1, 1) + f (-1, -1, 0) +$

$f (0, -1, 0) + f (1, -1, 0) + f (-1, -1, -1) + f (0, -1, -1) + f (1, -1, -1) + f (-1, 1, 1) + f (0, 1, 1)$

$+ f (1, 1, 1) + f (-1, 1, 0) + f (0, 1, 0) + f (1, 1, 0) + f (-1, 1, -1) + f (0, 1, -1) + f (1, 1, -1)]$

$* [a (x + y + z)^2 - (2a + 1/2) (x + y + z) + 3/4 (a+1)]$ (1)

The illustration of the neighborhood of $f (0, 0, 0)$, which is not included in (1), is given in figures 2, 3 and 4 in chapter 2.

Let us posit:

$\omega_1 = [f (-1, 0, 1) + f (-1, 0, 0) + f (1, 0, 0) + f (1, 0, -1) + f (-1, -1, 1) + f (-1, -1, 0) + f (1,$

$-1, 0) + f (1, -1, -1) + f (-1, 1, 1) + f (-1, 1, 0) + f (1, 1, 0) + f (1, 1, -1)]$ (2)

$\omega_2 = [f (-1, 0, 1) + f (0, 0, 1) + f (1, 0, 1) + f (-1, 0, 0) + f (1, 0, 0) + f (-1, 0, -1) + f (0, 0,$

$-1) + f (1, 0, -1) + f (-1, -1, 1) + f (0, -1, 1) + f (1, -1, 1) + f (-1, -1, 0) + f (0, -1, 0) + f (1,$

$-1, 0) + f (-1, -1, -1) + f (0, -1, -1) + f (1, -1, -1) + f (-1, 1, 1) + f (0, 1, 1) + f (1, 1, 1) +$

$f (-1, 1, 0) + f (0, 1, 0) + f (1, 1, 0) + f (-1, 1, -1) + f (0, 1, -1) + f (1, 1, -1)]$ (3)

So that equation (1) is written as:

$h_3(x, y, z) = \omega_1 * [- 2a (x + y + z)^2 + 1/2 (a+1)] + \omega_2 * [a (x + y + z)^2 - (2a + 1/2) (x + y +$

$z) + 3/4 (a+1)]$ (4)

CALCULATION OF THE PARTIAL SECOND ORDER DERIVATIVES: TRIVARIATE QUADRATIC B-SPLINE

First Order Partial Derivative with Respect to the x Variable

$(\partial \, (\, h_3(x, y, z) \,) \,) / \partial x = \partial \, \{ \, \omega_1 * [- 2a \, (x + y + z)^2 + 1/2 \, (a+1)] + \omega_2 * [a \, (x + y + z)^2 - (2a$

$+ 1/2) \, (x + y + z) + 3/4 \, (a+1)] \} \, / \partial x = \{ \, \omega_1 * [- 2a \, \partial \, (x + y + z)^2 / \partial x] + \omega_2 * [a \, \partial \, (x + y$

$+ z)^2 / \partial x - (2a + 1/2) \, \partial \, (x + y + z) / \partial x] \} =$

$\{ \, \omega_1 * [- 4a \, (x + y + z)] + \omega_2 * [2a \, (x + y + z) - (2a + 1/2)] \}$ (5)

First Order Partial Derivative with Respect to the y Variable

$(\partial \, (\, h_3(x, y, z) \,) \,) / \partial y = \partial \, \{ \, \omega_1 * [- 2a \, (x + y + z)^2 + 1/2 \, (a+1)] + \omega_2 * [a \, (x + y + z)^2 - (2a$

$+ 1/2) \, (x + y + z) + 3/4 \, (a+1)] \} \, / \partial y = \{ \, \omega_1 * [- 2a \, \partial \, (x + y + z)^2 / \partial y] + \omega_2 * [a \, \partial \, (x + y$

$+ z)^2 / \partial y - (2a + 1/2) \, \partial \, (x + y + z) / \partial y] \} =$

$\{ \, \omega_1 * [- 4a \, (x + y + z)] + \omega_2 * [2a \, (x + y + z) - (2a + 1/2)] \}$ (6)

First Order Partial Derivative with Respect to the z Variable

$(\partial \, (\, h_3(x, y, z) \,) \,) / \partial z = \partial \, \{ \, \omega_1 * [- 2a \, (x + y + z)^2 + 1/2 \, (a+1)] + \omega_2 * [a \, (x + y + z)^2 - (2a$

$- 1/2) \, (x + y + z) + 3/4 \, (a+1)] \} \, / \partial z = \{ \, \omega_1 * [- 2a \, \partial \, (x + y + z)^2 / \partial z] + \omega_2 * [a \, \partial \, (x + y +$

$z)^2 / \partial z - (2a + 1/2) \, \partial \, (x + y + z) / \partial z] \} =$

$\{ \, \omega_1 * [- 4a \, (x + y + z)] + \omega_2 * [2a \, (x + y + z) - (2a + 1/2)] \}$ (7)

Therefore:

$(\partial \, (\, h_3(x, y, z) \,) \,) / \partial x = (\partial \, (\, h_3(x, y, z) \,) \,) / \partial y = (\partial \, (\, h_3(x, y, z) \,) \,) / \partial z =$

$\{ \, \omega_1 * [- 4a \, (x + y + z)] + \omega_2 * [2a \, (x + y + z) - (2a + 1/2)] \}$ (8)

Second Order Partial Derivative with Respect to the x Variable

$(\partial^2 \, (\, h_3(x, y, z) \,) \,) / \partial x^2 = \{ \partial \, \{ \partial \, (\, h_3(x, y, z) \,) \,) / \partial x \} / \partial x \} = \{ \partial \, \{ \, \omega_1 * [- 4a \, (x + y + z)] + \omega_2 *$

$[2a \, (x + y + z) - (2a + 1/2)] \} / \partial x \} = \{ - 4a \, \omega_1 + 2a \, \omega_2 \}$ (9)

Second Order Partial Derivative with Respect to the y Variable

$$(\partial^2 \, (\, h_3(x, y, z) \,) \, / \partial y^2) = \{\partial \, \{\partial \, (\, h_3(x, y, z) \,) \, / \partial y\} \, / \partial y\} = \{\partial \, \{ \, \omega_1 * [\text{-} \, 4a \, (x + y + z)] + \omega_2 *$$

$$[\, 2a \, (x + y + z) \, \text{-} \, (2a + 1/2) \,] \, \}/ \partial y\} = \{\text{-} \, 4a \, \omega_1 + 2a \, \omega_2 \} \tag{10}$$

Second Order Partial Derivative with Respect to the z Variable

$$(\partial^2 \, (\, h_3(x, y, z) \,) \, / \partial z^2) = \{\partial \, \{\partial \, (\, h_3(x, y, z) \,) \, / \partial z\} \, / \partial z\} = \{\partial \, \{ \, \omega_1 * [\text{-} \, 4a \, (x + y + z)] + \omega_2 *$$

$$[\, 2a \, (x + y + z) \, \text{-} \, (2a + 1/2) \,] \, \}/ \partial z\} = \{\text{-} \, 4a \, \omega_1 + 2a \, \omega_2 \} \tag{11}$$

Second Order Partial Derivatives with Respect to the x and y Variables

$$(\partial^2 \, (\, h_3(x, y, z) \,) \, / \partial x \partial y) = \{\partial \, (\partial \, (\, h_3(x, y, z) \,) \, / \partial x) \, / \partial y\} = \{\partial \, \{ \, \omega_1 * [\text{-} \, 4a \, (x + y + z)] + \omega_2$$

$$* [\, 2a \, (x + y + z) \, \text{-} \, (2a + 1/2) \,] \, \} \, / \partial y\} = \{ \, \text{-} \, 4a \, \omega_1 + 2a \, \omega_2 \} \tag{12}$$

$$(\partial^2 \, (\, h_3(x, y, z) \,) \, / \partial y \partial x) = \{\partial \, (\partial \, (\, h_3(x, y, z) \,) \, / \partial y) \, / \partial x\} = \{\partial \, \{ \, \omega_1 * [\text{-} \, 4a \, (x + y + z)] + \omega_2$$

$$* [\, 2a \, (x + y + z) \, \text{-} \, (2a + 1/2) \,] \, \} \, / \partial x\} = \{ \, \text{-} \, 4a \, \omega_1 + 2a \, \omega_2 \} \tag{13}$$

Therefore:

$$(\partial^2 \, (\, h_3(x, y, z) \,) \, / \partial x \partial y) = (\partial^2 \, (\, h_3(x, y, z) \,) \, / \partial y \partial x) = \{ \, \text{-} \, 4a \, \omega_1 + 2a \, \omega_2 \} \tag{14}$$

Second Order Partial Derivatives with Respect to the x and z Variables

$$(\partial^2 \, (\, h_3(x, y, z) \,) \, / \partial x \partial z) = \{\partial \, (\partial \, (\, h_3(x, y, z) \,) \, / \partial x) \, / \partial z\} = \{\partial \, \{ \, \omega_1 * [\text{-} \, 4a \, (x + y + z)] + \omega_2$$

$$* [\, 2a \, (x + y + z) \, \text{-} \, (2a + 1/2) \,] \, \} \, / \partial z\} = \{ \, \text{-} \, 4a \, \omega_1 + 2a \, \omega_2 \} \tag{15}$$

$$(\partial^2 \, (\, h_3(x, y, z) \,) \, / \partial z \partial x) = \{\partial \, (\partial \, (\, h_3(x, y, z) \,) \, / \partial z) \, / \partial x\} = \{\partial \, \{ \, \omega_1 * [\text{-} \, 4a \, (x + y + z)] + \omega_2$$

$$* [\, 2a \, (x + y + z) \, \text{-} \, (2a + 1/2) \,] \, \} \, / \partial x\} = \{ \, \text{-} \, 4a \, \omega_1 + 2a \, \omega_2 \} \tag{16}$$

Second Order Partial Derivatives with Respect to the y and z Variables

$$(\partial^2 \, (\, h_3(x, y, z) \,) \, / \partial y \partial z) = \{\partial \, (\partial \, (\, h_3(x, y, z) \,) \, / \partial y) \, / \partial z\} = \{\partial \, \{ \, \omega_1 * [\text{-} \, 4a \, (x + y + z)] + \omega_2$$

$$* [\, 2a \, (x + y + z) \, \text{-} \, (2a + 1/2) \,] \, \} \, / \partial z\} = \{ \, \text{-} \, 4a \, \omega_1 + 2a \, \omega_2 \} \tag{17}$$

$$(\partial^2 \, (\, h_3(x, y, z) \,) \, / \partial z \partial y) = \{\partial \, (\partial \, (\, h_3(x, y, z) \,) \, / \partial z) \, / \partial y\} = \{\partial \, \{ \, \omega_1 * [\text{-} \, 4a \, (x + y + z)] + \omega_2$$

$$* [\, 2a \, (x + y + z) \, \text{-} \, (2a + 1/2) \,] \, \} \, / \partial y\} = \{ \, \text{-} \, 4a \, \omega_1 + 2a \, \omega_2 \} \tag{18}$$

Therefore:

$(\partial^2\,(h_3(x, y, z))\,/\partial x^2)\ = (\partial^2\,(h_3(x, y, z))\,/\partial y^2) = (\partial^2\,(h_3(x, y, z))\,/\partial z^2) =$

$(\partial^2\,(h_3(x, y, z))\,/\partial x\partial y) = (\partial^2\,(h_3(x, y, z))\,/\partial y\partial x) = (\partial^2\,(h_3(x, y, z))\,/\partial x\partial z) =$

$(\partial^2\,(h_3(x, y, z))\,/\partial z\partial x) = (\partial^2\,(h_3(x, y, z))\,/\partial y\partial z) = (\partial^2\,(h_3(x, y, z))\,/\partial z\partial y) =$

$\{\,-\,4a\,\omega_1 + 2a\,\omega_2\,\}$ (19)

CALCULATION OF THE INTENSITY-CURVATURE TERMS: TRIVARIATE QUADRATIC B-SPLINE

Let the intensity-curvature term before interpolation be defined as:

$E_o = E_o\,(x, y, z) = \int\limits_0^x\int\limits_0^y\int\limits_0^z f\,(0, 0, 0)\,*\,\{\,(\partial^2\,(h_3(x, y, z))\,/\partial x^2) + (\partial^2\,(h_3(x, y, z))\,/\partial y^2) +$

$(\partial^2\,(h_3(x, y, z))\,/\partial z^2) + (\partial^2\,(h_3(x, y, z))\,/\partial x\partial y) + (\partial^2\,(h_3(x, y, z))\,/\partial y\partial x) + (\partial^2\,(h_3(x, y,$

$z))\,/\partial x\partial z) + (\partial^2\,(h_3(x, y, z))\,/\partial z\partial x) + (\partial^2\,(h_3(x, y, z))\,/\partial y\partial z) + (\partial^2\,(h_3(x, y, z))\,/\partial z\partial y)\}$

$(0, 0, 0)\ dx\ dy\ dz$ (20)

On the basis of equation (19), it can be written that:

$E_o = E_o\,(x, y, z) = \int\limits_0^x\int\limits_0^y\int\limits_0^z 9\,*\,f\,(0, 0, 0)\,*\,\{\,-\,4a\,\omega_1 + 2a\,\omega_2\,\}\ dx\ dy\ dz =$

$9\ xyz\,*\,f\,(0, 0, 0)\,*\,\{\,-\,4a\,\omega_1 + 2a\,\omega_2\,\}$ (21)

Let the intensity-curvature term after interpolation be defined as:

$E_{IN} = E_{IN}\,(x, y, z) = \int\limits_0^x\int\limits_0^y\int\limits_0^z h_3(x, y, z)\,*\,\{\,(\partial^2\,(h_3(x, y, z))\,/\partial x^2) + (\partial^2\,(h_3(x, y, z))\,/\partial y^2) +$

$(\partial^2\,(h_3(x, y, z))\,/\partial z^2) + (\partial^2\,(h_3(x, y, z))\,/\partial x\partial y) + (\partial^2\,(h_3(x, y, z))\,/\partial y\partial x) + (\partial^2\,(h_3(x, y,$

$z))\,/\partial x\partial z) + (\partial^2\,(h_3(x, y, z))\,/\partial z\partial x) + (\partial^2\,(h_3(x, y, z))\,/\partial y\partial z) + (\partial^2\,(h_3(x, y, z))\,/\partial z\partial y)\}$

$dx\ dy\ dz$ (22)

Consider the following calculations:

$(x + y + z)^2 = (x + y + z)\,*\,(x + y + z) = x^2 + xy + xz + xy + y^2 + yz + xz + yz + z^2 =$

$x^2 + y^2 + z^2 + 2\ xy + 2\ xz + 2\ yz$ (23)

$$\int_0^x \int_0^y \int_0^z (x + y + z)^2 \, dx \, dy \, dz = \int_0^x \int_0^y \int_0^z (x^2 + y^2 + z^2 + 2\,xy + 2\,xz + 2\,yz) \, dx \, dy \, dz =$$

$$\int_0^x \int_0^y (x^2z + y^2z + z^3/3 + 2\,xyz + xz^2 + yz^2) \, dx \, dy =$$

$$\int_0^x (x^2yz + y^3z/3 + z^3y/3 + xy^2z + xyz^2 + y^2z^2/2) \, dx =$$

$$(x^3yz/3 + xy^3z/3 + xz^3y/3 + x^2y^2z/2 + x^2yz^2/2 + xy^2z^2/2) \tag{24}$$

$$\int_0^x \int_0^y \int_0^z (x + y + z) \, dx \, dy \, dz = \int_0^x \int_0^y (xz + yz + z^2/2) \, dx \, dy = \int_0^x (xyz + y^2z/2 + yz^2/2) \, dx =$$

$$(x^2yz/2 + xy^2z/2 + xyz^2/2) \tag{25}$$

Let us proceed to the calculation of equation (22):

$$E_{IN} = E_{IN}(x, y, z) = \int_0^x \int_0^y \int_0^z 9 * \{ \omega_1 * [-2a\,(x + y + z)^2 + 1/2\,(a+1)] + \omega_2 * [a\,(x + y + z)^2 -$$

$$(2a + 1/2)\,(x + y + z) + 3/4\,(a+1)]\} * \{-4a\,\omega_1 + 2a\,\omega_2\} \, dx \, dy \, dz =$$

$$9 * \{-4a\,\omega_1 + 2a\,\omega_2\} * \{\omega_1 * [-2a\,(x^3yz/3 + xy^3z/3 + xz^3y/3 + x^2y^2z/2 + x^2yz^2/2 +$$

$$xy^2z^2/2) + 1/2\,(a+1)\,xyz] + \omega_2 * [a\,(x^3yz/3 + xy^3z/3 + xz^3y/3 + x^2y^2z/2 + x^2yz^2/2 +$$

$$xy^2z^2/2) - (2a + 1/2)\,(x^2yz/2 + xy^2z/2 + xyz^2/2) + 3/4(a+1)\,xyz]\} \tag{26}$$

EQUATING THE TWO INTENSITY-CURVATURE TERMS BEFORE AND AFTER INTERPOLATION: TRIVARIATE QUADRATIC B-SPLINE

Let us proceed to solve the following equation which is derived posing in equality the two intensity-curvature terms E_o (equation (21)) and E_{IN} (equation (26)).

$$E_o(x, y, z) = 9\,xyz * f(0, 0, 0) * \{-4a\,\omega_1 + 2a\,\omega_2\} = E_{IN}(x, y, z) = 9 * \{-4a\,\omega_1 + 2a$$

$$\omega_2\} * \{\omega_1 * [-2a\,(x^3yz/3 + xy^3z/3 + xz^3y/3 + x^2y^2z/2 + x^2yz^2/2 + xy^2z^2/2) + 1/2\,(a+1)$$

$$xyz] + \omega_2 * [a\,(x^3yz/3 + xy^3z/3 + xz^3y/3 + x^2y^2z/2 + x^2yz^2/2 + xy^2z^2/2) - (2a + 1/2)$$

$$(x^2yz/2 + xy^2z/2 + xyz^2/2) + 3/4\,(a+1)\,xyz]\} \tag{27}$$

Equation (27) yields:

$$f(0, 0, 0) = \{\omega_1 * [-2a\,(x^3yz/3 + xy^3z/3 + xz^3y/3 + x^2y^2z/2 + x^2yz^2/2 + xy^2z^2/2) + 1/2$$

(a+1) xyz] + ω_2 * [a ($x^2yz/3 + xy^3z/3 + xz^3y/3 + x^2y^2z/2 + x^2yz^2/2 + xy^2z^2/2$) - (2a + 1/2)

($x^2yz/2 + xy^2z/2 + xyz^2/2$) + 3/4 (a+1) xyz] } / { xyz }

Equation (28) hereto following furnishes the value of the signal resilient to the trivariate quadratic B-Spline interpolation function of equation (4).

f (0, 0, 0) = { ω_1 * [- 2a ($x^2/3 + y^2/3 + z^2/3 + xy/2 + xz/2 + yz/2$) + 1/2 (a+1)] + ω_2 * [a

($x^2/3 + y^2/3 + z^2/3 + xy/2 + xz/2 + yz/2$) - (2a + 1/2) (x/2 + y/2 + z/2) + 3/4 (a+1)] } $\qquad$ (28)

PART III - TRIVARIATE CUBIC B-SPLINE FUNCTION
WITHOUT EMBEDDING

TRIVARIATE CUBIC B-SPLINE

Let us consider the trivariate cubic B-Spline of the form:

$$h_4(x, y, z) = \omega_1 * [1/2 (x + y + z)^3 - (x + y + z)^2 + 2/3] + \omega_2 * [-1/6 (x + y + z)^3 + (x + y + z)^2 - 2 (x + y + z) + 4/3] \tag{1}$$

where the values of the positions ω_1 and ω_2 hold true as per equations hereto follow:

$$\omega_1 = [f (-1, 0, 1) + f (-1, 0, 0) + f (1, 0, 0) + f (1, 0, -1) + f (-1, -1, 1) + f (-1, -1, 0) + f (1, -1, 0) + f (1, -1, -1) + f (-1, 1, 1) + f (-1, 1, 0) + f (1, 1, 0) + f (1, 1, -1)] \tag{2}$$

$$\omega_2 = [f (-1, 0, 1) + f (0, 0, 1) + f (1, 0, 1) + f (-1, 0, 0) + f (1, 0, 0) + f (-1, 0, -1) + f (0, 0, -1) + f (1, 0, -1) + f (-1, -1, 1) + f (0, -1, 1) + f (1, -1, 1) + f (-1, -1, 0) + f (0, -1, 0) + f (1, -1, 0) + f (-1, -1, -1) + f (0, -1, -1) + f (1, -1, -1) + f (-1, 1, 1) + f (0, 1, 1) + f (1, 1, 1) + f (-1, 1, 0) + f (0, 1, 0) + f (1, 1, 0) + f (-1, 1, -1) + f (0, 1, -1) + f (1, 1, -1)] \tag{3}$$

CALCULATION OF THE PARTIAL SECOND ORDER DERIVATIVES: TRIVARIATE CUBIC B-SPLINE

First Order Partial Derivative with Respect to the x Variable

$$(\partial (h_4(x, y, z)) / \partial x) = \partial \{ \omega_1 * [1/2 (x + y + z)^3 - (x + y + z)^2 + 2/3] + \omega_2 * [-1/6 (x + y + z)^3 + (x + y + z)^2 - 2 (x + y + z) + 4/3] / \partial x \} = \{ \omega_1 * [1/2 \partial (x + y + z)^3 / \partial x - \partial (x + y + z)^2 / \partial x] + \omega_2 * [-1/6 \partial (x + y + z)^3 / \partial x + \partial (x + y + z)^2 / \partial x - 2 \partial (x + y + z) / \partial x] \} =$$

$$\{ \omega_1 * [3/2 (x + y + z)^2 - 2 (x + y + z)] + \omega_2 * [-3/6 \partial (x + y + z)^2 + 2 (x + y + z) - 2] \} \tag{4}$$

First Order Partial Derivative with Respect to the y Variable

$$(\partial (h_4(x, y, z)) / \partial y) = \partial \{ \omega_1 * [1/2 (x + y + z)^3 - (x + y + z)^2 + 2/3] + \omega_2 * [-1/6 (x + y + z)^3 + (x + y + z)^2 - 2 (x + y + z) + 4/3] / \partial y \} = \{ \omega_1 * [1/2 \partial (x + y + z)^3 / \partial y - \partial (x + y + z)^2 / \partial y] + \omega_2 * [-1/6 \partial (x + y + z)^3 / \partial y + \partial (x + y + z)^2 / \partial y - 2 \partial (x + y +$$

z) $/\partial y$] } =

$$\{ \omega_1 * [3/2 (x + y + z)^2 - 2 (x + y + z)] + \omega_2 * [-3/6 (x + y + z)^2 + 2 (x + y + z) - 2] \} \tag{5}$$

First Order Partial Derivative with Respect to the z Variable

$(\partial (h_4(x, y, z)) /\partial z) = \partial \{ \omega_1 * [1/2 (x + y + z)^3 - (x + y + z)^2 + 2/3] + \omega_2 * [-1/6 (x + y$

$- z)^3 + (x + y + z)^2 - 2 (x + y + z) + 4/3] /\partial z \} = \{ \omega_1 * [1/2 \partial (x + y + z)^3 /\partial z - \partial (x + y +$

$z)^2 /\partial z] + \omega_2 * [-1/6 \partial (x + y + z)^3 /\partial z + \partial (x + y + z)^2 /\partial z - 2 \partial (x + y + z) /\partial z] \} =$

$$\{ \omega_1 * [3/2 (x + y + z)^2 - 2 (x + y + z)] + \omega_2 * [-3/6 (x + y + z)^2 + 2 (x + y + z) - 2] \} \tag{6}$$

Therefore:

$(\partial (h_4(x, y, z)) /\partial x) = (\partial (h_4(x, y, z)) /\partial y) = (\partial (h_4(x, y, z)) /\partial z) =$

$$\{ \omega_1 * [3/2 (x + y + z)^2 - 2 (x + y + z)] + \omega_2 * [-3/6 (x + y + z)^2 + 2 (x + y + z) - 2] \} \tag{7}$$

Second Order Partial Derivatives

$(\partial^2 (h_4(x, y, z)) /\partial x^2) = \{\partial \{\partial (h_4(x, y, z)) /\partial x\} /\partial x\} = \{\partial \{ \omega_1 * [3/2 (x + y + z)^2 - 2 (x$

$+ y + z)] + \omega_2 * [-3/6 (x + y + z)^2 + 2 (x + y + z) - 2] \} /\partial x\} =$

$$\{ \omega_1 * [3 (x + y + z) - 2] + \omega_2 * [- (x + y + z) + 2] \} \tag{8}$$

Because of equations (4) through (5) it is true that:

$(\partial^2 (h_4(x, y, z)) /\partial x^2) = \{\partial \{\partial (h_4(x, y, z)) /\partial x\} /\partial x\} =$

$(\partial^2 (h_4(x, y, z)) /\partial y^2) = \{\partial \{\partial (h_4(x, y, z)) /\partial y\} /\partial y\} =$

$(\partial^2 (h_4(x, y, z)) /\partial z^2) = \{\partial \{\partial (h_4(x, y, z)) /\partial z\} /\partial z\} =$

$$\{ \omega_1 * [3 (x + y + z) - 2] + \omega_2 * [- (x + y + z) + 2] \} \tag{9}$$

Also it is true that:

$(\partial^2 (h_4(x, y, z)) /\partial x \partial y) = \{\partial (\partial (h_4(x, y, z)) /\partial x) /\partial y\} = \{\partial \{ \omega_1 * [3/2 (x + y + z)^2 - 2 (x$

$+ y + z)] + \omega_2 * [-3/6 (x + y + z)^2 + 2 (x + y + z) - 2] \} /\partial y\} = \{ \omega_1 * [3 (x + y + z) - 2]$

$$+ \omega_2 * [- (x + y + z) + 2] \} \tag{10}$$

$(\partial^2 (h_4(x, y, z)) /\partial x \partial z) = \{\partial (\partial (h_4(x, y, z)) /\partial x) /\partial z\} = \{\partial \{ \omega_1 * [3/2 (x + y + z)^2 - 2 (x$

$+ y + z)] + \omega_2 * [-3/6 (x + y + z)^2 + 2 (x + y + z) - 2] \} / \partial z\} = \{ \omega_1 * [3 (x + y + z) - 2]$

$+ \omega_2 * [- (x + y + z) + 2] \}$ (11)

$(\partial^2 (h_4(x, y, z)) / \partial y \partial z) = \{\partial (\partial (h_4(x, y, z)) / \partial y) / \partial z\} = \{\partial \{ \omega_1 * [3/2 (x + y + z)^2 - 2 (x$

$+ y + z)] + \omega_2 * [-3/6 (x + y + z)^2 + 2 (x + y + z) - 2] \} / \partial z\} = \{ \omega_1 * [3 (x + y + z) - 2]$

$+ \omega_2 * [- (x + y + z) + 2] \}$ (12)

$(\partial^2 (h_4(x, y, z)) / \partial x \partial y) = \{\partial (\partial (h_4(x, y, z)) / \partial y) / \partial x\} =$

$(\partial^2 (h_4(x, y, z)) / \partial x \partial z) = \{\partial (\partial (h_4(x, y, z)) / \partial z) / \partial x\} =$

$(\partial^2 (h_4(x, y, z)) / \partial y \partial z) = \{\partial (\partial (h_4(x, y, z)) / \partial z) / \partial y\} =$

$\{ \omega_1 * [3 (x + y + z) - 2] + \omega_2 * [- (x + y + z) + 2] \}$ (13)

CALCULATION OF THE INTENSITY-CURVATURE TERMS: TRIVARIATE CUBIC B-SPLINE

The intensity-curvature term before interpolation is defined as:

$E_o = E_o (x, y, z) = \int_0^x \int_0^y \int_0^z f (0, 0, 0) * \{ (\partial^2 (h_4(x, y, z)) / \partial x^2) + (\partial^2 (h_4(x, y, z)) / \partial y^2) +$

$(\partial^2 (h_4(x, y, z)) / \partial z^2) + (\partial^2 (h_4(x, y, z)) / \partial x \partial y) + (\partial^2 (h_4(x, y, z)) / \partial y \partial x) + (\partial^2 (h_4(x, y,$

$z)) / \partial x \partial z) + (\partial^2 (h_4(x, y, z)) / \partial z \partial x) + (\partial^2 (h_4(x, y, z)) / \partial y \partial z) + (\partial^2 (h_4(x, y, z)) / \partial z \partial y)\}$

$(0, 0, 0) \, dx \, dy \, dz$ (14)

On the basis of equations (9) and (13), it can be written that:

$(\partial^2 (h_4(x, y, z)) / \partial x^2) (0, 0, 0) = (\partial^2 (h_4(x, y, z)) / \partial y^2) (0, 0, 0) =$

$(\partial^2 (h_4(x, y, z)) / \partial z^2) (0, 0, 0) = \{ - 2 \omega_1 + 2 \omega_2 \}$ (15)

$(\partial^2 (h_4(x, y, z)) / \partial x \partial y) (0, 0, 0) = (\partial^2 (h_4(x, y, z)) / \partial x \partial z) (0, 0, 0) =$

$(\partial^2 (h_4(x, y, z)) / \partial y \partial z) (0, 0, 0) = \{ - 2 \omega_1 + 2 \omega_2 \}$ (16)

Therefore:

$E_o = E_o (x, y, z) = \int_0^x \int_0^y \int_0^z 9 * f (0, 0, 0) * \{ - 2 \omega_1 + 2 \omega_2 \} \, dx \, dy \, dz =$

$9 \, xyz * f (0, 0, 0) * \{ - 2 \omega_1 + 2 \omega_2 \}$ (17)

The intensity-curvature term after interpolation is defined as:

$$E_{IN} = E_{IN}(x, y, z) = \int_0^x \int_0^y \int_0^z h_4(x, y, z) * \{ (\partial^2 (h_4(x, y, z)) / \partial x^2) + (\partial^2 (h_4(x, y, z)) / \partial y^2 +$$

$$(\partial^2 (h_4(x, y, z)) / \partial z^2) + (\partial^2 (h_4(x, y, z)) / \partial x \partial y) + (\partial^2 (h_4(x, y, z)) / \partial y \partial x) + (\partial^2 (h_4(x, y,$$

$$z)) / \partial x \partial z) + (\partial^2 (h_4(x, y, z)) / \partial z \partial x) + (\partial^2 (h_4(x, y, z)) / \partial y \partial z) + (\partial^2 (h_4(x, y, z)) / \partial z \partial y)\}$$

$$dx\, dy\, dz \tag{18}$$

On the basis of equations (1), (9) and (13), it can be written that:

$$E_{IN} = E_{IN}(x, y, z) = \int_0^x \int_0^y \int_0^z 9 * \{ \omega_1 * [1/2 (x + y + z)^3 - (x + y + z)^2 + 2/3] + \omega_2$$

$$* [-1/6 (x + y + z)^3 + (x + y + z)^2 - 2 (x + y + z) + 4/3]\} * \{ \omega_1 * [3 (x + y + z) - 2] + \omega_2$$

$$* [- (x + y + z) + 2]\}\, dx\, dy\, dz \tag{19}$$

Equation (19) is solved with the following procedure:

$$E_{IN} = E_{IN}(x, y, z) = 9 * \int_0^x \int_0^y \int_0^z \omega_1^2 * [1/2 (x + y + z)^3 - (x + y + z)^2 + 2/3][3 (x + y + z) -$$

$$2] + \omega_1 \omega_2 * [1/2 (x + y + z)^3 - (x + y + z)^2 + 2/3][- (x + y + z) + 2] + \omega_2 \omega_1 * [-1/6 (x$$

$$+ y + z)^3 + (x + y + z)^2 - 2 (x + y + z) + 4/3][3 (x + y + z) - 2] + \omega_2^2 * [-1/6 (x + y + z)^3$$

$$+ (x + y + z)^2 - 2 (x + y + z) + 4/3][- (x + y + z) + 2]\, dx\, dy\, dz =$$

$$9 * \int_0^x \int_0^y \int_0^z \omega_1^2 * [3/2 (x + y + z)^4 - 3 (x + y + z)^3 + 2 (x + y + z) - (x + y + z)^3 + 2 (x + y +$$

$$z)^2 - 4/3] + \omega_1 \omega_2 * [-1/2 (x + y + z)^4 + (x + y + z)^3 - 2/3 (x + y + z) + (x + y + z)^3 - 2 (x$$

$$+ y + z)^2 + 4/3] + \omega_2 \omega_1 * [-3/6 (x + y + z)^4 + 3 (x + y + z)^3 - 6 (x + y + z)^2 + 4 (x + y + z)$$

$$+ 2/6 (x + y + z)^3 - 2 (x + y + z)^2 + 4 (x + y + z) - 8/3] + \omega_2^2 [1/6 (x + y + z)^4 - (x + y +$$

$$z)^3 + 2 (x + y + z)^2 - 4/3 (x + y + z) - 2/6 (x + y + z)^3 + 2 (x + y + z)^2 - 4 (x + y + z) + 8/3]$$

$$dx\, dy\, dz =$$

$$9 * \int_0^x \int_0^y \int_0^z \omega_1^2 * [3/2 (x + y + z)^4 - 4 (x + y + z)^3 + 2 (x + y + z) + 2 (x + y + z)^2 - 4/3] +$$

$\omega_1 \omega_2 * [-1/2 (x + y + z)^4 + 2 (x + y + z)^3 - 2/3 (x + y + z) - 2 (x + y + z)^2 + 4/3] + \omega_2 \omega_1 *$

$[-3/6 (x + y + z)^4 + 4/3 (x + y + z)^3 - 8 (x + y + z)^2 + 8 (x + y + z) - 8/3] + \omega_2{}^2 [1/6 (x + y$

$+ z)^4 - 4/3 (x + y + z)^3 + 4 (x + y + z)^2 - 16/3 (x + y + z) + 8/3] \, dx \, dy \, dz =$

$9 * \int\limits_0^x \int\limits_0^y \int\limits_0^z [3/2 \, \omega_1{}^2 - 1/2 \, \omega_1 \, \omega_2 - 3/6 \, \omega_2 \, \omega_1 + 1/6 \, \omega_2{}^2] (x + y + z)^4 + [- 4 \, \omega_1{}^2 + 2 \, \omega_1 \, \omega_2 + 4/3 \, \omega_2$

$\omega_1 - 4/3 \, \omega_2{}^2] (x + y + z)^3 + [2 \, \omega_1{}^2 - 2 \, \omega_1 \, \omega_2 - 8 \, \omega_2 \, \omega_1 + 4 \, \omega_2{}^2] (x + y + z)^2 + [2 \, \omega_1{}^2 - 2/3$

$\omega_1 \, \omega_2 + 8 \, \omega_2 \, \omega_1 - 16/3 \, \omega_2{}^2] (x + y + z) + [- 4/3 \, \omega_1{}^2 + 4/3 \, \omega_1 \, \omega_2 - 8/3 \, \omega_2 \, \omega_1 + 8/3 \, \omega_2{}^2] \, dx$

$dy \, dz$ (20)

Now consider the following equation:

$(x + y + z)^2 = (x + y + z) * (x + y + z) = x^2 + xy + xz + xy + y^2 + yz + xz + yz + z^2 =$

$x^2 + y^2 + z^2 + 2 \, xy + 2 \, xz + 2 \, yz$ (21)

On the basis of (21), it is true that:

$(x + y + z)^4 = (x + y + z)^2 (x + y + z)^2 =$

$(x^2 + y^2 + z^2 + 2 \, xy + 2 \, xz + 2 \, yz) (x^2 + y^2 + z^2 + 2 \, xy + 2 \, xz + 2 \, yz) =$

$(x^4 + x^2y^2 + x^2z^2 + 2 \, x^3y + 2 \, x^3z + 2 \, x^2yz + y^4 + x^2y^2 + y^2z^2 + 2 \, xy^3 + 2 \, xzy^2 + 2 \, y^3z +$

$x^2z^2 + z^2y^2 + z^4 + 2 \, xyz^2 + 2 \, xz^3 + 2 \, yz^3 + 2 \, x^3y + 2 \, xy^3 + 2 \, xyz^2 + 4 \, x^2y^2 + 4 \, x^2yz + 4$

$xy^2z + 2 \, x^3z + 2 \, xy^2z + 2 \, xz^3 + 4 \, x^2yz + 4x^2z^2 + 4 \, xyz^2 + 2 \, x^2yz + 2 \, y^3z + 2 \, yz^3 + 4 \, xy^2z +$

$4 \, xyz^2 + 4 \, y^2z^2)$ (22)

$(x + y + z)^3 = (x + y + z)^2 (x + y + z) = (x^2 + y^2 + z^2 + 2 \, xy + 2 \, xz + 2 \, yz) (x + y + z) =$

$(x^3 + xy^2 + xz^2 + 2 \, x^2y + 2 \, x^2z + 2 \, xyz + x^2y + y^3 + yz^2 + 2 \, xy^2 + 2 \, xyz + 2 \, y^2z + x^2z +$

$y^2z + z^3 + 2 \, xyz + 2 \, xz^2 + 2 \, yz^2)$ (23)

Therefore, on the basis of equation (22) and (23) let us posit:

$\Pi^{(3)} = \int\limits_0^x \int\limits_0^y \int\limits_0^z (x + y + z)^3 \, dx \, dy \, dz = \int\limits_0^x \int\limits_0^y \int\limits_0^z (x^3 + xy^2 + xz^2 + 2 \, x^2y + 2 \, x^2z + 2 \, xyz + x^2y + y^3 +$

$yz^2 + 2xy^2 + 2 \, xyz + 2 \, y^2z + x^2z + y^2z + z^3 + 2 \, xyz + 2 \, xz^2 + 2 \, yz^2) \, dx \, dy \, dz =$

$$\int_0^x \int_0^y (x^3z + xy^2z + xz^3/3 + 2\,x^2yz + x^2z^2 + xyz^2 + x^2yz + y^3z + yz^3/3 + 2xy^2z + xyz^2 + y^2z^2$$

$$+ x^2z^2/2 + y^2z^2/2 + z^4/4 + xyz^2 + 2/3\,xz^3 + 2/3\,yz^3)\,dx\,dy =$$

$$\int_0^x (x^3yz + xy^3z/3 + xyz^3/3 + x^2y^2z + x^2yz^2 + xy^2z^2/2 + x^2y^2z/2 + y^4z/4 + y^2z^3/6 + 2/3\,xy^3z$$

$$+ xy^2z^2/2 + y^3z^2/3 + x^2yz^2/2 + y^3z^2/6 + yz^4/4 + xy^2z^2/2 + 2/3\,xyz^3 + 1/3\,y^2z^3)\,dx =$$

$$(x^4yz/4 + x^2y^3z/6 + x^2yz^2/6 + x^3y^2z/3 + 1/3\,x^3yz^2 + x^2y^2z^2/4 + x^3y^2z/6 + xy^4z/4 - xy^2z^3/6 +$$

$$1/3\,x^2y^3z + x^2y^2z^2/4 + xy^3z^2/3 + x^3yz^2/6 + xy^3z^2/6 + xyz^4/4 + x^2y^2z^2/4 + 1/3\,x^2yz^3 + 1/3$$

$$xy^2z^3) \tag{24}$$

$$\Pi^{(4)} = \int_0^x \int_0^y \int_0^z (x+y+z)^4\,dx\,dy\,dz = \int_0^x \int_0^y \int_0^z (x^4 + x^2y^2 + x^2z^2 + 2\,x^3y + 2\,x^3z + 2\,x^2yz + y^4 +$$

$$x^2y^2 + y^2z^2 + 2\,xy^3 + 2\,xzy^2 + 2\,y^3z + x^2z^2 + z^2y^2 + z^4 + 2\,xyz^2 + 2\,xz^3 + 2\,yz^3 + 2\,x^3y + 2$$

$$xy^3 + 2\,xyz^2 + 4\,x^2y^2 + 4\,x^2yz + 4\,xy^2z + 2\,x^3z + 2\,xy^2z + 2\,xz^3 + 4\,x^2yz + 4x^2z^2 + 4\,xyz^2$$

$$+ 2\,x^2yz + 2\,y^3z + 2\,yz^3 + 4\,xy^2z + 4\,xyz^2 + 4\,y^2z^2)\,dx\,dy\,dz =$$

$$\int_0^x \int_0^y (x^4z + x^2y^2z + x^2z^3/3 + 2\,x^3yz + x^3z^2 + x^2yz^2 + y^4z + x^2y^2z + y^2z^3/3 + 2\,xy^3z + xz^2y^2$$

$$+ y^3z^2 + x^2z^3/3 + z^3y^2/3 + z^5/5 + 2/3\,xyz^3 + 1/2\,xz^4 + 1/2\,yz^4 + 2\,x^3yz + 2\,xy^3z + 2/3\,xyz^3$$

$$+ 4\,x^2y^2z + 2\,x^2yz^2 + 2\,xy^2z^2 + x^3z^2 + xy^2z^2 + 1/2\,xz^4 + 2\,x^2yz^2 + 4/3\,x^2z^3 + 4/3\,xyz^3 +$$

$$x^2yz^2 + y^3z^2 + 1/2\,yz^4 + 2\,xy^2z^2 + 4/3\,xyz^3 + 4/3\,y^2z^3)\,dx\,dy =$$

$$\int_0^x (x^4zy + x^2y^3z/3 + x^2yz^3/3 + x^3y^2z + x^3z^2y + x^2y^2z^2/2 + y^5z/5 + x^2y^3z/3 + y^3z^3/9 + 1/2$$

$$xy^4z + xz^2y^3/3 + y^4z^2/4 + x^2yz^3/3 - z^3y^3/9 + z^5y/5 + 1/3\,xy^2z^3 + 1/2\,xyz^4 + 1/4\,y^2z^4 + x^3y^2z$$

$$+ 1/2\,xy^4z + 1/3\,xy^2z^3 + 4/3\,x^2y^3z + x^2y^2z^2 + 2/3\,xy^3z^2 + x^3z^2y + xy^3z^2/3 + 1/2\,xyz^4 +$$

$$x^2y^2z^2 + 4/3\,x^2yz^3 + 4/6\,xy^2z^3 + x^2y^2z^2/2 + y^4z^2/4 + 1/4\,y^2z^4 + 2/3\,xy^3z^2 + 4/6\,xy^2z^3 + 4/9$$

$$y^3z^3)\,dx$$

$$= (x^5zy/5 + x^3y^3z/9 + x^3yz^3/9 + x^4y^2z/4 + x^4z^2y/4 + x^3y^2z^2/6 + xy^5z/5 + x^3y^3z/9 + xy^3z^3/9 +$$

$$x^2y^4z + x^2z^2y^3/6 + xy^4z^2/4 + x^3yz^3/9 + xz^3y^3/9 + xz^5y/5 + 1/6\ x^2y^2z^3 + 1/4\ x^2yz^4 + 1/4$$

$$xy^2z^4 + x^4y^2z/4 + 1/4\ x^2y^4z + 1/6\ x^2y^2z^3 + 4/9\ x^3y^3z + x^3y^2z^2/3 + 1/3\ x^2y^3z^2 + x^4z^2y/4 +$$

$$x^2y^3z^2/6 + x^2yz^4 + x^3y^2z^2/3 + 4/9\ x^3yz^3 + 4/12\ x^2y^2z^3 + x^3y^2z^2/6 + xy^4z^2/4 + 1/4\ xy^2z^4 +$$

$$1/3\ x^2y^3z^2 + 4/12\ x^2y^2z^3 + 4/9\ xy^3z^3) \tag{25}$$

It can be calculated that:

$$\int_0^x\int_0^y\int_0^z (x + y + z)^2\ dx\ dy\ dz = \int_0^x\int_0^y\int_0^z (x^2 + y^2 + z^2 + 2\ xy + 2\ xz + 2\ yz)\ dx\ dy\ dz =$$

$$\int_0^x\int_0^y (x^2z + y^2z + z^3/3 + 2\ xyz + xz^2 + yz^2)\ dx\ dy =$$

$$\int_0^x (x^2yz + y^3z/3 + z^3y/3 + xy^2z + xyz^2 + y^2z^2/2)\ dx =$$

$$(x^3yz/3 + xy^3z/3 + xz^3y/3 + x^2y^2z/2 + x^2yz^2/2 + xy^2z^2/2) \tag{26}$$

$$\int_0^x\int_0^y\int_0^z (x + y + z)\ dx\ dy\ dz = \int_0^x\int_0^y (xz + yz + z^2/2)\ dx\ dy = \int_0^x (xyz + y^2z/2 + yz^2/2)\ dx =$$

$$(x^2yz/2 + xy^2z/2 + xyz^2/2) \tag{27}$$

Let us posit:

$$\Pi^{(2)} = (x^3yz/3 + xy^3z/3 + xz^3y/3 + x^2y^2z/2 + x^2yz^2/2 + xy^2z^2/2) \tag{28}$$

$$\Pi^{(1)} = (x^2yz/2 + xy^2z/2 + xyz^2/2) \tag{29}$$

Now equation (20) can be solved considering the knowledge provided in equations (24), (25), (28) and (29).

$$E_{IN}(x, y, z) = 9 * \{ [3/2\ \omega_1^2 - 1/2\ \omega_1\omega_2 - 3/6\ \omega_2\ \omega_1 + 1/6\ \omega_2^2]\ \Pi^{(4)} + [-4\ \omega_1^2 + 2\ \omega_1\ \omega_2 + 4/3$$

$$\omega_2\ \omega_1 - 4/3\ \omega_2^2]\ \Pi^{(3)} + [2\ \omega_1^2 - 2\ \omega_1\ \omega_2 - 8\ \omega_2\ \omega_1 + 4\ \omega_2^2]\ \Pi^{(2)} + [2\ \omega_1^2 - 2/3\ \omega_1\ \omega_2 + 8\ \omega_2$$

$$\omega_1 - 16/3\ \omega_2^2]\ \Pi^{(1)} + [-4/3\ \omega_1^2 + 4/3\ \omega_1\ \omega_2 - 8/3\ \omega_2\ \omega_1 + 8/3\ \omega_2^2]\ xyz \} \tag{30}$$

EQUATING THE TWO INTENSITY-CURVATURE TERMS BEFORE AND AFTER INTERPOLATION: TRIVARIATE CUBIC B-SPLINE

Let us proceed to equate the two intensity-curvature terms (before and after interpolation) in order to derive the signal resilient to the cubic trivariate B-Spline interpolation function of equation (1). The two intensity curvature terms $E_o(x, y, z)$ and $E_{IN}(x, y, z)$ are given in equations (17) and (30).

$$E_o(x, y, z) = 9\ xyz * f(0, 0, 0) * \{ -2\ \omega_1 + 2\ \omega_2 \} = E_{IN}(x, y, z) = 9 * \{ [3/2\ \omega_1^2 - 1/2$$

$$\omega_1\omega_2 - 3/6\ \omega_2\ \omega_1 + 1/6\ \omega_2^2]\ \Pi^{(4)} + [-4\ \omega_1^2 + 2\ \omega_1\ \omega_2 + 4/3\ \omega_2\ \omega_1 - 4/3\ \omega_2^2]\ \Pi^{(3)} + [\ 2\ \omega_1^2 -$$

$$2\ \omega_1\ \omega_2 - 8\ \omega_2\ \omega_1 + 4\ \omega_2^2]\ \Pi^{(2)} + [2\ \omega_1^2 - 2/3\ \omega_1\ \omega_2 + 8\ \omega_2\ \omega_1 - 16/3\ \omega_2^2]\ \Pi^{(1)} + [-4/3\ \omega_1^2$$

$$+ 4/3\ \omega_1\ \omega_2 - 8/3\ \omega_2\ \omega_1 - 8/3\ \omega_2^2]\ xyz \}$$

$$\tag{31}$$

Equation (31) can be re-written as:

$$xyz * f(0, 0, 0) * \{ -2\ \omega_1 + 2\ \omega_2 \} = \{ [3/2\ \omega_1^2 - 1/2\ \omega_1\omega_2 - 3/6\ \omega_2\ \omega_1 + 1/6\ \omega_2^2]\ \Pi^{(4)} +$$

$$[-4\ \omega_1^2 + 2\ \omega_1\ \omega_2 + 4/3\ \omega_2\ \omega_1 - 4/3\ \omega_2^2]\ \Pi^{(3)} + [\ 2\ \omega_1^2 - 2\ \omega_1\ \omega_2 - 8\ \omega_2\ \omega_1 + 4\ \omega_2^2]\ \Pi^{(2)} +$$

$$[2\ \omega_1^2 - 2/3\ \omega_1\ \omega_2 + 8\ \omega_2\ \omega_1 - 16/3\ \omega_2^2]\ \Pi^{(1)} + [-4/3\ \omega_1^2 + 4/3\ \omega_1\ \omega_2 - 8/3\ \omega_2\ \omega_1 + 8/3\ \omega_2^2]$$

$$xyz \}$$

$$\tag{32}$$

Let us posit:

$$\Lambda_\gamma = \{ [3/2\ \omega_1^2 - 1/2\ \omega_1\omega_2 - 3/6\ \omega_2\ \omega_1 + 1/6\ \omega_2^2]\ \Pi^{(4)} + [-4\ \omega_1^2 + 2\ \omega_1\ \omega_2 + 4/3\ \omega_2\ \omega_1 - 4/3$$

$$\omega_2^2]\ \Pi^{(3)} + [\ 2\ \omega_1^2 - 2\ \omega_1\ \omega_2 - 8\ \omega_2\ \omega_1 + 4\ \omega_2^2]\ \Pi^{(2)} + [2\ \omega_1^2 - 2/3\ \omega_1\ \omega_2 + 8\ \omega_2\ \omega_1 - 16/3$$

$$\omega_2^2]\ \Pi^{(1)} + [-4/3\ \omega_1^2 + 4/3\ \omega_1\ \omega_2 - 8/3\ \omega_2\ \omega_1 + 8/3\ \omega_2^2]\ xyz \}$$

$$\tag{33}$$

Equation (32) becomes:

$$xyz * f(0, 0, 0) * \{ -2\ \omega_1 + 2\ \omega_2 \} = \Lambda_\gamma$$

$$\tag{34}$$

And is solved hereto:

$$f(0, 0, 0) = \Lambda_\gamma / \{ xyz * [-2\ \omega_1 + 2\ \omega_2] \}$$

$$\tag{35}$$

Where $f(0, 0, 0)$ of equation (35) is the signal resilient to the cubic trivariate B-Spline interpolation function.

PART IV - TRIVARIATE CUBIC LAGRANGE FUNCTION WITHOUT EMBEDDING

TRIVARIATE LAGRANGE INTERPOLATION FUNCTION

The formula hereto reported is the trivariate Lagrange interpolation function having third degree order:

$LGR_3(x, y, z) = \omega_1 * [(1/2) (x + y + z)^3 - (x + y + z)^2 - 1/2 (x + y + z) + 1] +$

$\omega_2 * [-(1/6) (x + y + z)^3 + (x + y + z)^2 - (11/6) (x + y + z) + 1]$ (1)

This function has the same $((3 \times 3 \times 3) - 1)$ pixels' neighborhood of $f(0, 0, 0)$ (not included) as the one adopted for the quadratic and cubic B-Splines and this is illustrated in figures 2, 3 and 4 in chapter 2.

The values of ω_1 and ω_2 are:

$\omega_1 = [f(-1, 0, 1) + f(-1, 0, 0) + f(1, 0, 0) + f(1, 0, -1) + f(-1, -1, 1) + f(-1, -1, 0) + f(1,$

$-1, 0) + f(1, -1, -1) + f(-1, 1, 1) + f(-1, 1, 0) + f(1, 1, 0) + f(1, 1, -1)]$ (2)

$\omega_2 = [f(-1, 0, 1) + f(0, 0, 1) + f(1, 0, 1) + f(-1, 0, 0) + f(1, 0, 0) + f(-1, 0, -1) + f(0, 0,$

$-1) + f(1, 0, -1) + f(-1, -1, 1) + f(0, -1, 1) + f(1, -1, 1) + f(-1, -1, 0) + f(0, -1, 0) + f(1,$

$-1, 0) + f(-1, -1, -1) + f(0, -1, -1) + f(1, -1, -1) + f(-1, 1, 1) + f(0, 1, 1) + f(1, 1, 1) +$

$f(-1, 1, 0) + f(0, 1, 0) + f(1, 1, 0) + f(-1, 1, -1) + f(0, 1, -1) + f(1, 1, -1)]$ (3)

CALCULATION OF THE PARTIAL SECOND ORDER DERIVATIVES: TRIVARIATE CUBIC LAGRANGE

First Order and Second Order Partial Derivatives of the Trivariate Cubic Lagrange

For convenience, let us calculate the first and second order partial derivatives of the trivariate Lagrange function with respect to the x variable, also the second order partial derivatives with respect to x and y variables. Then, through the knowledge provided while studying the partial second order derivatives of the trivariate cubic B-Spline, let us demonstrate the assertions given hereto follow in equation (4).

$(\partial^2 (LGR_3(x, y, z)) / \partial x^2) = (\partial^2 (LGR_3(x, y, z)) / \partial y^2) = (\partial^2 (LGR_3(x, y, z)) / \partial z^2) =$

$(\partial^2 (LGR_3(x, y, z)) / \partial x \partial y) = (\partial^2 (LGR_3(x, y, z)) / \partial y \partial x) =$

$(\partial^2 (LGR_3(x, y, z)) / \partial x \partial z) = (\partial^2 (LGR_3(x, y, z)) / \partial z \partial x) =$

$$(\partial^2 (LGR_3(x, y, z)) / \partial y \partial z) = (\partial^2 (LGR_3(x, y, z)) / \partial z \partial y) \tag{4}$$

$$(\partial (LGR_3(x, y, z)) / \partial x) = \partial \{ \omega_1 * [(1/2) (x + y + z)^3 - (x - y + z)^2 - 1/2 (x +$$

$$y + z) + 1] + \omega_2 * [-(1/6) (x + y + z)^3 + (x + y + z)^2 - (11/6) (x + y + z) + 1] \} / \partial x =$$

$$\{ \omega_1 * [(1/2) \partial (x + y + z)^3 / \partial x - \partial (x + y + z)^2 / \partial x - 1/2 \partial (x + y + z) / \partial x] + \omega_2 * [-(1/6)$$

$$\partial (x + y + z)^3 / \partial x + \partial (x + y + z)^2 / \partial x - (11/6) \partial (x + y + z) / \partial x] \} =$$

$$\{ \omega_1 * [(3/2) (x + y + z)^2 - 2 (x + y + z) - 1/2] + \omega_2 * [-(3/6) (x + y + z)^2 + 2 (x + y + z)$$

$$- (11/6)] \} \tag{5}$$

$$(\partial^2 (LGR_3(x, y, z)) / \partial x^2) = (\partial (\partial (LGR_3(x, y, z)) / \partial x) / \partial x) = \partial \{ \omega_1 * [(3/2) (x + y + z)^2$$

$$- 2 (x + y + z) - 1/2] + \omega_2 * [-(3/6) (x + y + z)^2 + 2 (x + y + z) - (11/6)] \} / \partial x =$$

$$\{ \omega_1 * [3 (x + y + z) - 2] + \omega_2 * [- (x + y + z) + 2] \} \tag{6}$$

$$(\partial^2 (LGR_3(x, y, z)) / \partial x \partial y) = \partial \{ \omega_1 * [(3/2) (x + y + z)^2 - 2 (x + y + z) - 1/2] + \omega_2 * [-$$

$$(3/6) (x + y + z)^2 + 2 (x + y + z) - (11/6)] \} / \partial y =$$

$$\{ \omega_1 * [3 (x + y + z) - 2] + \omega_2 * [- (x + y + z) + 2] \} \tag{7}$$

Equation (4) is therefore true because of equations (8) and (9) hereto follow, which were determined for the the trivariate cubic B-Spline interpolation function.

$$(\partial^2 (h_4(x, y, z)) / \partial x^2) = \{ \partial \{ \partial (h_4(x, y, z)) / \partial x \} / \partial x \} =$$

$$(\partial^2 (h_4(x, y, z)) / \partial y^2) = \{ \partial \{ \partial (h_4(x, y, z)) / \partial y \} / \partial y \} =$$

$$(\partial^2 (h_4(x, y, z)) / \partial z^2) = \{ \partial \{ \partial (h_4(x, y, z)) / \partial z \} / \partial z \} =$$

$$\{ \omega_1 * [3 (x + y + z) - 2] + \omega_2 * [- (x + y + z) + 2] \} \tag{8}$$

$$(\partial^2 (h_4(x, y, z)) / \partial x \partial y) = \{ \partial (\partial (h_4(x, y, z)) / \partial y) / \partial x \} =$$

$$(\partial^2 (h_4(x, y, z)) / \partial x \partial z) = \{ \partial (\partial (h_4(x, y, z)) / \partial z) / \partial x \} =$$

$$(\partial^2 (h_4(x, y, z)) / \partial y \partial z) = \{ \partial (\partial (h_4(x, y, z)) / \partial z) / \partial y \} =$$

$$\{ \omega_1 * [3 (x + y + z) - 2] + \omega_2 * [- (x + y + z) + 2] \} \tag{9}$$

CALCULATION OF THE INTENSITY-CURVATURE TERMS: TRIVARIATE CUBIC LAGRANGE

The intensity-curvature term E_o of the trivariate cubic Lagrange function is:

$$E_o = E_o (x, y, z) = \int_0^x \int_0^y \int_0^z f (0, 0, 0) * \{ (\partial^2 (LGR_3 (x, y, z)) / \partial x^2) + (\partial^2 (LGR_3 (x, y, z)) / \partial y^2)$$

$$+ (\partial^2 (LGR_3 (x, y, z)) / \partial z^2) + (\partial^2 (LGR_3 (x, y, z))) / \partial x \partial y) + (\partial^2 (LGR_3 (x, y, z))) / \partial y \partial x) +$$

$$(\partial^2 (LGR_3 (x, y, z))) / \partial x \partial z) + (\partial^2 (LGR_3 (x, y, z))) / \partial z \partial x) + (\partial^2 (LGR_3 (x, y, z))) / \partial y \partial z) +$$

$$(\partial^2 (LGR_3 (x, y, z))) / \partial z \partial y) \} (0, 0, 0) \, dx \, dy \, dz \tag{10}$$

On the basis of equation (4), it can be written that:

$$(\partial^2 (LGR_3(x, y, z))) / \partial x^2) (0, 0, 0) = (\partial^2 (LGR_3(x, y, z))) / \partial y^2) (0, 0, 0) =$$

$$(\partial^2 (LGR_3(x, y, z))) / \partial z^2) (0, 0, 0) = (\partial^2 (LGR_3(x, y, z))) / \partial x \partial y) (0, 0, 0) =$$

$$(\partial^2 (LGR_3(x, y, z))) / \partial y \partial x) (0, 0, 0) = (\partial^2 (LGR_3(x, y, z))) / \partial x \partial z) (0, 0, 0) =$$

$$(\partial^2 (LGR_3(x, y, z))) / \partial z \partial x) (0, 0, 0) = (\partial^2 (LGR_3(x, y, z))) / \partial y \partial z) (0, 0, 0) =$$

$$(\partial^2 (LGR_3(x, y, z))) / \partial z \partial y) (0, 0, 0) =$$

$$\{ \omega_1 * [3 (x + y + z) - 2] + \omega_2 * [- (x + y + z) + 2] \} (0, 0, 0) = \{ - 2 \omega_1 + 2 \omega_2 \} \tag{11}$$

It follows that:

$$E_o (x, y, z) = \int_0^x \int_0^y \int_0^z 9 * f (0, 0, 0) * \{ - 2 \omega_1 + 2 \omega_2 \} =$$

$$= 9 xyz * f (0, 0, 0) * \{ - 2 \omega_1 + 2 \omega_2 \} \tag{12}$$

The intensity-curvature term E_{IN} of the trivariate cubic Lagrange function is:

$$E_{IN} = E_{IN} (x, y, z) = \int_0^x \int_0^y \int_0^z LGR_3(x, y, z) * \{ (\partial^2 (LGR_3(x, y, z)) / \partial x^2) + (\partial^2 (LGR_3(x, y, z))$$

$$/ \partial y^2) + (\partial^2 (LGR_3(x, y, z)) / \partial z^2) + (\partial^2 (LGR_3(x, y, z))) / \partial x \partial y) + (\partial^2 (LGR_3(x, y, z))) / \partial y \partial x)$$

$$+ (\partial^2 (LGR_3(x, y, z))) / \partial x \partial z) + (\partial^2 (LGR_3(x, y, z))) / \partial z \partial x) + (\partial^2 (LGR_3(x, y, z))) / \partial y \partial z) +$$

$$(\partial^2 (LGR_3(x, y, z))) / \partial z \partial y)\} \, dx \, dy \, dz \tag{13}$$

And because of equations (1) and (4):

$E_{IN} = E_{IN}(x, y, z) = \int_0^x\int_0^y\int_0^z 9 * \{ \omega_1 * [(1/2)(x+y+z)^3 - (x+y+z)^2 - 1/2(x+y+z) + 1] + \omega_2 * [-(1/6)(x+y+z)^3 + (x+y+z)^2 - (11/6)(x+y+z) + 1] \} * \{ \omega_1 * [3(x+y+z) - 2] + \omega_2 * [-(x+y+z) + 2] \} \, dx\, dy\, dz$ (14)

Let us solve equation (14):

$E_{IN}(x, y, z) = \int_0^x\int_0^y\int_0^z 9 * \{ \omega_1^2 * [(1/2)(x+y+z)^3 - (x+y+z)^2 - 1/2(x+y-z) + 1] * [3(x+y+z) - 2] + \omega_1\,\omega_2 * [-(1/6)(x+y+z)^3 + (x+y+z)^2 - (11/6)(x+y+z) + 1] * [3(x+y+z) - 2] + \omega_1\omega_2 * [(1/2)(x+y+z)^3 - (x+y+z)^2 - 1/2(x+y+z) + 1] * [-(x+y+z) + 2] + \omega_2^2 * [-(1/6)(x+y+z)^3 + (x+y+z)^2 - (11/6)(x+y+z) + 1] * [-(x+y+z) + 2] \} \, dx\, dy\, dz =$

$\int_0^x\int_0^y\int_0^z 9 * \{ \omega_1^2 * [(3/2)(x+y+z)^4 - 3(x+y+z)^3 - 3/2(x+y+z)^2 + 3(x+y+z)] + \omega_1^2 * [-(x+y+z)^3 + 2(x+y+z)^2 + (x+y+z) - 2] + \omega_1\,\omega_2 * [-(1/2)(x+y+z)^4 + 3(x+y+z)^3 - (33/6)(x+y+z)^2 + 3(x+y+z)] + \omega_1\,\omega_2 * [(1/3)(x+y+z)^3 - 2(x+y+z)^2 + (22/6)(x+y+z) - 2] + \omega_1\omega_2 * [-(1/2)(x+y+z)^4 + (x+y+z)^3 + 1/2(x+y+z)^2 - (x+y+z)] + \omega_1\omega_2 * [(x+y+z)^3 - 2(x+y+z)^2 - (x+y+z) + 2] + \omega_2^2 * [(1/6)(x+y+z)^4 - (x+y+z)^3 + (11/6)(x+y+z)^2 - (x+y+z)] + \omega_2^2 * [-(2/6)(x+y+z)^3 - 2(x+y+z)^2 - (22/6)(x+y+z) + 2] \} =$

$\int_0^x\int_0^y\int_0^z 9 * \{ [(3/2)\omega_1^2 - (1/2)\omega_1\omega_2 - (1/2)\omega_1\omega_2 + (1/6)\omega_2^2] (x+y+z)^4 + [-3\omega_1^2 - \omega_1^2 + 3\omega_1\omega_2 + (1/3)\omega_1\omega_2 + 2\omega_1\omega_2 - \omega_2^2 - (2/6)\omega_2^2] (x+y+z)^3 + [-3/2\omega_1^2 + 2\omega_1^2 - (33/6)\omega_1\omega_2 - 2\omega_1\omega_2 + 1/2\omega_1\omega_2 - 2\omega_1\omega_2 + (11/6)\omega_2^2 + 2\omega_2^2] (x+y+z)^2 + [3\omega_1^2 + \omega_1^2 + 3\omega_1\omega_2 + (22/6)\omega_1\omega_2 - \omega_1\omega_2 - \omega_1\omega_2 - \omega_2^2 - (22/6)\omega_2^2] (x+y+z) + [-2\omega_1^2 - 2\omega_1\omega_2 + 2\omega_1\omega_2 + 2\omega_2^2] \} =$

$$\int_0^x \int_0^y \int_0^z 9 * \{ [(3/2) \omega_1^2 - \omega_1 \omega_2 + (1/6) \omega_2^2] (x + y + z)^4 + [- 4 \omega_1^2 + (16/3) \omega_1 \omega_2 - (8/6)$$

$$\omega_2^2] (x + y + z)^3 + [1/2 \omega_1^2 - (54/6) \omega_1 \omega_2 + (23/6) \omega_2^2] (x + y + z)^2 + [4 \omega_1^2 + (28/6)$$

$$\omega_1 \omega_2 - (28/6) \omega_2^2] (x + y + z) + [- 2 \omega_1^2 + 2 \omega_2^2] \} \tag{15}$$

And equation (15) can be solved considering the knowledge provided in equations (24), (25), (26), (27), (28), (29) for the trivariate cubic B-Spline, which furnishes the values of $\Pi^{(4)}$ $\Pi^{(3)}$ $\Pi^{(2)}$ and $\Pi^{(1)}$. Therefore:

$$E_{IN} (x, y, z) = 9 * \{ [(3/2) \omega_1^2 - \omega_1 \omega_2 + (1/6) \omega_2^2] \Pi^{(4)} + [- 4 \omega_1^2 + (16/3) \omega_1 \omega_2 - (8/6)$$

$$\omega_2^2] \Pi^{(3)} + [1/2 \omega_1^2 - (54/6) \omega_1 \omega_2 + (23/6) \omega_2^2] \Pi^{(2)} + [4 \omega_1^2 + (28/6) \omega_1 \omega_2 - (28/6)$$

$$\omega_2^2] \Pi^{(1)} + [- 2 \omega_1^2 + 2 \omega_2^2] xyz \} \tag{16}$$

EQUATING THE TWO INTENSITY-CURVATURE TERMS BEFORE AND AFTER INTERPOLATION: TRIVARIATE CUBIC LAGRANGE

Let us posit: $E_o (x, y, z) = E_{IN} (x, y, z)$ from equations (12) and (16):

$$E_o (x, y, z) = E_{IN} (x, y, z) = 9 \, xyz * f (0, 0, 0) * \{ - 2 \omega_1 + 2 \omega_2 \} = 9 * \{ [(3/2) \omega_1^2 - \omega_1$$

$$\omega_2 + (1/6) \omega_2^2] \Pi^{(4)} + [- 4 \omega_1^2 + (16/3) \omega_1 \omega_2 - (8/6) \omega_2^2] \Pi^{(3)} + [1/2 \omega_1^2 - (54/6) \omega_1 \omega_2 +$$

$$(23/6) \omega_2^2] \Pi^{(2)} + [4 \omega_1^2 + (28/6) \omega_1 \omega_2 - (28/6) \omega_2^2] \Pi^{(1)} + [- 2 \omega_1^2 + 2 \omega_2^2] xyz \} \tag{17}$$

Let us posit:

$$\Lambda_L = 9 * \{ [(3/2) \omega_1^2 - \omega_1 \omega_2 + (1/6) \omega_2^2] \Pi^{(4)} + [- 4 \omega_1^2 + (16/3) \omega_1 \omega_2 - (8/6) \omega_2^2] \Pi^{(3)}$$

$$+ [1/2 \omega_1^2 - (54/6) \omega_1 \omega_2 + (23/6) \omega_2^2] \Pi^{(2)} + [4 \omega_1^2 + (28/6) \omega_1 \omega_2 - (28/6) \omega_2^2] \Pi^{(1)} +$$

$$[- 2 \omega_1^2 + 2 \omega_2^2] xyz \} \tag{18}$$

So that equation (17) is written as:

$$9 \, xyz * f (0, 0, 0) * \{ - 2 \omega_1 + 2 \omega_2 \} = \Lambda_L \tag{19}$$

And solved hereto:

$$f (0, 0, 0) = \Lambda_L / \{ 9 \, xyz * \{ - 2 \omega_1 + 2 \omega_2 \} \} \tag{20}$$

Where equation (20) gives the value of the signal resilient to the trivariate cubic Lagrange interpolation function.

SUMMARY

This chapter has presented the math developments to obtain the formulas of the signal resilient to interpolation for bivariate and trivariate polynomials: B-Spline (quadratic parametric and cubic), and Lagrange (cubic) without embedding the pixel to re-sample. Thus, two new math models are presented. Specifically, equations that calculate the signal resilient to interpolation with embedding (in chapter 2) and without embedding in chapter 3. The difference in the results obtained quantitatively and qualitatively from the four math models outlined in chapters 2 and 3 will be presented in subsequent chapters of the book. The methodological approach is the same seen in chapter 2. The interpolator needs to have second order derivative not null, and when the second order partial derivatives of the interpolation function are calculated, the equation of the two intensity-curvature terms furnishes through its solution the desired signal resilient to interpolation.

REFERENCES

Ciulla, C. (2009). *Improved Signal and Image Interpolation in Biomedical Applications: The Case of Magnetic Resonance Imaging (MRI)* – Medical Information Science Reference - IGI Global Publisher. Hershey, PA, U.S.A.

CHAPTER 4

PRELIMINARY INVESTIGATION WITH BIVARIATE POLYNOMIALS: QUADRATIC AND CUBIC, WITH AND WITHOUT EMBEDDING

INTRODUCTION

This chapter introduces concepts that are useful to the understanding of the book as a whole. The concepts dealt with are: (i) virtual-shift rotations: they relate to the validation paradigm and discuss the logic behind its rationale, (ii) convolution, (iii) proof of the concept relating to the significance of the signal resilient to interpolation, (iv) curvature, (v) transfer functions, and also along with these, (vi) a new conception of the Pixel Intensity Correction (PIC). This is introduced and validated with a planned experimental session of a quantitative nature, whereas the qualitative nature is presented in chapters 10 and 12 of this book.

As far as the experiments are concerned, in the first instance, this chapter intends to clarify that the transfer function is necessary to scale the convolution of the pixels' intensity and misplacement, which otherwise would make signal reconstruction awkward, if not impossible; moreover it introduces arbitration in the choice of its math form and thus bias in the interpolation error. Thus, in the second instance, we explain, following on along the exploratory research path of the chapter, that the calculation of the transfer function as derived from the curvature of the resilient interpolation formula is capable of removing bias and arbitration, so as to produce interpolation error improvement. This last set of experiments is labelled in table I "Curvature".

The experimental sessions conducted in this chapter shall be three types of signal reconstruction through interpolation: (i) without the use of the transfer function (noTF), (ii) with the use of the transfer function (TF), and (iii) with the use of the transfer function derived from the total curvature of the specific interpolation formula (Curvature). Each type will address classic interpolation and resilient interpolation with and without embedding ('WE').

The transfer functions (TF) are: cos(p), sin(p) and log($|1.0 / p|$); where log is the natural logarithm. Table I summarizes the experimental sessions.

Classic	noTF	TF	Curvature
Resilient	noTF	TF	Curvature
Resilient WE	noTF	TF	Curvature

Table I: Summary of the experimental sessions.

THE VALIDATION PARADIGM

Within the context of the preliminary investigation, the validation of the effectiveness of resilient interpolation, which is herein defined as signal reconstruction using the resilient formula (and therefore the estimation through the signal resilient to interpolation), is carried out versus classic interpolation polynomials: bivariate quadratic, bivariate cubic and bivariate cubic Lagrange formulae. By way of a general statement on the symbols adopted here to identify bivariate classic interpolation formulae let us use the following equation:

$$I(x, y) = f(0, 0) + \xi(x, y) \tag{1}$$

Where $f(0, 0)$ is the notation given to the pixel at the centre of the neighbourhood, which given the

Cartesian location (x, y), is estimated to be the value $I(x, y)$, and $\xi(x, y)$ is the notation adopted for the convolution made of the polynomial coefficients (neighbouring pixels) and polynomial powers.

In this chapter $\xi(x, y)$ can be of three mathematical forms:

Quadratic:

$$\xi(x, y) = \alpha_1 * [- 2a(x + y)^2 + 1/2(a+1)] +$$

$$\alpha_2 * [a(x + y)^2 - (2a + 1/2)(x + y) + 3/4(a+1)] \tag{2}$$

Cubic:

$$\xi(x, y) = \alpha_3 * [1/2(x + y)^3 - (x + y)^2 + 2/3] +$$

$$\alpha_2 * [-1/6(x + y)^3 + (x - y)^2 - 2(x + y) + 4/3] \tag{3}$$

Cubic Lagrange:

$$\xi(x, y) = \alpha_3 * [(1/2)(x + y)^3 - (x + y)^2 - 1/2(x + y) + 1] -$$

$$\alpha_2 * [-(1/6)(x + y)^3 + (x + y)^2 - (11/6)(x + y) + 1] \tag{4}$$

By way of a general statement on the symbols used to identify bivariate resilient interpolation formulae herein, let us use the following equation:

$$R(x, y) = f(0, 0) + \varrho(x, y) \tag{5}$$

Where $f(0, 0)$ indicates the signal at the centre of the neighbourhood, which is estimated through the resilient formula through $R(x, y)$, and $\varrho(x, y)$ is the term that results from the solution of the equation between the two intensity-curvature terms $E_o(x, y)$ (before interpolation) and $E_{IN}(x, y)$ (after interpolation), which is: $E_o(x, y) = E_{IN}(x, y)$.

Classic and resilient interpolation formulae shown in (1) and (5) incorporate the value of the signal which is being reconstructed ($f(0, 0)$). The term embedding is hereonin defined as the property of the interpolation function to incorporate in its math form the value of the signal which is being reconstructed: $f(0, 0)$.

The value $\varrho(x, y)$ results from the solution of the equation $E_o(x, y) = E_{IN}(x, y)$ As shall be clarified in the section "CONCEPTUALIZATION OF THE ISSUES IN THE VALIDATION PARADIGM", $\varrho(x, y)$ is the numerical value of the fraction of pixel intensity to be estimated at (x, y).

The value $f(0, 0)$ ($\xi(x, y)$ and/or $\varrho(x, y)$) will be seen in equations (16) with reference to the formulation with embedding, and in (20) and (21) with reference to the formulation without embedding. It is named as $f(0, 0)$ because it is obtained from the solution of the equation $E_o(x, y) = E_{IN}(x, y)$, nevertheless it will be treated as a fraction of pixel intensity and equated to $\xi(x, y)$ in order to calculate the estimate $I(x, y) = f(0, 0) + \xi(x, y)$ (classic interpolation) or equated to $\varrho(x, y)$ in order to calculate the estimate $R(x, y) = f(0, 0) + \varrho(x, y)$ (resilient interpolation). The reader will appreciate that there is a neat difference between the value $f(0, 0)$ in equations (1) and (5) and the value $f(0, 0)$ in equations (16), (20) and (21). The value of $f(0, 0)$ in equations (1) and (5) is the term embedded in the math form of the classic and resilient interpolation formulae respectively. The value $f(0, 0)$ in equations (16) (classic), and (20) and (21) (resilient) is the fraction of pixel intensity resulting from the solution of equation $E_o(x, y) = E_{IN}(x, y)$ and equated to $\xi(x, y)$ of $I(x, y) = f(0, 0) + \xi(x, y)$ (classic interpolation) or equated to $\varrho(x, y)$ of $R(x, y) = f(0, 0) + \varrho(x, y)$ (resilient interpolation). The concept also applies to 3D for the value $f(0, 0, 0)$. With the aim of determining $\varrho(x, y)$ with embedding, the interpolation formulae employed when solving the equation $E_o(x, y) = E_{IN}(x, y)$ are of the form identified in (1); quadratic, cubic and cubic Lagrange formulae respectively:

$$h_3(x, y) = f(0, 0) + \alpha_1 * [- 2a(x + y)^2 + 1/2(a+1)] + \alpha_2 * [a(x + y)^2 - (2a + 1/2)(x + y)$$

$$+ 3/4(a+1)] \tag{6}$$

$$h_4(x, y) = f(0, 0) + \alpha_3 * [1/2(x + y)^3 - (x + y)^2 + 2/3] + \alpha_2 * [-1/6(x + y)^3 + (x + y)^2 - 2$$

$$(x + y) + 4/3] \tag{7}$$

$$LGR_3(x, y) = f(0, 0) + \alpha_3 * [(1/2)(x + y)^3 - (x + y)^2 - 1/2(x + y) + 1] + \alpha_2 * [-(1/6)(x$$

$$+ y)^3 + (x + y)^2 - (11/6)(x + y) + 1] \tag{8}$$

Whereas, with the aim of determining $\varrho(x, y)$ without embedding, the interpolation formulae employed when solving the equation $E_o(x, y) = E_{IN}(x, y)$ are:

$$h_3(x, y) = \alpha_1 * [- 2a(x + y)^2 + 1/2(a+1)] +$$

$$\alpha_2 * [a(x + y)^2 - (2a + 1/2)(x + y) + 3/4(a+1)] \tag{9}$$

$$h_4(x, y) = \alpha_3 * [1/2(x + y)^3 - (x + y)^2 + 2/3] +$$

$$\alpha_2 * [-1/6(x + y)^3 + (x + y)^2 - 2(x + y) + 4/3] \tag{10}$$

$$LGR_3(x, y) = \alpha_3 * [(1/2)(x + y)^3 - (x + y)^2 - 1/2(x + y) + 1] +$$

$$\alpha_2 * [-(1/6)(x + y)^3 + (x + y)^2 - (11/6)(x + y) + 1] \tag{11}$$

INTRODUCTORY QUESTIONS

The signal resilient to the bivariate quadratic and cubic polynomial interpolation functions is convoluted as per defining formulae, which are derived through the math processes detailed in chapters 2 and 3. However the signal results from convolutions that do or do not embed the value of the signal which is being reconstructed and this establishes the difference between resilient interpolation with and without embedding. In the specific case of the two bivariate interpolation functions which will be the object of discussion, the term embedded is indicated with $f(0, 0)$ (see equations (6), (7) and (8)).

Negative values of the ratio (1-Classic/Resilient) indicate the percentage improvement of the resilient paradigm over the classic paradigm, and when positive, the ratio indicates the opposite. The ratio is calculated on the basis of the two values of Mean Absolute Error (MAE). Thus, in the formula (1-Classic/Resilient), 'Classic' means the MAE of the classic formula, and 'Resilient' means the MAE of the resilient formula.

The bivariate quadratic formula which calculates the signal resilient to interpolation with embedding is:

$$f(1/2, 1/2) = - \{ \beta_1 * \gamma_1 + \beta_2 * \gamma_2 \} / \{ \gamma_1 + \gamma_2 \} \tag{12}$$

The questions that relate to the approach to signal reconstruction herein called resilient interpolation are: *(i) what is the meaning of the signal resilient to interpolation?* And *(ii) is the signal resilient to interpolation related to the estimation of the true value of the unknown signal at the location (x, y)?*

While validating the concept of resilient interpolation questions (i) and (ii) require an answer. The validation methodology employs the concept of virtual shift-rotation as explained in the paragraph entitled: "VIRTUAL SHIFT-ROTATION". Through an investigative process that poses the problem on the basis of the results obtained and questions the logical root of the concepts behind the results, questions (i) and (ii) are addressed in the paragraph entitled: "CONCEPTUALIZATION OF THE ISSUES IN THE VALIDATION PARADIGM" and in subsequent sections of the book

Now, do we need to virtually shift back the signal after the virtual shift forwards? If we do so, then interpolation is subject to interpolation error twice: virtually shifting forwards and virtually shifting backwards. Is the resilient interpolation formula producing a virtual shift? The answer to this question is "yes" because the resilient interpolation formula determines signal convolution.

And since the convolution alone produces virtual shift-rotation, as stated in the paragraph entitled: "VIRTUAL SHIFT-ROTATION", this means that resilient interpolation is virtually shifting the signal. Therefore we do not need to virtually shift back the signal after the virtual shift forwards while employing the resilient interpolation formula. Neither is it necessary to virtually shift back the signal while employing classic interpolation because it determines signal convolution too.

An overview of the evolution of the theory and the applications of interpolation functions is provided in Ciulla (2009) and a list of authors is given there (Agarwal & Wong, 1993; Blu et al., 2001, 2004; Blu & Unser, 1999; De Boor, 1978; Grevera & Udupa, 1996; Herman et al., 1992; 1998; Newton & Huygens, trans. 1934; Raya & Udupa, 1990; Schoenberg, 1946a, 1946b, 1969; Unser et al., 1993a, 1993b; Waldron, 1998). The aims and scopes of this book diverge from previous research in this field because of the effort provided here to detach interpolation techniques from the use of the arbitrary transfer function to process the convolution produced through the interpolation formula.

In conclusion, the validation paradigm is acceptable when virtually shifting forwards while employing both classic and resilient interpolation formulae, and avoids virtually shifting backwards in either of the two formulations. This way, the advantage is two-fold: (a) to avoid interpolation error twice: the first time while virtually shifting forwards and the second time while virtually shifting backwards; and (b) to avoid the shift (or the shift-rotation) of the grid. The interpolation error determined therefore only accounts for the virtual shift (or the virtual shift-rotation) forwards.

The validation paradigm adopted in this book constitutes of a conceptual improvement of that one used to validate the unifying theory (Ciulla, 2009) because it does not use the virtual shift backward. The choice made to use the validation paradigm which uses both of virtual shift forward and virtual shift backward is justified through the its admissible congruency which is the admissibility provided through the rationality of the underlying logic.

VIRTUAL SHIFT-ROTATION

Polynomials and trigonometric convolutions (bivariate linear, trivariate linear, B-Splines, Lagrange, Sinc) do not produce any real shift-rotation. In order to shift-rotate signals the following process is needed: (i) shift-rotation of the grid, (ii) interpolation. If the grid is not shifted-rotated, then interpolation alone, through convolutions, does not suffice to produce any real shift-rotations. Therefore, convolutions do provide the interpolated signal at locations (x, y) or (x, y, z), however this signal is not shifted-rotated. *Thus, the convolution produces virtual shift-rotation onto locations (x, y) or (x, y, z), where the signal assumes the value obtained through its calculation.*

The measure of the interpolation error, called in this book the Mean Absolute Error (MAE), is defined as the absolute value of the difference between the value of the signal at the grid point minus the value of the estimate being made (the reconstructed signal, which is still not necessarily the true value of the signal because such a value has not been sampled and thus remains unknown). Because of this definition of the MAE, the assumption that the true value of the signal which is being reconstructed is the one signal value at the grid point is explicitly demanded. Why should this assumption be so, in explicit terms?

To illustrate the practical implications of the assumption let us consider a single dimensional signal, for instance in the variable x. The assumption demands that in between two adjacent nodes, the true value of the signal is exactly the same as the value of the signal at the grid point from which the misplacement included in the math form of the convolution which is determinant of the virtual shift departs.

Is this assumption made in such a way as to have a logical rationale? Yes - the rationale is of a theoretical nature, given that interpolation, as seen in chapter 1, is employed in up-sampling. Additionally, as seen in chapter 1, the assumption allows the use of the Mean Absolute Error (MAE) as the measure of the interpolation error and the use of the MAE is related both to the search for minimal energy change and so the search for minimal change in curvature. In other words, the MAE provides with the assumption of the logical rationale. The advantage of the virtual shift-rotation is that of avoiding the processing needed to shift-rotate the grid of the signal (image). The virtual shift-rotation was the choice made to validate the SRE interpolation functions and this choice is now made once again in order to validate resilient interpolation.

What is to be ascertained is the following: which of the virtually shifted-rotated signals estimated through classic interpolation or resilient interpolation at the location (x, y) (or (x, y, z) in three dimensions) is more accurate (the closest possible to the true and unknown value).

CONCEPTUALIZATION OF THE ISSUES IN THE VALIDATION PARADIGM

This discussion starts to address the issue of the transfer functions within the context of some signal reconstruction techniques relevant to the topics covered in this book and also introduces the effect of embedding the pixel to be re-sampled.

The transfer function is usually employed to process the convolution resulting from the interpolation formula. The effect of the transfer function is that of improving signal reconstruction through the interpolation formula and is visible in the residual images where the interpolation error is clearly minimized. Since the transfer function can be arbitrary in its mathematical form, it certainly affects the results of signal reconstruction, and thus it might become a bias when comparing the results of two interpolation formulae.

Removing the transfer function from the interpolation formula produces the reconstructed signal, convoluted as per the equation defining the formula. In the specific case of the classic bivariate quadratic and cubic polynomial interpolation functions the reconstructed signal is obtained from equations (13) and (14), whose math form is consistent with the generalized form of equation (1).

$$h_3(x, y) = f(0, 0) + \alpha_1 * [- 2a(x + y)^2 + 1/2(a+1)] +$$

$$\alpha_2 * [a(x + y)^2 - (2a + 1/2)(x + y) + 3/4(a+1)] \qquad (13)$$

$$h_4(x, y) = f(0, 0) + \alpha_3 * [1/2(x + y)^3 - (x + y)^2 + 2/3] +$$

$$\alpha_2 * [-1/6 (x + y)^3 + (x + y)^2 - 2 (x + y) + 4/3] \tag{14}$$

The effect of removing the transfer function to reconstruct the signal through polynomial interpolation functions is visible in the error images where the residual is substantially high.

When approaching the validation of resilient signal reconstruction with embedding, let us recall that from equation (1): $I (x, y) = f (0, 0) + \xi (x, y)$, the term $\xi (x, y)$ (called f (0, 0 in equation (16)) indicates the convolution made of the polynomial coefficients (neighbouring pixels) and polynomial powers in classic interpolation.

Also, from equation (5): $R (x, y) = f (0, 0) + \varrho (x, y)$, the term $\varrho (x, y)$ (called f (0, 0 in equations (20) and (21)) is derived from the solution of the equation which poses the equality between the two intensity-curvature terms: $E_o (x, y)$ (before interpolation) and $E_{IN} (x, y)$ (after interpolation).

Now, both $\xi (x, y)$ and $\varrho (x, y)$ are convolutions and this can be seen in equations (2), (3) and (4) ($\xi (x, y)$) and (15) and (16) ($\varrho (x, y) = f (1/2, 1/2)$ and $\varrho (x, y) = f (0, 0)$ respectively), thus they all share the same nature. The species of $\xi (x, y)$ is not different from the species of $\varrho (x, y)$. This is because $\xi (x, y)$ and $\varrho (x, y)$ are both additive to the pixel intensity to be estimated through the calculation of I (x, y) and R (x, y) respectively (see equations (1) and (5)), they are thus numerical values of the fraction of pixel intensity to be estimated at (x, y).

Equations (15) and (16) furnish the bivariate quadratic and cubic polynomial resilient signals obtained through the math process that includes in $E_o (x, y) = E_{IN} (x, y)$ equations (13) and (14) respectively.

$$f (1/2, 1/2) = - \{ \beta_1 * \gamma_1 + \beta_2 * \gamma_2 \} / \{ \gamma_1 + \gamma_2 \} \tag{15}$$

$$f (0, 0) = \Lambda / \{ 4xy * [- 2 \alpha_3 + 2 \alpha_2] - 4\alpha_3 * [3 (x^2y/2 + xy^2/2) - 2xy] + 4\alpha_2 * [- (x^2y/2 -$$

$$xy^2/2) + 2xy] \} \tag{16}$$

Where Λ is defined as

$$\Lambda = 4 * \{ [3/2 \alpha_3^2 - 1/2 \alpha_3 \alpha_2 - 3/5 \alpha_3 \alpha_2 + 1/6 \alpha_2^2] [x^5y/5 + 3/8 x^4y^2 + x^3y^3/3 + x^2y^4/8 +$$

$$x^4y^2/8 + x^3y^3/3 + 3/8 x^2y^4 + xy^5/5] + [4 \alpha_3^2 + 2 \alpha_3 \alpha_2 + 4/3 \alpha_3 \alpha_2 - 4/3 \alpha_2^2] [x^4y/4 +$$

$$x^3y^2/2 + x^2y^3/2 + xy^4/4] + [2 \alpha_3^2 - 2 \alpha_3 \alpha_2 - 8 \alpha_3 \alpha_2 + 4 \alpha_2^2] [x^3y/3 + x^2y^2/2 + xy^3/3] +$$

$$[2 \alpha_3^2 - 2/3 \alpha_3 \alpha_2 + 8 \alpha_3 \alpha_2 - 16/3 \alpha_2^2] [x^2y/2 + xy^2/2] + [- 4/3 \alpha_3^2 + 4/3 \alpha_3 \alpha_2 - 8/3 \alpha_3 \alpha_2$$

$$+ 8/3 \alpha_2^2] [xy] \} \tag{17}$$

The fact that $\xi (x, y)$ and $\varrho (x, y)$ are both fractions of the pixel intensity to be estimated at (x, y), allows classic and resilient interpolation to share the same basis of thought. This issue needs to be addressed when validating resilient signal reconstruction. Figure 1 shows an illustration of the geometry of $\xi (x, y)$ and $\varrho (x, y)$, the line of the square triangle opposite to the angle called θ.

Another way of thinking of the aforementioned concept, which is shared in both classic and resilient interpolation is described as follows, and remains valid also when classic signal reconstruction is performed without embedding the pixel to re-sample f (0, 0). The concept is valid and shared in both classic and resilient interpolation because: (a) the signal reconstructed through the classic polynomial interpolation function is convoluted and at the same time embeds the signal which is being reconstructed (which is f (0, 0) and is termed Signal(x, y) in equations (18) and (19)).

In the specific case of the bivariate quadratic and cubic polynomials, equations (13) and (14) are calculated as:

$$\text{shifted\_Signal(x, y)} = \text{Signal(x, y)} + H3\_2D\_1 + H3\_2D\_2 \tag{18}$$

$$\text{shifted\_Signal(x, y)} = \text{Signal(x, y)} + H4\_2D\_1 + H4\_2D\_2 \tag{19}$$

Where the terms H3_2D_1 and H3_2D_2 are those seen after f (0, 0) in equation (13); and terms H4_2D_1 and H4_2D_2 are those seen after f (0, 0) in equation (14). Terms H3_2D_1, H3_2D_2, H4_2D_1 and H4_2D_2 are convolved, that is to say: processed (scaled) with the transfer function.

In the classic formulae (18) and (19) the signal estimate I(x, y) is a function of f (0, 0) (Signal(x, y)) and a fraction value furnished through the convolution terms (H3_2D_1, H3_2D_2 in (18) H4_2D_1, H4_2D_2 in (19)). Thus, for a matter of clarity as much as formalism equations (18) and (19) can be generalized with:

$$I (x, y) = f (0, 0) + TF(\xi (x, y)) \tag{19.a}$$

The concept is valid also because (b): the bivariate resilient signals are obtained through the math process, which includes into the equation $E_o (x, y) = E_{IN} (x, y)$ equations (13) and (14) in the case of quadratic and cubic polynomials respectively, and arrives at the formulations (15) and (16). In the resilient formulae given in equations (15) and (16) the estimations f(1/2, 1/2) and f(0, 0) are fractions of pixel intensity and they are called from now on $\varrho (x, y)$, which is the term to be added to the value of the signal that is being reconstructed (f(0, 0) as shown in (5)).

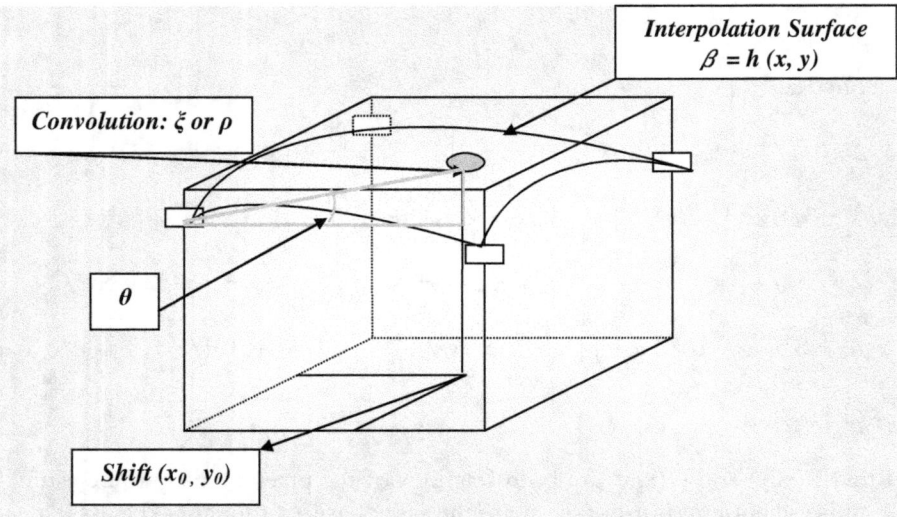

Figure 1: The geometrical meaning of the fractions (convolutions: $\xi (x, y)$ or $\varrho (x, y)$) of pixel intensity estimated through classic and resilient interpolation formulae.

Let us see now why the concept is still valid when the resilient signal reconstruction is calculated without embedding the pixel to re-sample $f(0, 0)$, however does not share the same conceptual basis with signal reconstruction performed with embedding. When calculating $\varrho(x, y)$ without embedding, the interpolation formulae employed while solving the equation $E_o(x, y) = E_{IN}(x, y)$ are given in equations (9) and (10), quadratic and cubic respectively, which are repeated below:

$$h_3(x, y) = \alpha_1 * [- 2a (x + y)^2 + 1/2 (a+1)] +$$

$$\alpha_2 * [a (x + y)^2 - (2a + 1/2) (x + y) + 3/4 (a+1)] \tag{20.a}$$

$$h_4(x, y) = \alpha_3 * [1/2 (x + y)^3 - (x + y)^2 + 2/3] +$$

$$\alpha_2 * [-1/6 (x + y)^3 + (x + y)^2 - 2 (x + y) + 4/3] \tag{20.b}$$

Equations (9) and (10) are the formulae from which we derive the signal resilient to interpolation, as per equations (21.a) and (21.b) repeated below (quadratic and cubic polynomial respectively):

$$f(0, 0) = \{\alpha_1 * [- 2a (x^2/3 + xy/2 + y^2/3) + \frac{1}{2} (a+1)] +$$

$$\alpha_2 * [a (x^2/3 + xy/2 + y^2/3) - (2a + 1/2) (x/2 + y/2) + 3/4 (a+1)]\} \tag{21.a}$$

$$f(0, 0) = \Lambda / \{4xy * [- 2 \alpha_3 + 2 \alpha_2]\} \tag{21.b}$$

Where:

$$\Lambda = 4 * \{[3/2 \alpha_3^2 - 1/2 \alpha_3 \alpha_2 - 3/6 \alpha_3 \alpha_2 + 1/6 \alpha_2^2] [x^5y/5 + 3/8 x^4y^2 + x^3y^3/3 + x^2y^4/8 +$$

$$x^4y^2/8 + x^3y^3/3 + 3/8 x^2y^4 + xy^5/5] + [-4 \alpha_3^2 + 2 \alpha_3 \alpha_2 + 4/3 \alpha_3 \alpha_2 - 4/3 \alpha_2^2] [x^4y/4 +$$

$$x^3y^2/2 + x^2y^3/2 + xy^4/4] + [2 \alpha_3^2 - 2 \alpha_3 \alpha_2 - 8 \alpha_3 \alpha_2 + 4 \alpha_2^2] [x^3y/3 + x^2y^2/2 + xy^3/3] +$$

$$[2 \alpha_3^2 - 2/3 \alpha_3 \alpha_2 + 8 \alpha_3 \alpha_2 - 16/3 \alpha_2^2] [x^2y/2 + xy^2/2] + [-4/3 \alpha_3^2 + 4/3 \alpha_3 \alpha_2 - 8/3 \alpha_3 \alpha_2$$

$$+ 8/3 \alpha_2^2] [xy]\} \tag{22}$$

Both equations (21.a) and (21.b) are derived from the math deduction which starts in either of the two cases from the classic forms of quadratic and cubic polynomial interpolation functions without embedding (equations (9) and (10)). The resilient formulae (21.a) and (21.b) provide in either case fractional values of the signal estimate $R(x, y)$, which are termed $\varrho(x, y)$ in equation (5): $R(x, y) = f(0, 0) + \varrho(x, y)$. As mentioned earlier, $\varrho(x, y)$ is of the same species as $\xi(x, y)$, which is the fraction value of the pixel intensity estimated through the classic interpolation paradigm.

However, equations (20.a) and (20.b) does not share the same conceptual basis with equations (13) and (14) because they do not have $f(0, 0)$ embedded in them. Thus, the math deduction of both classic and resilient interpolation without embedding does not share the same conceptual basis with the relevant interpolation paradigms with embedding. Nevertheless, the validation of classic and resilient signal reconstruction techniques with and without embedding can be theoretically consistent (sharing the same basis of thought) because of the generalized expressions of classic and resilient signal reconstruction given through the equations: $I(x, y) = f(0, 0) + TF(\xi(x, y))$ and $R(x, y) = f(0, 0)$

+ ϱ (x, y) respectively. To summarize, the following two issues are raised: (i) The transfer function might become a bias because of the arbitration of its mathematical form. (ii) Removing the transfer function determines signal reconstruction through polynomial interpolation with a substantially increased residual (high interpolation error).

PROOF OF CONCEPT

Quadratic classic and resilient signal reconstruction formulae were the object of study in this section, which shows quantitatively the following findings relevant to the bivariate quadratic paradigms: (i) It is possible to obtain signal reconstruction with resilient interpolation (with and without embedding). (ii) It is highly recommended that one should employ the transfer function in order to reduce considerably the interpolation error. Results presented in figures 2 through 4 had the value *'the_A_const'* set equal to -0.32 (with embedding) and equal to -0.254 (without embedding); *'the_A_const'* is the constant parameter of classic and resilient quadratic polynomials.

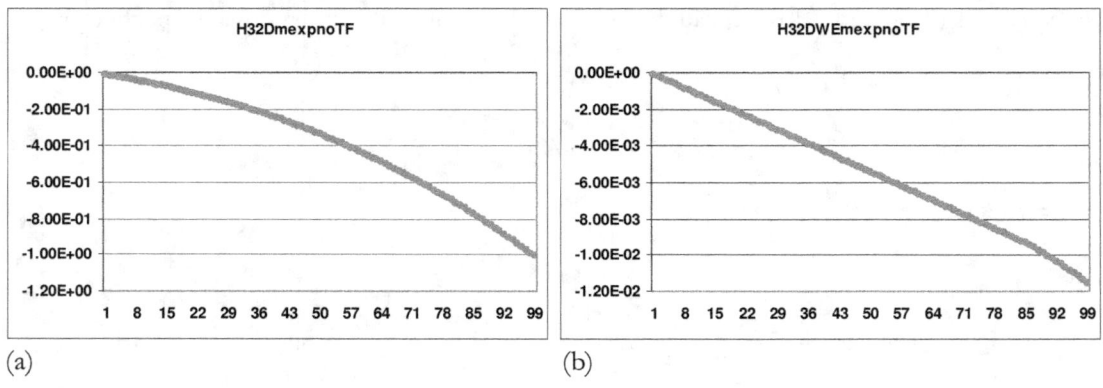

(a)   (b)

Figure 2: Ratio (1- Classic/ Resilient) relevant to the comparison of the classic quadratic formula versus resilient interpolation: (a) with embedding, (b) without embedding ('WE'). The signal resilient to interpolation was calculated as R (x, y) = f (0, 0) + ϱ (x, y). The transfer function was not employed.

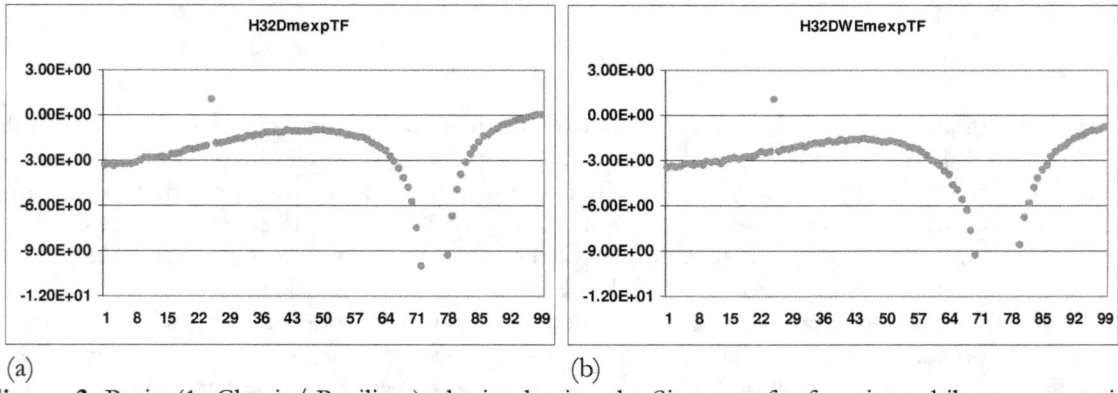

(a)   (b)

Figure 3: Ratio (1- Classic/ Resilient) obtained using the Sinc transfer function while reconstructing the signal with the classic formula and also with the resilient formula calculated as R (x, y) = f (0, 0) + ϱ (x, y). (a) With embedding, and (b) without embedding ('WE').

The ratio (1-Classic/Resilient) numerically equates to one minus the MAE of the classic formula over the MAE of the resilient formula. Data in figure 2 show that the classic quadratic formula is less accurate than the resilient formula (with and without embedding, see (a) and (b) respectively) across the full range of misplacements: (x, y) from [0.01, 0.01] to [0.99, 0.99] (steps of [0.01, 0.01]).

Data in figure 3 also show the visible superiority of signal reconstruction with resilient interpolation compared to the classic formula. This superiority tends to diminish and almost vanish in (a) when the misplacement approaches the value of [0.99, 0.99]. In both graphs (a) and (b) there exists an odd-ball point corresponding to the misplacement [0.25, 0.25]. Tables II and III report values of the Ratio not shown in figure 3, in order to improve visibility.

H32DmexpTF	
(x, y)	(1-Classic/Resilient)
(0.73, 0.73)	-1.551 E+01
(0.74, 0.74)	-3.066 E+01
(0.75, 0.75)	-6.024 E−05
(0.76, 0.76)	-3.032 E−01
(0.77, 0.77)	-1.487 E+01

Table II: Values of the Ratio not shown in figure 3.a

H32DWEmexpTF	
(x, y)	(1-Classic/Resilient)
(0.71, 0.71)	-1.176 E+01
(0.72, 0.72)	-1.571 E+01
(0.73, 0.73)	-2.430 E+01
(0.74, 0.74)	-4.855 E+01
(0.75, 0.75)	-8.620 E+05
(0.76, 0.76)	-4.699 E+01
(0.77, 0.77)	-2.293 E+01
(0.78, 0.78)	-1.490 E+01
(0.79, 0.79)	-1.107 E+01

Table III: Values of the Ratio not shown in figure 3.b.

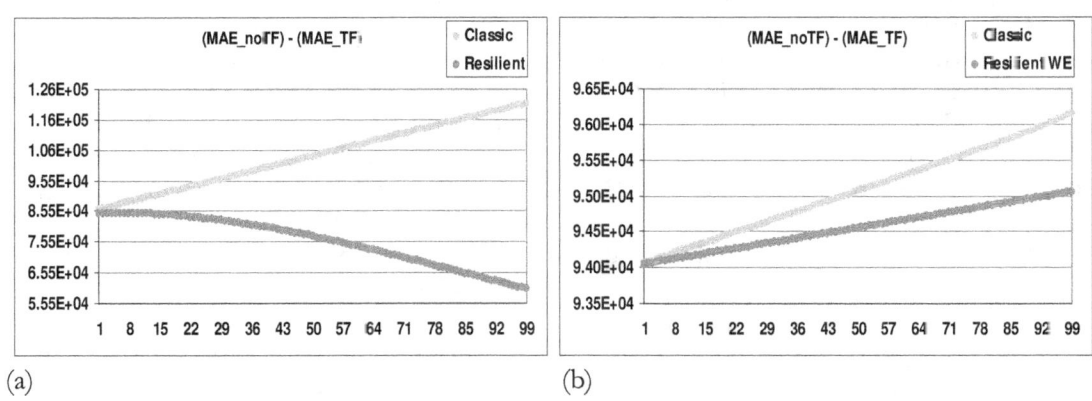

(a)　　　　　　　　　　　　　　(b)

Figure 4: Lines that indicate the value of ((MAE_noTF) - (MAE_TF)) versus the misplacement ranging from (x, y) = [0.01, 0.01] to (x, y) = [0.99, 0.99] (steps of [0.01, 0.01]). ((MAE_noTF) - (MAE_TF)) indicates the difference between the mean absolute error without the use of the transfer function (MAE_noTF) and the mean absolute error with the use of the transfer function (MAE_TF)). (a) With embedding, and (b) without embedding ('WE').

Data in figure 4 are clearly in favor of the use of the transfer function to process the convolution of the classic interpolation formula and also in favor of the signal resilient to interpolation calculated with: R (x, y) = f (0, 0) + ϱ (x, y); both with and without embedding. The transfer function employed in this set of experiments was the Sinc function.

STUDY OF TRANSFER FUNCTIONS

The convolutions of classic and resilient formulae calculated as ξ (x, y) and ϱ (x, y) respectively are processed through the use of the transfer function and so the signal is reconstructed. The residual observable is minimised, signal reconstruction is improved, while the transfer function tends to minimize the numerical value of the convolution calculated through the interpolation formula.

It would be reasonable to argue that this approach to signal reconstruction is accurate. There is need of further discussion though. There are two main reasons.

In the first instance the effect of the transfer function, which tends to make the numerical value of the convolution vanish, is debatable. This reasoning applies to both classic and resilient reconstruction approaches and raises the following two questions. What would signal reconstruction through interpolation without the transfer function be?

The answer is immediately provided in figure 4 where the non-beneficial effect of the lack of transfer function is visible. In other words, convolutions need to be made to behave in such a way that their resulting value acts in favour of improving the approximation property of the interpolation function.

The second question is: *what transfer function should be used?* This question raises an issue that relates to arbitration of the math form of the function, which might be solved with the use of the transfer function derived from the curvature of the specific interpolation formula. This possibility will be addressed in this chapter in the section entitled: "DERIVATION OF THE TRANSFER FUNCTION FROM THE MATH FORM OF THE INTERPOLATION FORMULA".

To discuss this concept further, consider three functions: cos(p), sin(p) and log($|$1.0 / p$|$). Given 'p' as the value of the numerical convolution which in relationship to the conceptualization proposed in what has been written here can either be ξ (x, y) or ϱ (x, y), classic and resilient formulae respectively (see equations (1) and (5)). Our task is to choose the suitable transfer function.

A natural answer to this question might be that of choosing the transfer function which is able to insure best signal reconstruction, which then corresponds to zero interpolation error. To have zero interpolation error is, however, limited to those cases which relate to interpolating f (0, 0) within a given neighbourhood with all of the pixels' intensities having the same value f (0, 0). In this case, regardless of the misplacement (x, y), the value of the pixel intensity ought to be estimated as f (0, 0).

It is then fair to assert that *an interpolation function, while interpolating f (0, 0) at (x, y) in a given neighbourhood with pixels having diverse intensity values ought not to release f (0, 0) as the estimate*, because regardless of the model interpolation function, the local curvature of the model function demands that the value of the intensity I (x, y) or R(x, y) to be estimated is not the same as f (0, 0).

The aforementioned assertion can then lead us to think carefully about how to choose a transfer function and the reason consists in the fact that the transfer function, while processing the convolution (which can be either ξ (x, y) or ϱ (x, y), already dealt with in this chapter), can be capable of almost vanishing or even nullifying its value and therefore releasing as an estimate I (x, y) = f(0, 0) (R(x, y) = f(0, 0)) in the presence of the neighbourhood with pixels having diverse intensity values, thus the practical effect of the transfer function in such a case would be that of neglecting the curvature of the model interpolation function. This would be a conceptual incongruence which leads to an incorrect signal reconstruction. The transfer function thus models the curvature of the interpolator.

The following set of experiments was conducted employing bivariate polynomials: B-Spline Quadratic and Cubic, and Cubic Lagrange. These three functions' convolutions, each in the form: classic and resilient (with embedding and without embedding), were processed with cos(p), sin(p) and log($|$1.0 / p$|$) (for a total of eighteen experimental sessions).

The image tested is shown in figure 5. The signal was reconstructed applying virtual shifts along X and Y concurrently in the range between [(0.01, 0,01), (0.01, 0.01)] and [0.99, 0.99), (0.99, 0.99)] along the diagonal of the pixel and the shifts are displayed on the abscissa of figures 6 through 9.

The ratio (1-Classic/Resilient) is displayed in the graphs in figures 6 through 9, showing that there exist differences between classic and resilient signal reconstruction, differences between resilient interpolation with and without embedding, and differences in the reconstructed signal depending on the chosen transfer function among the three studied: sin(p), cos(p) and log(|1.0 / p|).

Figure 5: In the image employed the pixel intensity values range in [0, 32757].

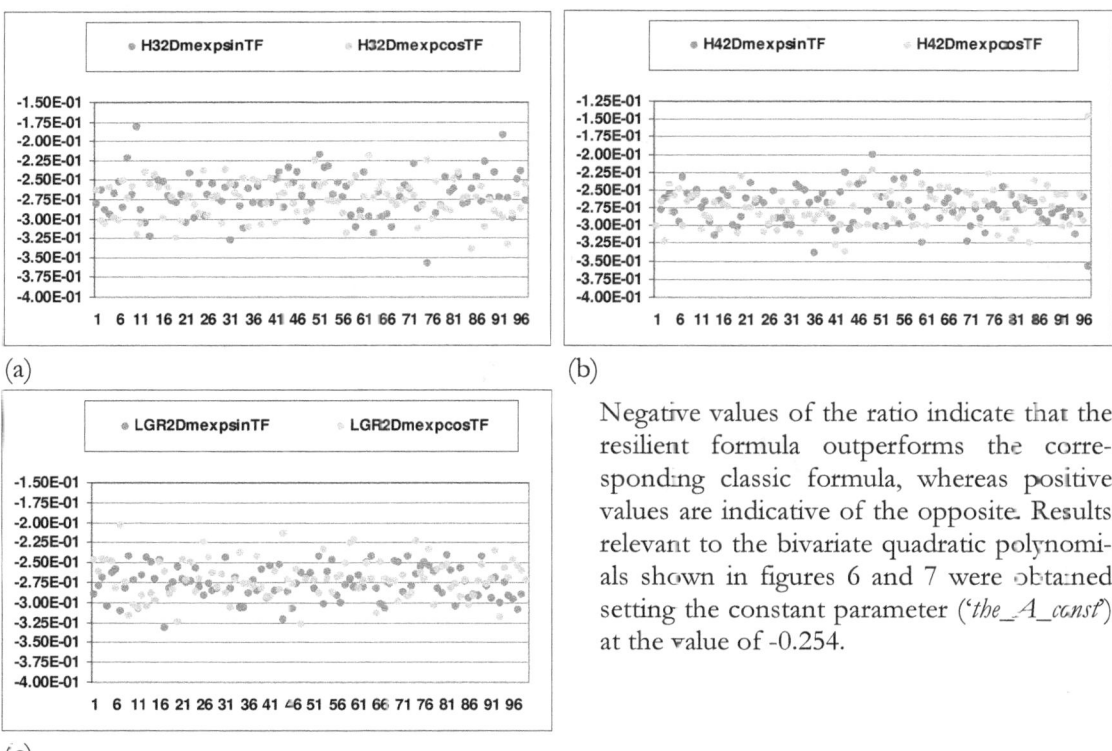

(a)

(b)

(c)

Negative values of the ratio indicate that the resilient formula outperforms the corresponding classic formula, whereas positive values are indicative of the opposite. Results relevant to the bivariate quadratic polynomials shown in figures 6 and 7 were obtained setting the constant parameter ('*the_A_const*') at the value of -0.254.

Figure 6: The ratio (1-Classic/Resilient) resulting from experimental sessions testing bivariate classic and resilient interpolation formulae. The ratio shown in (a), (b), and (c) is relevant to quadratic, cubic and cubic Lagrange formulae respectively. The transfer functions are sin(p) and cos(p).

Furthermore, the suffix 'WE' corresponds to the resilient functions calculated without embedding. In the absence of the suffix 'WE' the resilient function is calculated with embedding.

In what follows, each experiment will be labelled with an identifier. Bivariate quadratic, cubic and cubic Lagrange polynomials are identified with H32D, H42D and LGR2D respectively. The identifiers mexpsinTF, mexpcosTF, mexplogTF correspond to the transfer functions (TF): sin(p), cos(p) and log(|1.0 / p|), respectively. For instance the identifier H32DmexpsinTF labels the bivariate quadratic classic and resilient interpolation formulae processed with the sin transfer function.

In figures 6 and 7 the resilient formula was calculated with embedding. Results indicate that the resilient formula (R (x, y) = f (0, 0) + ϱ (x, y)) outperforms the classic formula (I (x, y) = f (0, 0) + ξ (x, y)). Some values of the ratio corresponding to misplacements along X and Y are given in tables IV, V and VI.

The values located in tables IV and V are relevant to the comparison of the bivariate quadratic classic interpolation formula with the resilient interpolation formulae calculated with embedding (table IV) and without embedding (table V).

Results shown in figures 8 and 9 are relevant to the ratio obtained with bivariate classic and resilient interpolation formulae calculated without embedding ('WE'). The value of the ratio indicates that the resilient formula produces a smaller interpolation error when compared with the classic formula, and these findings are consistent with those shown in figures 6 and 7.

In table VI the two polynomial comparisons are bivariate cubic classic Lagrange interpolation formulae versus resilient interpolation formulae calculated with embedding (LGR2D) and resilient interpolation formulae calculated without embedding (LGR2DWE).

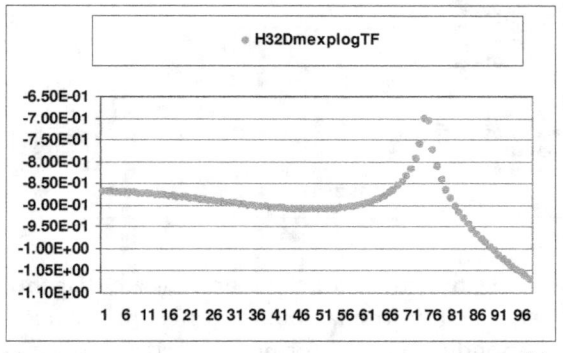

(a)

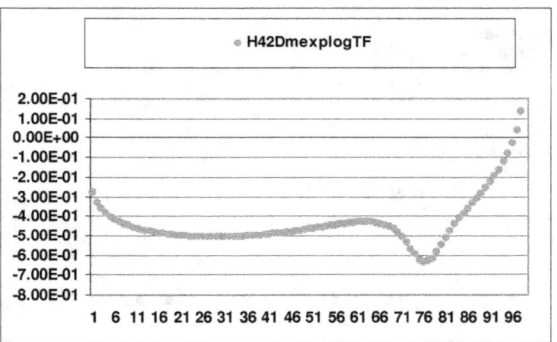

(b)

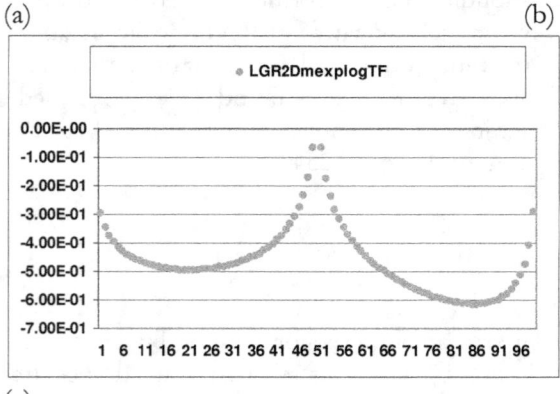

(c)

Figure 7: The ratio is relevant to quadratic classic and resilient polynomials that were processed with the transfer function log(|1.0 / p|). The ratio shown is relevant to quadratic (a), cubic (b) and cubic Lagrange (c) formulae.

(H32D)		
(x, y)	(1-Classic/Resilient)	TF
(0.25, 0.25)	9.999 E-01	sinTF
(0.75, 0.75)	7.298 E-03	sinTF
(0.25, 0.25)	-5.892 E-01	cosTF
(0.75, 0.75)	-5.798 E-01	cosTF
(0.75, 0.75)	-6.472 E-02	logTF

Table IV: Values of the ratio that are not displayed in figures 6 and 7 so as to improve the visibility of the remaining dots. From left to right: the value of the misplacement, the value of the ratio, and the type of transfer function (TF). The quadratic polynomials are: the classic and the resilient ones with embedding (H32D).

(H32DWE)		
(x, y)	(1-Classic/Resilient)	TF
(0.25, 0.25)	9.999 E-01	sinTF
(0.75, 0.75)	-8.256 E-03	sinTF
(0.25, 0.25)	-5.723 E-01	cosTF
(0.75, 0.75)	-5.580 E-01	cosTF
(0.75, 0.75)	-3.530 E-02	logTF

Table V: Values of the ratio that are not displayed in figures 8 and 9 so as to improve the visibility of the remaining dots. The layout of this table is the same as that of table IV. The quadratic polynomials are: the classic and the resilient ones without embedding (H32DWE).

(LGR2D)		
(x, y)	(1-Classic/Resilient)	TF
(0.50, 0.50)	1.000 E+00	sinTF
(0.50, 0.50)	-2.132 E+00	cosTF
(0.50, 0.50)	-2.755 E+00	logTF
(LGR2DWE)		
(x, y)	(1-Classic/Resilient)	TF
(0.50, 0.50)	1.000 E+01	sinTF

Table VI: Values of the ratio that are relevant to figures 8 and 9, removed from the graphs to improve the visibility. The transfer functions employed are: sin(p), cos(p), log(|1.0 / p|) (with embedding) and sin(p) (without embedding).

Error Images

In figure 10 the misplacement employed to calculate the virtual shift is (x, y) = (0.083, 0.838), and the value of the constant polynomial parameter 'the_A_const' is set at 0.54. Results are shown in the following order: error images obtained processing the convolutions with TF = sin(p), TF = cos(p) and TF = log(| 1.0 / p |) are placed in the first, fourth and seventh row from the top.

An observation that can naturally arise from figure 10 is that the pictures in (b) and (h) show a brighter image than the pictures in (a) and (g), notwithstanding the cumulative value of the MAE indicated in (s). The MAE is 0.814 in (a) and 0.637 in (b) (classic and resilient formulae respectively) and is 0.832 in (g) and 0.633 in (h) (classic and resilient formulae respectively). To explain the behavior in figures (a) and (b) let us look at the histograms in (c), (d) and (e), (f) and the corresponding values of the MAE average seen in (s), which is: 8.057 E+00 (c), 5.364 E+00 (e), 6.713 E+00 (d), 35.181 E+00 (f).

In both (c) (classic) and (d) (resilient), the cumulative error (MAE) seen in the histograms (higher in the case of the classic formula) is relevant to the interval [1, 244], whereas in (e) (classic) and (f) (resilient), the cumulative error is relevant to the interval [245, 255] (and the MAE is higher in the case of the resilient formula). Looking at the areas of the histograms in (c) and (d), the number of pixels in (c) appears larger (1.966 E+03 pixels out of a total of 2.209 E+03 across the image) than the number of pixels in (d) (1.638 E+03 pixels out of a total of 2.209 E-03 across the image). This demonstrates that in figure (a) the average error is higher than the average error in figure (b) and this happens in the majority of the pixels.

121

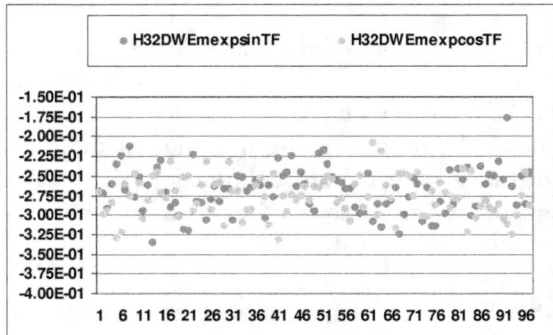

(a)

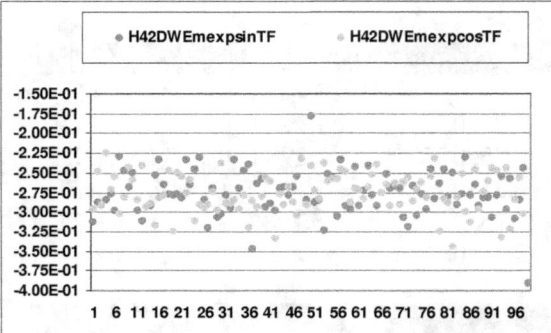

(b)

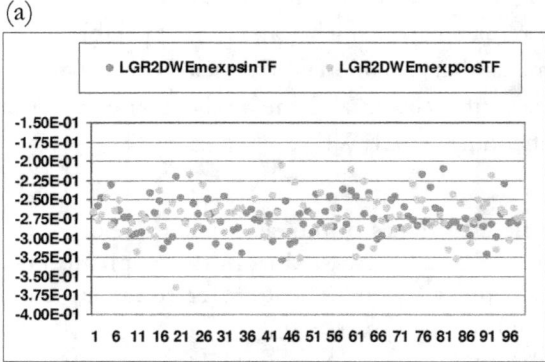

(c)

Figure 8: The values of the ratio (1-Classic/Resilient) obtained with two classes of polynomials: (i) bivariate classic and (ii) resilient ones calculated without embedding ('WE'). The polynomials were processed with the transfer functions: sin(p) and cos(p). (a) quadratic, (b) cubic, and (c) cubic Lagrange.

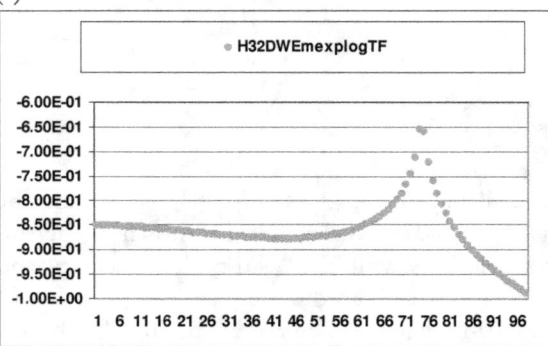

(a)

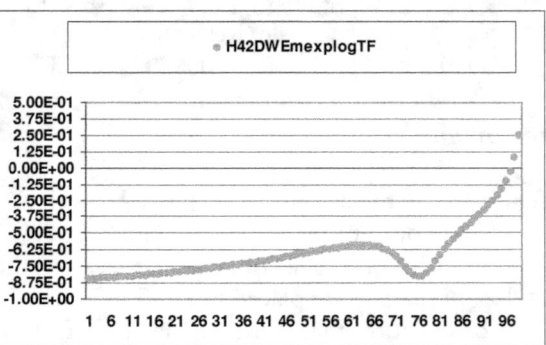

(b)

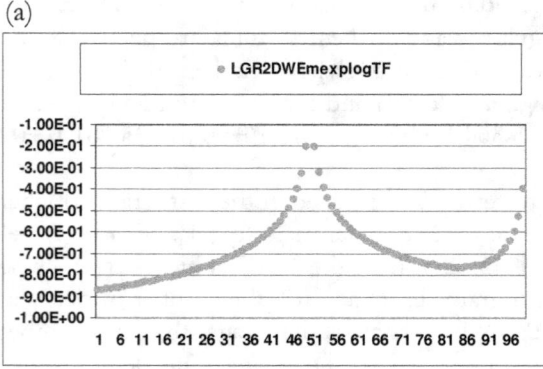

(c)

Figure 9: The ratio obtained with (i) bivariate classic and (ii) resilient formulae calculated without embedding ('WE'); in either case processing the interpolation formula with the transfer function log(|1.0 / p|). (a) Quadratic, (b) cubic, and (c) cubic Lagrange. The resilient signal was calculated as R (x, y) = f (0, 0) + ϱ (x, y) and the classic signal reconstruction was obtained with: I (x, y) = f (0, 0) + ξ(x, y).

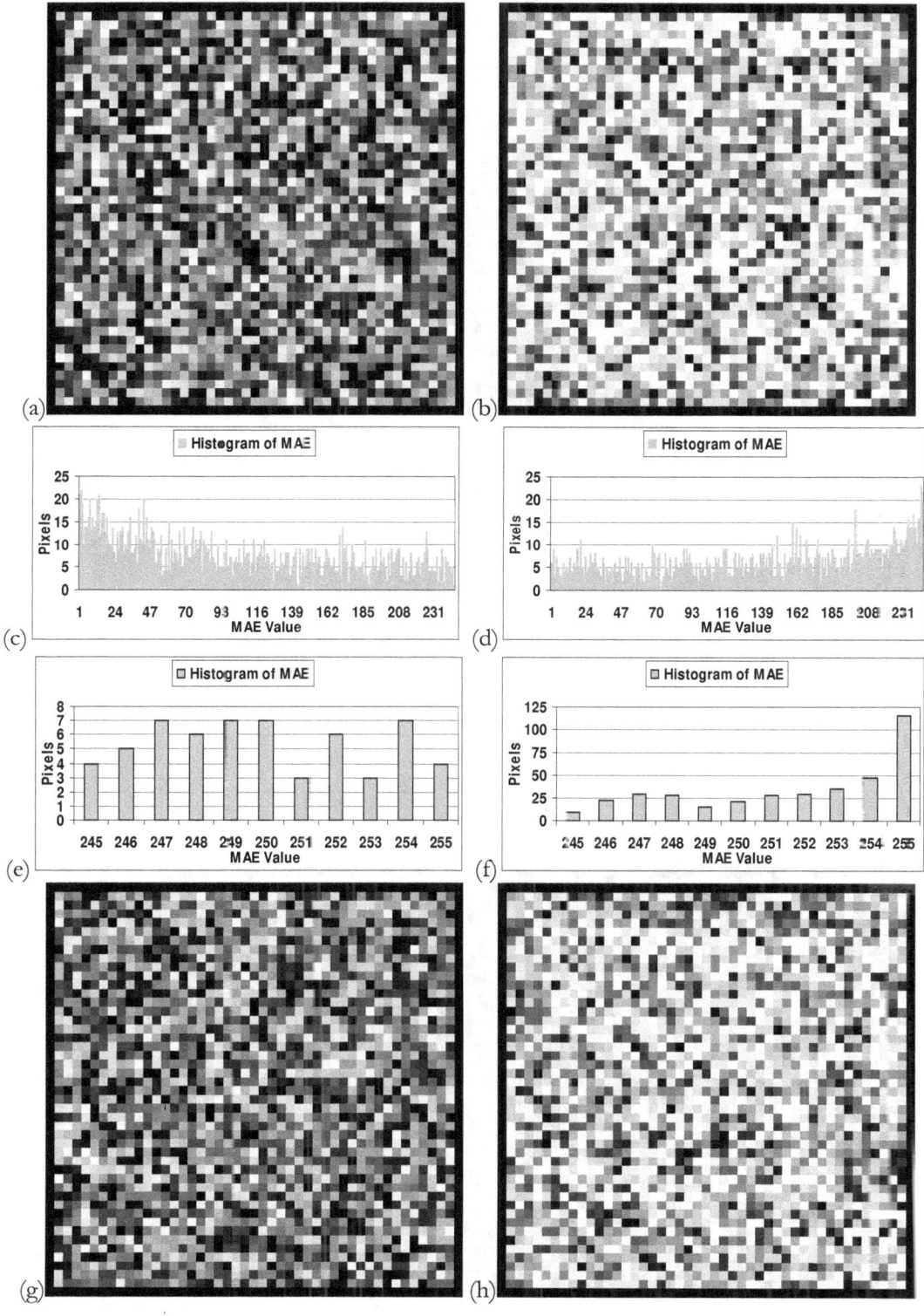

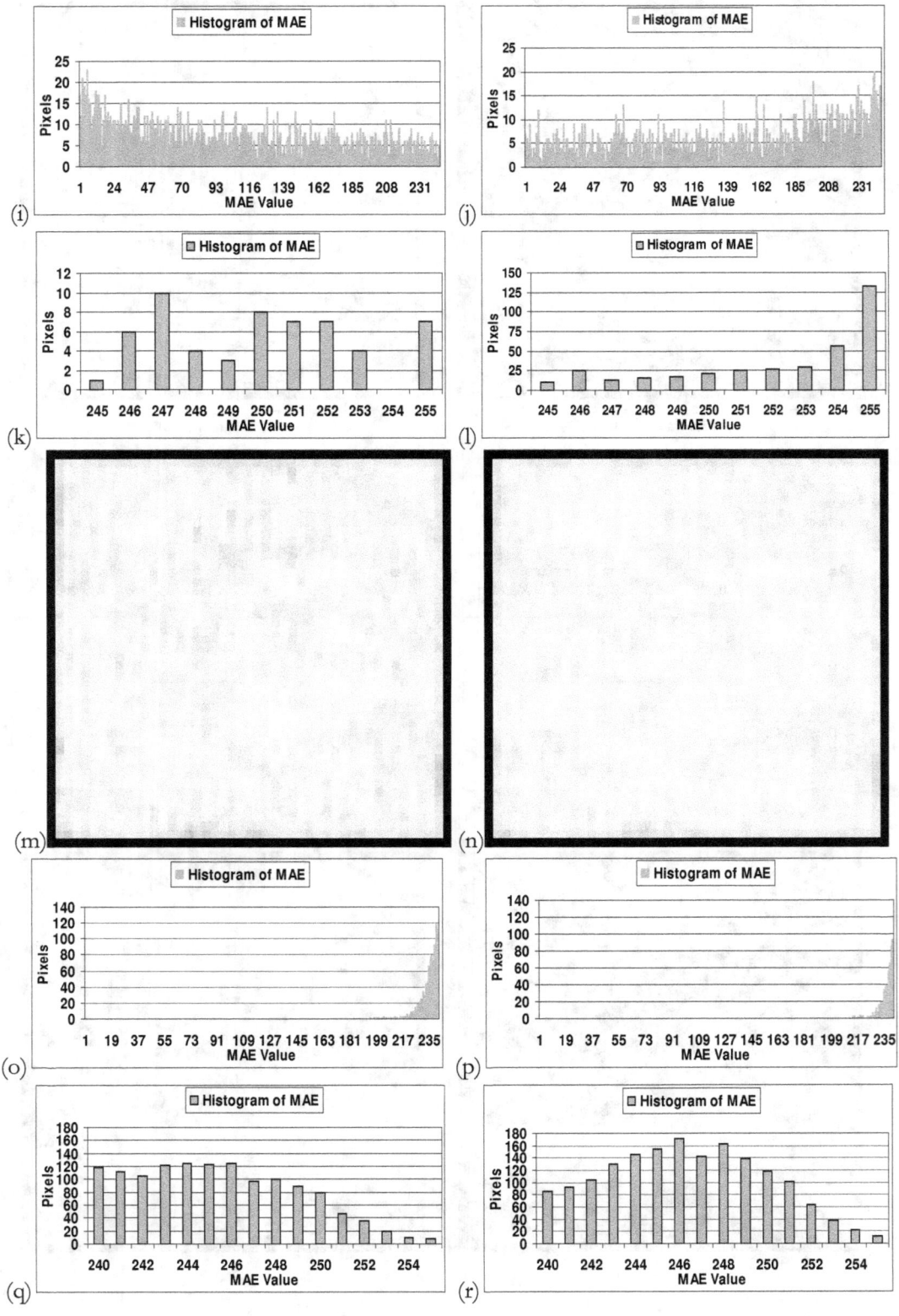

(H32DWE)						
MAE interval	MAE average	Pixels	Pixels	MAE Classic	MAE Resilient	TF
[1, 244]	8.057 E+00 (c)	1.966 E+03		0.814 (a)		sin(p)
[245, 255]	5.364 E+00 (e)	5.900 E+01				
[1, 244]	6.713 E+00 (d)		1.638 E+03		0.637 (b)	
[245, 255]	35.181 E+00 (f)		3.870 E+02			
[1, 244]	8.066 E +00 (i)	1.968 E+03		0.832 (g)		cos(p)
[245, 255]	5.182 E+00 (k)	5.700 E+01				
[1, 244]	6.754 E+00 (j)		1.648 E+03		0.633 (h)	
[245, 255]	34.272 E+00 (l)		3.770 E+02			
[1, 241]	1.991 E+01 (o)	1.289 E+03		18.943 (m)		$\log(|1.0 / p|)$
[242, 255]	7.779 E+01 (q)	7.360 E+02				
[1, 241]	11.021 E+00 (p)		8.990 E+02		11.501 (n)	
[242, 255]	1.076 E+02 (r)		1.126 E+03			

(s)

Figure 10: Error images obtained processing the test image with bivariate quadratic classic and resilient formulae (calculated without embedding) with three transfer functions. The residual is high in bright regions and low in dark regions (a), (g) and (m) are relevant to classic signal reconstruction. (b), (h) and (n) are relevant to resilient signal reconstruction. Each pixel in each of the error images had the intensity scaled in the range [0, 255]. Scaling was done consistently but separately from image to image. The quasi-uniform brightness in images (m) and (n) is indicative of pixel-by-pixel low variability of error within the image. The table in (s) is a summary of the values of the MAE across intervals, MAE average, number of pixels located in the histograms, MAE of classic and resilient formulae, transfer functions used, and is given as information which completes the information provided with both histograms and images.

On the other hand in the remaining minority of pixels (5.900 E+01, visualized in (e)) in figure (a), the average error is lower than the average error of the remaining minority of pixels (3.870 E+02 visualized in (f)) in figure (b). Clearly, (f) shows more pixels than (e) and a larger error than (e). Nevertheless (e) and (f) cover the interval [245, 255]. This analysis allows one to comprehend why the figure in (b) appears brighter than the figure in (a), even though the values of the MAE are 0.637 and 0.814 respectively.

Similar analysis can be synthesized as far as the figures shown in (g) and (h) are concerned, classic and resilient formulae respectively. The histograms of the classic formula are seen in (i) and (k) and those of the resilient formula in (j) and (l). Values from the table in (s) referring to the histograms indicate: 8.066 E +00 (i), 5.182 E+00, (k), 6.754 E+00 (j), 34.272 E+00 (l), whereas the value of the MAE across the full images is 0.832 in figure (g) and 0.633 in figure (h). The interval of the histograms in (i) and (j) is [1, 244] and the interval of the histogram in (k) and (l) is [245, 255]. There are 1.968 E+03 and 1.648 E+03 pixels in (i) and (j) out of a total of 2.209 E+03 across the image. Thus the majority of pixels in figure (g) has a higher error than the pixels in figure (h), and this explains why the values of the MAE are 0.832 (g), 0.633 (h) respectively. On the other hand, there are 5.700 E+01 and 3.770 E+02 pixels in (k) and (l) respectively and these are the minority of the image pixels with a corresponding MAE of 5.182 E+00 (k) and 34.272 E+00 (l), and this explains why the figure in (h) appears brighter than the figure in (g).

Finally the analysis can be extended to figures (m) and (n) having MAE values of 18.943 and 11.501, classic and resilient formulae respectively. The histograms in (o) and (p) show MAE values of 1.991 E+01 and 11.021 E+00 respectively, with the number of corresponding pixels being 1.289 E+03 (o) and 8.990 E+02 (p) (out of a total of 2.209 E+03 across the image); and this is clearly in favor of the image in (m) being brighter than the image in (n).

The histograms in (q) and (r) show MAE values of 7.779 E+01 and 1.076 E+02 with the number of pixels being 7.360 E+02 (q) and 1.126 E+03 (r); and this is clearly in favor of the image in (n) being brighter than the image in (m). The interval [1, 241] covered in (o) and (p) is much larger than the interval [242, 255] covered in (q) and (r), and this detail does not explain why the image in (m) is brighter than the image in (n). Let us then calculate the ratio of the MAE over pixels: 0.0154 in (o), 0.0122 in (p), 0.1056 in (q), 0.0955 in (r).

Even though the cumulative MAE is 11.501 in (n) and 18.943 in (m), these ratios explain why the image in (n) is brighter than the image in (m), and this is because the ratio in (o) is bigger than the ratio in (p) and the ratio in (q) is bigger than the ratio in (r).

Naturally, given that each of these error images are calculated as the difference between the reference image (see figure 5) and an image that has been virtually shifted-rotated and interpolated, and which on a pixel-by-pixel basis had been estimated in its amplitude, and knowing that because the model interpolation function has a given curvature behaviour which is dependent on the math form of the function, to admit zero interpolation error may lead one to think that it is admissible that the curvature of the interpolation function is null.

That is to say: the second order partial derivatives of the model interpolator are null. Looking at the image tested in figure 5, one can see that the neighbourhood of the pixels to be re-sampled is diverse in terms of pixel intensity; thus as far as regards figure 5, even if the interpolation error were zero, there would be an absurdity in such reasoning; in fact, in figure 5, the curvature of the interpolator is not null.

To think that it is admissible that the curvature of the interpolation function is null, at least conceptually, requires that the estimation of the pixel intensity should return to the same original value termed f (0, 0) when the neighbouring pixels taken into the convolution of the interpolation function are all of the same amplitude value, and this event corresponds to curvature (second order derivative) zero of the model interpolation function. That is to say: all pixels have the same value inside the chosen neighbourhood. This reasoning is logical, however it may not happen in practice because the convolution is processed through the transfer function.

(H32DWE)		
Classic	Resilient	TF
0.814	0.637	sin(p)
0.832	0.633	cos(p)
18.943	11.501	log(|1.0/p|)

Table VII: Mean Absolute Error (MAE) across the image. The values located in the first column from the left are relevant to the error images shown in figure 10 in (a), (c) and (e) (classic signal reconstruction), and those located in the second column from the left are relevant to the error images shown in figure 10 in (b), (d) and (f) (resilient signal reconstruction).

Furthermore, as shown in figures 5 and 10, even when the curvature is not null within the neighbourhood, the interpolation error is not null. So it follows that the transfer function (TF = sin(p), TF = cos(p) and TF = log(\| 1.0 / p \|)) does contribute to the interpolation error and this happens because of its processing of the numerical value of the convolution. Table VII gives the numerical values of the mean absolute error (MAE), which is the measure adopted here for the interpolation error across the image.

DERIVATION OF THE TRANSFER FUNCTION FROM THE MATH FORM OF THE INTERPOLATION FORMULA

The Pixel Intensity Correction (PIC)

This section will address the issue relating to the possibility as much as the necessity of removing the arbitration introduced through the choice of the transfer function employed to scale the convolutions in either classic or resilient interpolation.

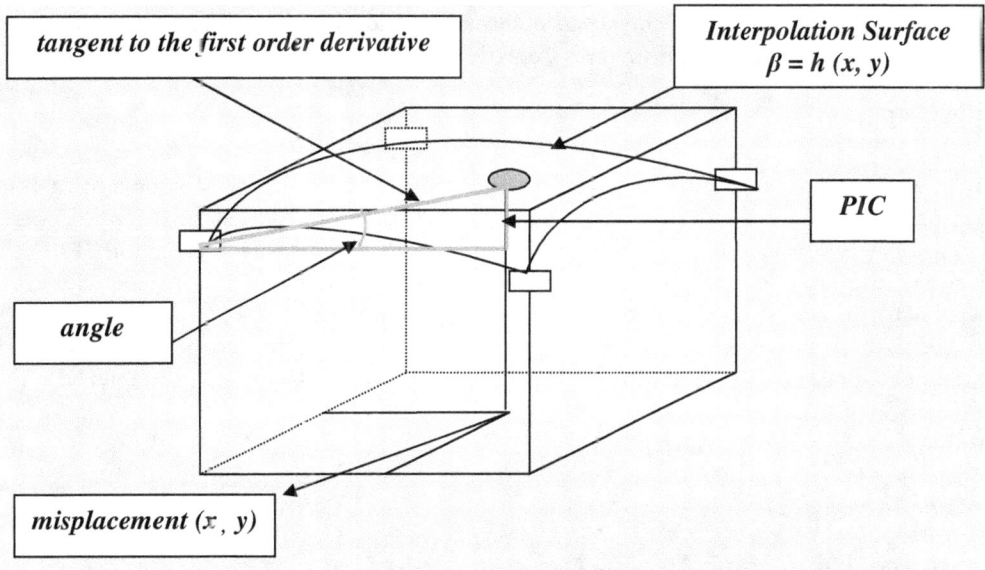

Figure 11: Pixel Intensity Correction (PIC).

One good way to address the aforementioned removal of the arbitration in scaling the convolutions is through experimental evidence. Nevertheless, experimental evidence is obtained and presented below along the theoretical grounds upon which it is based.

To the curvature of any interpolation functions which admit a non null second order derivative, one assigns the geometrical counterpart of its nature, which is that of the arc tangent of the angle between the horizontal and the tangent to the first order derivative of the curve. The geometrical counterpart is univocally defined through an infinitesimal (here called increment). The angle subtended from the tangent to the first order derivative and the horizontal (parallel to the axis of the abscissa) or the vertical (parallel to the axis of the ordinate) is independent of the increment and yet defined when the increment is zero. The cyclic function named tangent provides the measure of what is here called the Pixel Intensity Correction (PIC). Let us remind us the concept of the total curvature of a model interpolation function. It is defined to be the sum of second order derivatives of its math form respect to the dimensional variables. In other words, the tangent of the total curvature multiplied with the misplacement is equal to the PIC (see figure 11); the angle is the total curvature of the model interpolator. What, then, is the geometrical meaning of the PIC? It is that of the fraction of pixel intensity to add to the pixel intensity f (0, 0) located at the location of signal reconstruction to obtain the estimate of the signal's true unknown value.

SIGNAL RESILIENT TO INTERPOLATION

The model interpolator in 2D admits multiple partial second order derivatives, strictly depending in number on the dimensionality of the interpolator and each of the partial second order derivatives relating to the angle in one given dimension, thus implicitly referring to the PIC in the dimension. If it were admissible to consider multiple Pixel Intensity Corrections to the signal being reconstructed, it would be like admitting four dimensions at the same exact location, and this is an absurdity in space because the of the two spatial dimensions of the interpolation function. This absurdity provides justification as to why multiple second order partial derivatives of the interpolator are added together to obtain the total curvature of the model interpolator. To admit multiple PICs at the same exact location without this being an absurdity, would require time differences at the spaces where they are located. Thus the question arises as to whether it is admissible to consider time differences between the spaces where the PICs are located. The admissibility of considering time differences at the same space location in interpolation in a space domain is denied.

The Transfer Function

What has been illustrated in figure 1 conveys the geometrical meaning of the concept of both Pixel-Intensity Correction and convolution. At this stage it would seem natural and it would be advantageous to connect two concepts: (i) convolution and (ii) Pixel Intensity Correction (PIC). The two concepts are of the same nature because both the convolution and the PIC provide us with the fraction of signal to be added to the signal intensity being estimated through the model interpolation function. Let us now move on to consider equations (1) and (5), repeated below, which furnish the math expression of classic I (x, y) and resilient R (x, y) signal reconstructions respectively:

$$I (x, y) = f (0, 0) + \xi (x, y) \tag{23.a}$$

$$R (x, y) = f (0, 0) + \varrho (x, y) \tag{23.b}$$

To obtain the transfer function as derived from the math form of the model interpolation function which removes the arbitration introduced through transfer functions of the choice, such as those seen in the section: "STUDY OF TRANSFER FUNCTIONS", the PIC is added to the signal f(0, 0), which is being reconstructed at the spatial location (x, y) into:

$$I (x, y) = f (0, 0) + PIC (x, y) \tag{24.a}$$

$$R (x, y) = f (0, 0) + PIC (x, y) \tag{24.b}$$

It is worth noting that the PIC is divided with the value of maximum misplacement, formulated with:

$$A = [(n_1 * PSizeX)^2 + (n_2 * PSizeY)^2]^{1/2} \tag{25}$$

Where n_1 and n_2 are the number of pixels along the X and Y axis respectively, and PSizeX, PSizeY are the pixels' sizes. The math developments which formalize the calculation of the total curvature of resilient interpolation are given in the following chapters. Specifically, chapter 5 offers the treatise of bivariate quadratic formulae, with and without embedding; chapters 6 and 7 do the same to bivariate cubic signals resilient to interpolation. The calculation of the total curvature of classic interpolation was provided in chapters 2 and 3.

Results

The use of the total curvature to obtain the transfer function from the math form of the model interpolator - Classic and Resilient in the two variants, with and without embedding - has produced the results herein presented and pertaining to the third column to the right of table I ('Curvature') located in the section: "THE VALIDATION PARADIGM". The results are described in the following two paragraphs.

Transfer Function Derived From the Curvature of the Resilient Interpolation Formula

Results of the experiments conducted to compare signal reconstruction obtained with one transfer function chosen among cos(p), sin(p) and log(|1.0 / p|) versus the transfer function obtained through the total curvature are presented in figure 12 through figure 17.

The misplacement increases at steps of (x, y) = [(0.01, 0.01), (0.01, 0.01)] in the range (x, y) in [(0.01, 0.01), (0.99, 0.99)], thus moving along the diagonal of the pixel. Details regarding the nomenclature of the figures are given in Table VIII below.

Negative values of the ratio of improvement are indicative of superiority in signal reconstruction of resilient over classic interpolation.

Transfer Function	Comparison Between Classic and Resilient Formula	
	Quadratic	
	With E	WE
sin(p)	H32DCurvaturesinTF	H32DWECurvaturesinTF
cos(p)	H32DCurvaturecosTF	H32DWECurvaturecosTF
log(|1.0/p|)	H32DCurvaturelogTF	H32DWECurvaturelogTF
	Cubic	
	With E	WE
sin(p)	H42DCurvaturesinTF	H42DWECurvaturesinTF
cos(p)	H42DCurvaturecosTF	H42DWECurvaturecosTF
log(|1.0/p|)	H42DCurvaturelogTF	H42DWECurvaturelogTF
	Cubic Lagrange	
	With E	WE
sin(p)	LGR2DCurvaturesinTF	LGR2DWECurvaturesinTF
cos(p)	LGR2DCurvaturecosTF	LGR2DWECurvaturecosTF
log(|1.0/p|)	LGR2DCurvaturelogTF	LGR2DWECurvaturelogTF

Table VIII: Relevant to the nomenclature of figures 12 through 17. Quadratic, cubic and cubic Lagrange resilient interpolation formulas which make use of the transfer function. The resilient interpolation paradigm in its two variants: with and without embedding.

Table VIII reports the nomenclature of the figures 12 through 17. The four graphs shown in figure 12 show a common exponential behaviour which starts from high negative values of the Ratio and approaches, with increasing misplacement, the horizontal 'steady state' of improvement at approximately 2,000% (see figure 12 in (b) in the case of sin(p) and cos(p)). That practically means that resilient reconstruction performed with the custom transfer function obtained through the use of the total curvature offers the option of removing the arbitration of cos(p) and sin(p) transfer functions employed to process classic signal reconstruction, and such an option is readily provided as viable because of the ratio of improvement offered from the resilient formula. The horizontal 'steady state' of improvement approaches values of approximately 38,000% (see figure 12 in (d) in the case of log(|1.0 / p|)).

The graphs shown in figure 13 show behaviour similar to that seen in figure 12: the horizontal 'steady state' of improvement is reached at approximately 2,000% (see figure 13 in (b), sin(p) and cos(p)); and at approximately 75,000% (see figure 13 in (d), log(|1.0 / p|)). In both of the figures 12 and 13 the parametric constant *the_A_const* is set equal at -0.7045.

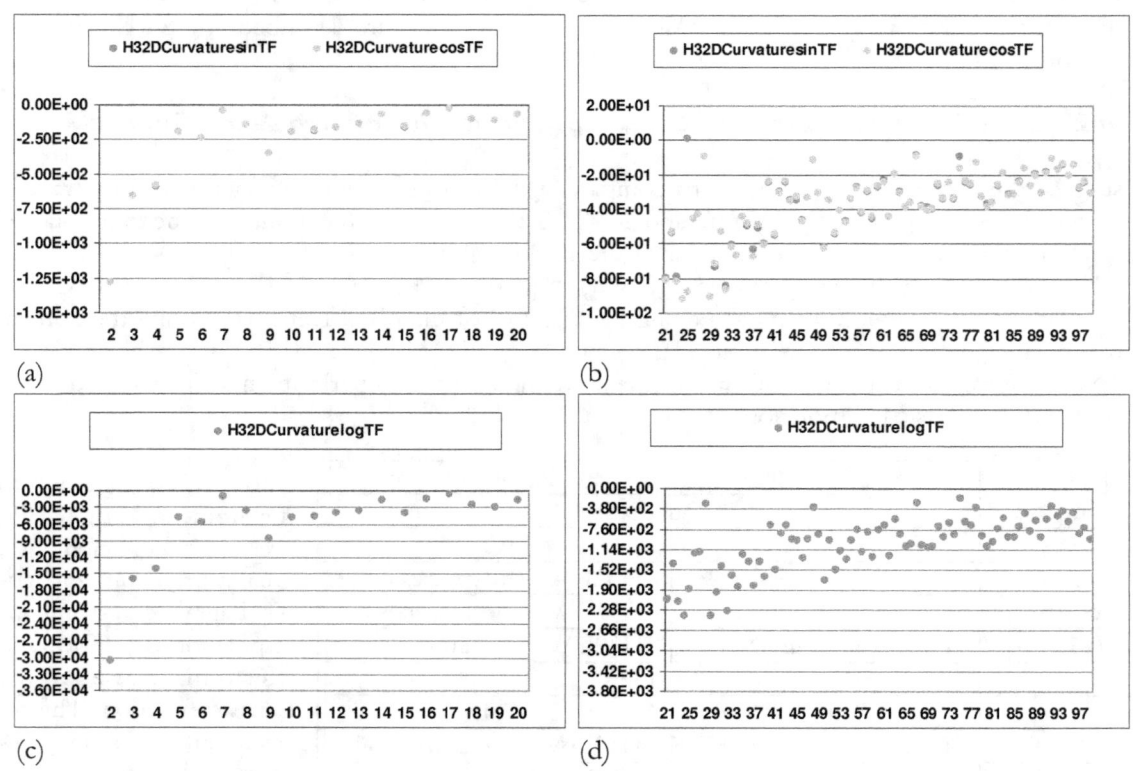

(a)

(b)

(c)

(d)

Figure 12: Ratio of improvement of resilient interpolation with embedding over classic interpolation. The polynomial is the bivariate quadratic. The transfer functions are: sin(p) and cos (p) as shown in (a) and (b), log(|1.0 / p|) as shown in (c) and (d). The graphs in (a) and (c) are relevant to misplacements inside the range (x, y) in [(0.02, 0.02), (0.20, 0.20)] and the graphs in (b) and (d) are relevant to misplacements inside the range (x, y) in [(0.21, 0.21), (0.99, 0.99)].

To improve visibility in figures 12, 13, 14, 15, 16 and 17 the value of the ratio of improvement has not been shown at the misplacement location (x, y) = (0.01, 0.01) but instead it has been placed in Tables IX, X and XI.

The behaviour observable in figure 14 is that of a decrease in the ratio of improvement with increasing misplacement. In the first portion of data seen in figure 14(a) and 14(b) (range (x, y) in [(0.02, 0.02), (0.49, 0.49)]) as relevant to the transfer functions of sin(p) and cos(p), the ratio approaches values of approximately 10,000% in (a), it then decreases further down to values that are approximately between 1,000% and 2,000% (range (x, y) in [(0.50, 0.50), (0.99, 0.99)] in (b)). In figure 14(c) and 14(d) (transfer function: log(|1.0 / p|), the ratio approaches values of approximately 200,000% (c) and 20,000% (d) (see middle portion of the graphs).

The behaviour of the ratio of improvement seen in figure 15 does not appear different from the behaviours seen in figures 12 through 14. The ratio decreases with the increase of misplacement in

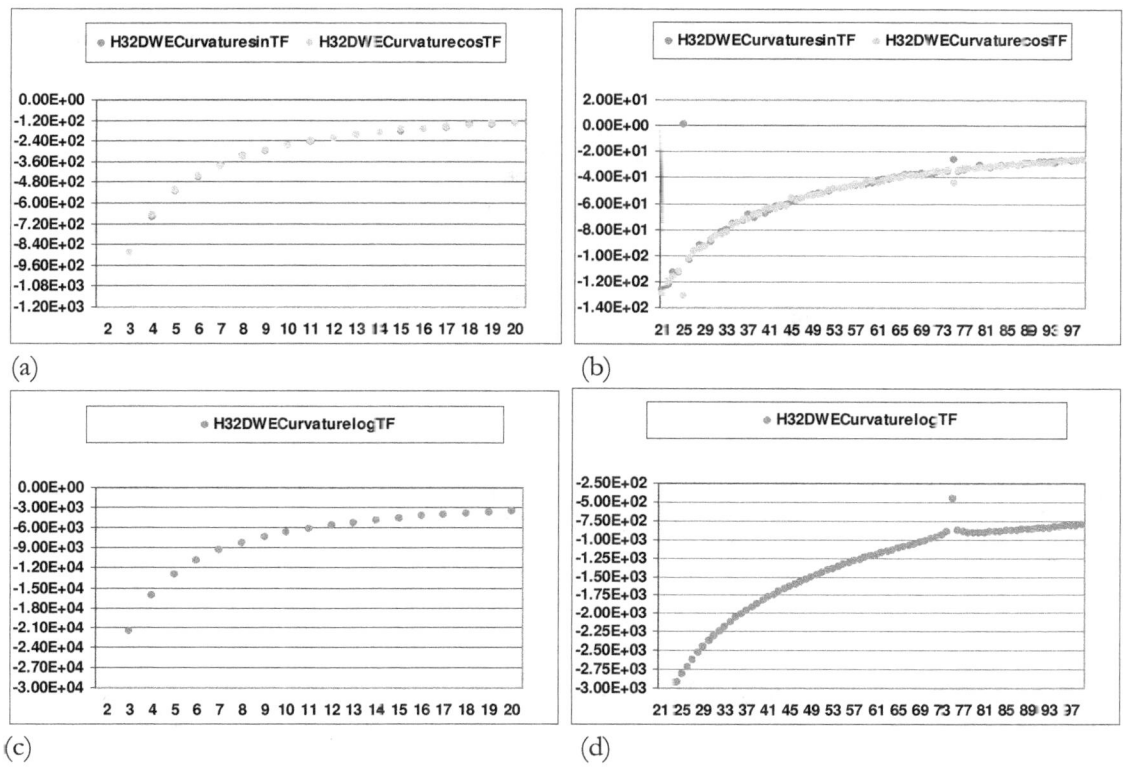

Figure 13: Ratio of improvement of resilient interpolation without embedding ('WE') over classic interpolation. The polynomial is the bivariate quadratic. Data are organized with the same layout and display as those seen in figure 12. The three transfer functions are the same: sin(p) and cos (p) (see (a) and (b)), log(|1.0 / p|) (see (c) and (d)). The graphs showing the ratio of improvement cover the ranges (x, y) in [(0.02, 0.02), (0.20, 0.20)] (see (a) and (c)), and (x, y) in [(0.21, 0.21), (0.99, 0.99)] (see (b) and (d)).

(H32DCurvature)			
Classic	Resilient	(1-Classic/Resilient)	TF
8.052 E-01	4.208 E-04	-1.879 E+03	sin(p)
8.191 E-01	4.208 E-04	-1.912 E+03	cos(p)
1.936 E+01	4.208 E-04	-4.522 E+04	log(|1.0/p|)
(H32DWECurvature)			
Classic	Resilient	(1-Classic/Resilient)	TF
8.052 E-01	3.030 E-04	-2.659 E+03	sin(p)
8.191 E-01	3.030 E-04	-2.704 E+03	cos(p)
1.936 E+01	3.030 E-04	-6.397 E+04	log(|1.0/p|)

Table IX: Bivariate quadratic polynomials: Classic and Resilient with and without embedding ('WE'). The resilient formula has the feature of employing the transfer function derived from its own math form through the curvature. The ratio of improvement is shown at the misplacement (x, y) = (0.01, 0.01) and is relevant to the three transfer functions sin(p), cos(p) and log(|1.0/p|) employed to process the classic formula to reconstruct the signal.

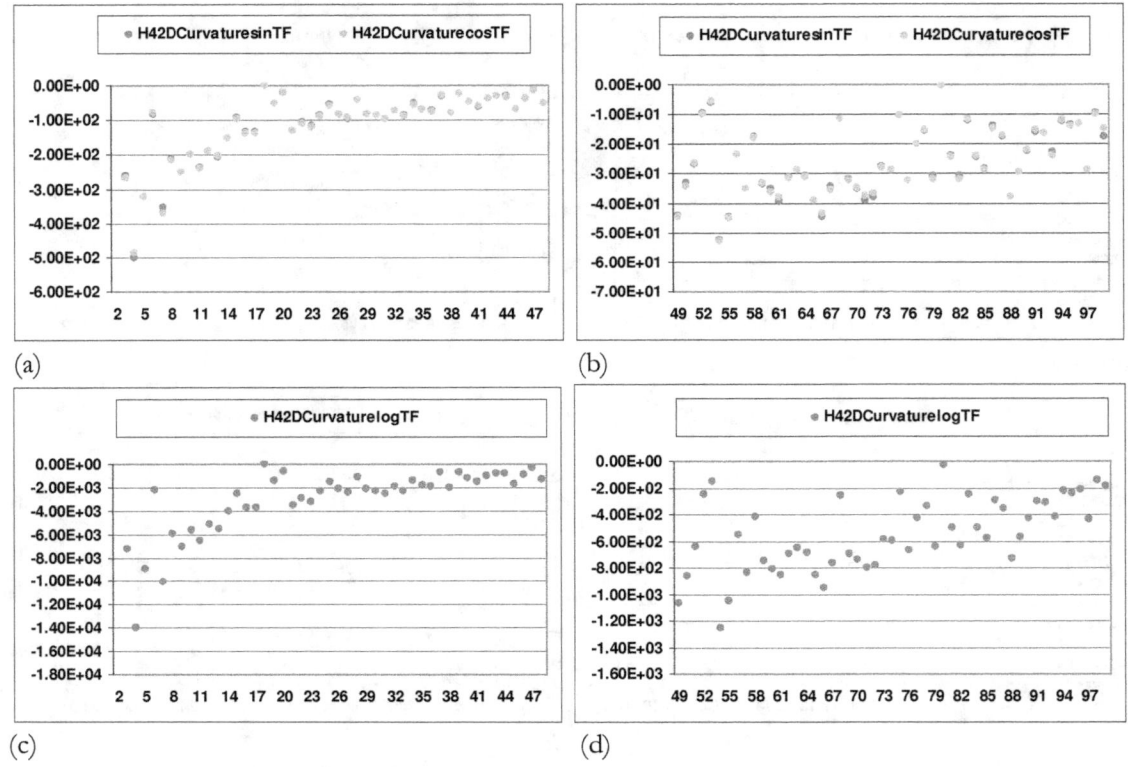

(a) (b) (c) (d)

Figure 14: Ratio of improvement of resilient interpolation with embedding versus classic interpolation which uses sin(p), cos(p) and log($|1.0 / p|$) as transfer functions. The graphs are relevant to the bivariate cubic polynomials. The graphs cover the ranges (x, y) in [(0.02, 0.02), (0.49, 0.49)] (see (a) and (c)), and (x, y) in [(0.50, 0.50), (0.99, 0.99)] (see (b) and (d)).

the graph seen in figure 15(a) down to approximately 10,000% and continues to decrease as shown in graph 15(b) up to almost 2,000% and decreases after that. In figures 15(c) and 15(d) the decrease goes down to approximately 500,000% (see graph in (c)), then it goes down to approximately 36,000% (see graph in (d)) and decreases after that.

The behaviour of the ratio of improvement seen in figure 17 is similar to the one seen in figure 12, where data decreases exponentially towards a 'steady state' of improvement. In the graph shown in figure 17(b), relating to the improvement of the resilient formula over the classic formula processed with sin(p) and cos(p), the ratio approaches approximately 2,000%; whereas in figure 17(d) (transfer function log (1/p)), the ratio approaches approximately 33,000%. The behaviour of the ratio seen in figure 17(b) is also similar to the one seen in figure 16(b) (2,000%), and the ratio seen in figure 17(d) is also similar to that one seen in figure 16(d) (35,000%).

The purpose of being able to produce a lower interpolation error through the total curvature, when compared to the classic interpolator with transfer function (see equation (19(a))), is supported from the empirical evidence presented here. It is thus possible through the total curvature to devise the novel interpolator called resilient interpolation with the transfer function derived from the total curvature (see equation (24(b))).

The results presented necessitate further commentary of a conceptual nature. *What is the meaning conveyed through the ratio of improvement of resilient versus classic signal reconstruction?* Figure 18 shows one admissible answer to the aforementioned question and at the same time raises another necessity, which is that of obtaining evidence which demonstrates the answer to the question.

In figure 18, the difference between resilient and classic interpolation functions is emphasized with the focus on the profile of the two model functions and such a difference of profiles is visually manifest through the curvature of the two which explains the numerical values of the ratio of improvement. Is such an explanation given in figure 18 logical, rational and exhaustive of the data presented in figures 12 through 17?

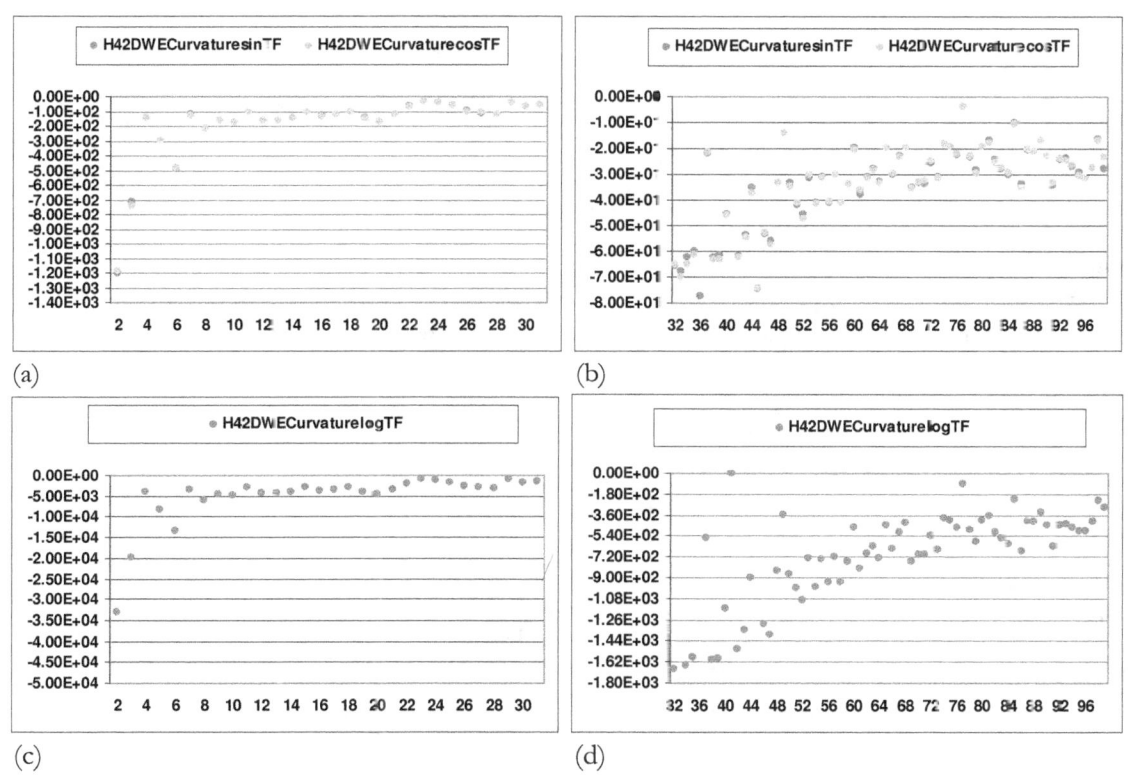

Figure 15: Ratio of improvement of resilient interpolation without embedding ('WE') versus classic interpolation. The graphs are relevant to the bivariate cubic polynomials, transfer functions are the same seen in figures 12, 13 and 14, whereas the ranges where data are displayed are: (x, y) in [(0.02, 0.02), (0.31, 0.31)] (see (a) and (c)) and (x, y) in [(0.32, 0.32), (0.99, 0.99)] (see (b) and (d)).

(H42DCurvature)			
Classic	Resilient	(1-Classic/Resilient)	TF
8.294 E-01	3.450 E-04	-2.404 E+03	sin(p)
8.305 E-01	3.450 E-04	-2.408 E+03	cos(p)
2.257 E+01	3.450 E-04	-6.548 E+04	log(| 1.0 / p |)
(H42DWECurvature)			
Classic	Resilient	(1-Classic/Resilient)	TF
8.294 E-01	6.710 E-04	-1.235 E+03	sin(p)
8.305 E-01	6.710 E-04	-1.237 E+03	cos(p)
2.257 E+01	6.710 E-04	-3.366 E+05	log(| 1.0 / p |)

Table X: Bivariate cubic polynomials: Classic and Resilient with and without embedding ('WE'). The ratio of improvement shown at the misplacement (x, y) = (0.01, 0.01).

This is logical and to understand why, simply compare equation 24(b), which gives the math expression of the resilient interpolator with the transfer function calculated from the total curvature: R (x, y) = f (0, 0) + PIC (x, y) with equation 19(a), which gives the math expression of the classic interpolator with an arbitrary transfer function: I (x, y) = f (0, 0) + TF(ξ (x, y)). Negative values of the ratio imply PIC (x, y) < TF(ξ (x, y)) given that the Mean Absolute Error (MAE) is defined as the absolute value of the difference between the original signal at the grid point (f (0, 0)) and the interpolated signal: either R (x, y) or I (x, y).

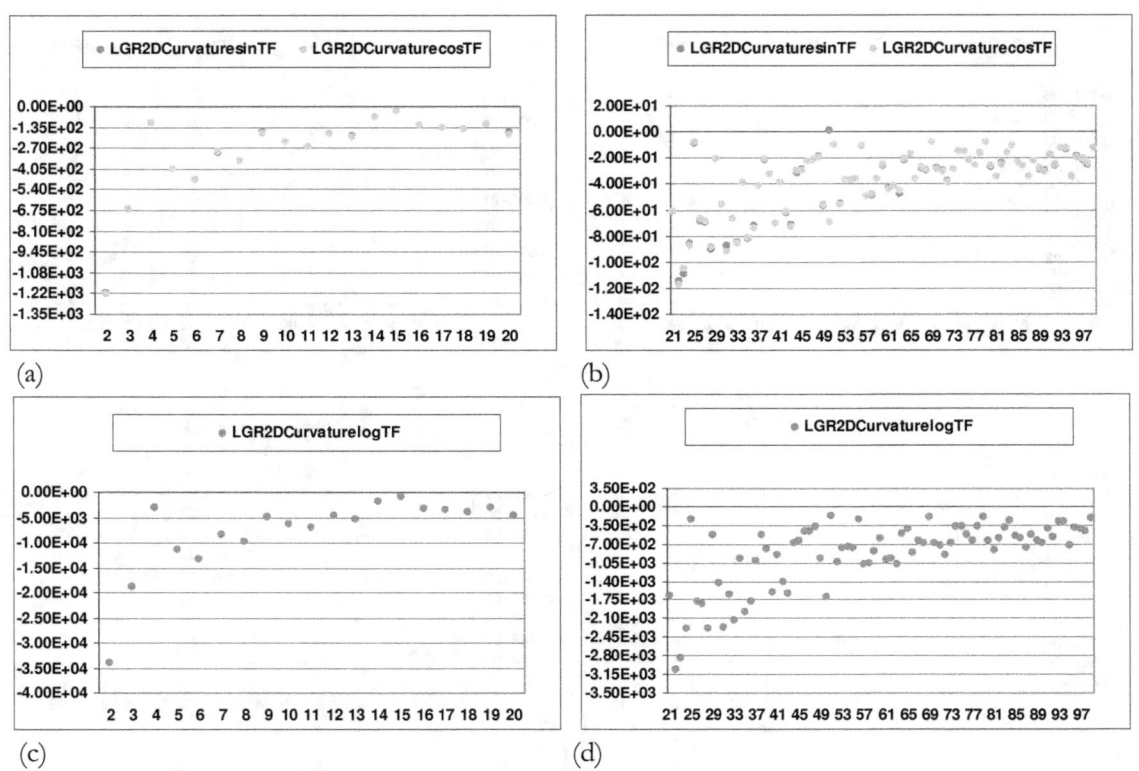

Figure 16: Ratio of improvement of resilient signal reconstruction versus classic signal reconstruction: cubic Lagrange polynomials. The resilient signal is calculated with embedding. As in figure 13, the graphs show data in two ranges: (x, y) in [(0.02, 0.02), (0.20, 0.20)] (see (a) and (c)), and (x, y) in [(0.21, 0.21), (0.99, 0.99)] (see (b) and (d)).

Is this rational? Consider moving from a pixel which is bright to a pixel which is dark and making an estimate between them. Which one of the two between, say, 15% grey and 25% grey would the one to be believed? When the estimate is made at a spatial location which is closer to the bright pixel than it is to the dark pixel, common sense suggests believing in 15% grey, whereas when the estimate is made at a spatial location which is closer to the dark pixel than it is to the bright pixel, 25% grey is more likely to be believed as the true estimate.

Then the natural question, when looking at the graphs of the ratio of improvement is: 'Why does the improvement keep decreasing with increasing misplacement across the full span of the pixel, from (x, y) = (0, 0) to (x, y) = (0.99, 0.99)?' In such a case it is true that PIC (x, y) < TF(ξ (x, y)) at each (x, y) in [(0. 0), (0.99, 0.99)].

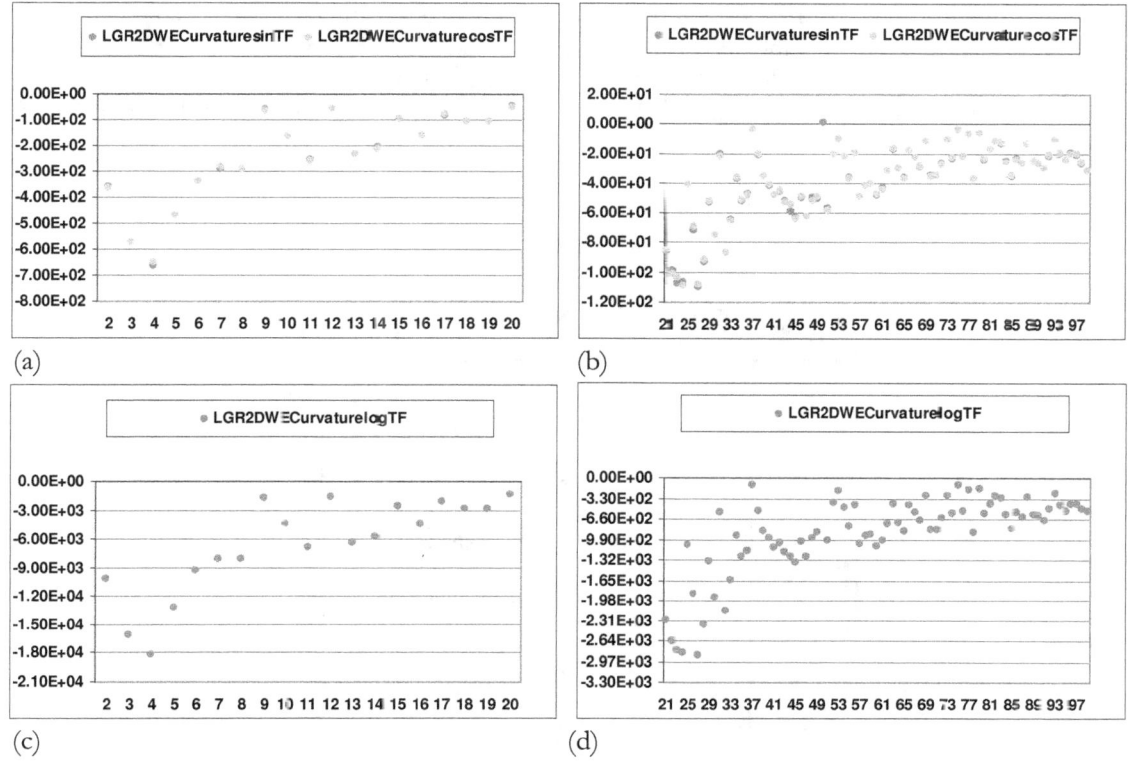

Figure 17: Ratio of improvement of resilient signal reconstruction without embedding ('WE') over classic signal reconstruction calculated with cubic Lagrange polynomials. Data have the same organization as figures 12 through 16. The ranges on the abscissa of each graph are the same as those seen in figures 12, 13 and 16, and the transfer functions of classic signal reconstruction are: sin(p), cos(p) (shown in (a) and (b)) and log(|1.0 / p|) (shown in (c) and (d)).

(LGR2DCurvature)			
Classic	Resilient	(1-Classic/Resilient)	TF
8.170 E-01	3.335 E-04	-2.439 E+03	sin(p)
7.944 E-01	3.335 E-04	-2.371 E+03	cos(p)
2.267 E+01	3.335 E-04	-6.772 E+05	log(|1.0/p|)
(LGR2DWECurvature)			
Classic	Resilient	(1-Classic/Resilient)	TF
8.170 E-01	3.410 E-04	-2.394 E+03	sin(p)
7.944 E-01	3.410 E-04	-2.327 E+03	cos(p)
2.267 E+01	3.410 E-04	-6.646 E+05	log(|1.0/p|)

Table XI: Cubic Lagrange polynomials: Classic and Resilient with and without embedding ('WE'). The ratio of improvement shown at the misplacement $(x, y) = (0.01, 0.01)$.

Consider the case that the curvatures of the two interpolators- classic and resilient- have a type of behaviour such that, with the increase of misplacement, one interpolator curvature decreases in its magnitude and then continues to decrease (the classic formula), and the other interpolator curvature (the resilient formula, which is related to PIC (x, y)) decreases less in its magnitude than the decrease of the curvature of the classic formula. With such behaviours, the ratio of improvement decreases in its magnitude with the increase of the misplacement (the PIC is given by the function tangent of the total curvature and thus it is proportional to the magnitude of its argument). Such behaviour is what we see in figure 18, when looking separately to the right side (when moving from right to left, the curvature decreases) and to the left side (when moving from left to right, the curvature decreases). Noteworthy in figure 18 is that the value of $R(x, y) - I(x, y)$ is larger in magnitude at the right than it is at the left.

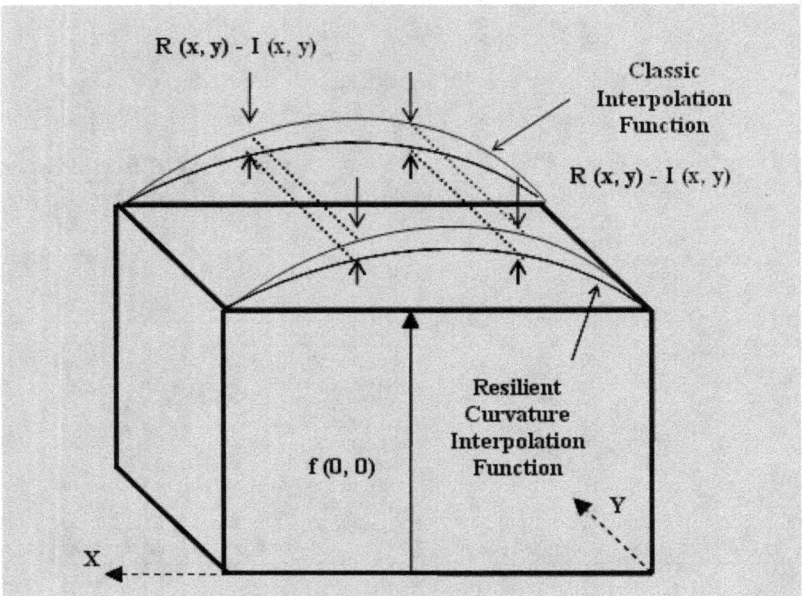

Figure 18: One sample admissible profiles of resilient curvature and classic interpolation functions.

X	Y	MAE Classic	MAE Resilient	Ratio
1.000 E-02	1.000 E-02	2.257 E+01	6.710 E-04	-3.366 E+05
2.000 E-02	2.000 E-02	2.254 E+01	6.800 E-04	-3.315 E+05
3.000 E-02	3.000 E-02	2.251 E+01	1.138 E-03	-1.977 E+05
4.000 E-02	4.000 E-02	2.247 E+01	5.545 E-03	-4.052 E+03
5.000 E-02	5.000 E-02	2.243 E+01	2.712 E-03	-8.272 E+03
6.000 E-02	6.000 E-02	2.240 E+01	1.671 E-03	-1.346 E+05
9.400 E-01	9.400 E-01	1.378 E+01	2.933 E-02	-4.689 E+03
9.500 E-01	9.500 E-01	1.330 E+01	2.667 E-02	-4.978 E+03
9.600 E-01	9.600 E-01	1.270 E+01	2.549 E-02	-4.971 E+03
9.700 E-01	9.700 E-01	1.190 E+01	2.842 E-02	-4.177 E+03
9.800 E-01	9.800 E-01	1.074 E+01	4.600 E-02	-2.326 E+03
9.900 E-01	9.900 E-01	8.731 E+00	2.999 E-02	-2.900 E+03

(a)

(b)

(c)

(d)

(e)

(f)

(g)

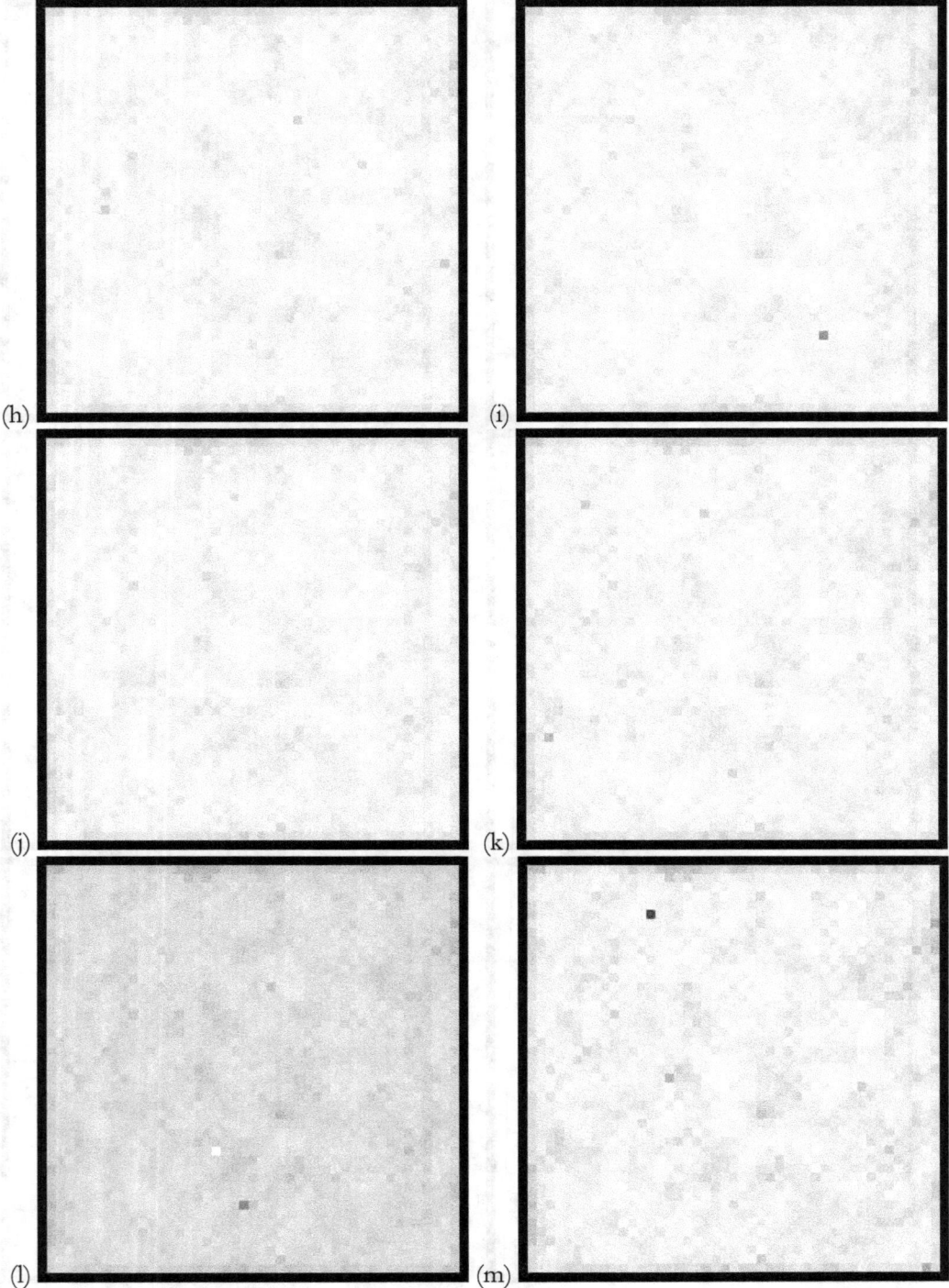

Figure 19: Maps of MAE absolute differences of data seen in figures 15(c) and 15(d) at the misplacements in the ranges (x, y) in [(0.01, 0.01), (0.06, 0.06)] shown in (b) through (g), and (x, y) in [(0.94, 0.94), (0.99, 0.99)] shown in (h) through (m) respectively. The table in (a) shows, from left to

right: x misplacement, y misplacement, the MAE of the classic formula, the MAE of the resilient formula, and the ratio of improvement.

Such behaviours is also reasonable as an explanation of why one interpolator (either classic or resilient) would be more accurate (at each (x, y) in [(0, 0), (0.99, 0.99)]) in the estimation of the grey level when moving (increase in the misplacement) from the bright pixel to the dark pixel.

This is certainly not exhaustive, because as shown in figures 12, 13, 16 and 17, the horizontal 'steady state' of improvement denies what figure 18 shows, which instead proposes a decrease from a maximum value of improvement when moving from right towards the middle of the pixel and then an increase towards a maximum value when moving from the middle of the pixel towards left.

The picture seen in figure 18 is therefore exhaustive of the data seen in figures 12, 13, 16 and 17 when considering only the first half of the pixel (from the far right to the middle), however not exhaustive of the overall behaviour of data within the full span of the pixel because of the constant decrease in the ratio of improvement seen in figures 12, 13, 16 and 17.

When looking at figure 18 from left to right, the value of R(x, y) – I(x, y) is larger in magnitude at the right than it is at the left. Such a difference between the two measurements of the value of R(x, y) – I(x, y) is seen in figure 15.

The evidence demonstrating the answer to the question: *"What is the meaning conveyed through the ratio of improvement of resilient versus classic signal reconstruction?"* shall be presented in visual form through the maps of the absolute value of the difference between the Mean Absolute Error (MAE) of the classic formula and the MAE of the resilient formula, at the intra-pixel locations (misplacements) in the ranges (x, y) in [(0.01, 0.01), (0.06, 0.06)] and (x, y) in [(0.94, 0.94), (0.99, 0.99)] of figures 15(c), 15(d) respectively.

The maps of MAE absolute differences are shown in figure 19 along with the table in (a) showing the details relevant to each figure shown from 19(b) through figure 19(m). With the increase of misplacement from (x, y) = (0.01, 0.01) to (x, y) = (0.06, 0.06), the table in (a) shows the MAE of the classic formula decreasing persistently, the MAE of the resilient formula almost increasing persistently and ratio of improvement almost decreasing persistently (this connects with the right side of figure 18 and the interval [(0.01, 0.01), (0.05, 0.05)] because the curvature of the two formulae tends not to be the same when moving towards the middle).

When the misplacement decreases from (x, y) = (0.99, 0.99) to (x, y) = (0.94, 0.94), the MAE of classic formula increases constantly and the MAE of the resilient formula shows oscillating behaviour concurrently with the quasi-constant increase of the ratio of improvement, and such increase is less in its magnitude than the decrease observed into the interval [(0.01, 0.01), (0.05, 0.05)], and so, the behaviour observed into the interval [(0.99, 0.99), (0.94, 0.94)] does not correlate with the left side of figure 18, because the ratio of improvement increases, when moving out from (0.99, 0.99), as shown in the table in (a).

In figures 19(b) through 19(m) the absolute difference between the Mean Square Error (MAE) of the classic formula and the MAE of the resilient formula, show images becoming darker consistently with the persistent decrease of the absolute difference between the MAE of the classic formula and MAE of the resilient formula, observable when looking at the third and fourth column (from left to right) of the table in (a) (the MAE Classic is persistently decreasing and the MAE Resilient quasi-constantly increasing). This behaviour correlates well with the quasi constant decrease of the ratio of improvement seen in the right most column of the table in (a). Therefore, figure 19 shows that the profiles of the resilient curvature and the classic interpolation functions are exhaustive of the data seen in figures 12, 13, 16 and 17, only in relationship to the first half of the pixel (from the far right to the middle).

SIGNAL RESILIENT TO INTERPOLATION

Transfer Function Derived from the Curvature of the Classic Interpolation Formula

In this section another option for conceiving an interpolation formula is shown, and this option is offered through the total curvature. To be specific, the total curvature of the classic interpolation formula allows the following argument to be made.

Equation 23(a) is pertinent to classic signal reconstruction and equation 24(a) is pertinent to classic signal reconstruction obtained through the PIC. The two equations differ because the convolution named ξ (x, y) in 23(a) is substituted with PIC (x, y) in 24(a). Conceptually, convolution and PIC are of the same nature because they both furnish that fraction of pixel intensity value which is necessary in order to reconstruct the signal at a given intra-pixel location (x, y) when the value of pixel intensity f (0, 0) is given. However, convolution and PIC are not the same because of a difference in their theoretical foundations and practical implications.

PIC is proportional to the tangent of the angle, subtended from the tangent to the first order derivative with the horizontal (parallel to the axis of the abscissa) or the vertical (parallel to the axis of the ordinate), times the misplacement; whereas the convolution is the fraction value of the pixel intensity. Thus, the difference consists in the way they are calculated.

The convolution needs scaling through the transfer function, which can be chosen arbitrarily, whereas the PIC is capable of being employed in order to derive the transfer function of a given interpolation formula and such a transfer function removes the arbitration introduced through transfer functions of the choice, such as those transfer functions studied in this chapter (sin(p), cos(p) and log($|1.0 / p|$)). As to how it is possible to invent the transfer function obtained from the PIC, the answer springs from the nature of the interpolator, and specifically from the math form which defines the total curvature of the interpolator. And the total curvature is sufficient to define the transfer function of the interpolation formula.

The math treatise of the total curvature of classic interpolation as pertinent to bivariate quadratic and cubic polynomials is included in part I of chapters 2 and 3 (while calculating the mathematical formulation of the signal resilient to interpolation), with and without embedding respectively.

The presentation of the results of the comparison of classic signal reconstruction (equation 23(a)), classic signal reconstruction obtained through the PIC (equation 24(a)) and resilient signal reconstruction obtained through the PIC (equation 24(b)), is given below through the graphs of the Mean Absolute Error (MAE). 'Without TF' signifies that interpolation is conducted with the transfer function derived from the total curvature of the math formulation of the interpolator. The interpolators are: bivariate quadratic, bivariate cubic and bivariate cubic Lagrange formula, and each of the three is calculated with embedding and without embedding ('WE'), giving a total of six interpolators. The nomenclature 'without TF' is abbreviated with 'WTF' (see Table XIII).

As far as regards the bivariate quadratic interpolation formulae, figure 20 shows the graphs of the MAE of classic interpolation in (a) with sin(p) as the transfer function (TF), in (b) with cos(p) as the transfer function, and in (c) with log($|1.0 / p|$) as the transfer function. The MAE of the classic and resilient interpolation formulae with the transfer function derived from the total curvature with embedding are shown in (d) and (e), and those without embedding are shown in (f) and (g). Naturally, the total curvature of classic and resilient interpolation formulae are not the same as it is shown in chapters 2, 3, 5, 6 and 7. The values of the MAE on the ordinate refer to pixel intensity. The pixel intensity values ranges in [0, 32757]. The behaviour of the MAE seen in the graphs (a) and (b) (sin(p) and cos(p)) is predominantly constant, the value is approximately 0.8. In (c), the MAE ranges predominantly between 20 and 25 (log($|1.0 / p|$)), in (d) it linearly increases with increasing misplacement (x, y) up to approximately below 3.00 E-02, in (e) it is predominantly below 6.00 E-02, in (f) it increases linearly up to approximately below 3.00 E-02 just as in (d), and finally in (g) its linear

Transfer Function	Classic with TF
Quadratic	
sin(p)	H32DsinTF
cos(p)	H32DcosTF
log(|1.0/p|)	H32DlogTF
Cubic	
sin(p)	H42DsinTF
cos(p)	H42DcosTF
log(|1.0/p|)	H42DlogTF
Cubic Lagrange	
sin(p)	LGR2DsinTF
cos(p)	LGR2DcosTF
log(|1.0/p|)	LGR2DlogTF

Table XII: Headings of the graphs shown in figures 20 through 23: classic interpolation with transfer function ('TF').

increase with increasing misplacement reaches up to approximately below 3.00 E-02. The misplacement (x, y) ranges in [(0.01, 0.01), (0.99, 0.99)] and is shown on the abscissa with the relationship 1 to 100. Tables XII and XIII present the nomenclature to be used to read figures 20 through 23. There are three classes of interpolation formulae: (i) classic with TF, (ii) classic without TF, and (iii) resilient without TF. With TF signifies that the interpolation is conducted with either of the three transfer functions: sin(p), cos(p) and log(\|1.0 / p\|) as shown in Table XII. The following remark is due: the graphs seen in (f) and (d) in figure 20 are exactly the same, and the reason is because the calculation of the total curvature of the

Classic without TF With E	Classic without TF WE	Resilient without TF With E	Resilient without TF WE
Quadratic			
H32DClassicWTF	H32DWEClassicWTF	H32DResilientWTF	H32DWEResilientWTF
Cubic			
H42DClassicWTF	H42DWEClassicWTF	H42DResilientWTF	H42DWEResilientWTF
Cubic Lagrange			
LGR2DClassicWTF	LGR2DWEClassicWTF	LGR2DResilientWTF	LGR2DWEResilientWTF

Table XIII: Headings of the graphs shown in figures 20 through 23: classic and resilient interpolation without TF ('WTF'), with embedding ('With E') and without embedding ('WE').

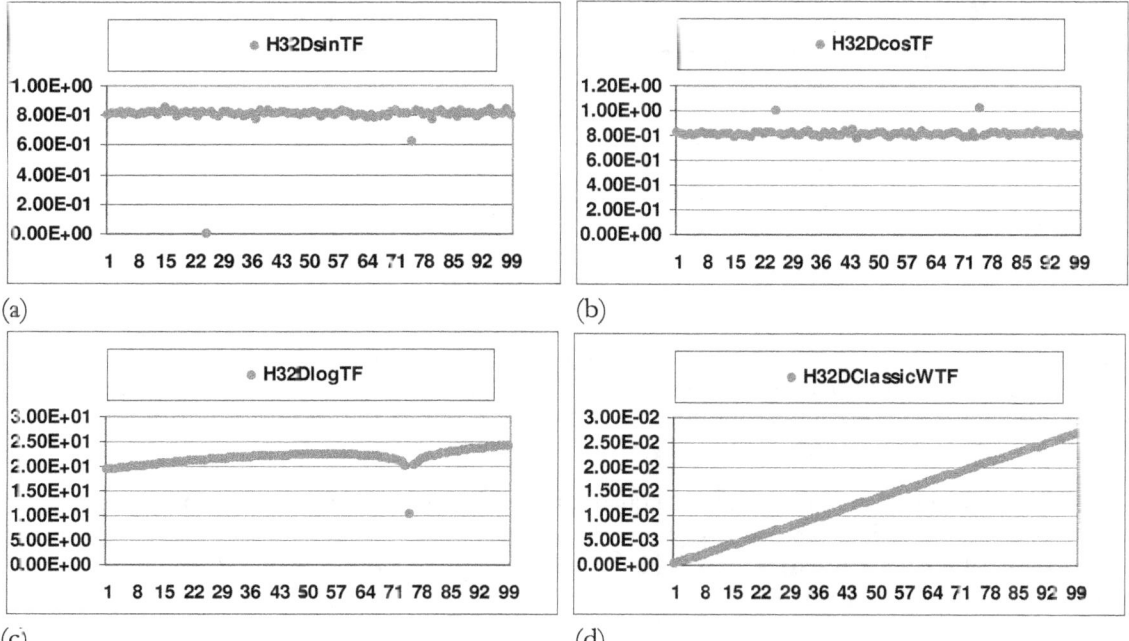

(a)   (b)   (c)   (d)

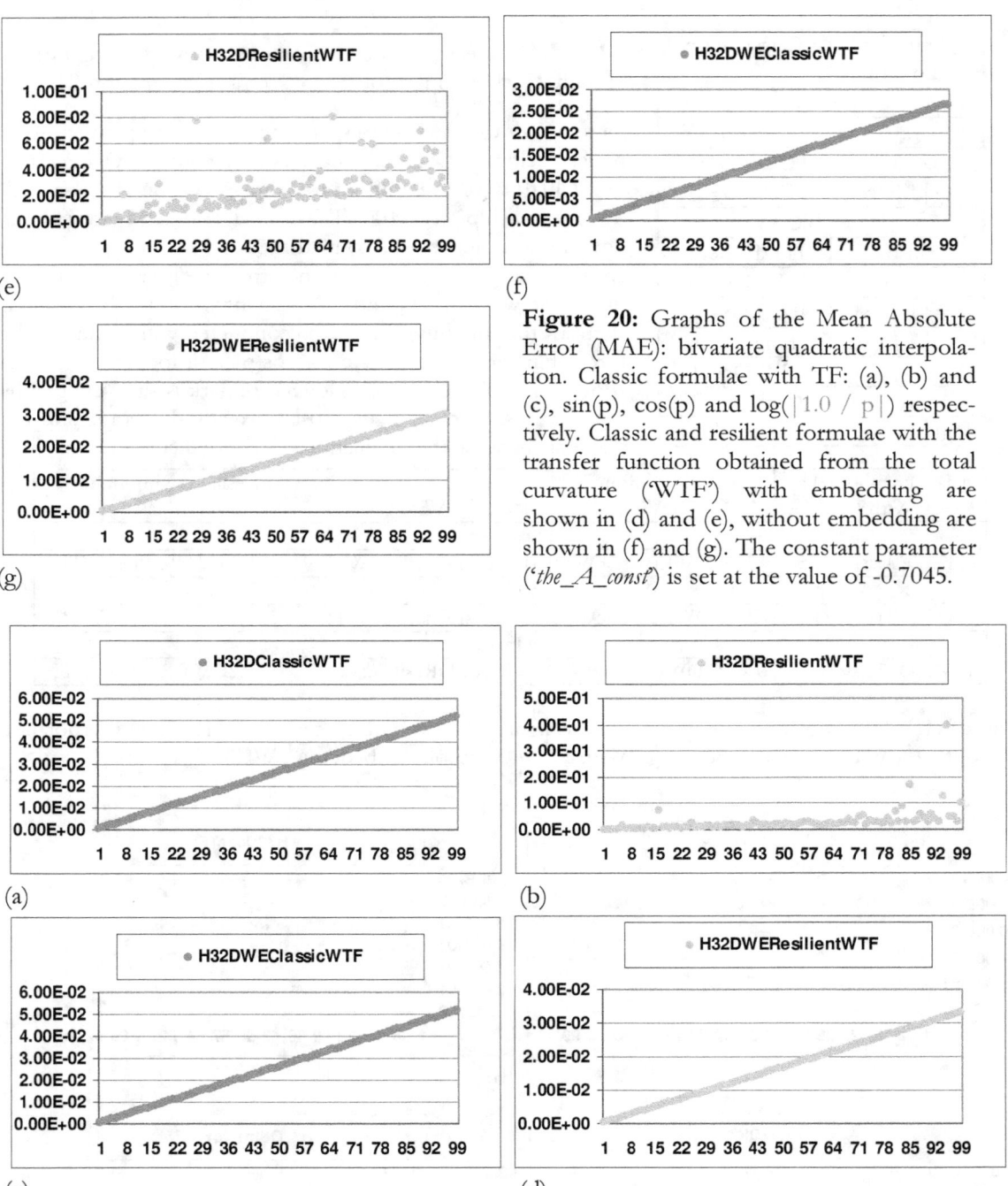

(e)

(f)

Figure 20: Graphs of the Mean Absolute Error (MAE): bivariate quadratic interpolation. Classic formulae with TF: (a), (b) and (c), sin(p), cos(p) and log($|1.0 / p|$) respectively. Classic and resilient formulae with the transfer function obtained from the total curvature ('WTF') with embedding are shown in (d) and (e), without embedding are shown in (f) and (g). The constant parameter ('*the_A_const*') is set at the value of -0.7045.

(g)

(a)

(b)

(c)

(d)

Figure 21: Behaviour of the MAE: bivariate quadratic classic and resilient interpolation formulae with the transfer function derived from the total curvature (the parameter '*the_A_const*' is set at -0.45). The graphs in (a) and (b) are relevant to formulae with embedding and the graphs in (c) and (d) are relevant to formulae without embedding.

classic interpolation formula with or without embedding results in the same math. This happens because f (0, 0) is constant and additive and thus removed when calculating any derivatives, and it can be understood through the comparison of equations (6), (7) and (8) (bivariate polynomials with embedding) with equations (9), (10) and (11) (bivariate polynomials without embedding). The remark is to be considered relevant to the graphs in figures 21, 22 and 23 also. It is not possible to infer general behaviours from figure 20. We can see, though, that the bivariate quadratic interpolation formula with the transfer function derived from the total curvature, both in its resilient and classic forms, outperforms the bivariate classic interpolation formula when the transfer function is sin(p), cos(p) and log(|1.0 / p|).

Such performance is seen also when changing the value of the constant parameter (*'the_A_const'*) to -0.45 as illustrated in figure 21. The behaviour is almost the same showing that the good performance is displayed in the graphs of both classic and resilient formulae with and without embedding. It is both informative and necessary that we should highlight the difference between the formulations obtained to calculate the total curvature in both bivariate quadratic classic and resilient interpolation formulae with the transfer function derived from the curvature.

The bivariate quadratic classic interpolation formula calculated either with embedding or without embedding has a total curvature which is independent of misplacements and heavily dependent on the value of the *'the_A_const'*:

$$(\partial^2 (h_3(x, y)) / \partial x^2) = - 4a\ \alpha_1 + 2a\ \alpha_2; (\partial^2 (h_3(x, y)) / \partial x \partial y) = - 4a\ \alpha_1 + 2a\ \alpha_2;$$

$$(\partial^2 (h_3(x, y)) / \partial y \partial x) = - 4a\ \alpha_1 + 2a\ \alpha_2; (\partial^2 (h_3(x, y)) / \partial y^2) = - 4a\ \alpha_1 + 2a\ \alpha_2;$$

(see chapter 2 Part I, with embedding and chapter 3 Part I, without embedding). Also, the bivariate quadratic resilient interpolation formula calculated without embedding is only dependent on the values of the *'the_A_const'* and signal intensity, as reported in equation (58) of chapter 5.

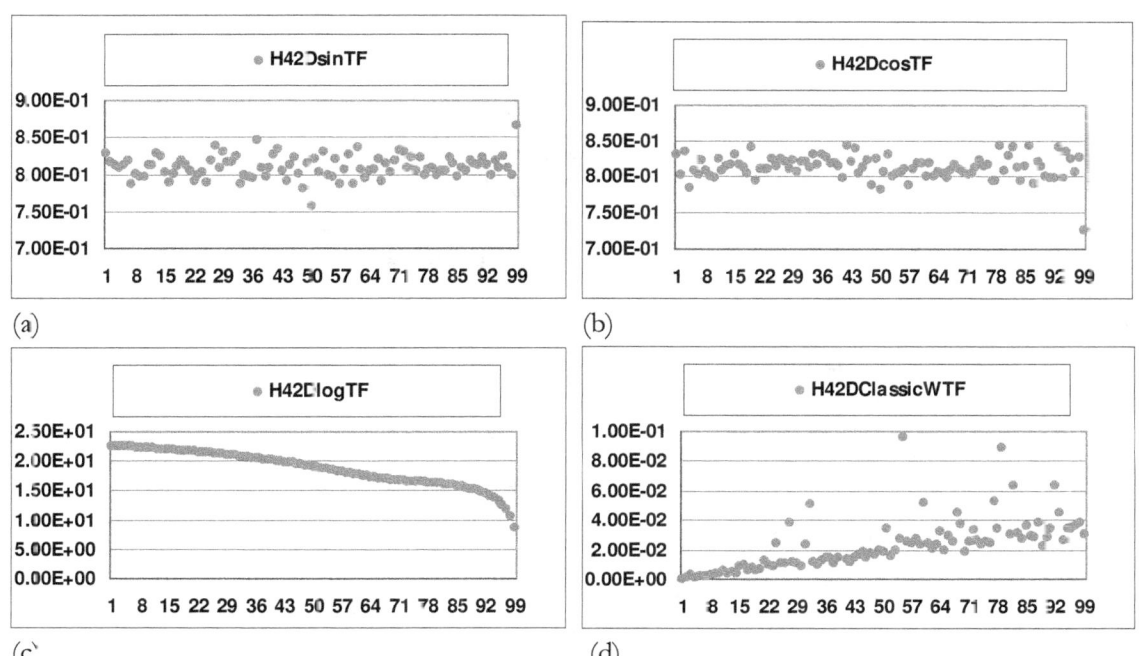

(a)

(b)

(c)

(d)

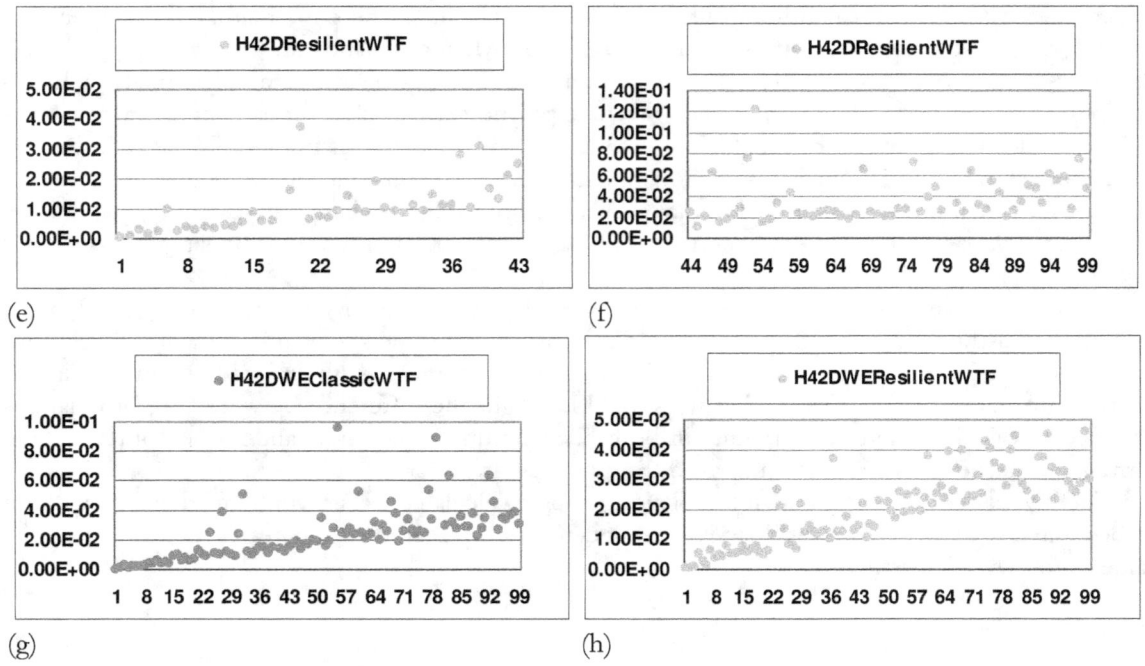

Figure 22: Graphs of the MAE: bivariate cubic interpolation formulae with transfer functions: sin(p) (a), cos(p) (b), log(| 1.0 / p |) (c). Classic and resilient formulae with the transfer function derived from the total curvature ('WTF'): with embedding are shown in (d) through (f), and without embedding in (g) and (h).

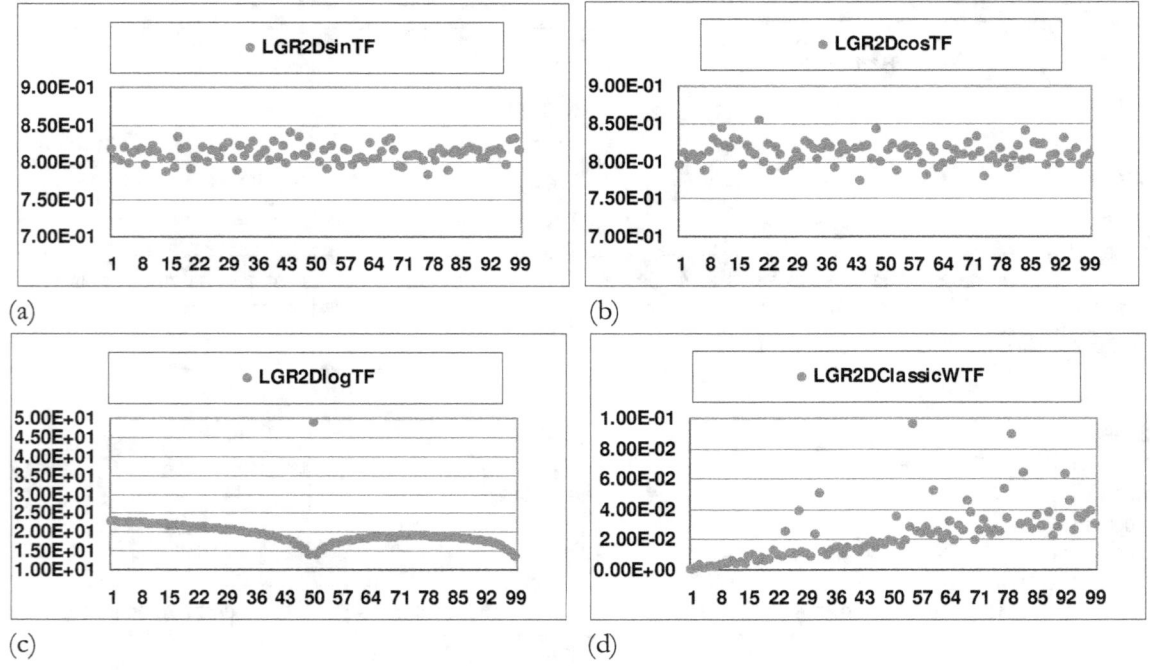

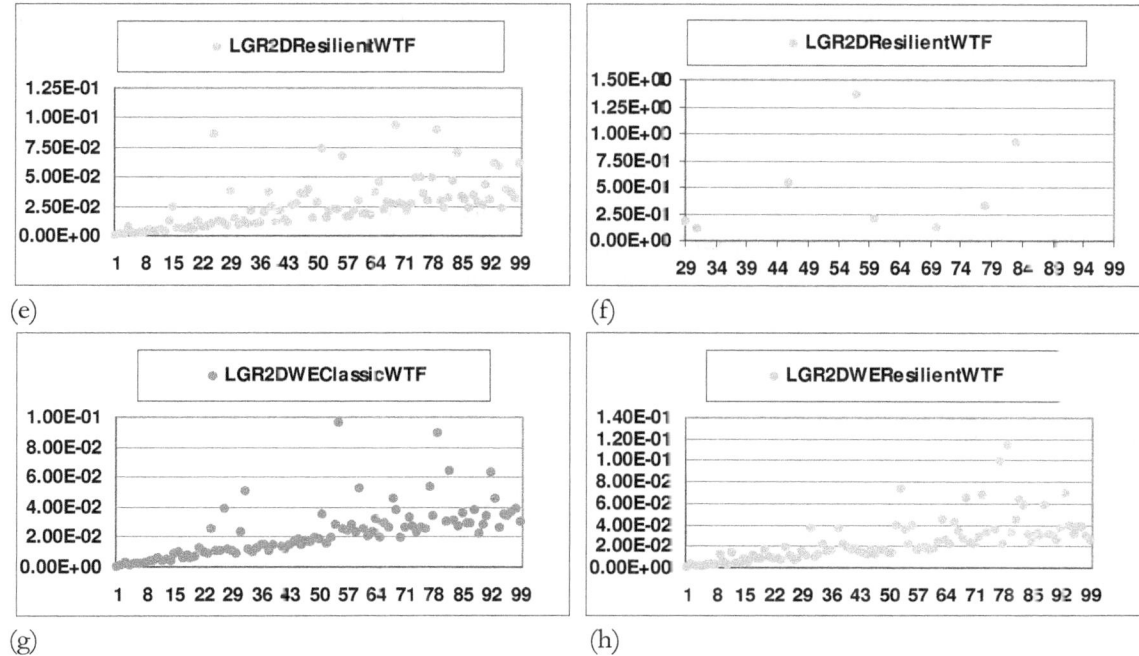

Figure 23: Graphs of MAE relevant to the bivariate cubic Lagrange formulae with transfer functions: (a), (b) and (c), referring to sin(p), cos(p) and log(|1.0 / p|) respectively. Classic and resilient formulae with the transfer function derived from their own math form through the use of the total curvature ('WTF') are shown in (d) and (e) and (f): with embedding; and in (g) and (h): without embedding.

Even though the bivariate quadratic resilient interpolation formula calculated with embedding has the total curvature C(R'(x, y), f)) given by adding up the equations: (26), (29), (31) and (39) of chapter 5, it is indeed dependent on a convolution made of signal values and misplacement.

The behaviour of the MAE seen in figure 22(a) and 22(o) shows the large majority of values to be in between 0.75 and 0.85 (respectively sin(p) and cos(p)), it decreases with increasing misplacement, with values in between approximately below 25 and above 5 (see figure 22(c), log(|1.0 / p|)).

Interpolation with the transfer function derived from the total curvature shows in (d) behaviour of the MAE which features a quasi linear increase with increasing misplacement up to approximately below 0.04 (classic with embedding), with some scattered values above 0.04.

In (e) and (f), the MAE shows a predominance of values below 0.03 (range of misplacement in [(0.01, 0.01), (0.43, 0.43)]) and 0.08 (range of misplacement in [(0.44, 0.44), (0.99, 0.99)]) respectively (resilient with embedding).

Graphs relevant to classic and resilient formulae with the transfer function derived from the total curvature ('WTF'), without embedding are shown in figure 22(g) and 22(h). In (g) the same results as in (d) are seen, because the total curvature is the same in both of the classic formulae: with and without embedding.

The graph in (h) is relevant to the resilient formula calculated without embedding with the transfer function derived from the total curvature ('WTF'). It shows values of the MAE below 0.05. Validation of the bivariate cubic interpolation formulae WTF need experimentation with a larger data set to conjecture general behaviours.

Figure 23 shows behaviour of the MAE in (a) and (b) with values predominantly in between 0.8 and 0.85; the transfer functions are sin(p) and cos(p) respectively. In (c) the transfer function $\log(|\,1.0\,/\,p\,|)$ shows values between approximately 10 and 25 with a peak seen at the location $(x, y) = (0.5, 0.5)$. Data relevant to the classic formula with transfer function derived from the total curvature are seen in (d) and (g). These results are the same as those of the classic bivariate cubic WTF seen in figure 22(d) and 22(g). The reason for this behaviour naturally derives from the math resulting from the calculation of the curvature of the two classic bivariate cubic formulae: $h_4(x, y)$ and $LGR_3(x, y)$ (in both of the cases: with and without embedding). The math of the total curvature of $h_4(x, y)$ and $LGR_3(x, y)$ is exactly the same.

The reader is referred to part I of chapters 2 and 3 where the total curvature is visible:

$$(\partial^2\,(\,h_4(x, y)\,)\,/\partial x^2) + (\partial^2\,(\,h_4(x, y)\,)\,/\partial y^2) + (\partial^2\,(\,h_4(x, y)\,)\,/\partial x \partial y) + (\partial^2\,(\,h_4(x, y)\,)\,/\partial y \partial x)\ \text{results in}$$

the same math form of the total curvature:

$$(\partial^2\,(LGR_3(x, y)\,)\,/\partial x^2) + (\partial^2\,(LGR_3(x, y)\,)\,/\partial y^2) + (\partial^2\,(LGR_3(x, y)\,)\,/\partial y \partial x) + (\partial^2\,(LGR_3(x, y)\,)\,/\partial x \partial y),$$

which is:

$$4 * \{\,\alpha_3 * [\,3\,(x + y) - 2\,] + \alpha_2 * [\,-\,(x + y) + 2\,]\,\}.$$

In (e), (f) and (h) in figure 23 the data are relevant to the resilient formulae with the transfer function derived from the total curvature with embedding: (e), (f); and without embedding (h). Data in (e) range across the span of misplacements (x, y) in $[(0.01, 0.01), (0.99, 0.99)]$ showing the MAE values below 0.1, except for those values seen in (f) which are below 1.5. Data in (h) show the MAE values below 0.12. Data seen in the graph of figure 23(f) are missing from figure 23(e), so as to improve the visibility of both (e) and (f).

SUMMARY

It is worthwhile to remove the arbitration introduced through the choice of the transfer function employed to scale the convolutions in either classic or resilient interpolation. Given an interpolation formula, to remove such arbitration gives control over the variability of the interpolation error produced through diverse transfer functions.

In general, a suitable transfer function is such that extreme filtering of the convolution does not happen, and is such that it does not ignore the relevance of the curvature of the model interpolation function within the neighbourhood of the pixel, otherwise signal reconstruction is incongruent.

To summarize, the use of the curvature of the resilient interpolation function in order to derive the transfer function can be beneficial in a two-fold manner: (i) removing the arbitration induced when using the transfer function of the choice to process the convolution of the classic interpolation function, and (ii) improving the performance of signal reconstruction in terms of reduced interpolation error.

To conclude, it can be argued that the total curvature of the interpolator is a viable option to obtain the transfer function from the math form of the interpolator, thus removing the arbitration induced in the results through transfer functions of the choice. When designing an interpolator, the interpolation error can be reduced using its own total curvature.

REFERENCES

Agarwal, R. P., & Wong, P. J. Y. (1993). *Error inequalities in polynomial interpolation and their applications. Mathematics and its applications.* Dordrecht, The Netherlands: Kluwer Academic Publishers.

Blu, T., & Unser M. (1999). *Quantitative Fourier analysis of approximation techniques: Part I – interpolators and projectors.* IEEE Transactions on Signal Processing, 47(10), 2783-2795.

Blu, T., Thevenaz, P. & Unser M. (2001). *MOMS: Maximal-order interpolation of minimal support.* IEEE Transactions on Image Processing, 10(7), 1069-1080.

Blu, T., Thevenaz, P. & Unser, M. (2003). *Complete parameterization of piecewise-polynomial interpolation kernels.* IEEE Transactions on Image Processing, 12(11), 1297-1309.

Blu, T., Thevenaz, P. & Unser, M. (2004). *Linear interpolation revitalized.* IEEE Transactions on Image Processing, 13(5), 710-719.

Ciulla, C. (2009). *Improved Signal and Image Interpolation in Biomedical Applications: The Case of Magnetic Resonance Imaging (MRI)* – Medical Information Science Reference - IGI Global Publisher. Hershey, PA, U.S.A.

De Boor, C. & Fix, G. J. (1973). *Spline approximation by quasi-interpolants.* Journal of Approximation Theory, 8, 19-45.

De Boor, C. (1978). *A practical guide to splines. Applied mathematical sciences.* New York, NY: Springer-Verlag.

Grevera, G. J. & Udupa, J. K. (1996). *Shape-based interpolation of multidimensional grey-level images* IEEE Transactions on Medical Imaging, 15(6), 881-892.

Grevera, G. J. & Udupa, J. K. (1998). *An objective comparison of 3-D image interpolation methods.* IEEE Transactions on Medical Imaging, 17(4), 642-652.

Hajnal, J. V., Saeed, N., Soar, E. J., Oatridge, A., Young, I. R. & Bydder, G. M. (1995). *A registration and interpolation procedure for subvoxel matching of serially acquired MR images.* Journal of Computer Assisted Tomography, 19(2), 289-296.

Herman, G. T., Zheng, J. & Bucholtz, C. A. (1992). *Shape-based interpolation.* IEEE Computer Graphics and Applications, 12(3), 69-79.

Lorentz, H. A., Einstein, A., Minkowski, H. & Weyl, H. (1952). *The principle of relativity: A collection of original memoirs on the special and general theory of relativity* (W. Perrett & G. B. Jeffery, Trans.). New York, NY: Dover Publications. (Original work published 1923).

Newton I. & Huygens, C. (1934). The motion of the moon's nodes. In R. M. Aynard Hutchins (Ed.), *Mathematical Principles of Natural Philosophy: Optics, Treatise on light* (A. Motte, Trans.). (pp. 338-339). William Benton.

Raya, S. P. & Udupa J. K. (1990). *Shape-based interpolation of multidimensional objects.* IEEE Transactions on Medical Imaging, 9(1), 32-42.

Schoenberg, I. J. (1946a). *Contributions to the problem of approximation of equidistant data by analytic functions. Part A. On the problem of smoothing or graduation. A first class of analytic approximation formulae* Quarterly of Applied Mathematics, 4, 45-99.

Schoenberg, I. J. (1946b). *Contributions to the problem of approximation of equidistant data by analytic functions. Part A. On the problem of osculatory interpolation. A second class of analytic approximation formulae.* Quarterly of Applied Mathematics, 4, 112-141.

Schoenberg, I. J. (1969). *Cardinal interpolation and spline functions.* Journal of Approximation Theory 2, 167-206.

Unser, M. & Daubechies, I. (1997). *On the approximation power of convolution-based least squares versus interpolation.* IEEE Transactions on Signal Processing, 45(7), 1697-1711.

Unser, M., Aldroubi, A. & Eden, M. (1993a). *B-spline signal processing: Part I – theory.* IEEE Transactions on Signal Processing, 41(2), 821-833.

Unser, M., Aldroubi, A. & Eden, M. (1993b). *B-spline signal processing: Part II – efficient design and applications*. IEEE Transactions on Signal Processing, 41(2), 834-848.

Waldron, S. (1998). *The error in linear interpolation at the vertices of a simplex*. SIAM Journal on Numerical Analysis 35(3), 1191-1200.

CHAPTER 5

CALCULATION OF THE CURVATURE OF THE BIVARIATE QUADRATIC SIGNAL RESILIENT TO INTERPOLATION: WITH AND WITHOUT EMBEDDING

INTRODUCTION

In this chapter the total curvature of the bivariate quadratic signal resilient to interpolation is calculated both with and without embedding and the math developments are oulined at each step. In chapter 4 the reader was informed how to obtain the custom transfer function from the math form of the interpolation formula: $I(x, y) = f(0, 0) + \xi(x, y)$ (classic signal, equation (23.a) in chapter 4) or $R(x, y) = f(0, 0) + \varrho(x, y)$ (signal resilient to interpolation, equation (23.b) in chapter 4). In the specifics of this chapter the classic signal is equation (3) in chapter 2: $I(x, y) = h_3(x, y) = f(0, 0) + \alpha_1 * [- 2a(x + y)^2 + 1/2(a+1)] + \alpha_2 * [a(x + y)^2 - (2a + 1/2)(x + y) + 3/4(a+1)]$, with embedding and $I(x, y) = h_3(x, y) = \alpha_1 * [- 2a(x + y)^2 + 1/2(a+1)] + \alpha_2 * [a(x + y)^2 - (2a + 1/2)(x + y) + 3/4(a-1)]$ without embedding. The signal resilient to interpolation is given in this chapter in equation (1): $\varrho(x, y) = f(1/2, 1/2) = - \{\beta_1 * \gamma_1 + \beta_2 * \gamma_2\} / \{\gamma_1 + \gamma_2\}$, with embedding and equation (40): $\varrho(x, y) = \{\alpha_1 * [- 2a(x^2/3 + xy/2 + y^2/3) + 1/2(a+1)] + \alpha_2 * [a(x^2/3 + xy/2 + y^2/3) - (2a + 1/2)(x/2 + y/2) + 3/4(a+1)]\}$, without embedding.

The Pixel Intensity Correction needs to be calculated through the formula PIC = Shift * tan(total curvature); the shift is obtained through the Pythagorean theorem and the total curvature of the classic signal is given through the sum of second order partial derivatives of $h_3(x, y)$, which is: $(\partial^2(h_3(x, y))/\partial x^2) + (\partial^2(h_3(x, y))/\partial y^2) + (\partial^2(h_3(x, y))/\partial x\partial y) + (\partial^2(h_3(x, y))/\partial y\partial x)$.

Given the math form of the classic signal $I(x, y)$, the total curvature with and without embedding is the same. The total curvature of the signal resilient to interpolation: $\varrho(x, y)$ changes with embedding in equation (1) and without embedding is in equation (40). In either case the total curvature is given through the sum: $(\partial^2(\varrho(x, y))/\partial x^2) + (\partial^2(\varrho(x, y))/\partial y^2) + (\partial^2(\varrho(x, y))/\partial x\partial y) + (\partial^2(\varrho(x, y))/\partial y\partial x)$; and it is calculated in this chapter adding together the equations (26), (29), (31) and (39) (with embedding) and in equation (58) (without embedding).

BIVARIATE QUADRATIC SIGNAL CALCULATED WITH EMBEDDING

The signal resilient to interpolation is given in the following equation:

$$f(1/2, 1/2) = - \{\beta_1 * \gamma_1 + \beta_2 * \gamma_2\} / \{\gamma_1 + \gamma_2\} \tag{1}$$

Where it is posited:

$$\beta_1 = f(-1/2, -1/2) + f(2/3, 2/3) + f(-2/3, -2/3) \tag{1.a}$$

$$\beta_2 = f(-1/2, -1/2) + f(2/3, 2/3) + f(-2/3, -2/3) + f(-1, -1) + f(1, 1) +$$

$$f(3/2, 3/2) + f(-3/2, -3/2) \tag{1.b}$$

$$\gamma_1 = [- 2a(x^3y/3 + x^2y^2/2 + xy^3/3) + 1/2 xy(a+1)] \tag{1.c}$$

$$\gamma_2 = [a(x^3y/3 + x^2y^2/2 + xy^3/3) - (2a + 1/2)(x^2y/2 + xy^2/2) + 3/4 xy(a+1)] \tag{1.d}$$

For convenience let us calculate the first order partial derivatives of γ_1 and γ_2 with respect to the variables x and y.

$$(\partial\,(\gamma_1)\,/\partial x) = (\partial\,[-\,2a\,(x^3y/3 + x^2y^2/2 + xy^3/3) + 1/2\,xy\,(a+1)\,]\,/\partial x) =$$

$$[-\,2a\,(x^2y + xy^2 + y^3/3) + 1/2\,y\,(a+1)\,] \tag{2}$$

$$(\partial\,(\gamma_1)\,/\partial y) = (\partial\,[-\,2a\,(x^3y/3 + x^2y^2/2 + xy^3/3) + 1/2\,xy\,(a+1)\,]\,/\partial y) =$$

$$[-\,2a\,(x^3/3 + x^2y + xy^2) + 1/2\,x\,(a+1)\,] \tag{3}$$

$$(\partial\,(\gamma_2)\,/\partial x) = (\partial\,[\,a\,(x^3y/3 + x^2y^2/2 + xy^3/3) - (2a + 1/2)\,(x^2y/2 + xy^2/2) + 3/4\,xy\,(a+1)\,]$$

$$/\partial x) = [\,a\,(x^2y + xy^2 + y^3/3) - (2a + 1/2)\,(xy + y^2/2) + 3/4\,y\,(a+1)\,] \tag{4}$$

$$(\partial\,(\gamma_2)\,/\partial y) = (\partial\,[\,a\,(x^3y/3 + x^2y^2/2 + xy^3/3) - (2a + 1/2)\,(x^2y/2 + xy^2/2) + 3/4\,xy\,(a+1)\,]$$

$$/\partial y) = [\,a\,(x^3/3 + x^2y + xy^2) - (2a + 1/2)\,(x^2/2 + xy) + 3/4\,x\,(a+1)\,] \tag{5}$$

Now let us calculate the second order partial derivatives of γ_1 and γ_2 with respect to the variables x and y.

$$(\partial^2\,(\gamma_1)\,/\partial x^2) = (\partial\,[-\,2a\,(x^2y + xy^2 + y^3/3) + 1/2\,y\,(a+1)\,]\,/\partial x) = [-\,2a\,(2xy + y^2)\,] \tag{6}$$

$$(\partial\,(\partial\,(\gamma_1)\,/\partial x)\partial y) = (\partial\,[-\,2a\,(x^2y + xy^2 + y^3/3) + 1/2\,y\,(a+1)\,]\,/\partial y) =$$

$$[-\,2a\,(x^2 + 2xy + y^2) + 1/2\,(a+1)\,] \tag{7}$$

$$(\partial^2\,(\gamma_1)\,/\partial y^2) = (\partial\,[-\,2a\,(x^3/3 + x^2y + xy^2) + 1/2\,x\,(a+1)\,]\,/\partial y) = [-\,2a\,(x^2 + 2xy)\,] \tag{8.a}$$

$$(\partial\,(\partial\,(\gamma_1)\,/\partial y)\partial x) = (\partial\,[-\,2a\,(x^3/3 + x^2y + xy^2) + 1/2\,x\,(a+1)\,]\,/\partial x) =$$

$$[-\,2a\,(x^2 + 2xy + y^2) + 1/2\,(a+1)\,] \tag{8.b}$$

Thus: $(\partial\,(\partial\,(\gamma_1)\,/\partial x)\partial y) = (\partial\,(\partial\,(\gamma_1)\,/\partial y)\partial x) = [-\,2a\,(x^2 + 2xy + y^2) + 1/2\,(a+1)\,]$

Additionally:

$$(\partial^2\,(\gamma_2)\,/\partial x^2) = (\partial\,[\,a\,(x^2y + xy^2 + y^3/3) - (2a + 1/2)\,(xy + y^2/2) + 3/4\,y\,(a+1)\,]\,/\partial x) =$$

$$[\,a\,(2xy + y^2) - (2a + 1/2)y\,] \tag{9}$$

$$(\partial\,(\partial\,(\gamma_2)\,/\partial x)\partial y) = (\partial\,[\,a\,(x^2y + xy^2 + y^3/3) - (2a + 1/2)\,(xy + y^2/2) + 3/4\,y\,(a+1)\,]\,/\partial y) =$$

$$[\,a\,(x^2 + 2xy + y^2) - (2a + 1/2)\,(x + y) + 3/4\,(a+1)\,] \tag{10}$$

$$(\partial\,(\partial\,(\gamma_2)\,/\partial y)\partial x) = (\partial\,[\,a\,(x^3/3 + x^2y + xy^2) - (2a + 1/2)\,(x^2/2 + xy) + 3/4\,x\,(a+1)\,]\,/\partial x) =$$

$$[a (x^2 + 2xy + y^2) - (2a + 1/2) (x + y) + 3/4 (a+1)] \tag{11}$$

Thus:

$$(\partial (\partial (\gamma_2) /\partial x)\partial y) = (\partial (\partial (\gamma_2) /\partial y)\partial x) = [a (x^2 + 2xy + y^2) - (2a + 1/2) (x + y) + 3/4 (a+1)]$$

$$(\partial^2 (\gamma_2) /\partial y^2) = (\partial [a (x^3/3 + x^2y + xy^2) - (2a + 1/2) (x^2/2 + xy) + 3/4 x (a+1)] /\partial y) =$$

$$[a (x^2 + 2xy) - (2a + 1/2) x] \tag{12}$$

From the generalized form of the bivariate resilient interpolation calculated with embedding:

$R (x, y) = f (0, 0) + \varrho (x, y)$, and letting $\varrho (x, y) = - \{ \beta_1 * \gamma_1 - \beta_2 * \gamma_2 \} / \{ \gamma_1 + \gamma_2 \}$, is obtained:

$$R((x, y), f) = f (1/2, 1/2) + f (1/2, 1/2) - \{ \beta_1 * \gamma_1 + \beta_2 * \gamma_2 \} / \{ \gamma_1 + \gamma_2 \} \tag{13}$$

Let the total curvature $C(R((x, y), f))$ be defined as:

$$C(R((x, y), f)) = (\partial^2 (R((x, y), f)) /\partial x^2) + (\partial (\partial (R((x, y), f)) /\partial y)\partial x) + (\partial (\partial (R((x, y),$$

$$f) /\partial x)\partial y) + (\partial^2 (R((x, y), f)) /\partial y^2) \tag{14}$$

Second Order Partial Derivatives of R((x, y), f)

Let us calculate the second order partial derivative of $R((x, y), f)$ with respect to the variable x:

$$(\partial^2 (R((x, y), f)) /\partial x^2) = - (\partial^2 \{ \{ \beta_1 * \gamma_1 + \beta_2 * \gamma_2 \} / \{ \gamma_1 + \gamma_2 \} \} /\partial x^2) =$$

$$- (\partial (\partial \{ \{ \beta_1 * \gamma_1 + \beta_2 * \gamma_2 \} / \{ \gamma_1 + \gamma_2 \} \} /\partial x) /\partial x) =$$

$$- (\partial \{ \{ \{ \beta_1 * (\partial (\gamma_1) /\partial x) + \beta_2 * (\partial (\gamma_2) /\partial x) \} * \{ \gamma_1 + \gamma_2 \} - \{ \beta_1 * \gamma_1 + \beta_2 * \gamma_2 \} * \{ (\partial (\gamma_1)$$

$$/\partial x) + (\partial(\gamma_2) /\partial x) \} \} / \{ \gamma_1 + \gamma_2 \}^2 \} /\partial x) \tag{15}$$

Let us posit: $\pi_1 = \{ \{ \beta_1 * (\partial (\gamma_1) /\partial x) + \beta_2 * (\partial (\gamma_2) /\partial x) \} * \{ \gamma_1 + \gamma_2 \} - \{ \beta_1 * \gamma_1 + \beta_2 * \gamma_2 \}$

$$* \{ (\partial (\gamma_1) /\partial x) + (\partial (\gamma_2) /\partial x) \} \} \tag{16}$$

Thus equation (15) is written as: $- (\partial \{ \pi_1 / \{ \gamma_1 + \gamma_2 \}^2 \} /\partial x)$, therefore:

$$(\partial^2 (R((x, y), f)) /\partial x^2) = - \{ \{ (\partial (\pi_1) /\partial x) * \{ \gamma_1 + \gamma_2 \}^2 - \pi_1 * (\partial \{ \gamma_1 + \gamma_2 \}^2 /\partial x) \} \} / \{ \gamma_1 + \gamma_2 \}^4 \tag{17}$$

Let us calculate the second order partial derivative of $R((x, y), f)$ with respect to the variable y:

$$(\partial^2 (R((x, y), f)) /\partial y^2) = - (\partial^2 \{ \{ \beta_1 * \gamma_1 + \beta_2 * \gamma_2 \} / \{ \gamma_1 + \gamma_2 \} \} /\partial y^2) =$$

$$- (\partial \{ \partial \{ \{ \beta_1 * \gamma_1 + \beta_2 * \gamma_2 \} / \{ \gamma_1 + \gamma_2 \} \} /\partial y \} /\partial y) =$$

$- (\partial \; \{ \; \{ \; \{ \; \beta_1 * (\partial (\gamma_1) /\partial y) + \beta_2 * (\partial (\gamma_2) /\partial y) \; \} * \{ \; \gamma_1 + \gamma_2 \} - \{ \; \beta_1 * \gamma_1 + \beta_2 * \gamma_2 \} * \{ \; (\partial (\gamma_1)$

$/\partial y) + (\partial (\gamma_2) /\partial y) \; \} \; \} \; / \; \{ \; \gamma_1 + \gamma_2 \}^2 \; \} \; /\partial y \;)$ \hfill (18)

Let us posit: $\pi_2 = \{ \; \{ \; \beta_1 * (\partial (\gamma_1) /\partial y) + \beta_2 * (\partial (\gamma_2) /\partial y) \; \} * \{ \; \gamma_1 + \gamma_2 \} - \{ \; \beta_1 * \gamma_1 + \beta_2 * \gamma_2 \}$

$* \; \{ \; (\partial (\gamma_1) /\partial y) + (\partial (\gamma_2) /\partial y) \; \} \; \}$ \hfill (19)

Thus equation (18) is written as: $- (\partial \; \{ \; \pi_2 / \; \{ \; \gamma_1 + \gamma_2 \}^2 \; \} \; /\partial y \;)$, therefore:

$(\partial^2 \; (\; R((x, y), f) \;) \; /\partial y^2) = - \{ \; \{ \; (\partial (\pi_2) /\partial y) * \{ \; \gamma_1 + \gamma_2 \}^2 - \pi_2 * (\partial \; \{ \; \gamma_1 + \gamma_2 \}^2 /\partial y) \; \} \; \} \; / \; \{ \; \gamma_1 + \gamma_2 \}^4$ \hfill (20)

Equation (17)

To solve equation (17), we may transcribe from equation (16):

$(\partial (\pi_1) /\partial x) = (\partial \; \{ \; \{ \; \beta_1 * (\partial (\gamma_1) /\partial x) + \beta_2 * (\partial (\gamma_2) /\partial x) \; \} * \{ \; \gamma_1 + \gamma_2 \} - \{ \; \beta_1 * \gamma_1 + \beta_2 * \gamma_2 \} *$

$\{ \; (\partial (\gamma_1) /\partial x) + (\partial (\gamma_2) /\partial x) \; \} \; \} \; /\partial x) =$

$\{ \; \{ \; \beta_1 * (\partial^2 (\gamma_1) /\partial x^2) + \beta_2 * (\partial^2 (\gamma_2) /\partial x^2) \; \} * \{ \; \gamma_1 + \gamma_2 \} + \{ \; \beta_1 * (\partial (\gamma_1) /\partial x) + \beta_2 * (\partial (\gamma_2)$

$/\partial x) \; \} * (\partial \{ \; \gamma_1 + \gamma_2 \} \; /\partial x) - (\partial \; \{ \; \beta_1 * \gamma_1 + \beta_2 * \gamma_2 \} \; /\partial x) * \{ \; (\partial (\gamma_1) /\partial x) + (\partial (\gamma_2) /\partial x) \; \} - \{ \; \beta_1$

$* \; \gamma_1 + \beta_2 * \gamma_2 \} * \{ \; (\partial^2 (\gamma_1) /\partial x^2) + (\partial^2 (\gamma_2) /\partial x^2) \; \} \; \}$ \hfill (21)

And making use of equations (2), (4), (6), (9) respectively:

$(\partial (\gamma_1) /\partial x) = (\partial \; [- 2a \; (x^3 y/3 + x^2 y^2/2 + xy^3/3) + 1/2 \; xy \; (a+1) \;] \; /\partial x) =$

$[- 2a \; (x^2 y + xy^2 + y^3/3) + 1/2 \; y \; (a+1) \;]$

$(\partial (\gamma_2) /\partial x) = (\partial \; [\; a \; (x^3 y/3 + x^2 y^2/2 + xy^3/3) - (2a + 1/2) \; (x^2 y/2 + xy^2/2) + 3/4 \; xy \; (a+1) \;]$

$/\partial x) = [\; a \; (x^2 y + xy^2 + y^3/3) - (2a + 1/2) \; (xy + y^2/2) + 3/4 \; y \; (a+1) \;]$

$(\partial^2 (\gamma_1) /\partial x^2) = (\partial \; [- 2a \; (x^2 y + xy^2 + y^3/3) + 1/2 \; y \; (a+1) \;] \; /\partial x) = [- 2a \; (2xy + y^2) \;]$

$(\partial^2 (\gamma_2) /\partial x^2) = (\partial \; [\; a \; (x^2 y + xy^2 + y^3/3) - (2a + 1/2) \; (xy + y^2/2) + 3/4 \; y \; (a+1) \;] \; /\partial x) =$

$[\; a \; (2xy + y^2) - (2a + 1/2)y \;]$

It is true that:

$(\partial \{ \; \gamma_1 + \gamma_2 \} \; /\partial x) = \{ \; (\partial (\gamma_1) /\partial x) + (\partial (\gamma_2) /\partial x) \; \} = \{ \; [- 2a \; (x^2 y + xy^2 + y^3/3) + 1/2 \; y \; (a+1) \;]$

$+ [\; a \; (x^2 y + xy^2 + y^3/3) - (2a + 1/2) \; (xy + y^2/2) + 3/4 \; y \; (a+1) \;] \; \}$ \hfill (22)

$(\partial\{\beta_1 * \gamma_1 + \beta_2 * \gamma_2\}/\partial x) = \{\beta_1 * (\partial(\gamma_1)/\partial x) + \beta_2 * (\partial(\gamma_2)/\partial x)\} = \{\beta_1 * [-2a(x^2y + xy^2 + y^3/3) + 1/2\, y(a+1)] + \beta_2 * [a(x^2y + xy^2 + y^3/3) - (2a + 1/2)(xy + y^2/2) + 3/4\, y(a+1)]\}$ (23)

Equation (21) becomes:

$(\partial(\pi_1)/\partial x) = \{\{\beta_1 * [-2a(2xy + y^2)] + \beta_2 * [a(2xy + y^2) - (2a + 1/2)y]\} * \{\gamma_1 + \gamma_2\} + \{\beta_1 * [-2a(x^2y + xy^2 + y^3/3) + 1/2\, y(a+1)] + \beta_2 * [a(x^2y + xy^2 + y^3/3) - (2a + 1/2)(xy + y^2/2) + 3/4\, y(a+1)]\} * \{[-2a(x^2y + xy^2 + y^3/3) + 1/2\, y(a+1)] + [a(x^2y + xy^2 + y^3/3) - (2a + 1/2)(xy + y^2/2) + 3/4\, y(a+1)]\}\} - \{\beta_1 * [-2a(x^2y + xy^2 + y^3/3) + 1/2\, y(a+1)] + \beta_2 * [a(x^2y + xy^2 + y^3/3) - (2a + 1/2)(xy + y^2/2) + 3/4\, y(a+1)]\} * \{[-2a(x^2y + xy^2 + y^3/3) + 1/2\, y(a+1)] + [a(x^2y + xy^2 + y^3/3) - (2a + 1/2)(xy + y^2/2) + 3/4\, y(a+1)]\} - \{\beta_1 * \gamma_1 + \beta_2 * \gamma_2\} * \{[-2a(2xy + y^2)] + [a(2xy + y^2) - (2a + 1/2)y]\}\} =$

$\{\{\beta_1 * [-2a(2xy + y^2)] + \beta_2 * [a(2xy + y^2) - (2a + 1/2)y]\} * \{\gamma_1 + \gamma_2\} - \{\beta_1 * \gamma_1 + \beta_2 * \gamma_2\} * \{[-2a(2xy + y^2)] + [a(2xy + y^2) - (2a + 1/2)y]\}\}$ (24)

Given the position: $(\partial(\pi_1)/\partial x) = \Pi_1$, and also given that:

$(\partial\{\gamma_1 + \gamma_2\}^2/\partial x) = 2\{\gamma_1 + \gamma_2\} * \{[-2a(x^2y + xy^2 + y^3/3) + 1/2\, y(a+1)] + [a(x^2y + xy^2 + y^3/3) - (2a + 1/2)(xy + y^2/2) + 3/4\, y(a+1)]\}$ (25)

Let us posit: $(\partial\{\gamma_1 + \gamma_2\}^2/\partial x) = \varkappa_1$

Equation (17) is written as:

$(\partial^2(R((x, y), f))/\partial x^2) = -\{\{\Pi_1 * \{\gamma_1 + \gamma_2\}^2 - \pi_1 * \varkappa_1\}\}/\{\gamma_1 + \gamma_2\}^4$ (26)

Equation (20)

From equation (19):

$\pi_2 = \{\{\beta_1 * (\partial(\gamma_1)/\partial y) + \beta_2 * (\partial(\gamma_2)/\partial y)\} * \{\gamma_1 + \gamma_2\} - \{\beta_1 * \gamma_1 + \beta_2 * \gamma_2\} * \{(\partial(\gamma_1)/\partial y) + (\partial(\gamma_2)/\partial y)\}\}$

$(\partial(\pi_2)/\partial y) = (\partial\{\{\beta_1 * (\partial(\gamma_1)/\partial y) + \beta_2 * (\partial(\gamma_2)/\partial y)\} * \{\gamma_1 + \gamma_2\} - \{\beta_1 * \gamma_1 + \beta_2 * \gamma_2\} *$

153

$\{ (\partial (\gamma_1) / \partial y) + (\partial (\gamma_2) / \partial y) \} \} / \partial y =$

$\{ \{ \beta_1 * (\partial^2 (\gamma_1) / \partial y^2) + \beta_2 * (\partial^2 (\gamma_2) / \partial y^2) \} * \{ \gamma_1 + \gamma_2 \} + \{ \beta_1 * (\partial (\gamma_1) / \partial y) + \beta_2 * (\partial (\gamma_2)$

$/ \partial y) \} * \{ (\partial (\gamma_1) / \partial y) + (\partial (\gamma_2) / \partial y) \} - \{ \beta_1 * (\partial (\gamma_1) / \partial y) + \beta_2 * (\partial (\gamma_2) / \partial y) \} * \{ (\partial (\gamma_1)$

$/ \partial y) + (\partial (\gamma_2) / \partial y) \} - \{ \beta_1 * \gamma_1 + \beta_2 * \gamma_2 \} * \{ (\partial^2 (\gamma_1) / \partial y^2) + (\partial^2 (\gamma_2) / \partial y^2) \} \}$ (27)

And making use of equations (3), (5), (8.a) and (12):

$(\partial (\gamma_1) / \partial y) = (\partial [- 2a (x^3 y / 3 + x^2 y^2 / 2 + xy^3 / 3) + 1/2 xy (a+1)] / \partial y) =$

$[- 2a (x^3 / 3 + x^2 y + xy^2) + 1/2 x (a+1)]$

$(\partial (\gamma_2) / \partial y) = (\partial [a (x^3 y / 3 + x^2 y^2 / 2 + xy^3 / 3) - (2a + 1/2) (x^2 y / 2 + xy^2 / 2) + 3/4 xy (a+1)]$

$/ \partial y) = [a (x^3 / 3 + x^2 y + xy^2) - (2a + 1/2) (x^2 / 2 + xy) + 3/4 x (a+1)]$

$(\partial^2 (\gamma_1) / \partial y^2) = (\partial [- 2a (x^3 / 3 + x^2 y + xy^2) + 1/2 x (a+1)] / \partial y) = [- 2a (x^2 + 2xy)]$

$(\partial^2 (\gamma_2) / \partial y^2) = (\partial [a (x^3 / 3 + x^2 y + xy^2) - (2a + 1/2) (x^2 / 2 + xy) + 3/4 x (a+1)] / \partial y) =$

$[a (x^2 + 2xy) - (2a + 1/2) x]$

Equation (27) becomes:

$(\partial (\pi_2) / \partial y) = \{ \{ \beta_1 * [- 2a (x^2 + 2xy)] + \beta_2 * [a (x^2 + 2xy) - (2a + 1/2) x] \} * \{ \gamma_1 + \gamma_2 \} +$

$\{ \beta_1 * [- 2a (x^3 / 3 + x^2 y + xy^2) + 1/2 x (a+1)] + \beta_2 * [a (x^3 / 3 + x^2 y + xy^2) - (2a + 1/2) (x^2 / 2$

$+ xy) + 3/4 x (a+1)] \} * \{ [- 2a (x^3 / 3 + x^2 y + xy^2) + 1/2 x (a+1)] + [a (x^3 / 3 + x^2 y + xy^2) -$

$(2a + 1/2) (x^2 / 2 + xy) + 3/4 x (a+1)] \} - \{ \beta_1 * [- 2a (x^3 / 3 + x^2 y + xy^2) + 1/2 x (a+1)] +$

$\beta_2 * [a (x^3 / 3 + x^2 y + xy^2) - (2a + 1/2) (x^2 / 2 + xy) + 3/4 x (a+1)] \} * \{ [- 2a (x^3 / 3 + x^2 y +$

$xy^2) + 1/2 x (a+1)] + [a (x^3 / 3 + x^2 y + xy^2) - (2a + 1/2) (x^2 / 2 + xy) + 3/4 x (a+1)] \} - \{ \beta_1$

$* \gamma_1 + \beta_2 * \gamma_2 \} * \{ [- 2a (x^2 + 2xy)] + [a (x^2 + 2xy) - (2a + 1/2) x] \} \} =$

$\{ \{ \beta_1 * [- 2a (x^2 + 2xy)] + \beta_2 * [a (x^2 + 2xy) - (2a + 1/2) x] \} * \{ \gamma_1 + \gamma_2 \} -$

$\{ \beta_1 * \gamma_1 + \beta_2 * \gamma_2 \} * \{ [- 2a (x^2 + 2xy)] + [a (x^2 + 2xy) - (2a + 1/2) x] \} \}$ (28)

Furthermore:

$(\partial \{ \gamma_1 + \gamma_2 \}^2 / \partial y) = 2 * \{ \gamma_1 + \gamma_2 \} * \{ (\partial (\gamma_1) / \partial y) + (\partial (\gamma_2) / \partial y) \} = 2 * \{ \gamma_1 + \gamma_2 \} * \{ [- 2a$

$(x^3/3 + x^2y + xy^2) + 1/2\, x\, (a+1) \,] + [\, a\, (x^3/3 + x^2y + xy^2) - (2a + 1/2)\, (x^2/2 + xy) + 3/4\, x$

$(a+1) \,] \}$

Let us posit: $(\partial\, (\pi_2)\, /\partial y) = \Pi_2$, $(\partial\, \{\, \gamma_1 + \gamma_2\, \}^2\, /\partial y) = \varkappa_2$ and now equation (20) can be written as:

$$(\partial^2\, (\, R((x, y), f)\,)\, /\partial y^2) = -\, \{\, \{\, \Pi_2 * \{\, \gamma_1 + \gamma_2\, \}^2 - \pi_2 * \varkappa_2\, \}\, \}/\, \{\, \gamma_1 + \gamma_2\, \}^4 \tag{29}$$

Below is the calculation of the second order partial derivatives of $R((x, y), f)$ with respect to the variables x and y:

$(\partial\, (\partial\, (\, R((x, y), f)\,)\, /\partial x)\partial y) = -\, (\partial\, (\partial\, \{\, \{\, \beta_1 * \gamma_1 + \beta_2 * \gamma_2\, \}\, /\, \{\, \gamma_1 + \gamma_2\, \}\, \}\, /\partial x)\partial y) =$

$-\, (\partial\, \{\, \{\, \{\, \beta_1 * (\partial\, (\gamma_1)\, /\partial x) + \beta_2 * (\partial\, (\gamma_2)\, /\partial x)\, \} * \{\, \gamma_1 + \gamma_2\, \} - \{\, \beta_1 * \gamma_1 + \beta_2 * \gamma_2\, \} * \{\, (\partial\, (\gamma_1)$

$/\partial x) + (\partial\, (\gamma_2)\, /\partial x)\, \}\, \}\, /\, \{\, \gamma_1 + \gamma_2\, \}^2\, \}\, /\partial y\,) =$

$-\, \{\, \{\, \{\, \beta_1 * (\partial\, (\partial\, (\gamma_1)\, /\partial x)\partial y) + \beta_2 * (\partial\, (\partial\, (\gamma_2)\, /\partial x)\partial y)\, \} * \{\, \gamma_1 + \gamma_2\, \} + \{\, \beta_1 * (\partial\, (\gamma_1)\, /\partial x) + \beta_2$

$* (\partial\, (\gamma_2)\, /\partial x)\, \} * \{\, (\partial\, (\gamma_1)\, /\partial y) + (\partial\, (\gamma_2)\, /\partial y)\, \} - \{\, \beta_1 * (\partial\, (\gamma_1)\, /\partial y) + \beta_2 * (\partial\, (\gamma_2)\, /\partial y)\, \} * \{\, (\partial$

$(\gamma_1)\, /\partial x) + (\partial\, (\gamma_2)\, /\partial x)\, \} - \{\, \beta_1 * \gamma_1 + \beta_2 * \gamma_2\, \} * \{\, (\partial\, (\partial\, (\gamma_1)\, /\partial x)\partial y) + (\partial\, (\partial\, (\gamma_2)\, /\partial x)\partial y)\, \}\, \} * \{\, \gamma_1 + \gamma_2\, \}^2$

$-\, \{\, \{\, \beta_1 * (\partial\, (\gamma_1)\, /\partial x) + \beta_2 * (\partial\, (\gamma_2)\, /\partial x)\, \} * \{\, \gamma_1 + \gamma_2\, \} - \{\, \beta_1 * \gamma_1 + \beta_2 * \gamma_2\, \} * \{\, (\partial\, (\gamma_1)$

$/\partial x) + (\partial\, (\gamma_2)\, /\partial x)\, \}\, \} * 2 * \{\, (\partial\, (\gamma_1)\, /\partial y) + (\partial\, (\gamma_2)\, /\partial y)\, \} * \{\, \gamma_1 + \gamma_2\, \}\, \}\, /\, \{\, \gamma_1 + \gamma_2\, \}^4 \tag{30}$

Making use of equations (2), (3), (4), (5), (7), (8.b), (10) and (11):

$(\partial\, (\gamma_1)\, /\partial x) = (\partial\, [-\, 2a\, (x^3y/3 + x^2y^2/2 + xy^3/3) + 1/2\, xy\, (a+1)\,]\, /\partial x) =$

$[-\, 2a\, (x^2y + xy^2 + y^3/3) + 1/2\, y\, (a+1)\,]$

$(\partial\, (\gamma_1)\, /\partial y) = (\partial\, [-\, 2a\, (x^3y/3 + x^2y^2/2 + xy^3/3) + 1/2\, xy\, (a+1)\,]\, /\partial y) =$

$[-\, 2a\, (x^3/3 + x^2y + xy^2) + 1/2\, x\, (a+1)\,]$

$(\partial\, (\gamma_2)\, /\partial x) = (\partial\, [\, a\, (x^3y/3 - x^2y^2/2 + xy^3/3) - (2a + 1/2)\, (x^2y/2 + xy^2/2) + 3/4\, xy\, (a+1)\,]$

$/\partial x) =$

$[\, a\, (x^2y + xy^2 + y^3/3) - (2a - 1/2)\, (xy + y^2/2) + 3/4\, y\, (a+1)\,]$

$(\partial\, (\gamma_2)\, /\partial y) = (\partial\, [\, a\, (x^3y/3 + x^2y^2/2 + xy^3/3) - (2a + 1/2)\, (x^2y/2 + xy^2/2) + 3/4\, xy\, (a-1)\,]$

$/\partial y) = [\, a\, (x^3/3 + x^2y + xy^2) - (2a + 1/2)\, (x^2/2 + xy) + 3/4\, x\, (a+1)\,]$

$(\partial\, (\partial\, (\gamma_1)\, /\partial x)\partial y) = (\partial\, [-\, 2a\, (x^2y + xy^2 + y^3/3) + 1/2\, y\, (a+1)\,]\, /\partial y) =$

$[- 2a (x^2 + 2xy + y^2) + 1/2 (a+1)]$

$(\partial (\partial (\gamma_1) /\partial y)\partial x) = (\partial [- 2a (x^3/3 + x^2y + xy^2) + 1/2 x (a+1)] /\partial x) =$

$[- 2a (x^2 + 2xy + y^2) + 1/2 (a+1)]$

$(\partial (\partial (\gamma_2) /\partial x)\partial y) = (\partial [a (x^2y + xy^2 + y^3/3) - (2a + 1/2) (xy + y^2/2) + 3/4 y (a+1)] /\partial y) =$

$[a (x^2 + 2xy + y^2) - (2a + 1/2) (x + y) + 3/4 (a+1)]$

$(\partial (\partial (\gamma_2) /\partial y)\partial x) = (\partial [a (x^3/3 + x^2y + xy^2) - (2a + 1/2) (x^2/2 + xy) + 3/4 x (a+1)] /\partial x) =$

$[a (x^2 + 2xy + y^2) - (2a + 1/2) (x + y) + 3/4 (a+1)]$

$\gamma_1 + \gamma_2 = \{ [- 2a (x^3y/3 + x^2y^2/2 + xy^3/3) + 1/2 xy (a+1)] + [a (x^3y/3 + x^2y^2/2 + xy^3/3) - (2a + 1/2)$

$(x^2y/2 + xy^2/2) + 3/4 xy (a+1)] \}$

It is calculated:

$(\partial (\partial (R((x, y), f)) /\partial x)\partial y) = - \{ \{ \{ \beta_1 * [- 2a (x^2 + 2xy + y^2) + 1/2 (a+1)] + \beta_2 * [a (x^2 +$

$2xy + y^2) - (2a + 1/2) (x + y) + 3/4 (a+1)] \} * \{ \gamma_1 + \gamma_2 \} + \{ \beta_1 * [- 2a (x^2y + xy^2 + y^3/3) +$

$1/2 y (a+1)] + \beta_2 * [a (x^2y + xy^2 + y^3/3) - (2a + 1/2) (xy + y^2/2) + 3/4 y (a+1)] \} * \{ [- 2a$

$(x^3/3 + x^2y + xy^2) + 1/2 x (a+1)] + [a (x^3/3 + x^2y + xy^2) - (2a + 1/2) (x^2/2 + xy) + 3/4 x$

$(a+1)] \} - \{ \beta_1 * [- 2a (x^3/3 + x^2y + xy^2) + 1/2 x (a+1)] + \beta_2 * [a (x^3/3 + x^2y + xy^2) - (2a$

$+ 1/2) (x^2/2 + xy) + 3/4 x (a+1)] \} * \{ [- 2a (x^2y + xy^2 + y^3/3) + 1/2 y (a+1)] + [a (x^2y +$

$xy^2 + y^3/3) - (2a + 1/2) (xy + y^2/2) + 3/4 y (a+1)] \} - \{ \beta_1 * \gamma_1 + \beta_2 * \gamma_2 \} * \{ [- 2a (x^2 +$

$2xy + y^2) + 1/2 (a+1)] + [a (x^2 + 2xy + y^2) - (2a + 1/2) (x + y) + 3/4 (a+1)] \} \} * \{ \gamma_1 + \gamma_2 \}^2 -$

$\{ \{ \beta_1 * [- 2a (x^2y + xy^2 + y^3/3) + 1/2 y (a+1)] + \beta_2 * [a (x^2y + xy^2 + y^3/3) - (2a + 1/2) (xy + y^2/2) +$

$3/4 y (a+1)] \} * \{ \gamma_1 + \gamma_2 \} - \{ \beta_1 * \gamma_1 + \beta_2 * \gamma_2 \} * \{ [- 2a (x^2y + xy^2 + y^3/3) + 1/2 y (a+1)] + [a (x^2y$

$+ xy^2 + y^3/3) - (2a + 1/2) (xy + y^2/2) + 3/4 y (a+1)] \} \} * 2 * \{ [- 2a (x^3/3 + x^2y + xy^2) + 1/2 x$

$(a+1)] + [a (x^3/3 + x^2y + xy^2) - (2a + 1/2) (x^2/2 + xy) + 3/4 x (a+1)] \} * \{ [- 2a (x^3y/3 + x^2y^2/2 +$

$xy^3/3) + 1/2 xy (a+1)] + [a (x^3y/3 + x^2y^2/2 + xy^3/3) - (2a + 1/2) (x^2y/2 + xy^2/2) + 3/4 xy (a+1)] \}$

$\} / \{ \gamma_1 + \gamma_2 \}^4$
(31)

Furthermore:

$$(\partial\,(\partial\,(\,R((x,\,y),\,f)\,)\,)\,/\partial y)\partial x) = -\,(\partial\,(\partial\,\{\,\{\,\beta_1 * \gamma_1 + \beta_2 * \gamma_2\,\}\,/\,\{\,\gamma_1 + \gamma_2\,\}\,\}\,/\partial y)\partial x) =$$

$$-\,(\partial\,\{\,\{\,\{\,\beta_1 * (\partial\,(\gamma_1)\,/\partial y) + \beta_2 * (\partial\,(\gamma_2)\,/\partial y)\,\} * \{\,\gamma_1 + \gamma_2\,\} - \{\,\beta_1 * \gamma_1 + \beta_2 * \gamma_2\,\} * \{\,(\partial\,(\gamma_1)$$

$$/\partial y) + (\partial\,(\gamma_2)\,/\partial y)\,\}\,\}\,/\,\{\,\gamma_1 + \gamma_2\,\}^2\,\}\,/\partial x\,) =$$

$$-\,\{\,\{\,\{\,\beta_1 * (\partial\,(\partial\,(\gamma_1)\,/\partial y)\partial x) + \beta_2 * (\partial\,(\partial\,(\gamma_2)\,/\partial y)\partial x)\,\} * \{\,\gamma_1 + \gamma_2\,\} + \{\,\beta_1 * (\partial\,(\gamma_1)\,/\partial y) + \beta_2$$

$$* (\partial\,(\gamma_2)\,/\partial y)\,\} * \{\,(\partial\,(\gamma_1)\,/\partial x) + (\partial\,(\gamma_2)\,/\partial x)\,\} - \{\,\beta_1 * (\partial\,(\gamma_1)\,/\partial x) + \beta_2 * (\partial\,(\gamma_2)\,/\partial x)\,\} * \{\,(\partial$$

$$(\gamma_1)\,/\partial y) + (\partial\,(\gamma_2)\,/\partial y)\,\} - \{\,\beta_1 * \gamma_1 + \beta_2 * \gamma_2\,\} * \{\,(\partial\,(\partial\,(\gamma_1)\,/\partial y)\partial x) + (\partial\,(\partial\,(\gamma_2)\,/\partial y)\partial x)\,\}\,\} * \{\,\gamma_1 + \gamma_2\,\}^2$$

$$-\,\{\,\{\,\beta_1 * (\partial\,(\gamma_1)\,/\partial y) + \beta_2 * (\partial\,(\gamma_2)\,/\partial y)\,\} * \{\,\gamma_1 + \gamma_2\,\} - \{\,\beta_1 * \gamma_1 + \beta_2 * \gamma_2\,\} * \{\,(\partial\,(\gamma_1)$$

$$/\partial y) + (\partial\,(\gamma_2)\,/\partial y)\,\}\,\} * 2 * \{\,(\partial\,(\gamma_1)\,/\partial x) + (\partial\,(\gamma_2)\,/\partial x)\,\} * \{\,\gamma_1 + \gamma_2\,\}\,\}\,/\,\{\,\gamma_1 + \gamma_2\,\}^4 \qquad (32)$$

Because of equations (2), (3), (4), (5), (7), (8.b), (10) and (11), it is true that:

$$(\partial\,(\gamma_1)\,/\partial x) = (\partial\,[-\,2a\,(x^3y/3 + x^2y^2/2 + xy^3/3) + 1/2\,xy\,(a+1)\,]\,/\partial x) =$$

$$[-\,2a\,(x^2y + xy^2 + y^3/3) + 1/2\,y\,(a+1)]$$

$$(\partial\,(\gamma_1)\,/\partial y) = (\partial\,[-\,2a\,(x^3y/3 + x^2y^2/2 + xy^3/3) + 1/2\,xy\,(a+1)\,]\,/\partial y) =$$

$$[-\,2a\,(x^3/3 + x^2y + xy^2) + 1/2\,x\,(a+1)]$$

$$(\partial\,(\gamma_2)\,/\partial x) = (\partial\,[\,a\,(x^3y/3 + x^2y^2/2 + xy^3/3) - (2a + 1/2)\,(x^2y/2 + xy^2/2) + 3/4\,xy\,(a+1)\,]$$

$$/\partial x) = [\,a\,(x^2y + xy^2 + y^3/3) - (2a + 1/2)\,(xy + y^2/2) + 3/4\,y\,(a+1)\,]$$

$$(\partial\,(\gamma_2)\,/\partial y) = (\partial\,[\,a\,(x^3y/3 + x^2y^2/2 + xy^3/3) - (2a + 1/2)\,(x^2y/2 + xy^2/2) + 3/4\,xy\,(a+1)\,]$$

$$/\partial y) = [\,a\,(x^3/3 + x^2y + xy^2) - (2a + 1/2)\,(x^2/2 + xy) + 3/4\,x\,(a+1)\,]$$

$$(\partial\,(\partial\,(\gamma_1)\,/\partial x)\partial y) = (\partial\,[-\,2a\,(x^2y + xy^2 + y^3/3) + 1/2\,y\,(a+1)\,]\,/\partial y) =$$

$$[-\,2a\,(x^2 + 2xy + y^2) + 1/2\,(a+1)]$$

$$(\partial\,(\partial\,(\gamma_1)\,/\partial y)\partial x) = (\partial\,[-\,2a\,(x^3/3 + x^2y + xy^2) + 1/2\,x\,(a+1)\,]\,/\partial x) =$$

$$[-\,2a\,(x^2 + 2xy + y^2) + 1/2\,(a+1)]$$

$$(\partial\,(\partial\,(\gamma_2)\,/\partial x)\partial y) = (\partial\,[\,a\,(x^2y + xy^2 + y^3/3) - (2a + 1/2)\,(xy - y^2/2) + 3/4\,y\,(a+1)\,]\,/\partial y) =$$

$$[\,a\,(x^2 + 2xy + y^2) - (2a + 1/2)\,(x + y) + 3/4\,(a+1)]$$

$(\partial\,(\partial\,(\gamma_2)\,/\partial y)\partial x) = (\partial\,[\,a\,(x^3/3 + x^2y + xy^2) - (2a + 1/2)\,(x^2/2 + xy) + 3/4\,x\,(a+1)\,]\,/\partial x) =$

$[\,a\,(x^2 + 2xy + y^2) - (2a + 1/2)\,(x + y) + 3/4\,(a+1)]$

$\{\,\beta_1 * (\partial\,(\partial\,(\gamma_1)\,/\partial y)\partial x) + \beta_2 * (\partial\,(\partial\,(\gamma_2)\,/\partial y)\partial x)\,\} = \{\,\beta_1 * [\,- 2a\,(x^2 + 2xy + y^2) + 1/2\,(a+1)$

$]+ \beta_2 * [\,a\,(x^2 + 2xy + y^2) - (2a + 1/2)\,(x + y) + 3/4\,(a+1)\,]\,\}$ (33)

$\{\,\beta_1 * (\partial\,(\gamma_1)\,/\partial y) + \beta_2 * (\partial\,(\gamma_2)\,/\partial y)\,\} = \{\,\beta_1 * [\,- 2a\,(x^3/3 + x^2y + xy^2) + 1/2\,x\,(a+1)\,] + \beta_2 *$

$[\,a\,(x^3/3 + x^2y + xy^2) - (2a + 1/2)\,(x^2/2 + xy) + 3/4\,x\,(a+1)\,]\,\}$ (34)

$\{\,(\partial\,(\gamma_1)\,/\partial x) + (\partial\,(\gamma_2)\,/\partial x)\,\} = \{\,[\,- 2a\,(x^2y + xy^2 + y^3/3) + 1/2\,y\,(a+1)\,] + [\,a\,(x^2y + xy^2 +$

$y^3/3) - (2a + 1/2)\,(xy + y^2/2) + 3/4\,y\,(a+1)\,]\,\}$ (35)

$\{\,\beta_1 * (\partial\,(\gamma_1)\,/\partial x) + \beta_2 * (\partial\,(\gamma_2)\,/\partial x)\,\} = \{\,\beta_1 * [\,- 2a\,(x^2y + xy^2 + y^3/3) + 1/2\,y\,(a+1)\,] + \beta_2 *$

$[\,a\,(x^2y + xy^2 + y^3/3) - (2a + 1/2)\,(xy + y^2/2) + 3/4\,y\,(a+1)\,]\,\}$ (36)

$\{\,(\partial\,(\gamma_1)\,/\partial y) + (\partial\,(\gamma_2)\,/\partial y)\,\} = \{\,[\,- 2a\,(x^3/3 + x^2y + xy^2) + 1/2\,x\,(a+1)\,] + [\,a\,(x^3/3 + x^2y +$

$xy^2) - (2a + 1/2)\,(x^2/2 + xy) + 3/4\,x\,(a+1)\,]\,\}$ (37)

$\{\,(\partial\,(\partial\,(\gamma_1)\,/\partial y)\partial x) + (\partial\,(\partial\,(\gamma_2)\,/\partial y)\partial x)\,\} = \{\,[\,- 2a\,(x^2 + 2xy + y^2) + 1/2\,(a+1)\,] + [\,a\,(x^2 +$

$2xy + y^2) - (2a + 1/2)\,(x + y) + 3/4\,(a+1)\,]\,\}$ (38)

Making use of equations (33) through (38), equation (32) becomes:

$(\partial\,(\partial\,(\,R((x, y), f)\,)\,)\,/\partial y)\partial x) = -\,\{\,\{\,\{\,\beta_1 * [\,- 2a\,(x^2 + 2xy + y^2) + 1/2\,(a+1)\,] + \beta_2 * [\,a\,(x^2 +$

$2xy + y^2) - (2a + 1/2)\,(x + y) + 3/4\,(a+1)\,]\,\} * \{\,\gamma_1 + \gamma_2\,\} + \{\,\beta_1 * [\,- 2a\,(x^3/3 + x^2y + xy^2) +$

$1/2\,x\,(a+1)\,] + \beta_2 * [\,a\,(x^3/3 + x^2y + xy^2) - (2a + 1/2)\,(x^2/2 + xy) + 3/4\,x\,(a+1)\,]\,\} * \{\,[\,- 2a$

$(x^2y + xy^2 + y^3/3) + 1/2\,y\,(a+1)\,] + [\,a\,(x^2y + xy^2 + y^3/3) - (2a + 1/2)\,(xy + y^2/2) + 3/4\,y$

$(a+1)\,]\,\} - \{\,\beta_1 * [\,- 2a\,(x^2y + xy^2 + y^3/3) + 1/2\,y\,(a+1)\,] + \beta_2 * [\,a\,(x^2y + xy^2 + y^3/3) - (2a$

$+ 1/2)\,(xy + y^2/2) + 3/4\,y\,(a+1)\,]\,\} * \{\,[\,- 2a\,(x^3/3 + x^2y + xy^2) + 1/2\,x\,(a+1)\,] + [\,a\,(x^3/3 +$

$x^2y + xy^2) - (2a + 1/2)\,(x^2/2 + xy) + 3/4\,x\,(a+1)\,]\,\} - \{\,\beta_1 * \gamma_1 + \beta_2 * \gamma_2\,\} * \{\,[\,- 2a\,(x^2 +$

$2xy + y^2) + 1/2\,(a+1)\,] + [\,a\,(x^2 + 2xy + y^2) - (2a + 1/2)\,(x + y) + 3/4\,(a+1)\,]\,\}\,\} * \{\,\gamma_1 + \gamma_2\,\}^2 -$

$\{\,\{\,\beta_1 * [\,- 2a\,(x^3/3 + x^2y + xy^2) + 1/2\,x\,(a+1)] + \beta_2 * [\,a\,(x^3/3 + x^2y + xy^2) - (2a + 1/2)\,(x^2/2 + xy) +$

$3/4 \, x \, (a+1)] \} * \{ \gamma_1 + \gamma_2 \} - \{ \beta_1 * \gamma_1 + \beta_2 * \gamma_2 \} * \{ [- 2a \, (x^3/3 + x^2y + xy^2) + 1/2 \, x \, (a+1)] + [a \, (x^3/3$

$+ x^2y + xy^2) - (2a + 1/2) \, (x^2/2 + xy) + 3/4 \, x \, (a+1)] \} \} * 2 * \{ [- 2a \, (x^2y + xy^2 + y^3/3) + 1/2 \, y$

$(a+1)] + [a \, (x^2y + xy^2 + y^3/3) - (2a + 1/2) \, (xy + y^2/2) + 3/4 \, y \, (a+1)] \} * \{ [- 2a \, (x^3y/3 + x^2y^2/2 +$

$xy^3/3) + 1/2 \, xy \, (a+1)] + [a \, (x^3y/3 + x^2y^2/2 + xy^3/3) - (2a + 1/2) \, (x^2y/2 + xy^2/2) + 3/4 \, xy \, (a+1)] \}$

$/ \{ \gamma_1 + \gamma_2 \}^4$
(39)

In summary, as per equation (14), the total curvature $C(R((x, y), f))$ of the quadratic bivariate signal resilient to interpolation is given adding together the following equations: (26), (29), (31) and (39).

BIVARIATE QUADRATIC SIGNAL CALCULATED WITHOUT EMBEDDING

The initial equation to consider is: $R (x, y) = f (0, 0) + \varrho (x, y)$, with $\varrho (x, y)$ given in equation (17) of chapter 3 given below as equation (40):

$\varrho (x, y) = \{ \alpha_1 * [- 2a \, (x^2/3 + xy/2 + y^2/3) + \frac{1}{2} \, (a+1)] + \alpha_2 * [a \, (x^2/3 +$

$xy/2 + y^2/3) - (2a + 1/2) \, (x/2 + y/2) + 3/4 \, (a+1)] \}$
(40)

Where:

$\alpha_1 = [f (1/2, 1/2) + f (-1/2, -1/2) + f (2/3, 2/3) + f (-2/3, -2/3)]$
(40.a)

$\alpha_2 = [f (1/2, 1/2) + f (-1/2, -1/2) + f (2/3, 2/3) + f (-2/3, -2/3) + f (-1, -1) + f (1, 1) +$

$f (3/2, 3/2) + f (-3/2, -3/2)]$
(40.b)

The initial equation becomes:

$R((x, y), f) = f (0, 0) + \{ \alpha_1 * [- 2a \, (x^2/3 + xy/2 + y^2/3) + \frac{1}{2} \, (a+1)] + \alpha_2 * [a \, (x^2/3 +$

$xy/2 + y^2/3) - (2a + 1/2) \, (x/2 + y/2) + 3/4 \, (a+1)] \} = f (0, 0) + \{ \alpha_1 * \theta_1 \} + \{ \alpha_2 * \theta_2 \}$
(41)

$\theta_1 = [- 2a \, (x^2/3 + xy/2 + y^2/3) + \frac{1}{2} \, (a+1)]$
(41.a)

$\theta_2 = [a \, (x^2/3 + xy/2 + y^2/3) - (2a + 1/2) \, (x/2 + y/2) + 3/4 \, (a+1)]$
(41.b)

Let us calculate the first and second order partial derivatives of θ_1 and θ_2 with respect to the variables x and y.

$(\partial (\theta_1) /\partial x) = [- 2a \, (2x/3 + y/2)]$
(42)

$(\partial (\theta_2) /\partial x) = [a \, (2x/3 + y/2) - (2a - 1/2) \, (1/2)]$
(43)

$(\partial (\theta_1) /\partial y) = [- 2a \, (x/2 + 2y/3)]$
(44)

$$(\partial(\theta_2)/\partial y) = [\, a\,(x/2 + 2y/3) - (2a + 1/2)\,(1/2)\,] \tag{45}$$

$$(\partial^2(\theta_1)/\partial x^2) = [-\,2a\,(2/3)\,] \tag{46}$$
$$(\partial^2(\theta_1)/\partial y^2) = [-\,2a\,(2/3)\,] \tag{47}$$
$$(\partial^2(\theta_2)/\partial x^2) = [\, a\,(2/3)\,] \tag{48}$$
$$(\partial^2(\theta_2)/\partial y^2) = [\, a\,(2/3)\,] \tag{49}$$

$$(\partial(\partial(\theta_1)/\partial x)\partial y) = [-\,2a\,(1/2)\,] \tag{50}$$
$$(\partial(\partial(\theta_2)/\partial x)\partial y) = [\, a\,(1/2)\,] \tag{51}$$
$$(\partial(\partial(\theta_1)/\partial y)\partial x) = [-\,2a\,(1/2)\,] \tag{52}$$
$$(\partial(\partial(\theta_2)/\partial y)\partial x) = [\, a\,(1/2)\,] \tag{53}$$

Second Order Partial Derivatives and Curvature of R((x, y), f)

$$(\partial^2(R((x, y), f))/\partial x^2) = (\partial^2\{\{\alpha_1 * \theta_1\} + \{\alpha_2 * \theta_2\}\}/\partial x^2) = (\partial\{\{\alpha_1 * (\partial(\theta_1)/\partial x)\} +$$

$$\{\alpha_2 * (\partial(\theta_2)/\partial x)\}\}/\partial x) = \{\{\alpha_1 * (\partial^2(\theta_1)/\partial x^2)\} + \{\alpha_2 * (\partial^2(\theta_2)/\partial x^2)\}\}$$

$$= \{\alpha_1 * [-\,2a\,(2/3)\,]\} + \{\alpha_2 * [\, a\,(2/3)\,]\} \tag{54}$$

Analogously:

$$(\partial^2(R((x, y), f))/\partial y^2) = (\partial^2\{\{\alpha_1 * \theta_1\} + \{\alpha_2 * \theta_2\}\}/\partial y^2) = (\partial\{\{\alpha_1 * (\partial(\theta_1)/\partial y)\} +$$

$$\{\alpha_2 * (\partial(\theta_2)/\partial y)\}\}/\partial y) = \{\{\alpha_1 * (\partial^2(\theta_1)/\partial y^2)\} + \{\alpha_2 * (\partial^2(\theta_2)/\partial y^2)\}\}$$

$$= \{\alpha_1 * [-\,2a\,(2/3)\,]\} + \{\alpha_2 * [\, a\,(2/3)\,]\} \tag{55}$$

$$(\partial(\partial(R((x, y), f))/\partial x)\partial y) = (\partial(\partial\{\{\alpha_1 * \theta_1\} + \{\alpha_2 * \theta_2\}\}/\partial x)\partial y) = (\partial\{\{\alpha_1 * (\partial(\theta_1)$$

$$/\partial x)\} + \{\alpha_2 * (\partial(\theta_2)/\partial x)\}\}/\partial y) = \{\{\alpha_1 * (\partial(\partial(\theta_1)/\partial x)\partial y)\} + \{\alpha_2 * (\partial(\partial(\theta_2)/\partial x)\partial y)$$

$$\}\} = \{\alpha_1 * [-\,2a\,(1/2)\,]\} + \{\alpha_2 * [\, a\,(1/2)\,]\} \tag{56}$$

$$(\partial(\partial(R((x, y), f))/\partial y)\partial x) = (\partial(\partial\{\{\alpha_1 * \theta_1\} + \{\alpha_2 * \theta_2\}\}/\partial y)\partial x) = (\partial\{\{\alpha_1 * (\partial(\theta_1)$$

$$/\partial y)\} + \{\alpha_2 * (\partial(\theta_2)/\partial y)\}\}/\partial x) = \{\{\alpha_1 * (\partial(\partial(\theta_1)/\partial y)\partial x)\} + \{\alpha_2 * (\partial(\partial(\theta_2)/\partial y)\partial x)$$

$$\}\} = \{\alpha_1 * [-\,2a\,(1/2)\,]\} + \{\alpha_2 * [\, a\,(1/2)\,]\} \tag{57}$$

Therefore the total curvature is defined in equation (14) as:

$$C(R((x, y), f)) = (\partial^2(R((x, y), f))/\partial x^2) + (\partial(\partial(R((x, y), f))/\partial y)\partial x) + (\partial(\partial(R((x, y),$$

$$f))/\partial x)\partial y) + (\partial^2(R((x, y), f))/\partial y^2) = \{\alpha_1 * [-\,2a\,(2/3)\,]\} + \{\alpha_2 * [\, a\,(2/3)\,]\} + \{\alpha_1 *$$

$$[-\,2a\,(1/2)\,]\} + \{\alpha_2 * [\, a\,(1/2)\,]\} + \{\alpha_1 * [-\,2a\,(1/2)\,]\} + \{\alpha_2 * [\, a\,(1/2)\,]\} + \{\alpha_1 * [-\,2a$$

(2/3)] } + { α_2 * [a (2/3)] } = 2 * { { α_1 * [- 2a (2/3)] } + { α_2 * [a (2/3)] } + { α_1 * [-

2a (1/2)] }+ { α_2 * [a (1/2)] } }   (58)

SUMMARY

Given the interpolation function $h_3(x, y)$, which we call classic signal I $(x, y) = f (0, 0) + \xi (x, y)$, there are two types of total curvature that can be calculated: classic curvature and resilient curvature. Given its definition: $(\partial^2 (h_3(x, y)) /\partial x^2) + (\partial^2 (h_3(x, y)) /\partial y^2) + (\partial^2 (h_3(x, y)) /\partial x\partial y) + (\partial^2 (h_3(x, y)) /\partial y\partial x)$; the classic curvature is the same in both of the cases of signal calculated with embedding and without embedding. From the preceding chapters, it was shown that in order to calculate the signal resilient to the interpolation function, the equation of the two intensity-curvature terms needs to be resolved in f (0, 0), which is the resilient signal, and conceptually it is also that fraction of pixel intensity to be added to the original signal, which we call $\varrho (x, y)$. *Thus, there is conceptual equality between $\varrho (x, y)$ and f (0, 0) (which is not the original signal seen as first term of the sum in resilient signal: R $(x, y) = f (0, 0) + \varrho (x, y)$).* R (x, y) is the sum of the original signal f (0, 0) (seen as the left term of the sum) and the resilient signal $\varrho (x, y) = f (0, 0)$, where f (0, 0) derives from the solution of $E_o (x, y) = E_{IN}(x, y)$ Having found $\varrho (x, y) = f (0, 0)$, as the fraction of pixel intensity to be added to the original signal allows to set the basis to calculate the resilient curvature (total curvature of the resilient signal) through the sum of the second order partial derivatives of $\varrho (x, y)$.

Since the math form of the resilient signal with embedding differs from the math form of the resilient signal without embedding, the same may said to be true for the derived resilient curvature. Thus, in this chapter the resilient curvature with and without embedding is calculated with different math formulations.

REFERENCES

Ciulla, C. (2009). *Improved Signal and Image Interpolation in Biomedical Applications: The Case of Magnetic Resonance Imaging (MRI)* – Medical Information Science Reference - IGI Global Publisher Hershey, PA, U.S.A.

CALCULATION OF THE CURVATURE OF THE BIVARIATE CUBIC SIGNAL RESILIENT TO INTERPOLATION: WITH AND WITHOUT EMBEDDING

INTRODUCTION

In this chapter is calculated the total curvature of the signal resilient to the bivariate cubic interpolation function given as equation (4) in chapter 2: $h_4(x, y) = f(0, 0) + \alpha_3 * [1/2 (x + y)^3 - (x + y)^2 + 2/3] + \alpha_2 * [-1/6 (x + y)^3 + (x + y)^2 - 2 (x + y) + 4/3]$. The classic signal is: $I(x, y) = h_4(x, y) = f(0, 0) + \alpha_3 * [1/2 (x + y)^3 - (x + y)^2 + 2/3] + \alpha_2 * [-1/6 (x + y)^3 + (x + y)^2 - 2 (x + y) + 4/3]$; with embedding and: $I(x, y) = \alpha_3 * [1/2 (x + y)^3 - (x + y)^2 + 2/3] + \alpha_2 * [-1/6 (x + y)^3 + (x + y)^2 - 2 (x + y) + 4/3]$, without embedding. The total curvature is calculated in either case through the sum: $(\partial^2 (h_4(x, y))) /\partial x^2) + (\partial^2 (h_4(x, y))) /\partial y^2) + (\partial^2 (h_4(x, y))) /\partial x \partial y) + (\partial^2 (h_4(x, y))) /\partial y \partial x)$. The signal resilient to interpolation is given in this chapter in equations (1) and (2), with and without embedding respectively and the total curvature is calculated through the sum: $(\partial^2 (\varrho(x, y))) /\partial x^2) + (\partial^2 (\varrho(x, y))) /\partial y^2) + (\partial^2 (\varrho(x, y))) /\partial x \partial y) + (\partial^2 (\varrho(x, y))) /\partial y \partial x)$ given in equation (90). To obtain signal reconstruction, the Pixel Intensity Correction needs to be calculated through the formula PIC = Shift * tan(total curvature), consistently with what has been seen previously.

CALCULATION OF THE TOTAL CURVATURE

The total curvature to calculate is:

$C(R((x, y), f)) = (\partial^2 (R((x, y), f)) /\partial x^2) + (\partial (\partial (R((x, y), f)) /\partial y)\partial x) + (\partial (\partial (R((x, y),$

$f)) /\partial x)\partial y) + (\partial^2 (R((x, y), f)) /\partial y^2)$

$R(x, y) = f(0, 0) + \varrho(x, y)$, with $\varrho(x, y)$ given in equation (34) of chapter 2 where the bivariate cubic signal is calculated with embedding and given below as equation (1):

$\varrho(x, y) = \Lambda / \{ 4xy * [-2\alpha_3 + 2\alpha_2] - 4\alpha_3 * [3(x^2y/2 + xy^2/2) - 2xy] - 4\alpha_2 * [-(x^2y/2 +$

$xy^2/2) + 2xy] \}$ (1)

And with $\varrho(x, y)$ given in equation (34) of chapter 3 where the bivariate cubic signal is calculated without embedding and given below as equation (2):

$\varrho(x, y) = \Lambda / \{ 4xy * [-2\alpha_3 + 2\alpha_2] \}$ (2)

$\Lambda = 4 * \{ [3/2 \alpha_3^2 - 1/2 \alpha_3 \alpha_2 - 3/6 \alpha_3 \alpha_2 + 1/6 \alpha_2^2] [x^5y/5 + 3/8 x^4y^2 + x^3y^3/3 + x^2y^4/8 +$

$x^4y^2/8 + x^3y^3/3 + 3/8 x^2y^4 + xy^5/5] + [-4 \alpha_3^2 + 2 \alpha_3 \alpha_2 + 4/3 \alpha_3 \alpha_2 - 4/3 \alpha_2^2] [x^4y/4 +$

$x^3y^2/2 + x^2y^3/2 + xy^4/4] + [2 \alpha_3^2 - 2 \alpha_3 \alpha_2 - 8 \alpha_3 \alpha_2 + 4 \alpha_2^2] [x^3y/3 + x^2y^2/2 + xy^3/3] +$

$[2 \alpha_3^2 - 2/3 \alpha_3 \alpha_2 + 8 \alpha_3 \alpha_2 - 16/3 \alpha_2^2] [x^2y/2 + xy^2/2] + [-4/3 \alpha_3^2 + 4/3 \alpha_3 \alpha_2 - 8/3 \alpha_3 \alpha_2$

$+ 8/3 \alpha_2{}^2] [xy] \}$

$\alpha_2 = [f (1/2, 1/2) + f (-1/2, -1/2) + f (2/3, 2/3) + f (-2/3, -2/3) + f (-1, -1) + f (1, 1) +$

$f (3/2, 3/2) + f (-3/2, -3/2)]$

$\alpha_3 = [f (1/2, 1/2) + f (-1/2, -1/2) + f (-1, -1) + f (1, 1)]$

Calculation of the Second Order Partial Derivatives of Λ

In reference to Λ let us posit:

$$\tau_1 = [x^5y/5 + 3/8\ x^4y^2 + x^3y^3/3 + x^2y^4/8 + x^4y^2/8 + x^3y^3/3 + 3/8\ x^2y^4 + xy^5/5] \tag{3}$$

$$\tau_2 = [x^4y/4 + x^3y^2/2 + x^2y^3/2 + xy^4/4] \tag{4}$$

$$\tau_3 = [x^3y/3 + x^2y^2/2 + xy^3/3] \tag{5}$$

$$\tau_4 = [x^2y/2 + xy^2/2] \tag{6}$$

$$\tau_5 = [xy] \tag{7}$$

$(\partial \{ \tau_1 \} / \partial x) = (\partial [x^5y/5 + 3/8\ x^4y^2 + x^3y^3/3 + x^2y^4/8 + x^4y^2/8 + x^3y^3/3 + 3/8\ x^2y^4 + xy^5/5]$

$$/\partial x) = [x^4y + (12/8)\ x^3y^2 + x^2y^3 + xy^4/4 + x^3y^2/2 + x^2y^3 + (6/8)\ xy^4 + y^5/5] \tag{8}$$

$(\partial \{ \tau_2 \} / \partial x) = (\partial[x^4y/4 - x^3y^2/2 + x^2y^3/2 + xy^4/4] /\partial x) =$

$$[x^3y + (3/2)\ x^2y^2 + xy^3 + y^4/4] \tag{9}$$

$$(\partial \{ \tau_3 \} / \partial x) = (\partial[x^3y/3 + x^2y^2/2 + xy^3/3] /\partial x) = [x^2y + xy^2 + y^3/3] \tag{10}$$

$$(\partial \{ \tau_4 \} / \partial x) = (\partial [x^2y/2 + xy^2/2] /\partial x) = [xy + y^2/2] \tag{11}$$

$$(\partial \{ \tau_5 \} / \partial x) = (\partial[xy] /\partial x) = y \tag{12}$$

$(\partial^2 \{ \tau_1 \} / \partial x^2) = (\partial (\partial \{ \tau_1 \} / \partial x) \partial x) = (\partial (x^4y + (12/8)\ x^3y^2 + x^2y^3 + xy^4/4 + x^3y^2/2 + x^2y^3$

$$+ (6/8)\ xy^4 + y^5/5)\ \partial x) = [4x^3y + (36/8)\ x^2y^2 + 2xy^3 + y^4/4 - (3/2)\ x^2y^2 + 2xy^3 + (6/8)y^4] \tag{13}$$

$(\partial^2 \{ \tau_2 \} / \partial x^2) = (\partial (\partial \{ \tau_2 \} / \partial x)\ \partial x) = (\partial (x^3y + (3/2)\ x^2y^2 + xy^3 + y^4/4)\ \partial x) =$

$$[3x^2y + 3xy^2 + y^3] \tag{14}$$

$$(\partial^2 \{ \tau_3 \} / \partial x^2) = (\partial (\partial \{ \tau_3 \} / \partial x)\ \partial x) = (\partial (x^2y + xy^2 + y^3/3)\ \partial x) = [2xy + y^2] \tag{15}$$

$$(\partial^2 \{ \tau_4 \} / \partial x^2) = (\partial (\partial \{ \tau_4 \} / \partial x) \, \partial x) = (\partial (xy + y^2/2) \, \partial x) = [\, y \,] \tag{16}$$

$$(\partial^2 \{ \tau_5 \} / \partial x^2) = (\partial (\partial \{ \tau_5 \} / \partial x) \, \partial x) = (\partial (y) \partial x) = 0 \tag{17}$$

$$(\partial \{ \tau_1 \} / \partial y) = (\partial [\, x^5 y/5 + 3/8 \, x^4 y^2 + x^3 y^3/3 + x^2 y^4/8 + x^4 y^2/8 + x^3 y^3/3 + 3/8 \, x^2 y^4 + x y^5/5 \,]$$

$$/\partial y) = [\, x^5/5 + 6/8 \, x^4 y + x^3 y^2 + x^2 y^3/2 + x^4 y/4 + x^3 y^2 + 12/8 \, x^2 y^3 + x y^4 \,] \tag{18}$$

$$(\partial \{ \tau_2 \} / \partial y) = (\partial [\, x^4 y/4 + x^3 y^2/2 + x^2 y^3/2 + x y^4/4 \,] / \partial y) = [x^4/4 + x^3 y + (3/2) \, x^2 y^2 + x y^3] \tag{19}$$

$$(\partial \{ \tau_3 \} / \partial y) = (\partial [\, x^3 y/3 + x^2 y^2/2 + x y^3/3 \,] / \partial y) = [\, x^3/3 + x^2 y + x y^2 \,] \tag{20}$$

$$(\partial \{ \tau_4 \} / \partial y) = (\partial [\, x^2 y/2 + x y^2/2 \,] / \partial y) = [\, x^2/2 + xy \,] \tag{21}$$

$$(\partial \{ \tau_5 \} / \partial y) = (\partial [\, xy \,] / \partial y) = x \tag{22}$$

$$(\partial^2 \{ \tau_1 \} / \partial y^2) = (\partial (\partial \{ \tau_1 \} / \partial y) \, \partial y) = (\partial (x^5/5 + 6/8 \, x^4 y + x^3 y^2 + x^2 y^3/2 + x^4 y/4 + x^3 y^2 +$$

$$12/8 \, x^2 y^3 + x y^4) \, \partial y) = [\, 6/8 \, x^4 + 2x^3 y + (3/2) x^2 y^2 + x^4/4 + 2x^3 y + (36/8) \, x^2 y^2 + 4x y^3 \,] \tag{23}$$

$$(\partial^2 \{ \tau_2 \} / \partial y^2) = (\partial (\partial \{ \tau_2 \} / \partial y) \, \partial y) = (\partial (x^4/4 + x^3 y + (3/2) \, x^2 y^2 + x y^3) \, \partial y) =$$

$$[\, x^3 + 3x^2 y + 3x y^2] \tag{24}$$

$$(\partial^2 \{ \tau_3 \} / \partial y^2) = (\partial (\partial \{ \tau_3 \} / \partial y) \, \partial y) = (\partial (x^3/3 + x^2 y + x y^2) \, \partial y) = [\, x^2 + 2xy \,] \tag{25}$$

$$(\partial^2 \{ \tau_4 \} / \partial y^2) = (\partial (\partial \{ \tau_4 \} / \partial y) \, \partial y) = (\partial (x^2/2 + xy) \, \partial y) = [\, x \,] \tag{26}$$

$$(\partial^2 \{ \tau_5 \} / \partial y^2) = (\partial (\partial \{ \tau_5 \} / \partial y) \, \partial y) = (\partial (x) \, \partial y) = 0 \tag{27}$$

$$(\partial (\partial \{ \tau_1 \} / \partial x) \, \partial y) = (\partial (x^4 y + (12/8) \, x^3 y^2 + x^2 y^3 + x y^4/4 + x^3 y^2/2 + x^2 y^3 + (6/8) \, x y^4 +$$

$$y^5/5) \partial y) = [\, x^4 + (24/8) \, x^3 y + 3x^2 y^2 + x y^3 + x^3 y + 3x^2 y^2 + (24/8) \, x y^3 + y^4 \,] \tag{28}$$

$$(\partial (\partial \{ \tau_2 \} / \partial x) \, \partial y) = (\partial (x^3 y + (3/2) \, x^2 y^2 + x y^3 + y^4/4) \, \partial y) = [\, x^3 + 3x^2 y + 3x y^2 + y^3 \,] \tag{29}$$

$$(\partial (\partial \{ \tau_3 \} / \partial x) \, \partial y) = (\partial (x^2 y + x y^2 + y^3/3) \, \partial y) = [\, x^2 + 2xy + y^2 \,] \tag{30}$$

$$(\partial (\partial \{ \tau_4 \} / \partial x) \, \partial y) = (\partial (xy + y^2/2) \, \partial y) = [\, x + y \,] \tag{31}$$

$$(\partial (\partial \{ \tau_5 \} / \partial x) \, \partial y) = (\partial (y) \, \partial y) = 1 \tag{32}$$

$$(\partial (\partial \{ \tau_1 \} / \partial y) \, \partial x) = (\partial (x^5/5 + 6/8 \, x^4 y + x^3 y^2 + x^2 y^3/2 + x^4 y/4 + x^3 y^2 + 12/8 \, x^2 y^3 + x y^4)$$

$$\partial x) = [\, x^4 + (24/8) \, x^3 y + 3x^2 y^2 + x y^3 + x^3 y + 3x^2 y^2 + (24/8) \, x y^3 + y^4 \,] \tag{33}$$

$$(\partial (\partial \{ \tau_2 \} / \partial y) \, \partial x) = (\partial (x^4/4 + x^3 y + (3/2) x^2 y^2 + xy^3) \, \partial x) = [x^3 + 3x^2 y + 3xy^2 + y^3] \tag{34}$$

$$(\partial (\partial \{ \tau_3 \} / \partial y) \, \partial x) = (\partial (x^3/3 + x^2 y + xy^2) \, \partial x) = [x^2 + 2xy + y^2] \tag{35}$$

$$(\partial (\partial \{ \tau_4 \} / \partial y) \, \partial x) = (\partial (x^2/2 - xy) \, \partial x) = [x + y] \tag{36}$$

$$(\partial (\partial \{ \tau_5 \} / \partial y) \, \partial x) = (\partial (x) \, \partial x) = 1 \tag{37}$$

In view of equations (3) through (7), equation (1) (with embedding) and equation (2) (without embedding) become (38) (with embedding) and (39) (without embedding) respectively:

$$\varrho (x, y) = 4 * \{ [3/2 \, \alpha_3{}^2 - 1/2 \, \alpha_3 \, \alpha_2 - 3/6 \, \alpha_3 \, \alpha_2 + 1/6 \, \alpha_2{}^2] \, [\tau_1] + [-4 \, \alpha_3{}^2 + 2 \, \alpha_3 \, \alpha_2 + 4/3 \, \alpha_3 \, x_2$$

$$- 4/3 \, \alpha_2{}^2] \, [\tau_2] + [2 \, \alpha_3{}^2 - 2 \, \alpha_3 \, \alpha_2 - 8 \, \alpha_3 \, \alpha_2 + 4 \, \alpha_2{}^2] \, [\tau_3] + [2 \, \alpha_3{}^2 - 2/3 \, \alpha_3 \, \alpha_2 + 8 \, \alpha_3 \, \alpha_2 - 16/3$$

$$x_2{}^2] \, [\tau_4] + [- 4/3 \, \alpha_3{}^2 + 4/3 \, \alpha_3 \, \alpha_2 - 8/3 \, \alpha_3 \, \alpha_2 + 8/3 \, \alpha_2{}^2] \, [\tau_5] \} / \{ 4xy * [- 2 \, \alpha_3 + 2 \, \alpha_2] -$$

$$4\alpha_3 * [3 \, (x^2 y/2 + xy^2/2) - 2xy] - 4\alpha_2 * [- (x^2 y/2 + xy^2/2) + 2xy] \} \tag{38}$$

$$\varrho (x, y) = 4 * \{ [3/2 \, \alpha_3{}^2 - 1/2 \, \alpha_3 \, \alpha_2 - 3/6 \, \alpha_3 \, \alpha_2 + 1/6 \, \alpha_2{}^2] \, [\tau_1] + [-4 \, \alpha_3{}^2 + 2 \, \alpha_3 \, \alpha_2 + 4/3 \, \alpha_3 \, \alpha_2$$

$$- 4/3 \, \alpha_2{}^2] \, [\tau_2] + [2 \, \alpha_3{}^2 - 2 \, \alpha_3 \, \alpha_2 - 8 \, \alpha_3 \, \alpha_2 + 4 \, \alpha_2{}^2] \, [\tau_3] + [2 \, \alpha_3{}^2 - 2/3 \, \alpha_3 \, \alpha_2 + 8 \, \alpha_3 \, \alpha_2 - 16/3$$

$$\alpha_2{}^2] \, [\tau_4] + [- 4/3 \, \alpha_3{}^2 + 4/3 \, \alpha_3 \, \alpha_2 - 8/3 \, \alpha_3 \, \alpha_2 + 8/3 \, \alpha_2{}^2] \, [\tau_5] \} / \{ 4xy * [- 2 \, \alpha_3 + 2 \, \alpha_2] \} \tag{39}$$

Let us calculate the second order partial derivatives of the denominator of equations (38) and (39) given the positions:

$$\eta_6 = \{ 4xy * [- 2 \, \alpha_3 + 2 \, \alpha_2] - 4\alpha_3 * [3 \, (x^2 y/2 + xy^2/2) - 2xy] - 4\alpha_2 * [- (x^2 y/2 + xy^2/2) + 2xy] \} \tag{40}$$

$$\tau_5 = \{ 4xy * [- 2 \, \alpha_3 + 2 \, \alpha_2] \} \tag{41}$$

$$(\partial \{ \eta_6 \} / \partial x) = (\partial \{ 4xy * [- 2 \, \alpha_3 + 2 \, \alpha_2] - 4\alpha_3 * [3 \, (x^2 y/2 + xy^2/2) - 2xy] - 4\alpha_2 * [- (x^2 y/2$$

$$+ xy^2/2) + 2xy] \} / \partial x) = \{ 4y * [- 2 \, \alpha_3 + 2 \, \alpha_2] - 4\alpha_3 * [3 \, (xy + y^2/2) - 2y] - 4\alpha_2 * [- (xy + y^2/2) + 2y] \} \tag{42}$$

$$(\partial \{ \eta_6 \} / \partial y) = (\partial \{ 4xy * [- 2 \, \alpha_3 + 2 \, \alpha_2] - 4\alpha_3 * [3 \, (x^2 y/2 + xy^2/2) - 2xy] - 4\alpha_2 * [- (x^2 y/2$$

$$+ xy^2/2) + 2xy] \} / \partial y) = \{ 4x * [- 2 \, \alpha_3 + 2 \, \alpha_2] - 4\alpha_3 * [3 \, (x^2/2 + xy) - 2x] - 4\alpha_2 * [- (x^2/2 +$$

$$xy) + 2x] \} \tag{43}$$

$$(\partial^2 \{ \eta_6 \} / \partial x^2) = (\partial (\partial \{ r_6 \} / \partial x) \, \partial x) = (\partial \{ 4y * [- 2 \, \alpha_3 + 2 \, \alpha_2] - 4\alpha_3 * [3 \, (xy + y^2/2) - 2y]$$

$$- 4\alpha_2 * [- (xy + y^2/2) + 2y] \} / \partial x) = \{ - 4\alpha_3 * 3y + 4\alpha_2 * y \} \tag{44}$$

$$(\partial^2 \{ \eta_6 \} / \partial y^2) = (\partial (\partial \{ \eta_5 \} / \partial y) \, \partial y) = (\partial \{ 4x * [- 2 \, \alpha_3 + 2 \, \alpha_2] - 4\alpha_3 * [3 \, (x^2/2 + xy) - 2x]$$

$$- 4\alpha_2 * [- (x^2/2 + xy) + 2x] \} /\partial y) = \{- 4\alpha_3 * 3x + 4\alpha_2 * x \} \tag{45}$$

$$(\partial (\partial \{ \eta_6 \} /\partial x) \partial y) = (\partial \{ 4y * [- 2 \alpha_3 + 2 \alpha_2] - 4\alpha_3 * [3 (xy + y^2/2) - 2y] - 4\alpha_2 * [- (xy +$$

$$y^2/2) + 2y]\} \partial y) = \{ 4 * [- 2 \alpha_3 + 2 \alpha_2] - 4\alpha_3 * [3 (x + y) - 2] - 4\alpha_2 * [- (x + y) + 2] \} \tag{46}$$

$$(\partial (\partial \{ \eta_6 \} /\partial y) \partial x) = (\partial \{ 4x * [- 2 \alpha_3 + 2 \alpha_2] - 4\alpha_3 * [3 (x^2/2 + xy) - 2x] - 4\alpha_2 * [- (x^2/2 +$$

$$xy) + 2x]\} \partial x) = \{ 4 * [- 2 \alpha_3 + 2 \alpha_2] - 4\alpha_3 * [3 (x + y) - 2] - 4\alpha_2 * [- (x + y) + 2] \} \tag{47}$$

$$(\partial \{ \tau_6 \} /\partial x) = (\partial \{ 4xy * [- 2 \alpha_3 + 2 \alpha_2] \}/\partial x) = \{ 4y * [- 2 \alpha_3 + 2 \alpha_2] \} \tag{48}$$

$$(\partial \{ \tau_6 \} /\partial y) = (\partial \{ 4xy * [- 2 \alpha_3 + 2 \alpha_2] \}/\partial y) = \{ 4x * [- 2 \alpha_3 + 2 \alpha_2] \} \tag{49}$$

$$(\partial^2 \{ \tau_6 \} /\partial x^2) = (\partial (\partial \{ \tau_6 \} /\partial x) \partial x) = (\partial (4y * [- 2 \alpha_3 + 2 \alpha_2]) \partial x) = 0 \tag{50}$$

$$(\partial^2 \{ \tau_6 \} /\partial y^2) = (\partial (\partial \{ \tau_6 \} /\partial y) \partial y) = (\partial (4x * [- 2 \alpha_3 + 2 \alpha_2]) \partial y) = 0 \tag{51}$$

$$(\partial (\partial \{ \tau_6 \} /\partial x) \partial y) = (\partial (4y * [- 2 \alpha_3 + 2 \alpha_2]) \partial y) = \{ 4 * [- 2 \alpha_3 + 2 \alpha_2] \} \tag{52}$$

$$(\partial (\partial \{ \tau_6 \} /\partial y) \partial x) = (\partial (4x * [- 2 \alpha_3 + 2 \alpha_2]) \partial x) = \{ 4 * [- 2 \alpha_3 + 2 \alpha_2] \} \tag{53}$$

Let us posit:

$$\omega_1 = [3/2 \alpha_3^2 - 1/2 \alpha_3 \alpha_2 - 3/6 \alpha_3 \alpha_2 + 1/6 \alpha_2^2] \tag{54}$$

$$\omega_2 = [-4 \alpha_3^2 + 2 \alpha_3 \alpha_2 + 4/3 \alpha_3 \alpha_2 - 4/3 \alpha_2^2] \tag{55}$$

$$\omega_3 = [2 \alpha_3^2 - 2 \alpha_3 \alpha_2 - 8 \alpha_3 \alpha_2 + 4 \alpha_2^2] \tag{56}$$

$$\omega_4 = [2 \alpha_3^2 - 2/3 \alpha_3 \alpha_2 + 8 \alpha_3 \alpha_2 - 16/3 \alpha_2^2] \tag{57}$$

$$\omega_5 = [- 4/3 \alpha_3^2 + 4/3 \alpha_3 \alpha_2 - 8/3 \alpha_3 \alpha_2 + 8/3 \alpha_2^2] \tag{58}$$

Calculation of the Second Order Partial Derivatives of ϱ (x, y)

The signal resilient to interpolation is calculated with embedding in equation (1) and without embedding in equation (2). These two equations differ because of the denominator which has been posited as η_6 (with embedding) and τ_6 (without embedding).

Let us posit:

$$\lambda = \{ \omega_1 \tau_1 + \omega_2 \tau_2 + \omega_3 \tau_3 + \omega_4 \tau_4 + \omega_5 \tau_5 \} \tag{59}$$

In what follows, the symbol Φ will signify either η_6 to relate to the resilient signal calculated with embedding, or τ_6 to relate to the resilient signal calculated without embedding. Thus, the generalized expression of $\varrho(x, y)$, as relevant to the bivariate cubic signal resilient to interpolation, is now written as equation (60):

$$\varrho(x, y) = 4 * \lambda / \Phi \tag{60}$$

Given below is the calculation of the second order partal derivatives of the generalized expression of $\varrho(x, y)$.

$$(\partial(\varrho(x, y)) / \partial x) =$$

$$(\partial(4 * \lambda / \Phi) / \partial x) = \{\{(\partial(4 * \lambda) / \partial x) * \Phi\} - \{4 * \lambda * (\partial(\Phi) / \partial x)\}\} / \{\Phi\}^2 \tag{61}$$

$$(\partial^2(\varrho(x, y)) / \partial x^2) =$$

$$(\partial(\partial(\varrho(x, y)) / \partial x) \partial x) = (\partial(\{\{\{(\partial(4 * \lambda) / \partial x) * \Phi\} - \{4 * \lambda * (\partial(\Phi) / \partial x)\}\} / \{\Phi$$

$$\}^2) \partial x) = \{\{\{(\partial^2(4 * \lambda) / \partial x^2) * \Phi\} + \{(\partial(4 * \lambda) / \partial x) \times (\partial(\Phi) / \partial x)\} - \{(\partial(4 * \lambda)$$

$$/ \partial x) * (\partial(\Phi) / \partial x)\} - \{4 * \lambda * (\partial^2(\Phi) / \partial x^2)\}\} * \{\Phi\}^2 - \{\{(\partial(4 * \lambda) / \partial x) * \Phi\} - \{4 * \lambda * (\partial(\Phi)$$

$$/ \partial x)\}\} * \{2 * \Phi * (\partial(\Phi) / \partial x)\}\} / \{\Phi\}^4 \tag{62}$$

$$(\partial(\varrho(x, y)) / \partial y) =$$

$$(\partial(4 * \lambda / \Phi) / \partial y) = \{\{(\partial(4 * \lambda) / \partial y) * \Phi\} - \{4 * \lambda * (\partial(\Phi) / \partial y)\}\} / \{\Phi\}^2 \tag{63}$$

$$(\partial^2(\varrho(x, y)) / \partial y^2) =$$

$$(\partial(\partial(\varrho(x, y)) / \partial y) \partial y) = (\partial(\{\{\{(\partial(4 * \lambda) / \partial y) * \Phi\} - \{4 * \lambda * (\partial(\Phi) / \partial y)\}\} / \{\Phi$$

$$\}^2) \partial y) = \{\{\{(\partial^2(4 * \lambda) / \partial y^2) * \Phi\} + \{(\partial(4 * \lambda) / \partial y) * (\partial(\Phi) / \partial y)\} - \{(\partial(4 * \lambda)$$

$$/ \partial y) * (\partial(\Phi) / \partial y)\} - \{4 * \lambda * (\partial^2(\Phi) / \partial y^2)\}\} * \{\Phi\}^2 - \{\{(\partial(4 * \lambda) / \partial y) * \Phi\} - \{4 * \lambda * (\partial(\Phi)$$

$$/ \partial y)\}\} * \{2 * \Phi * (\partial(\Phi) / \partial y)\}\} / \{\Phi\}^4 \tag{64}$$

$$(\partial(\partial(\varrho(x, y)) / \partial x) \partial y) = (\partial(\{\{\{(\partial(4 * \lambda) / \partial x) * \Phi\} - \{4 * \lambda * (\partial(\Phi) / \partial x)\}\} / \{\Phi\}^2$$

$$) \partial y) = \{\{\{(\partial(\partial(4 * \lambda) / \partial x) \partial y) * \Phi\} + \{(\partial(4 * \lambda) / \partial x) * (\partial(\Phi) / \partial y)\} - \{(\partial(4 * \lambda)$$

$$/ \partial y) * (\partial(\Phi) / \partial x)\} - \{4 * \lambda * (\partial(\partial(\Phi) / \partial x) \partial y)\}\} * \{\Phi\}^2 - \{\{(\partial(4 * \lambda) / \partial x) * \Phi\} - \{4 * \lambda *$$

$\partial (\Phi) / \partial x) \} \} * \{ 2 * \Phi * (\partial (\Phi) / \partial y) \} \} / \{ \Phi \}^4$ (65)

$(\partial (\partial (\varrho (x, y)) / \partial y) \partial x) = (\partial (\{ \{ \{ (\partial (4 * \lambda) / \partial y) * \Phi \} - \{ 4 * \lambda * (\partial (\Phi) / \partial y) \} \} / \{ \Phi \}^2$

$) \partial x) = \{ \{ \{ (\partial (\partial (4 * \lambda) / \partial y) \partial x) * \Phi \} + \{ (\partial (4 * \lambda) / \partial y) * (\partial (\Phi) / \partial x) \} - \{ (\partial (4 * \lambda)$

$/ \partial x) * (\partial (\Phi) / \partial y) \} - \{ 4 * \lambda * (\partial (\partial (\Phi) / \partial y) \partial x) \} \} * \{ \Phi \}^2 - \{ \{ (\partial (4 * \lambda) / \partial y) * \Phi \} - \{ 4 * \lambda *$

$(\partial (\Phi) / \partial y) \} \} * \{ 2 * \Phi * (\partial (\Phi) / \partial x) \} \} / \{ \Phi \}^4$ (66)

In view of equation (59): $\lambda = \{ \omega_1 \tau_1 + \omega_2 \tau_2 + \omega_3 \tau_3 + \omega_4 \tau_4 + \omega_5 \tau_5 \}$, let us calculate:

$(\partial (\omega_1 \tau_1 + \omega_2 \tau_2 + \omega_3 \tau_3 + \omega_4 \tau_4 + \omega_5 \tau_5) / \partial x) = \omega_1 (\partial \{ \tau_1 \} / \partial x) + \omega_2 (\partial \{ \tau_2 \} / \partial x) + \omega_3 (\partial$

$\{ \tau_3 \} / \partial x) + \omega_4 (\partial \{ \tau_4 \} / \partial x) + \omega_5 (\partial \{ \tau_5 \} / \partial x) =$

$\omega_1 [x^4 y + (12/8) x^3 y^2 + x^2 y^3 + xy^4/4 + x^3 y^2/2 + x^2 y^3 + (6/8) xy^4 + y^5/5] + \omega_2 [x^3 y + (3/2)$

$x^2 y^2 + xy^3 + y^4/4] + \omega_3 [x^2 y + xy^2 + y^3/3] + \omega_4 [xy + y^2/2] + \omega_5 y$ (67)

$(\partial (\omega_1 \tau_1 + \omega_2 \tau_2 + \omega_3 \tau_3 + \omega_4 \tau_4 + \omega_5 \tau_5) / \partial y) = \omega_1 (\partial \{ \tau_1 \} / \partial y) + \omega_2 (\partial \{ \tau_2 \} / \partial y) + \omega_3 (\partial$

$\{ \tau_3 \} / \partial y) + \omega_4 (\partial \{ \tau_4 \} / \partial y) + \omega_5 (\partial \{ \tau_5 \} / \partial y) =$

$\omega_1 [x^5/5 + 6/8 x^4 y + x^3 y^2 + x^2 y^3/2 + x^4 y/4 + x^3 y^2 + 12/8 x^2 y^3 + xy^4] + \omega_2 [x^4/4 + x^3 y +$

$(3/2) x^2 y^2 + xy^3] + \omega_3 [x^3/3 + x^2 y + xy^2] + \omega_4 [x^2/2 + xy] + \omega_5 x$ (68)

$(\partial^2 (\omega_1 \tau_1 + \omega_2 \tau_2 + \omega_3 \tau_3 + \omega_4 \tau_4 + \omega_5 \tau_5) / \partial x^2) = \omega_1 (\partial^2 \{ \tau_1 \} / \partial x^2) + \omega_2 (\partial^2 \{ \tau_2 \} / \partial x^2) +$

$\omega_3 (\partial^2 \{ \tau_3 \} / \partial x^2) + \omega_4 (\partial^2 \{ \tau_4 \} / \partial x^2) + \omega_5 (\partial^2 \{ \tau_5 \} / \partial x^2) =$

$\omega_1 [4x^3 y + (36/8) x^2 y^2 + 2xy^3 + y^4/4 + (3/2) x^2 y^2 + 2xy^3 + (6/8)y^4] + \omega_2 [3x^2 y + 3xy^2 + y^3$

$] + \omega_3 [2xy + y^2] + \omega_4 y$ (69)

$(\partial^2 (\omega_1 \tau_1 + \omega_2 \tau_2 + \omega_3 \tau_3 + \omega_4 \tau_4 + \omega_5 \tau_5) / \partial y^2) = \omega_1 (\partial^2 \{ \tau_1 \} / \partial y^2) + \omega_2 (\partial^2 \{ \tau_2 \} / \partial y^2) +$

$\omega_3 (\partial^2 \{ \tau_3 \} / \partial y^2) + \omega_4 (\partial^2 \{ \tau_4 \} / \partial y^2) + \omega_5 (\partial^2 \{ \tau_5 \} / \partial y^2) = \omega_1 [6/8 x^4 + 2x^3 y + (3/2)x^2 y^2$

$+ x^4/4 + 2x^3 y + (36/8) x^2 y^2 + 4xy^3] + \omega_2 [x^3 + 3x^2 y + 3xy^2] + \omega_3 [x^2 + 2xy] + \omega_4 x$ (70)

$(\partial (\partial(\omega_1 \tau_1 + \omega_2 \tau_2 + \omega_3 \tau_3 + \omega_4 \tau_4 + \omega_5 \tau_5) / \partial x) \partial y) = \omega_1 (\partial (\partial \{ \tau_1 \} / \partial x) \partial y) + \omega_2 (\partial (\partial \{$

τ_2 } $/\partial x$) ∂y) + ω_3 (∂ (∂ { τ_3 } $/\partial x$) ∂y) + ω_4 (∂ (∂ { τ_4 } $/\partial x$) ∂y) + ω_5 (∂ (∂ { τ_5 } $/\partial x$) ∂y) =

ω_1 [x^4 + (24/8) x^3y + $3x^2y^2$ + xy^2 + x^3y + $3x^2y^2$ + (24/8) xy^3 + y^4] + ω_2 [x^3 + $3x^2y$ + $3xy^2$

+ y^3] + ω_3 [x^2 + $2xy$ + y^2] + ω_4 [x + y] + ω_5 (71)

(∂ (∂($\omega_1 \tau_1$ + $\omega_2 \tau_2$ + $\omega_3 \tau_3$ + $\omega_4 \tau_4$ + $\omega_5 \tau_5$) $/\partial y$) ∂x) = ω_1 (∂ (∂ { τ_1 } $/\partial y$) ∂x) + ω_2 (∂ (∂ {

τ_2 } $/\partial y$) ∂x) + ω_3 (∂ (∂ { τ_3 } $/\partial y$) ∂x) + ω_4 (∂ (∂ { τ_4 } $/\partial y$) ∂x) + ω_5 (∂ (∂ { τ_5 } $/\partial y$) ∂x) =

ω_1 [x^4 + (24/8)x^3y + $3x^2y^2$ + xy^3 + x^3y + $3x^2y^2$ + (24/8) xy^3 + y^4] + ω_2 [x^3 + $3x^2y$ − $3xy^2$ +

y^3] + ω_3 [x^2 + $2xy$ + y^2] + ω_4 [x + y] + ω_5 (72)

The Total Curvature C(R((x, y), f))

Given the formulation R (x, y) = f (0, 0) + ϱ (x, y), and since f(0, 0) is costant, the total curvature C(R((x, y), f)) is given in equation (121):

C(R((x, y), f)) = (∂^2 (R((x, y), f)) $/\partial x^2$) + (∂ (∂ (R((x, y), f)) $/\partial y$)∂x) + (∂ (∂ (R((x, y),

f)) $/\partial x$)∂y) + (∂^2 (R((x, y), f)) $/\partial y^2$) = (∂^2 (ϱ (x, y)) $/\partial x^2$) + (∂ (∂ (ϱ (x, y)) $/\partial y$) ∂x) + (∂ (∂

(ϱ (x, y)) $/\partial x$) ∂y) + (∂^2 (ϱ (x, y)) $/\partial y^2$) (73)

In view of the fact that the symbol Φ signifies η_6 when referring to the resilient signal calculated with embedding, or signifies τ_6 when referring to the resilient signal calculated without embedding, hereto equations (42) through (47) and (48) through (53) are posited as equations (74) through (79). Thus, $\Psi_x, \Psi_y, \Psi_{xx}, \Psi_{yy}, \Psi_{xy}$ and Ψ_{yx} below will be pertinent to the resilient signal calculated with embedding when Φ signifies η_6, and the resilient signal calculated without embedding when Φ signifies τ_6. Equation (74) identifies Ψ_x in the following two cases: (i) Φ signifying η_6 and (ii) Φ signifies τ_6. Analogously equations (75) through (79) identify $\Psi_y, \Psi_{xx}, \Psi_{yy}, \Psi_{xy}$ and Ψ_{yx}.

Ψ_x = (∂ { η_6 } $/\partial x$) = { $4y$ * [- $2\alpha_3$ + $2\alpha_2$] - $4\alpha_3$ * [3 (xy + $y^2/2$) − 2y] - $4\alpha_2$ * [- (xy + $y^2/2$) + 2y] }

Ψ_x = (∂ { τ_6 } $/\partial x$) = { $4y$ * [- $2\alpha_3$ + $2\alpha_2$] } (74)

Ψ_y = (∂ { η_6 } $/\partial y$) = { $4x$ * [- $2\alpha_3$ + $2\alpha_2$] - $4\alpha_3$ * [3 ($x^2/2$ + xy) − 2x] - $4\alpha_2$ * [- ($x^2/2$ + xy) + 2x] }

Ψ_y = (∂ { τ_6 } $/\partial y$) = { $4x$ * [- $2\alpha_3$ + $2\alpha_2$] } (75)

Ψ_{xx} = (∂^2 { η_6 } $/\partial x^2$) = { - $4\alpha_3$ * $3y$ + $4\alpha_2$ * y }

Ψ_{xx} = (∂^2 { τ_6 } $/\partial x^2$) = (∂ ($4y$ * [- $2\alpha_3$ + $2\alpha_2$]) ∂x) = 0 (76)

$\Psi_{yy} = (\partial^2 \{ \eta_6 \} / \partial y^2) = \{ - 4\alpha_3 * 3x + 4\alpha_2 * x \}$

$\Psi_{yy} = (\partial^2 \{ \tau_6 \} / \partial y^2) = (\partial (4x * [- 2 \alpha_3 + 2 \alpha_2]) \partial y) = 0$ \hfill (77)

$\Psi_{xy} = (\partial (\partial \{ \eta_6 \} / \partial x) \partial y) = (\partial \{ 4y * [- 2 \alpha_3 + 2 \alpha_2] - 4\alpha_3 * [3 (xy + y^2/2) - 2y] + 4\alpha_2 * [-$

$(xy + y^2/2) + 2y] \} \partial y) = \{ 4 * [- 2 \alpha_3 + 2 \alpha_2] - 4\alpha_3 * [3 (x + y) - 2] - 4\alpha_2 * [- (x + y) + 2] \}$

$\Psi_{xy} = (\partial (\partial \{ \tau_6 \} / \partial x) \partial y) = \{ 4 * [- 2 \alpha_3 + 2 \alpha_2] \}$ \hfill (78)

$\Psi_{yx} = (\partial (\partial \{ \eta_6 \} / \partial y) \partial x) = (\partial \{ 4x * [- 2 \alpha_3 + 2 \alpha_2] - 4\alpha_3 * [3 (x^2/2 + xy) - 2x] + 4\alpha_2 * [-$

$(x^2/2 + xy) + 2x] \} \partial x) = \{ 4 * [- 2 \alpha_3 + 2 \alpha_2] - 4\alpha_3 * [3 (x + y) - 2] - 4\alpha_2 * [- (x + y) + 2] \}$

$\Psi_{yx} = (\partial (\partial \{ \tau_6 \} / \partial y) \partial x) = \{ 4 * [- 2 \alpha_3 + 2 \alpha_2] \}$ \hfill (79)

And equations (67) through (72) are posited as:

$\Gamma_x = (\partial (\lambda) / \partial x) = (\partial (\omega_1 \tau_1 + \omega_2 \tau_2 + \omega_3 \tau_3 + \omega_4 \tau_4 + \omega_5 \tau_5) / \partial x) = \omega_1 [x^4y + (12/8) x^3y^2 +$

$x^2y^3 + xy^4/4 + x^3y^2/2 + x^2y^3 + (6/8) xy^4 + y^5/5] + \omega_2 [x^3y + (3/2) x^2y^2 + xy^3 + y^4/4] + \omega_3$

$[x^2y + xy^2 + y^3/3] + \omega_4 [xy + y^2/2] + \omega_5 y$ \hfill (80)

$\Gamma_y = (\partial (\lambda) / \partial y) = (\partial (\omega_1 \tau_1 + \omega_2 \tau_2 + \omega_3 \tau_3 + \omega_4 \tau_4 + \omega_5 \tau_5) / \partial y) = \omega_1 [x^5/5 + 6/8 x^4y +$

$x^3y^2 + x^2y^3/2 + x^4y/4 + x^3y^2 + 12/8 x^2y^3 + xy^4] + \omega_2 [x^4/4 + x^3y + (3/2) x^2y^2 + xy^3] + \omega_3 [$

$x^3/3 + x^2y + xy^2] + \omega_4 [x^2/2 + xy] + \omega_5 x$ \hfill (81)

$\Gamma_{xx} = (\partial^2 (\lambda) / \partial x^2) = (\partial^2 (\omega_1 \tau_1 + \omega_2 \tau_2 + \omega_3 \tau_3 + \omega_4 \tau_4 + \omega_5 \tau_5) / \partial x^2) = \omega_1 [4x^3y + (36/8)$

$x^2y^2 + 2xy^3 + y^4/4 + (3/2) x^2y^2 + 2xy^3 + (6/8)y^4] + \omega_2 [3x^2y + 3xy^2 + y^3] + \omega_3 [2xy + y^2]$

$+ \omega_4 y$ \hfill (82)

$\Gamma_{yy} = (\partial^2 (\lambda) / \partial y^2) = (\partial^2 (\omega_1 \tau_1 + \omega_2 \tau_2 + \omega_3 \tau_3 + \omega_4 \tau_4 + \omega_5 \tau_5) / \partial y^2) = \omega_1 [6/8 x^4 + 2x^3y +$

$(3/2)x^2y^2 + x^4/4 + 2x^3y + (36/8) x^2y^2 + 4xy^3] + \omega_2 [x^3 + 3x^2y + 3xy^2] + \omega_3 [x^2 + 2xy] +$

$\omega_4 x$ \hfill (83)

$\Gamma_{xy} = (\partial (\partial(\lambda) / \partial x) \partial y) = (\partial (\partial(\omega_1 \tau_1 + \omega_2 \tau_2 + \omega_3 \tau_3 + \omega_4 \tau_4 + \omega_5 \tau_5) / \partial x) \partial y) = \omega_1 [x^4 +$

$(24/8)\, x^3y + 3x^2y^2 + xy^3 - x^3y + 3x^2y^2 + (24/8)\, xy^3 + y^4] + \omega_2\, [\, x^3 + 3x^2y + 3xy^2 + y^3\,] +$

$\omega_3\, [\, x^2 + 2xy + y^2\,] + \omega_4\, [\, x + y\,] - \omega_5$ \hfill (84)

$\Gamma_{yx} = (\partial\, (\partial(\lambda)\, /\partial y)\, \partial x) = (\partial\, (\partial(\, \omega_1\, \tau_1 + \omega_2\, \tau_2 + \omega_3\, \tau_3 + \omega_4\, \tau_4 + \omega_5\, \tau_5)\, /\partial y)\, \partial x) = \omega_1\, [\, x^4 +$

$(24/8)\, x^3y + 3x^2y^2 + xy^3 + x^3y + 3x^2y^2 + (24/8)\, xy^3 + y^4] + \omega_2\, [\, x^3 + 3x^2y + 3xy^2 + y^3\,] +$

$\omega_3\, [\, x^2 + 2xy + y^2] + \omega_4\, [\, x + y\,] + \omega_5$ \hfill (85)

Where $\tau_1, \tau_2, \tau_3, \tau_4, \tau_5$ are given in equations (3) through (7), $\eta_6 = \{\, 4xy * [- 2\, \alpha_3 + 2\, \alpha_2] - 4\alpha_3 * [3\, (x^2y/2 + xy^2/2) - 2xy] - 4\alpha_2 * [- (x^2y/2 + xy^2/2) + 2xy]\, \}$ is the denominator of equation (38), $\tau_6 = \{\, 4xy * [- 2\, \alpha_3 + 2\, \alpha_2\,]\, \}$ is the denominator of equation (39). Also, $\omega_1, \omega_2, \omega_3, \omega_4, \omega_5$ are given in equations (54) through (59) and $\lambda = \{\, \omega_1\, \tau_1 + \omega_2\, \tau_2 + \omega_3\, \tau_3 + \omega_4\, \tau_4 + \omega_5\, \tau_5\, \}$ is given in equation (59).

In view of the above positions, equations (62), (64), (65) and (66) are made explicit as follows:

$(\partial^2\, (\varrho\, (x, y))\, /\partial x^2) =$

$(\partial\, (\partial\, (\varrho\, (x, y))\, /\partial x)\, \partial x) = (\partial\, (\, \{\, \{\, \{\, (\partial\, (4 * \lambda)\, /\partial x) * \Phi\} - \{4 * \lambda * (\partial\, (\Phi)\, /\partial x)\, \}\, \}\, /\, \{\Phi\}^2)\, \partial x) = \{\, \{\, \{\, (\partial^2\, (4 * \lambda)\, /\partial x^2) * \Phi\} + \{(\partial\, (4 * \lambda)\, /\partial x) * (\partial\, (\Phi)\, /\partial x)\, \} - \{(\partial\, (4 * \lambda)\, /\partial x) * (\partial\, (\Phi)\, /\partial x)\, \} - \{4 * \lambda * (\partial^2\, (\Phi)\, /\partial x^2)\, \}\, \} * \{\Phi\}^2 - \{\, \{\, (\partial\, (4 * \lambda)\, /\partial x) * \Phi\} - \{4 * \lambda * (\partial\, (\Phi)\, /\partial x)\, \}\, \} * \{2 * \Phi * (\partial\, (\Phi)\, /\partial x)\, \}\, \}\, /\, \{\Phi\}^4 = \{\, \{\, \{4\, \Gamma_{xx}\, \Phi\} + \{4\, \Gamma_x\, \Psi_x\} - \{4\, \Gamma_x\, \Psi_x\} - \{4\, \lambda\, \Psi_{xx}\}\, \} * \{\Phi\}^2 - \{\, \{4 * \Gamma_x * \Phi\} - \{4 * \lambda * \Psi_x\}\, \} * \{2 * \Phi * \Psi_x\}\, \}\, /\, \{\Phi\}^4$ \hfill (86)

$(\partial^2\, (\varrho\, (x, y))\, /\partial y^2) =$

$(\partial\, (\partial\, (\varrho\, (x, y))\, /\partial y)\, \partial y) = (\partial\, (\, \{\, \{\, \{\, (\partial\, (4 * \lambda)\, /\partial y) * \Phi\} - \{4 * \lambda * (\partial\, (\Phi)\, /\partial y)\, \}\, \}\, /\, \{\Phi\}^2)\, \partial y) = \{\, \{\, \{\, (\partial^2\, (4 * \lambda)\, /\partial y^2) * \Phi\} + \{(\partial\, (4 * \lambda)\, /\partial y) * (\partial\, (\Phi)\, /\partial y)\, \} - \{(\partial\, (4 * \lambda)\, /\partial y) * (\partial\, (\Phi)\, /\partial y)\, \} - \{4 * \lambda * (\partial^2\, (\Phi)\, /\partial y^2)\, \}\, \} * \{\Phi\}^2 - \{\, \{\, (\partial\, (4 * \lambda)\, /\partial y) * \Phi\} - \{4 * \lambda * (\partial\, (\Phi)\, /\partial y)\, \}\, \} * \{2 * \Phi * (\partial\, (\Phi)\, /\partial y)\, \}\, \}\, /\, \{\Phi\}^4 = \{\, \{\, \{4\, \Gamma_{yy}\, \Phi\} + \{4\, \Gamma_y\, \Psi_y\} - \{4\, \Gamma_y\, \Psi_y\} - \{4\, \lambda\, \Psi_{yy}\}\, \} * \{\Phi\}^2 - \{\, \{4 * \Gamma_y * \Phi\} - \{4 * \lambda * \Psi_y\}\, \} * \{2 * \Phi * \Psi_y\}\, \}\, /\, \{\Phi\}^4$ \hfill (87)

$(\partial\, (\partial\, (\varrho\, (x, y))\, /\partial x)\, \partial y) = (\partial\, (\, \{\, \{\, (\partial\, (4 * \lambda)\, /\partial x) * \Phi\} - \{4 * \lambda * (\partial\, (\Phi)\, /\partial x)\, \}\, \}\, /\, \{\Phi\}^2)\, \partial y) = \{\, \{\, \{\, (\partial\, (\partial\, (4 * \lambda)\, /\partial x)\, \partial y) * \Phi\} + \{(\partial\, (4 * \lambda)\, /\partial x) * (\partial\, (\Phi)\, /\partial y)\, \} - \{(\partial\, (4 * \lambda)$

$/\partial y) * (\partial (\Phi) / \partial x) \} - \{ 4 * \lambda * (\partial (\partial (\Phi) / \partial x) \partial y) \} \} * \{ \Phi \}^2 - \{ \{ (\partial (4 * \lambda) / \partial x) * \Phi \} - \{ 4 * \lambda *$

$(\partial (\Phi) / \partial x) \} \} * \{ 2 * \Phi * (\partial (\Phi) / \partial y) \} \} / \{ \Phi \}^4 = \{ 4 * \{ \Gamma_{xy} \Phi + \Gamma_x \Psi_y - \Gamma_y \Psi_x - \lambda \Psi_{xy} \} * \{ \Phi \}^2$

$- \{ \{ 4 * \Gamma_x * \Phi \} - \{ 4 * \lambda * \Psi_x \} \} * \{ 2 * \Phi * \Psi_y \} \} / \{ \Phi \}^4$ (88)

$(\partial (\partial (\varrho (x, y)) / \partial y) \partial x) = (\partial (\{ \{ (\partial (4 * \lambda) / \partial y) * \Phi \} - \{ 4 * \lambda * (\partial (\Phi) / \partial y) \} \} / \{ \Phi \}^2$

$) \partial x) = \{ \{ \{ (\partial (\partial (4 * \lambda) / \partial y) \partial x) * \Phi \} + \{ (\partial (4 * \lambda) / \partial y) * (\partial (\Phi) / \partial x) \} - \{ (\partial (4 * \lambda)$

$/ \partial x) * (\partial (\Phi) / \partial y) \} - \{ 4 * \lambda * (\partial (\partial (\Phi) / \partial y) \partial x) \} \} * \{ \Phi \}^2 - \{ \{ (\partial (4 * \lambda) / \partial y) * \Phi \} - \{ 4 * \lambda *$

$(\partial (\Phi) / \partial y) \} \} * \{ 2 * \Phi * (\partial (\Phi) / \partial x) \} \} / \{ \Phi \}^4 = \{ 4 * \{ \Gamma_{yx} \Phi + \Gamma_y \Psi_x - \Gamma_x \Psi_y - \lambda \Psi_{yx} \} * \{ \Phi$

$\}^2 - \{ \{ 4 * \Gamma_y * \Phi \} - \{ 4 * \lambda * \Psi_y \} \} * \{ 2 * \Phi * \Psi_x \} \} / \{ \Phi \}^4$ (89)

From equation (73), adding together equations (86), (87), (88) and (89), the total curvature is given as:

$C(R((x, y), f)) = \{ \{ \{ \{ 4 \Gamma_{xx} \Phi \} + \{ 4 \Gamma_x \Psi_x \} - \{ 4 \Gamma_x \Psi_x \} - \{ 4 \lambda \Psi_{xx} \} \} * \{ \Phi \}^2 - \{ \{ 4 * \Gamma_x * \Phi \}$

$- \{ 4 * \lambda * \Psi_x \} \} * \{ 2 * \Phi * \Psi_x \} \} + \{ \{ \{ 4 \Gamma_{yy} \Phi \} + \{ 4 \Gamma_y \Psi_y \} - \{ 4 \Gamma_y \Psi_y \} - \{ 4 \lambda \Psi_{yy} \} \} * \{ \Phi$

$\}^2 - \{ \{ 4 * \Gamma_y * \Phi \} - \{ 4 * \lambda * \Psi_y \} \} * \{ 2 * \Phi * \Psi_y \} \} + \{ 4 * \{ \Gamma_{xy} \Phi + \Gamma_x \Psi_y - \Gamma_y \Psi_x - \lambda \Psi_{xy} \} * \{$

$\Phi \}^2 - \{ \{ 4 * \Gamma_x * \Phi \} - \{ 4 * \lambda * \Psi_x \} \} * \{ 2 * \Phi * \Psi_y \} \} + \{ 4 * \{ \Gamma_{yx} \Phi + \Gamma_y \Psi_x - \Gamma_x \Psi_y - \lambda \Psi_{yx} \} *$

$\{ \Phi \}^2 - \{ \{ 4 * \Gamma_y * \Phi \} - \{ 4 * \lambda * \Psi_y \} \} * \{ 2 * \Phi * \Psi_x \} \} \} / \{ \Phi \}^4 =$

$\{ \{ \{ \{ 4 \Gamma_{xx} \Phi \} - \{ 4 \lambda \Psi_{xx} \} \} * \{ \Phi \}^2 - \{ \{ 4 * \Gamma_x * \Phi \} - \{ 4 * \lambda * \Psi_x \} \} * \{ 2 * \Phi * \Psi_x \} \} + \{ \{$

$\{ 4 \Gamma_{yy} \Phi \} - \{ 4 \lambda \Psi_{yy} \} \} * \{ \Phi \}^2 - \{ \{ 4 * \Gamma_y * \Phi \} - \{ 4 * \lambda * \Psi_y \} \} * \{ 2 * \Phi * \Psi_y \} \} + \{ 4 * \{ \Gamma_{xy}$

$\Phi + \Gamma_x \Psi_y - \Gamma_y \Psi_x - \lambda \Psi_{xy} \} * \{ \Phi \}^2 - \{ \{ 4 * \Gamma_x * \Phi \} - \{ 4 * \lambda * \Psi_x \} \} * \{ 2 * \Phi * \Psi_y \} \} + \{ 4 * \{$

$\Gamma_{yx} \Phi + \Gamma_y \Psi_x - \Gamma_x \Psi_y - \lambda \Psi_{yx} \} * \{ \Phi \}^2 - \{ \{ 4 * \Gamma_y * \Phi \} - \{ 4 * \lambda * \Psi_y \} \} * \{ 2 * \Phi * \Psi_x \} \} \} /$

$\{ \Phi \}^4$ (90)

Equation (90) furnishes $C(R((x, y), f))$ in either case of the following cases: (i) the resilient signal calculated with embedding and in such case the numerical values assigned to the symbols Φ, Ψ_x, Ψ_y, $\Psi_{xx}, \Psi_{yy}, \Psi_{xy}$ and Ψ_{yx} are derived with η_6, and (ii) the resilient signal calculated without embedding and in such case the numerical values assigned to the symbols Φ, $\Psi_x, \Psi_y, \Psi_{xx}, \Psi_{yy}, \Psi_{xy}$ and Ψ_{yx} are derived with τ_6.

SUMMARY

What was said about the two types of curvature (classic and resilient), that the classic-curvature is the same $((\partial^2 \, (\, h_4(x, y) \,) \, /\partial x^2) + (\partial^2 \, (\, h_4(x, y) \,) \, /\partial y^2) + (\partial^2 \, (\, h_4(x, y) \,) \, /\partial x \partial y) + (\partial^2 \, (\, h_4(x, y) \,) \, /\partial y \partial x))$ in both of the cases of signal calculated with embedding and without embedding, and that the resilient curvature is not the same with and without embeding given that the resilient signal assumes two different math forms, applies to this chapter also.

It can be added at this point as an anticipation of what will be seen as the results of the experiments in chapter 8, that the two types of curvature (classic and resilient) allow us to derive two Pixel Intensity Corrections (PICs). One should remember that the Pixel Intensity Correction is the transfer function customized, or better to say, derived from the math form of the interpolation function.

We now have I $(x, y) = f \, (0, 0) + $ PIC (x, y) where f $(0, 0)$ is the original signal, and PIC (x, y) is derived from the multiplication of the shift (misplacement) times tan(total classic curvature); and R $(x, y) = f \, (0, 0) + $ PIC (x, y), where f $(0, 0)$ is the original signal, and PIC (x, y) is derived from the multiplication of the shift (misplacement) times tan(total resilient curvature).

The reader may wonder then what the difference between I $(x, y) = f \, (0, 0) + $ PIC (x, y) and I $(x, y) = f \, (0, 0) + \xi \, (x, y)$ is, and what the difference between R $(x, y) = f \, (0, 0) + $ PIC (x, y) and R $(x, y) = f \, (0, 0) + \varrho \, (x, y)$ is. The classic signal is $\xi \, (x, y)$, the resilient signal is $\varrho \, (x, y)$, the transfer function is PIC (x, y), which is derived from the total curvature either classic: $((\partial^2 \, (\, h_4(x, y) \,) \, /\partial x^2) + (\partial^2 \, (\, h_4(x, y) \,) \, /\partial y^2) + (\partial^2 \, (\, h_4(x, y) \,) \, /\partial x \partial y) + (\partial^2 \, (\, h_4(x, y) \,) \, /\partial y \partial x))$, which furnishes the reconstructed signal I $(x, y) = f \, (0, 0) + $ PIC (x, y); or resilient: $(\partial^2 \, (\varrho \, (x, y) \,) \, /\partial x^2) + (\partial^2 \, (\varrho \, (x, y) \,) \, /\partial y^2) + (\partial^2 \, (\varrho \, (x, y) \,) \, /\partial x \partial y) + (\partial^2 \, (\varrho \, (x, y) \,) \, /\partial y \partial x)$, which furnishes the reconstructed signal R $(x, y) = f \, (0, 0) + $ PIC (x, y)

REFERENCES

Ciulla, C. (2009). *Improved Signal and Image Interpolation in Biomedical Applications: The Case of Magnetic Resonance Imaging (MRI)* – Medical Information Science Reference - IGI Global Publisher. Hershey, PA, U.S.A.

CALCULATION OF THE CURVATURE OF THE BIVARIATE CUBIC LAGRANGE SIGNAL RESILIENT TO INTERPOLATION: WITH AND WITHOUT EMBEDDING

INTRODUCTION

This chapter presents the calculation of the total curvature of the signal resilient to the bivariate cubic Lagrange interpolation function given in chapter 2 in equation (36):

$$LGR_3(x, y) = f(0, 0) + \alpha_3 * [\, (1/2)(x + y)^3 - (x + y)^2 - 1/2(x + y) + 1\,] + \alpha_2 * [\, -(1/6)(x + y)^3 + (x + y)^2 - (11/6)(x + y) + 1\,].$$

The classic signal is $I(x, y) = f(0, 0) + \xi(x, y)$, with: $\xi(x, y) = \alpha_3 * [\, (1/2)(x + y)^3 - (x + y)^2 - 1/2(x + y) + 1\,] + \alpha_2 * [\, -(1/6)(x + y)^3 + (x + y)^2 - (11/6)(x + y) + 1\,]$, with embedding and is $I(x, y) = \xi(x, y)$, without embedding. The total curvature is calculated through the sum: $(\partial^2 (LGR_3(x, y)))/\partial x^2) + (\partial^2 (LGR_3(x, y)))/\partial y^2) + (\partial^2 (LGR_3(x, y)))/\partial x \partial y) + (\partial^2 (LGR_3(x, y)))/\partial y \partial x)$.

The signal resilient to interpolation is given in this chapter in equations (1) and (51), with and without embedding respectively. The total curvature of the resilient signal through the sum: $(\partial^2 (\varrho(x, y)))/\partial x^2) + (\partial^2 (\varrho(x, y)))/\partial y^2) + (\partial^2 (\varrho(x, y)))/\partial x \partial y) + (\partial^2 (\varrho(x, y)))/\partial y \partial x)$; and it is calculated in this chapter in equation (74), with and without embedding. To finalize the signal reconstruction, the Pixel Intensity Correction is calculated through the formula: PIC = Shift * tan(total curvature).

BIVARIATE CUBIC LAGRANGE CALCULATED WITH EMBEDDING

The formulation $R(x, y) = f(0, 0) + \varrho(x, y)$ requires $\varrho(x, y)$ as per equation (51) of chapter 2 Part I:

$$\varrho(x, y) = \Lambda_{LRG} / \{\, 4xy * [-2\alpha_3 + 2\alpha_2] - 4 * \alpha_3 [\, 3(x^2y/2 + xy^2/2) - 2xy\,] - 4 * \alpha_2 [\, -(x^2y/2 + xy^2/2) + 2xy\,] \,\} \tag{1}$$

The value $\varrho(x, y)$ results from the solution of the equation: $E_o(x, y) = E_{IN}(x, y)$, where:

$$\Lambda_{LRG} = 4* \{\, [\, (3/2)\alpha_3^2 - (1/2)\alpha_3\alpha_2 - (3/6)\alpha_2\alpha_3 + (1/6)\alpha_2^2\,] (x^5y/5 + 3/8\, x^4y^2 + x^3y^3/3 +$$

$$x^2y^4/8 + x^4y^2/8 + x^3y^3/3 + 3/8\, x^2y^4 + xy^5/5) + [\, -4\alpha_3^2 + 2\alpha_3\alpha_2 + (10/3)\alpha_2\alpha_3 - (8/6)\alpha_2^2\,]$$

$$(x^4y/4 + x^3y^2/2 + x^2y^3/2 + xy^4/4) + [\, 1/2\,\alpha_3^2 - 3/2\,\alpha_3\alpha_2 - (45/6)\alpha_2\alpha_3 + (23/6)\alpha_2^2\,] (x^3y/3 +$$

$$x^2y^2/2 + xy^3/3) + [\, 4\alpha_3^2 - 2\alpha_3\alpha_2 + (40/6)\alpha_2\alpha_3 - (28/6)\alpha_2^2\,] (x^2y/2 + xy^2/2) +$$

$$[\, -2\alpha_3^2 + 2\alpha_2^2\,] (xy) \,\} \tag{2}$$

Let us posit:

$$\Phi_1 = [\, (3/2)\alpha_3^2 - (1/2)\alpha_3\alpha_2 - (3/6)\alpha_2\alpha_3 + (1/6)\alpha_2^2\,] \tag{3}$$

$$\Phi_2 = [-4\,\alpha_3{}^2 + 2\,\alpha_3\,\alpha_2 + (10/3)\,\alpha_2\,\alpha_3 - (8/6)\,\alpha_2{}^2] \tag{4}$$

$$\Phi_3 = [1/2\,\alpha_3{}^2 - 3/2\,\alpha_3\,\alpha_2 - (45/6)\,\alpha_2\,\alpha_3 + (23/6)\,\alpha_2{}^2] \tag{5}$$

$$\Phi_4 = [4\,\alpha_3{}^2 - 2\,\alpha_3\,\alpha_2 + (40/6)\,\alpha_2\,\alpha_3 - (28/6)\,\alpha_2{}^2] \tag{6}$$

$$\Phi_5 = [-2\,\alpha_3{}^2 + 2\,\alpha_2{}^2] \tag{7}$$

$$\varphi_1 = (x^5y/5 + 3/8\,x^4y^2 + x^3y^3/3 + x^2y^4/8 + x^4y^2/8 + x^3y^3/3 + 3/8\,x^2y^4 + xy^5/5) \tag{8}$$

$$\varphi_2 = (x^4y/4 + x^3y^2/2 + x^2y^3/2 + xy^4/4) \tag{9}$$

$$\varphi_3 = (x^3y/3 + x^2y^2/2 + xy^3/3) \tag{10}$$

$$\varphi_4 = (x^2y/2 + xy^2/2) \tag{11}$$

$$\varphi_5 = (xy) \tag{12}$$

Calculation of First and Second Order Partial Derivatives

$$(\partial\,(\varphi_1)\,/\partial x) = (x^4y + (3/2)\,x^3y^2 + x^2y^3 + xy^4/4 + x^3y^2/2 + x^2y^3 + (3/4)\,xy^4 + y^5/5) \tag{13}$$

$$(\partial\,(\varphi_2)\,/\partial x) = (x^3y + (3/2)\,x^2y^2 + xy^3 + y^4/4) \tag{14}$$

$$(\partial\,(\varphi_3)\,/\partial x) = (x^2y + xy^2 + y^3/3) \tag{15}$$

$$(\partial\,(\varphi_4)\,/\partial x) = (xy + y^2/2) \tag{16}$$

$$(\partial\,(\varphi_5)\,/\partial x) = y \tag{17}$$

$$(\partial^2\,(\varphi_1)\,/\partial x^2) = (4\,x^3y + (9/2)\,x^2y^2 - 2\,xy^3 + y^4/4 + (3/2)\,x^2y^2 + 2\,xy^3 + (3/4)\,y^4) \tag{18}$$

$$(\partial^2\,(\varphi_2)\,/\partial x^2) = (3\,x^2y + 3\,xy^2 + y^3) \tag{19}$$

$$(\partial^2\,(\varphi_3)\,/\partial x^2) = (2\,xy + y^2) \tag{20}$$

$$(\partial^2\,(\varphi_4)\,/\partial x^2) = y \tag{21}$$

$$(\partial^2\,(\varphi_5)\,/\partial x^2) = 0 \tag{22}$$

$$(\partial^2\,(\varphi_1)\,/\partial x\partial y) = (x^4 + 3\,x^3y + 3\,x^2y^2 + xy^3 + x^3y + 3\,x^2y^2 + 3\,xy^3 + y^4) \tag{23}$$

$$(\partial^2\,(\varphi_2)\,/\partial x\partial y) = (x^3 + 3\,x^2y + 3\,xy^2 + y^3) \tag{24}$$

$$(\partial^2\,(\varphi_3)\,/\partial x\partial y) = (x^2 + 2\,xy + y^2) \tag{25}$$

$$(\partial^2\,(\varphi_4)\,/\partial x\partial y) = (x + y) \tag{26}$$

$$(\partial^2 (\varphi_5) / \partial x \partial y) = 1 \tag{27}$$

$$(\partial (\varphi_1) / \partial y) = (x^5/5 + (3/4) x^4y + x^3y^2 + x^2y^3/2 + x^4y/4 + x^3y^2 + (3/2) x^2y^3 + xy^4) \tag{28}$$

$$(\partial (\varphi_2) / \partial y) = (x^4/4 + x^3y + (3/2) x^2y^2 + xy^3) \tag{29}$$

$$(\partial (\varphi_3) / \partial y) = (x^3/3 + x^2y + xy^2) \tag{30}$$

$$(\partial (\varphi_4) / \partial y) = (x^2/2 + xy) \tag{31}$$

$$(\partial (\varphi_5) / \partial y) = x \tag{32}$$

$$(\partial^2 (\varphi_1) / \partial y \partial x) = (x^4 + 3 x^3y + 3 x^2y^2 + xy^3 + x^3y + 3 x^2y^2 + 3 xy^3 + y^4) \tag{33}$$

$$(\partial^2 (\varphi_2) / \partial y \partial x) = (x^3 + 3 x^2y + 3 xy^2 + y^3) \tag{34}$$

$$(\partial^2 (\varphi_3) / \partial y \partial x) = (x^2 + 2 xy + y^2) \tag{35}$$

$$(\partial^2 (\varphi_4) / \partial y \partial x) = (x + y) \tag{36}$$

$$(\partial^2 (\varphi_5) / \partial y \partial x) = 1 \tag{37}$$

$$(\partial^2 (\varphi_1) / \partial y^2) = ((3/4) x^4 + 2 x^3y + (3/2) x^2y^2 + x^4/4 + 2 x^3y + (9/2) x^2y^2 + 4 xy^3) \tag{38}$$

$$(\partial^2 (\varphi_2) / \partial y^2) = (x^3 + 3 x^2y + 3 xy^2) \tag{39}$$

$$(\partial^2 (\varphi_3) / \partial y^2) = (x^2 + 2 xy) \tag{40}$$

$$(\partial^2 (\varphi_4) / \partial y^2) = x \tag{41}$$

$$(\partial^2 (\varphi_5) / \partial y^2) = 0 \tag{42}$$

From equation (51) of chapter 2 Part I let us posit:

$$\Lambda D_{LRG} = \{ 4xy * [- 2 \alpha_3 + 2 \alpha_2] - 4 * \alpha_3 [3 (x^2y/2 + xy^2/2) - 2xy]$$

$$- 4 * \alpha_2 [- (x^2y/2 + xy^2/2) + 2xy] \} \tag{43}$$

Calculation of First and Second Order Partial Derivatives of ΛD_{LRG}

$$(\partial (\Lambda D_{LRG}) / \partial x) = \{ 4y * [- 2 \alpha_3 + 2 \alpha_2] - 4 * \alpha_3 [3 (xy + y^2/2) - 2y]$$

$$- 4 * \alpha_2 [- (xy + y^2/2) + 2y] \} \tag{44}$$

$$(\partial^2 (\Lambda D_{LRG}) / \partial x^2) = \{ - 4 * \alpha_3 [3 y] - 4 * \alpha_2 [(- y)] \} = -12 \alpha_3 y + 4 \alpha_2 y \tag{45}$$

$$(\partial^2 (\Lambda D_{LRG}) / \partial x \partial y) = \{ 4 [- 2 \alpha_3 + 2 \alpha_2] - 4 * \alpha_3 [3 (x + y) - 2] - 4 * \alpha_2 [- (x + y) + 2] \} \tag{46}$$

$$(\partial\,(\Lambda D_{LRG})\,/\partial y) = \{\; 4x * [-2\,\alpha_3 + 2\,\alpha_2] - 4 * \alpha_3\,[\,3\,(x^2/2 + xy) - 2x\,]$$

$$- 4 * \alpha_2\,[\,-(x^2/2 + xy) + 2x\,]\;\} \tag{47}$$

$$(\partial^2\,(\Lambda D_{LRG})\,/\partial y\partial x) = \{\,4\,[-2\,\alpha_3 + 2\,\alpha_2] - 4 * \alpha_3\,[\,3\,(x + y) - 2\,] - 4 * \alpha_2\,[\,-(x + y) + 2\,]\,\} \tag{48}$$

$$(\partial^2\,(\Lambda D_{LRG})\,/\partial y^2) = \{\,-4 * \alpha_3\,[\,3\,x\,] - 4 * \alpha_2\,[\,(-x)\,]\,\} = -12\,\alpha_3\,x + 4\,\alpha_2\,x \tag{49}$$

Equation (51) of chapter 2 Part I can be written as:

$$\varrho\,(x, y) = \Lambda_{LRG}\,/\,\Lambda D_{LRG} = 4 * \{\;\Phi_1 * \varphi_1 + \Phi_2 * \varphi_2 + \Phi_3 * \varphi_3 + \Phi_4 * \varphi_4 + \Phi_5 * \varphi_5\;\}\,/\,\Lambda D_{LRG} \tag{50}$$

BIVARIATE CUBIC LAGRANGE CALCULATED WITHOUT EMBEDDING

Equation (51) of chapter 3 Part I is given below as (51):

$$f\,(0, 0) = \Lambda_{LRG}\,/\,\{\,4xy * [-2\,\alpha_3 + 2\,\alpha_2]\,\} \tag{51}$$

With:

$$\Lambda_{LRG} = 4 * \{\;[\,(3/2)\,\alpha_3{}^2 - (1/2)\,\alpha_3\alpha_2 - (3/6)\,\alpha_2\,\alpha_3 + (1/6)\,\alpha_2{}^2\,]\,(x^5 y/5 + 3/8\,x^4 y^2 + x^3 y^3/3 +$$

$$x^2 y^4/8 + x^4 y^2/8 + x^3 y^3/3 + 3/8\,x^2 y^4 + xy^5/5) + [-4\,\alpha_3{}^2 + 2\,\alpha_3\alpha_2 + (10/3)\,\alpha_2\,\alpha_3 - (8/6)\,\alpha_2{}^2]$$

$$(x^4 y/4 + x^3 y^2/2 + x^2 y^3/2 + xy^4/4) + [1/2\,\alpha_3{}^2 - 3/2\,\alpha_3\alpha_2 - (45/6)\,\alpha_2\,\alpha_3 + (23/6)\,\alpha_2{}^2]\,(x^3 y/3 +$$

$$x^2 y^2/2 + xy^3/3) + [4\,\alpha_3{}^2 - 2\,\alpha_3\alpha_2 + (40/6)\,\alpha_2\,\alpha_3 - (28/6)\,\alpha_2{}^2]\,(x^2 y/2 + xy^2/2) + [-2\,\alpha_3{}^2 + 2$$

$$\alpha_2{}^2]\,(xy)\;\} \tag{52}$$

In view of equations (2) through (12), equation (52), which is the same as equation (2), is reduced to the form of equation (53):

$$\Lambda_{LRG} = 4 * \{\;\Phi_1 * \varphi_1 + \Phi_2 * \varphi_2 + \Phi_3 * \varphi_3 + \Phi_4 * \varphi_4 + \Phi_5 * \varphi_5\;\} \tag{53}$$

From equation (51) let us posit:

$$\Omega_{LRG} = \{\,4xy * [-2\,\alpha_3 + 2\,\alpha_2]\,\} \tag{54}$$

As per equation (51) of chapter 3 Part I, the formulation R (x, y) = f (0, 0) + ϱ (x, y) requires:

$$\varrho\,(x, y) = \Lambda_{LRG}\,/\,\Omega_{LRG} \tag{55}$$

Calculation of First and Second Order Partial Derivatives of Ω_{LRG}

On the basis of equation (54) the following is true:

$$(\partial\,(\Omega_{LRG})\,/\partial x) = \{\,4y * [-2\,\alpha_3 + 2\,\alpha_2]\,\} \tag{56}$$

$$(\partial \, (\Omega_{LRG}) \, / \partial y) = \{ \, 4x * [-2\,\alpha_3 + 2\,\alpha_2] \, \} \tag{57}$$

$$(\partial^2 \, (\Omega_{LRG}) \, / \partial x^2) = (\partial^2 \, (\Omega_{LRG}) \, / \partial y^2) = 0 \tag{58}$$

$$(\partial^2 \, (\Omega_{LRG}) \, / \partial x \partial y) = \{ \, 4 * [-2\,\alpha_3 + 2\,\alpha_2] \, \} \tag{59}$$

$$(\partial^2 \, (\Omega_{LRG}) \, / \partial y \partial x) = \{ \, 4 * [-2\,\alpha_3 + 2\,\alpha_2] \, \} \tag{60}$$

CALCULATION OF FIRST AND SECOND ORDER PARTIAL DERIVATIVES OF ϱ (x, y): BIVARIATE CUBIC LAGRANGE SIGNAL WITH AND WITHOUT EMBEDDING

Equation (50) relevant to the signal calculated with embedding and equation (55) relevant to the signal calculated without embedding, can both be grouped in equation (61) given below, where $\Gamma_{LGR} = \Lambda D_{LRG}$ (signal calculated with embedding) or $\Gamma_{LGR} = \Omega_{LRG}$ (signal calculated without embedding).

$$\varrho \, (x, y) = \Lambda_{LRG} \, / \, \Gamma_{LGR} \tag{61}$$

It follows that the calculation of the first and second order partial derivatives of ϱ (x, y) applies to both of the cases of the signal calculated with embedding ($\Gamma_{LGR} = \Lambda D_{LRG}$) and the signal calculated without embedding ($\Gamma_{LGR} = \Omega_{LRG}$).

$$(\partial \, (\Lambda_{LRG} \, / \, \Gamma_{LGR}) \, / \partial x) = 4 * \{ \, [\, \Phi_1 \, (\partial \, (\varphi_1) \, / \partial x) + \Phi_2 \, (\partial \, (\varphi_2) \, / \partial x) + \Phi_3 \, (\partial \, (\varphi_3) \, / \partial x) + \Phi_4 \, (\partial \, (\varphi_4)$$

$$/ \partial x) + \Phi_5 \, (\partial \, (\varphi_5) \, / \partial x) \,] * \Gamma_{LGR} - [\, \Phi_1 * \varphi_1 + \Phi_2 * \varphi_2 + \Phi_3 * \varphi_3 + \Phi_4 * \varphi_4 + \Phi_5 * \varphi_5] * (\partial \, (\Gamma_{LGR}) \, / \partial x) \, \} \, /$$

$$[\, \Gamma_{LGR} \,]^2 \tag{62}$$

$$(\partial \, (\Lambda_{LRG} \, / \, \Gamma_{LGR}) \, / \partial y) = 4 * \{ \, [\, \Phi_1 \, (\partial \, (\varphi_1) \, / \partial y) + \Phi_2 \, (\partial \, (\varphi_2) \, / \partial y) + \Phi_3 \, (\partial \, (\varphi_3) \, / \partial y) + \Phi_4 \, (\partial \, (\varphi_4)$$

$$/ \partial y) + \Phi_5 \, (\partial \, (\varphi_5) \, / \partial y) \,] * \Gamma_{LGR} - [\, \Phi_1 * \varphi_1 + \Phi_2 * \varphi_2 + \Phi_3 * \varphi_3 + \Phi_4 * \varphi_4 + \Phi_5 * \varphi_5] * (\partial \, (\Gamma_{LGR}) \, / \partial y) \, \} \, /$$

$$[\, \Gamma_{LGR} \,]^2 \tag{63}$$

$$(\partial^2 \, (\Lambda_{LRG} \, / \, \Gamma_{LGR}) \, / \partial x^2) = 4 * \{ \, \{ \, [\, \Phi_1 \, (\partial^2 \, (\varphi_1) \, / \partial x^2) + \Phi_2 \, (\partial^2 \, (\varphi_2) \, / \partial x^2) + \Phi_3 \, (\partial^2 \, (\varphi_3) \, / \partial x^2) +$$

$$\Phi_4 \, (\partial^2 \, (\varphi_4) \, / \partial x^2) + \Phi_5 \, (\partial^2 \, (\varphi_5) \, / \partial x^2) \,] * \Gamma_{LGR} + [\, \Phi_1 \, (\partial \, (\varphi_1) \, / \partial x) + \Phi_2 \, (\partial \, (\varphi_2) \, / \partial x) + \Phi_3 \, (\partial \, (\varphi_3)$$

$$/ \partial x) + \Phi_4 \, (\partial \, (\varphi_4) \, / \partial x) + \Phi_5 \, (\partial \, (\varphi_5) \, / \partial x) \,] * (\partial \, (\Gamma_{LGR}) \, / \partial x) - [\, \Phi_1 * \varphi_1 + \Phi_2 * \varphi_2 + \Phi_3 * \varphi_3 + \Phi_4$$

$$* \varphi_4 + \Phi_5 * \varphi_5] * (\partial^2 \, (\Gamma_{LGR}) \, / \partial x^2) - [\, \Phi_1 \, (\partial \, (\varphi_1) \, / \partial x) + \Phi_2 \, (\partial \, (\varphi_2) \, / \partial x) + \Phi_3 \, (\partial \, (\varphi_3) \, / \partial x) + \Phi_4$$

$$(\partial \, (\varphi_4) \, / \partial x) + \Phi_5 \, (\partial \, (\varphi_5) \, / \partial x) \,] * (\partial \, (\Gamma_{LGR}) \, / \partial x) \, \} * [\, \Gamma_{LGR}]^2 - \{ \, [\, \Phi_1 \, (\partial \, (\varphi_1) \, / \partial x) + \Phi_2 \, (\partial \, (\varphi_2)$$

$$/ \partial x) + \Phi_3 \, (\partial \, (\varphi_3) \, / \partial x) + \Phi_4 \, (\partial \, (\varphi_4) \, / \partial x) + \Phi_5 \, (\partial \, (\varphi_5) \, / \partial x) \,] * [\, \Gamma_{LGR} \,] - [\, \Phi_1 * \varphi_1 + \Phi_2 * \varphi_2 +$$

$$\Phi_3 * \varphi_3 + \Phi_4 * \varphi_4 + \Phi_5 * \varphi_5] * (\partial \, (\Gamma_{LGR}) \, / \partial x) \, \} * [\, 2 \, \Gamma_{LGR} \, (\partial \, (\Gamma_{LGR}) \, / \partial x)] \, \} \, / \, [\, \Gamma_{LGR}]^4 \tag{64}$$

$(\partial^2 (\Lambda_{LRG} / \Gamma_{LGR}) / \partial y^2) = 4 * \{ \{ [\Phi_1 (\partial^2 (\varphi_1) / \partial y^2) + \Phi_2 (\partial^2 (\varphi_2) / \partial y^2) + \Phi_3 (\partial^2 (\varphi_3) / \partial y^2) +$

$\Phi_4 (\partial^2 (\varphi_4) / \partial y^2) + \Phi_5 (\partial^2 (\varphi_5) / \partial y^2)] * \Gamma_{LGR} + [\Phi_1 (\partial (\varphi_1) / \partial y) + \Phi_2 (\partial (\varphi_2) / \partial y) + \Phi_3 (\partial (\varphi_3)$

$/ \partial y) + \Phi_4 (\partial (\varphi_4) / \partial y) + \Phi_5 (\partial (\varphi_5) / \partial y)] * (\partial (\Gamma_{LGR}) / \partial y) - [\Phi_1 * \varphi_1 + \Phi_2 * \varphi_2 + \Phi_3 * \varphi_3 + \Phi_4$

$* \varphi_4 + \Phi_5 * \varphi_5] * (\partial^2 (\Gamma_{LGR}) / \partial y^2) - [\Phi_1 (\partial (\varphi_1) / \partial y) + \Phi_2 (\partial (\varphi_2) / \partial y) + \Phi_3 (\partial (\varphi_3) / \partial y) + \Phi_4$

$(\partial (\varphi_4) / \partial y) + \Phi_5 (\partial (\varphi_5) / \partial y)] * (\partial \Gamma_{LGR}) / \partial y) \} * [\Gamma_{LGR}]^2 - \{ [\Phi_1 (\partial (\varphi_1) / \partial y) + \Phi_2 (\partial (\varphi_2)$

$/ \partial y) + \Phi_3 (\partial (\varphi_3) / \partial y) + \Phi_4 (\partial (\varphi_4) / \partial y) + \Phi_5 (\partial (\varphi_5) / \partial y)] * [\Gamma_{LGR}] - [\Phi_1 * \varphi_1 + \Phi_2 * \varphi_2 +$

$\Phi_3 * \varphi_3 + \Phi_4 * \varphi_4 + \Phi_5 * \varphi_5] * (\partial (\Gamma_{LGR}) / \partial y) \} * [2 \Gamma_{LGR} (\partial (\Gamma_{LGR}) / \partial y)] \} / [\Gamma_{LGR}]^4$ (65)

$(\partial^2 (\Lambda_{LRG} / \Gamma_{LGR}) / \partial x \partial y) = 4 * \{ \{ [\Phi_1 (\partial^2 (\varphi_1) / \partial x \partial y) + \Phi_2 (\partial^2 (\varphi_2) / \partial x \partial y) + \Phi_3 (\partial^2 (\varphi_3)$

$/ \partial x \partial y) + \Phi_4 (\partial^2 (\varphi_4) / \partial x \partial y) + \Phi_5 (\partial^2 (\varphi_5) / \partial x \partial y)] * \Gamma_{LGR} + [\Phi_1 (\partial (\varphi_1) / \partial x) + \Phi_2 (\partial (\varphi_2) / \partial x)$

$+ \Phi_3 (\partial (\varphi_3) / \partial x) + \Phi_4 (\partial (\varphi_4) / \partial x) + \Phi_5 (\partial (\varphi_5) / \partial x)] * (\partial (\Gamma_{LGR}) / \partial y) - [\Phi_1 * \varphi_1 + \Phi_2 * \varphi_2 +$

$\Phi_3 * \varphi_3 + \Phi_4 * \varphi_4 + \Phi_5 * \varphi_5] * (\partial^2 (\Gamma_{LGR}) / \partial x \partial y) - [\Phi_1 (\partial (\varphi_1) / \partial y) + \Phi_2 (\partial (\varphi_2) / \partial y) + \Phi_3 (\partial$

$(\varphi_3) / \partial y) + \Phi_4 (\partial (\varphi_4) / \partial y) + \Phi_5 (\partial (\varphi_5) / \partial y)] * (\partial (\Gamma_{LGR}) / \partial x) \} * [\Gamma_{LGR}]^2 - \{ [\Phi_1 (\partial (\varphi_1)$

$/ \partial x) + \Phi_2 (\partial (\varphi_2) / \partial x) + \Phi_3 (\partial (\varphi_3) / \partial x) + \Phi_4 (\partial (\varphi_4) / \partial x) + \Phi_5 (\partial (\varphi_5) / \partial x)] * [\Gamma_{LGR}] - [\Phi_1$

$* \varphi_1 + \Phi_2 * \varphi_2 + \Phi_3 * \varphi_3 + \Phi_4 * \varphi_4 + \Phi_5 * \varphi_5] * (\partial (\Gamma_{LGR}) / \partial x) \} * [2 \Gamma_{LGR} (\partial (\Gamma_{LGR}) / \partial y)] \} / [\Gamma_{LGR}]^4$

    (66)

$(\partial^2 (\Lambda_{LRG} / \Gamma_{LGR}) / \partial y \partial x) = 4 * \{ \{ [\Phi_1 (\partial^2 (\varphi_1) / \partial y \partial x) + \Phi_2 (\partial^2 (\varphi_2) / \partial y \partial x) + \Phi_3 (\partial^2 (\varphi_3)$

$/ \partial y \partial x) + \Phi_4 (\partial^2 (\varphi_4) / \partial y \partial x) + \Phi_5 (\partial^2 (\varphi_5) / \partial y \partial x)] * \Gamma_{LGR} + [\Phi_1 (\partial (\varphi_1) / \partial y) + \Phi_2 (\partial (\varphi_2) / \partial y)$

$+ \Phi_3 (\partial (\varphi_3) / \partial y) + \Phi_4 (\partial (\varphi_4) / \partial y) + \Phi_5 (\partial (\varphi_5) / \partial y)] * (\partial (\Gamma_{LGR}) / \partial x) - [\Phi_1 * \varphi_1 + \Phi_2 * \varphi_2 +$

$\Phi_3 * \varphi_3 + \Phi_4 * \varphi_4 + \Phi_5 * \varphi_5] * (\partial^2 (\Gamma_{LGR}) / \partial y \partial x) - [\Phi_1 (\partial (\varphi_1) / \partial x) + \Phi_2 (\partial (\varphi_2) / \partial x) + \Phi_3 (\partial$

$(\varphi_3) / \partial x) + \Phi_4 (\partial (\varphi_4) / \partial x) + \Phi_5 (\partial (\varphi_5) / \partial x)] * (\partial (\Gamma_{LGR}) / \partial y) \} * [\Gamma_{LGR}]^2 - \{ [\Phi_1 (\partial (\varphi_1)$

$/ \partial y) + \Phi_2 (\partial (\varphi_2) / \partial y) + \Phi_3 (\partial (\varphi_3) / \partial y) + \Phi_4 (\partial (\varphi_4) / \partial y) + \Phi_5 (\partial (\varphi_5) / \partial y)] * [\Gamma_{LGR}] - [\Phi_1$

$* \varphi_1 + \Phi_2 * \varphi_2 + \Phi_3 * \varphi_3 + \Phi_4 * \varphi_4 + \Phi_5 * \varphi_5] * (\partial (\Gamma_{LGR}) / \partial y) \} * [2 \Gamma_{LGR} (\partial (\Gamma_{LGR}) / \partial x)] \} / [\Gamma_{LGR}]^4$

    (67)

Where it is true to state:

$[\Phi_1 (\partial (\varphi_1) / \partial x) + \Phi_2 (\partial (\varphi_2) / \partial x) + \Phi_3 (\partial (\varphi_3) / \partial x) + \Phi_4 (\partial (\varphi_4) / \partial x) + \Phi_5 (\partial (\varphi_5) / \partial x)] =$

$$[\Phi_1 * (x^4y + (3/2) x^3y^2 + x^2y^3 + xy^4/4 + x^3y^2/2 + x^2y^3 + (3/4) xy^4 + y^5/5) + \Phi_2 * (x^3y +$$

$$(3/2) x^2y^2 + xy^3 + y^4/4) + \Phi_3 * (x^2y + xy^2 + y^3/3) + \Phi_4 * (xy + y^2/2) + \Phi_5 * y] \tag{68}$$

$$[\Phi_1 (\partial (\varphi_1) /\partial y) + \Phi_2 (\partial (\varphi_2) /\partial y) + \Phi_3 (\partial (\varphi_3) /\partial y) + \Phi_4 (\partial (\varphi_4) /\partial y) + \Phi_5 (\partial (\varphi_5) /\partial y)] =$$

$$[\Phi_1 * (x^5/5 + (3/4) x^4y + x^3y^2 + x^2y^3/2 + x^4y/4 + x^3y^2 + (3/2) x^2y^3 + xy^4) + \Phi_2 * (x^4/4 + x^3y$$

$$+ (3/2) x^2y^2 + xy^3) + \Phi_3 * (x^3/3 + x^2y + xy^2) + \Phi_4 * (x^2/2 + xy) + \Phi_5 * x] \tag{69}$$

$$[\Phi_1 (\partial^2 (\varphi_1) /\partial x^2) + \Phi_2 (\partial^2 (\varphi_2) /\partial x^2) + \Phi_3 (\partial^2 (\varphi_3) /\partial x^2) + \Phi_4 (\partial^2 (\varphi_4) /\partial x^2) + \Phi_5 (\partial^2 (\varphi_5)$$

$$/\partial x^2)] = [\Phi_1 * (4 x^3y + (9/2) x^2y^2 + 2 xy^3 + y^4/4 + (3/2) x^2y^2 + 2 xy^3 + (3/4) y^4) + \Phi_2 * (3$$

$$x^2y + 3 xy^2 + y^3) + \Phi_3 * (2 xy + y^2) + \Phi_4 * y] \tag{70}$$

$$[\Phi_1 (\partial^2 (\varphi_1) /\partial x \partial y) + \Phi_2 (\partial^2 (\varphi_2) /\partial x \partial y) + \Phi_3 (\partial^2 (\varphi_3) /\partial x \partial y) + \Phi_4 (\partial^2 (\varphi_4) /\partial x \partial y) + \Phi_5 (\partial^2$$

$$(\varphi_5) /\partial x \partial y)] = [\Phi_1 * (x^4 + 3 x^3y + 3 x^2y^2 + xy^3 + x^3y + 3 x^2y^2 + 3 xy^3 + y^4) + \Phi_2 * (x^3 + 3$$

$$x^2y + 3 xy^2 + y^3) + \Phi_3 * (x^2 + 2 xy + y^2) + \Phi_4 * (x + y) + \Phi_5] \tag{71}$$

$$[\Phi_1 (\partial^2 (\varphi_1) /\partial y \partial x) + \Phi_2 (\partial^2 (\varphi_2) /\partial y \partial x) + \Phi_3 (\partial^2 (\varphi_3) /\partial y \partial x) + \Phi_4 (\partial^2 (\varphi_4) /\partial y \partial x) + \Phi_5 (\partial^2$$

$$(\varphi_5) /\partial y \partial x)] = [\Phi_1 * (x^4 + 3 x^3y + 3 x^2y^2 + xy^3 + x^3y + 3 x^2y^2 + 3 xy^3 + y^4) + \Phi_2 * (x^3 + 3$$

$$x^2y + 3 xy^2 + y^3) + \Phi_3 * (x^2 + 2 xy + y^2) + \Phi_4 * (x + y) + \Phi_5] \tag{72}$$

$$[\Phi_1 (\partial^2 (\varphi_1) /\partial y^2) + \Phi_2 (\partial^2 (\varphi_2) /\partial y^2) + \Phi_3 (\partial^2 (\varphi_3) /\partial y^2) + \Phi_4 (\partial^2 (\varphi_4) /\partial y^2) + \Phi_5 (\partial^2 (\varphi_5)$$

$$/\partial y^2)] = [\Phi_1 * ((3/4) x^4 + 2 x^3y + (3/2) x^2y^2 + x^4/4 + 2 x^3y + (9/2) x^2y^2 + 4 xy^3) + \Phi_2 * (x^3$$

$$+ 3 x^2y + 3 xy^2) + \Phi_3 * (x^2 + 2 xy) + \Phi_4 * x] \tag{73}$$

CALCULATION OF THE TOTAL CURVATURE OF THE RESILIENT SIGNAL: BIVARIATE CUBIC LAGRANGE

Given equation (61), let us posit:

$$(\partial^2 (R((x, y), f)) /\partial x^2) = (\partial^2 (\Lambda_{LRG} / \Gamma_{LGR}) /\partial x^2)$$

$$(\partial (\partial (R((x, y), f)) /\partial y) \partial x) = (\partial^2 (\Lambda_{LRG} / \Gamma_{LGR}) /\partial y \partial x)$$

$(\partial \, (\partial \, (\, R((x, y), f) \,) \, / \partial x) \partial y) = (\partial^2 \, (\Lambda_{LRG} \, / \, \Gamma_{LGR}) \, / \partial x \partial y)$

$(\partial^2 \, (\, R((x, y), f) \,) \, / \partial y^2) = (\partial^2 \, (\Lambda_{LRG} \, / \, \Gamma_{LGR}) \, / \partial y^2)$

The total curvature $C(\, R((x, y), f) \,)$ of the resilient signal is:

$C(\, R((x, y), f) \,) = (\partial^2 \, (\, R((x, y), f) \,) \, / \partial x^2) + (\partial \, (\partial \, (\, R((x, y), f) \,) \, / \partial y) \partial x) + (\partial \, (\partial \, (\, R((x, y),$

$f) \,) \, / \partial x) \partial y) + (\partial^2 \, (\, R((x, y), f) \,) \, / \partial y^2)$ 　　　　　　　　　　　　　　(74)

SUMMARY

To conclude the chapter and to make way for the results of chapter 8 it is necessary to explain that the two signal reconstruction formulas derived from classic and resilient curvature: $I(x, y) = f(0, 0) + PIC(x, y)$ and $R(x, y) = f(0, 0) + PIC(x, y)$; make it possible to devise the so called classic-resilient-curvature hybrid formula or hybrid formula ($HYB(x, y)$) in short.

The hybrid formula is defined as $I(x, y) = f(0, 0) + PIC(x, y)$ or $R(x, y) = f(0, 0) + PIC(x, y)$ depending as to which one of the two mean absolute errors is the smaller. Specifically, on a pixel-by-pixel basis, $HYB(x, y) = I(x, y)$ if $I(x, y) < R(x, y)$ and $HYB(x, y) = R(x, y)$ if $I(x, y) \geq R(x, y)$. Clearly, this third new interpolation formula will have the smallest error between the two formulas: classic-curvature ($I(x, y) = f(0, 0) + PIC(x, y)$), resilient-curvature ($R(x, y) = f(0, 0) + PIC(x, y)$).

REFERENCES

Ciulla, C. (2009). *Improved Signal and Image Interpolation in Biomedical Applications: The Case of Magnetic Resonance Imaging (MRI)* – Medical Information Science Reference - IGI Global Publisher, Hershey, PA, U.S.A.

CHAPTER 8

CLASSIC-RESILIENT-CURVATURE HYBRID INTERPOLATION: BIVARIATE QUADRATIC AND CUBIC, WITH AND WITHOUT EMBEDDING

INTRODUCTION

The interpolation formulae that have been the object of study so far can be reformulated on the basis of the concept of curvature and this was seen in chapter 4, where the results for the resilient formula $R(x, y) = f(0, 0) + \varrho(x, y)$ have been presented and also revised so as to incorporate the transfer function derived from the curvature of the interpolator.

This approach has made it possible to derive the interpolation formula: $R(x, y) = f(0, 0) + PIC(x, y)$ (resilient interpolation without the transfer function), in such a way as to be uncompromised with the arbitration of the transfer function of the choice. The transfer function is instead derived from the math form of the resilient interpolation formula $R(x, y) = f(0, 0) + \varrho(x, y)$. The notation employed is recalled below to remind us that such a process of obtainment of the transfer function from the curvature is also possible in the case of the classic interpolation formula. In fact, from the formulation: $I(x, y) = f(0, 0) + \xi(x, y)$ it is possible to arrive at the formulation: $I(x, y) = f(0, 0) + PIC(x, y)$, which may be denominated: classic interpolation without the transfer function.

In both of the cases of the classic interpolation formula without the transfer function: $I(x, y) = f(0, 0) + PIC(x, y)$, and the resilient interpolation formula without the transfer function: $R(x, y) = f(0, 0) + PIC(x, y)$, it is possible to obtain an improvement in signal reconstruction when making a comparison with the interpolation formulae: $I(x, y) = f(0, 0) + \xi(x, y)$ (classic) and $R(x, y) = f(0, 0) + \varrho(x, y)$ (resilient), from which the math form of the transfer functions $PIC(x, y)$ was derived (in either case). It is also true that from neither of the two classic and resilient formulae:

$$I(x, y) = f(0, 0) + PIC(x, y) \tag{1}$$

$$R(x, y) = f(0, 0) + PIC(x, y) \tag{2}$$

(which incorporate the transfer function derived from their math form) is it possible to achieve interpolation error improvement, which is absolute in its spatial extent so as to be on a pixel-by-pixel basis, and therefore to be found across the full image. There exist pixels that, when interpolated with the classic formula, furnish an interpolation error which is lower than the interpolation error obtained with the resilient formula, and vice versa. Thus, the possibility of devising the classic-resilient curvature hybrid interpolation formula $HYB(x, y)$ or also called hybrid interpolation formula $HYB(x, y)$ arises. This is constructed so as to be classic or resilient depending on which one of the two paradigms reveals the lower interpolation error. In an algorithmic approach $HYB(x, y) = f(0, 0) + PIC(x, y)$, where $PIC(x, y)$ is the smaller value obtained from equations (1) and (2).

This chapter reports on the investigation of the comparison of the performance of the hybrid interpolation formula $HYB(x, y)$, and thus reveals to what extent, in terms of ratio of improvement (interpolation error improvement) either the classic formula (equation (1)) or the resilient formula (equation (2)) can be beneficial to the other formula. For instance, if equation (1) furnishes a higher interpolation error than equation (2) at a given pixel, $HYB(x, y)$ becomes equation (2), thus the classic formula (1) benefits from the resilient formula (2) in the specific task of interpolating the given pixel. Conversely, if at the same pixel, equation (1) furnishes a lower interpolation error than equation (2), then $HYB(x, y)$ becomes equation (1) thus the resilient formula (2) benefits from the classic formula (1). This reasoning takes place on a pixel-by-pixel basis.

It will be useful to report here work on the of B-Spline interpolation functions (De Boor, 1978; Hecklin et al., 2008; Jonic et al., 2006; Keys, 1981; Ledesma-Carbayo et al., 2005; Meijering, 1999; Precioso et al., 2003, 2005; Schoenberg, 1946a, 1946b, 1969, 1971; Sorzano et al., 2006; Unser et al., 1993a; Unser, et al., 1993b; Unser, et al., 1995; Unser, 1999; Unser, and Blu, 2005; Unser, 2005; Unser and Blu, 2006). The purpose of this chapter will be that of devising an hybrid interpolation formula, which differs in essence from previous works reported in the literature, where the aforementioned list of authors have provided excellent contributions to the panorama of investigations into B-Spline interpolation functions.

PRELIMINARY INVESTIGATION

In this paragraph the preliminary investigation of the classic-resilient curvature hybrid interpolation formula HYB (x, y) is performed on the image seen in figure 1 using bivariate quadratic polynomials with embedding. The image in figure 1 is a rectangle of 8x16 pixels oriented perfectly parallel to the X and Y axis of the image coordinate system and placed with the origin at the centre of the image plane.

The pixel intensity is zero outside the rectangle and quasi constant (0.99 pixel intensity) inside the rectangle. In the remainder of the chapter the following glossary will be used: Classic-curvature (Classic) signal reconstruction means interpolation through equation (1), resilient-curvature (Resilient) signal reconstruction means interpolation through equation (2), classic-resilient-curvature (Hybrid) reconstruction means interpolation through HYB (x, y).

Figure 1: The image employed in the preliminary investigation: 47x47 pixels.

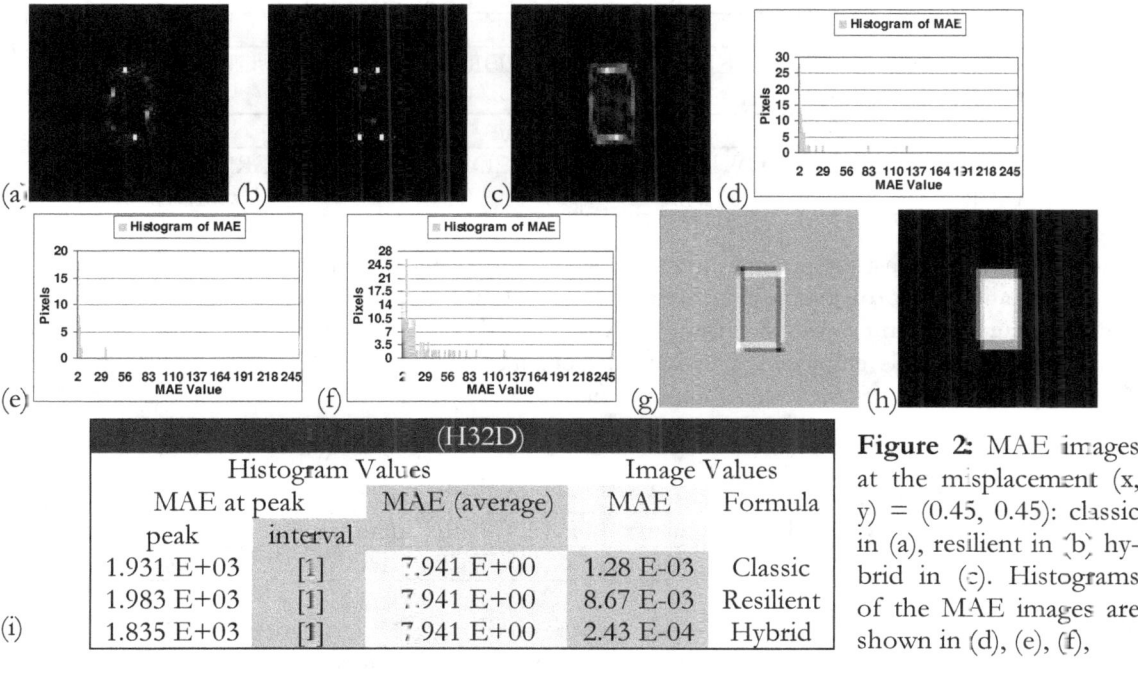

(H32D)				
Histogram Values			Image Values	
MAE at peak		MAE (average)	MAE	Formula
peak	interval			
1.931 E+03	[1]	7.941 E+00	1.28 E-03	Classic
1.983 E+03	[1]	7.941 E+00	8.67 E-03	Resilient
1.835 E+03	[1]	7.941 E+00	2.43 E-04	Hybrid

Figure 2: MAE images at the misplacement (x, y) = (0.45, 0.45): classic in (a), resilient in (b) hybrid in (c). Histograms of the MAE images are shown in (d), (e), (f),

classic, resilient and hybrid respectively. Classic curvature and resilient curvature maps of the image are shown in (g) and (h) respectively. The table in (i) shows values which serve to complete the information provided in both images and histograms.

Among the MAE images shown in figures 2(a), 2(b) and 2(c) the one with the greatest residual was obtained by employing the resilient-curvature interpolation formula (a). As is visible in (i), the MAE is minimal in all of the three formulae: classic-curvature (1.28 E-03), resilient-curvature (8.67 E-03) and hybrid (2.43 E-04). Also, each pixel of the MAE images had the intensity scaled in the range [0, 255] and such is the range of values of the MAE covered in each histogram. The average is the same (7.941 E+00), and what makes most of the difference are the peak values of the histogram, which are visible in (i) with a magnitude of the order 10^3 (see left-most column in (i)). These three values are not displayed in the histograms shown in (d), (e) and (f), so as to increase the visibility.

It is worth noticing the maps of the curvature of the two formulae: classic in (g) and resilient in (h), which show that the resilient curvature is prone to being sensitive in showing opposite values between the inside and the outside of the 8x16 rectangle (object), whereas this difference in not noticeable in the map of the classic formula (g) where there appear not to be any differences between the two aforementioned regions. At the edges of the object, it is remarkable that we can see the change in curvature of the model function, and this is clearly visible in (g) of figure 2, and in (h) of figure 2 with the difference that both of the upper right and lower left edges (corners) show almost null curvature

Classic-curvature With E	Classic-curvature WE	Resilient-curvature With E	Resilient-curvature WE
Quadratic			
H32DRCCurvature C.CR	H32DWERCCurvature C.CR	H32DRCCurvature R.CR	H32DWERCCurvature R.CR
Cubic			
H42DRCCurvature C.CR	H42DWERCCurvature C.CR	H42DRCCurvature R.CR	H42DWERCCurvature R.CR
Cubic Lagrange			
LGR2DRCCurvature C.CR	LGR2DWERCCurvature C.CR	LGR2DRCCurvature R.CR	LGR2DWERCCurvature R.CR

Table I: The headings of the graphs shown in figures 3 through 11 and figures 16, 17, 22 and 23, to indicate the ratio of improvement of the hybrid interpolation formula over either the classic-curvature or resilient-curvature interpolation formulae, with embedding ('With E') and without embedding ('WE'). This notation is employed with bivariate quadratic, bivariate cubic and bivariate cubic Lagrange formulae to identify the ratio of improvement of the hybrid formula in the various cases.

With E	WE
H32DRCCurvature (MAE ClassicResilient)	H32DWERCCurvature (MAE ClassicResilient)
H42DRCCurvature (MAE ClassicResilient)	H42DWERCCurvature (MAE ClassicResilient)
LGR2DRCCurvature (MAE ClassicResilient)	LGR2DWERCCurvature (MAE ClassicResilient)

Table II: The headings of the graphs shown in figures 3 through 11 and figures 16, 17, 22 and 23, to indicate the notation used to identify the MAE of the hybrid signal reconstruction throughout the three types of interpolation formulae: bivariate quadratic, bivariate cubic and bivariate cubic Lagrange, with embedding ('With E') and without embedding ('WE').

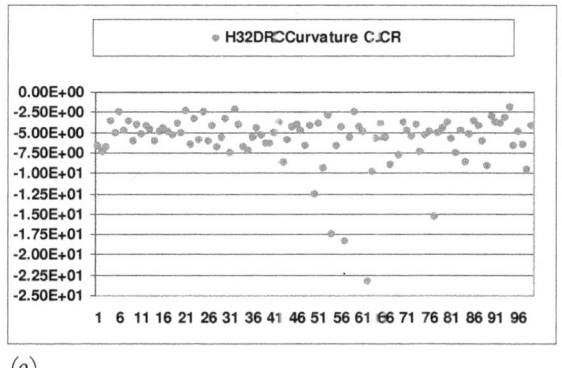

(a)

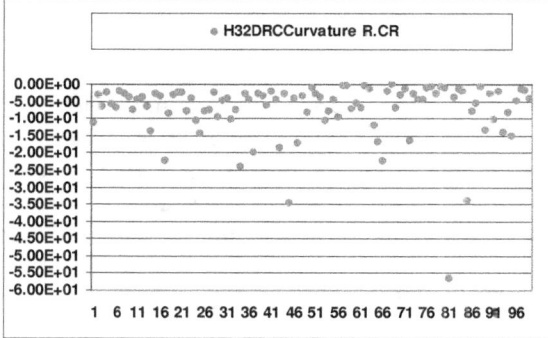

(b)

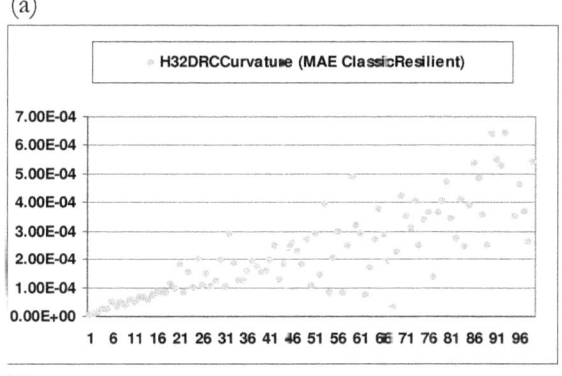

(c)

Figure 3: The ratio of improvement of the hybrid formula over the classic-curvature formula (a), the hybrid formula over resilient-curvature formula (b); and the MAE of the hybrid interpolation formula (c). The bivariate quadratic polynomials are calculated with embedding ('With E').

suggesting (without proof) that the change in curvature along X and Y directions void each other at the corners.

As far as regards the notation employed in some of the graphs of the figures of this chapter, the term 'H32DRCCurvature C.CR' refers to the ratio of improvement of the hybrid signal reconstruction (CR) over the bivariate classic-curvature signal reconstruction formula (C). Lastly: 'H32DRCCurvature (MAE ClassicResilient)' refers to the MAE of the Hybrid signal reconstruction formula, which thus employs classic-curvature or resilient-curvature signal reconstruction formulae on a pixel-by-pixel basis depending on which one of the two formulae give the smaller interpolation error (MAE). Details of such notation referring to the headings of the graphs are given in tables I and II. Throughout, the term: 'H32DRCCurvature R.CR' refers to the ratio of improvement of the hybrid signal reconstruction (CR) over the bivariate resilient-curvature signal reconstruction formula (R).

In figure 3(a) we can see that the ratio of improvement of the hybrid signal reconstruction over the classic-curvature signal reconstruction ranges between approximately 250 % and approximately 2,250 %, whereas in figure 3(b) the ratio of improvement of the hybrid signal reconstruction over the resilient-curvature signal reconstruction ranges between approximately 0 % and approximately 5,500 %, in the range of misplacements (x, y) = [(0.01, 0.01), (0.99, 0.99)] and each point in the abscissa corresponds to one point along the diagonal of the pixel that goes from (x, y) = (0.01, 0.01) to (x, y) = (0.99, 0.99). The value *the_A_const* is set equal to -0.450. The value of the misplacement is the same along x and y and appears multiplied 100 in the graphs of figure 3 as well as those of all the other similar graphs in the present chapter. It is worth noticing in figure 3 that the interpolation formula that most benefits from being merged into the hybrid signal reconstruction is the resilient-curvature signal reconstruction (higher ratio of improvement of the hybrid seen between (a) and (b)).

Generally, the more negative the ratio of improvement is, the more effective the gain in merging the classic-curvature and resilient-curvature interpolation formulae on a pixel-by-pixel basis is, and the higher the gain of the formula (classic or resilient) which transforms itself into the hybrid is. The value of the MAE of the hybrid signal reconstruction is shown in figure 3(c); it behaves almost scatter linearly, increasing with the increase of the misplacement up to values of approximately 6.0 E-04.

In the remaining sections of this chapter the image employed in the experimentations pertaining to bivariate quadratic, bivariate cubic and bivariate cubic Lagrange polynomials is shown in figure 5 of chapter 4, where the pixel intensity values ranges within [0, 32757].

BIVARIATE QUADRATIC POLYNOMIALS

In this section we give the result of the experiments that were conducted with the purpose of ascertaining which of the following- the classic-curvature formula and the resilient-curvature formula- benefits most from the transformation into the classic-resilient-curvature hybrid formula.

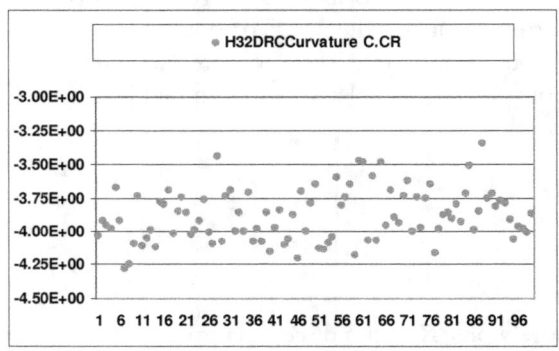

(a)

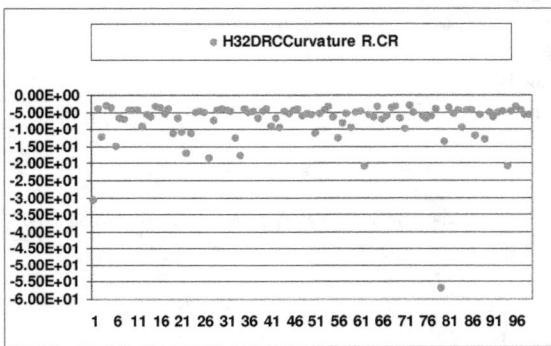

(b)

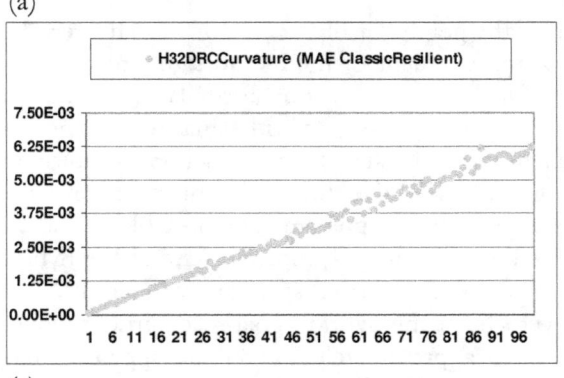

(c)

Figure 4: The ratio of improvement of: the hybrid formula over the classic-curvature formula (a), the hybrid formula over the resilient-curvature formula (b); and the MAE of the hybrid formula (c). The value *the_A_const* is set equal to -0.145. The bivariate quadratic polynomials are calculated with embedding ('With E').

Each of figures 4 through 11 is organized in a set of graphs composed of those relevant to the ratio of improvement of the hybrid formula over classic-curvature and resilient-curvature formulae respectively, and those showing the MAE of the classic-resilient-curvature hybrid formula.

In figure 4, the gain in quality of signal reconstruction is expressed in terms of the ratio of improvement of the hybrid formula: over the classic-curvature formula in (a), over the resilient-curvature formula in (b). In (a) values of improvement that are between approximately 325 % and approximately 425 % are shown. In (b) the values of improvement are between approximately 500 % and 2,000 %, and this demonstrates that on a pixel-by-pixel basis, in order to improve the MAE

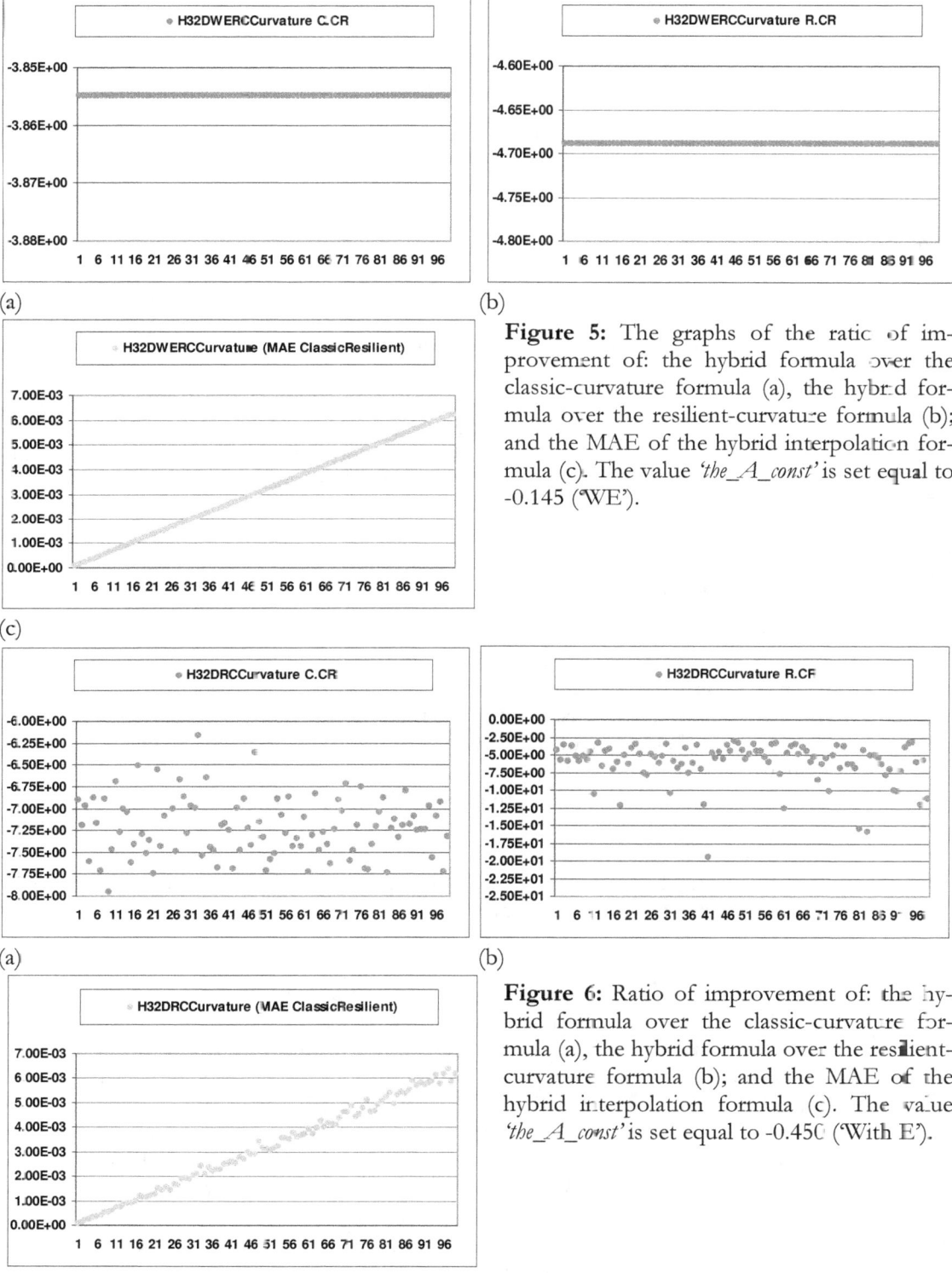

Figure 5: The graphs of the ratio of improvement of: the hybrid formula over the classic-curvature formula (a), the hybrid formula over the resilient-curvature formula (b); and the MAE of the hybrid interpolation formula (c). The value *'the_A_const'* is set equal to -0.145 ('WE').

Figure 6: Ratio of improvement of: the hybrid formula over the classic-curvature formula (a), the hybrid formula over the resilient-curvature formula (b); and the MAE of the hybrid interpolation formula (c). The value *'the_A_const'* is set equal to -0.450 ('With E').

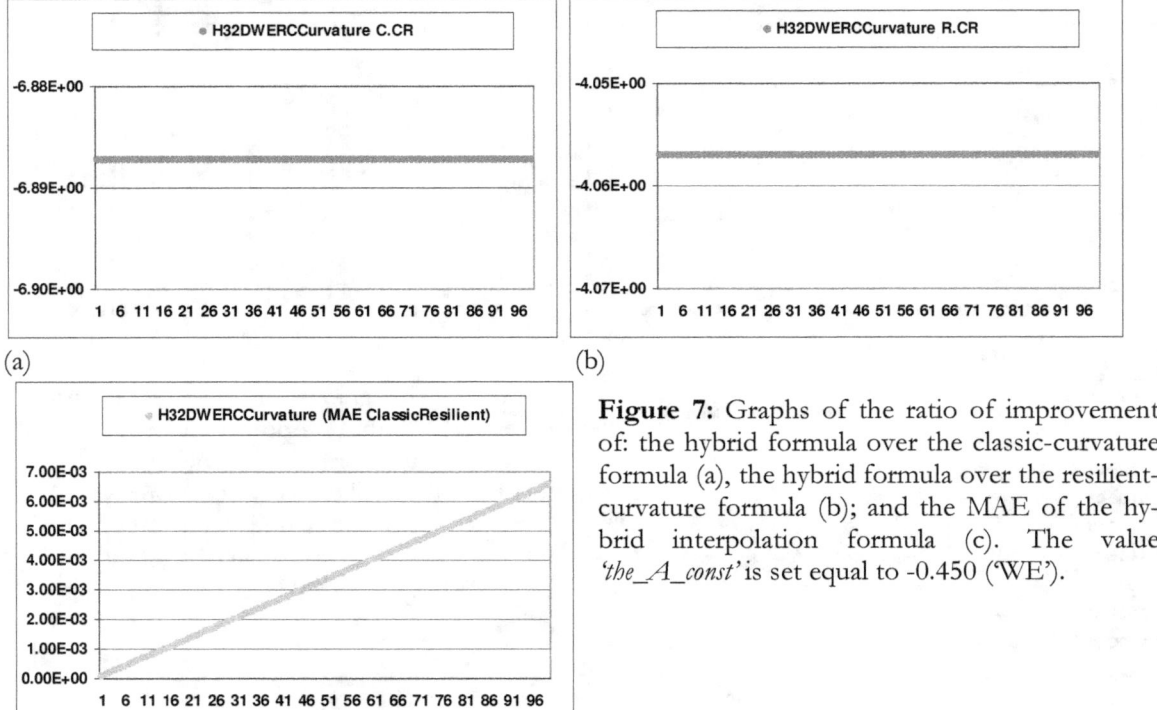

(a)   (b)

(c)

Figure 7: Graphs of the ratio of improvement of: the hybrid formula over the classic-curvature formula (a), the hybrid formula over the resilient-curvature formula (b); and the MAE of the hybrid interpolation formula (c). The value 'the_A_const' is set equal to -0.450 ('WE').

(to reduce the interpolation error), the resilient-curvature formula needs the classic-curvature formula more than the classic-curvature formula needs the resilient-curvature formula. In (c) the MAE of the resulting hybrid formula with behaviour of linear increase with increasing misplacements up to the value of approximately 6.25 E-03 is shown.

In figure 5, bivariate quadratic formulae were calculated without embedding and the results are such as to show quasi constant behaviour across the range of misplacements (x, y) = [(0.01, 0.01), (0.99, 0.99)] in both (a) and (b). The hybrid formula achieves a ratio of improvement over the classic-curvature formula to the order of approximately in between 385 % and 386 % (a), and over the resilient-curvature formula to the order of approximately 470 % (b). The Mean Absolute Error (MAE) of the hybrid formula linearly increases with increasing misplacements up to approximately 6.00 E-03 (see (c)).

In figures 6 and 7 the experiment illustrated in figures 4 and 5 is repeated changing the value of 'the_A_const' from -0.145 to -0.450. In figure 6(a) the hybrid formula achieved ratio of improvement over the classic-curvature formula between approximately 625 % and approximately 800 %. In (b) the hybrid formula achieved a ratio of improvement over the resilient formula of approximately 500 % (with some scattering between 250 % and 2000 %), showing that this time it is the classic-curvature formula that benefits more than the resilient-resilient formula from the transformation into the hybrid formula. The value of MAE reaches values below 6.00 E-03 in figure 6(c). Figure 7 shows a ratio of improvement of the hybrid in between 688 % and 689 % over the classic-curvature formula in (a), of between 405 % and 406 % over the resilient-curvature formula in (b), and the value of MAE that reaches values below 7.00 E-03 (c). This time it is the classic-curvature formula which benefits the most from the transformation into the hybrid formula.

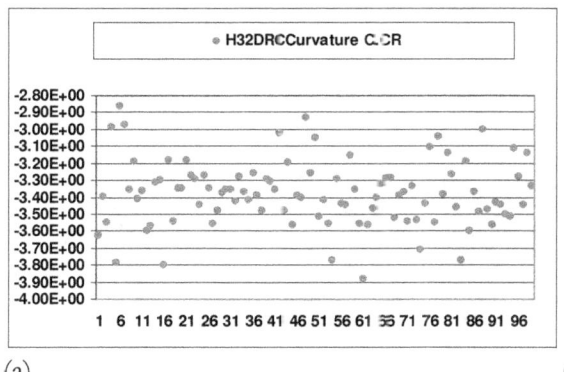

(a)

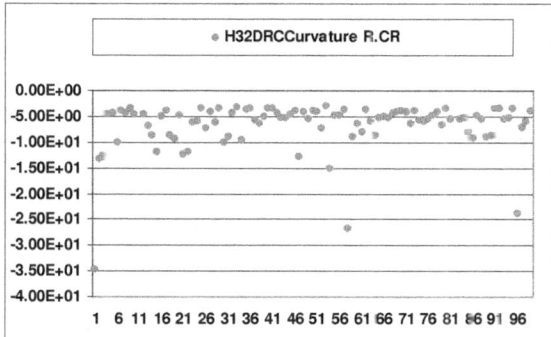

(b)

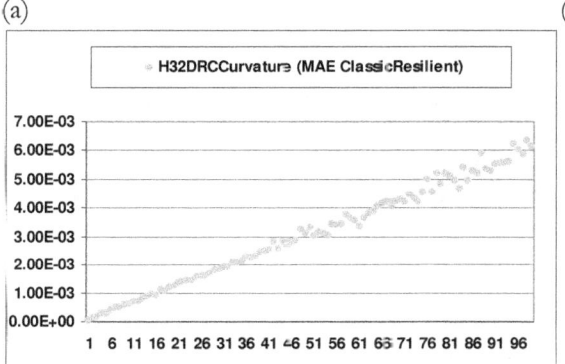

(c)

Figure 8: Ratio of improvement of: the hybrid formula over the classic-curvature formula (a), the hybrid formula over the resilient-curvature formula (b); and the MAE of the hybrid interpolation formula (c). The value *'the_A_const'* is set equal to -0.7045 ('With E').

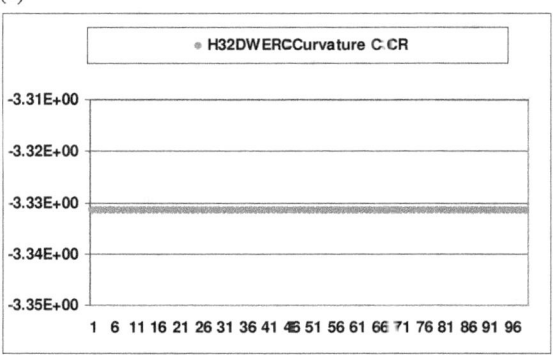

(a)

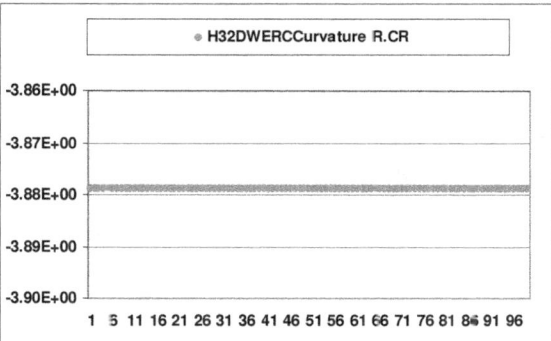

(b)

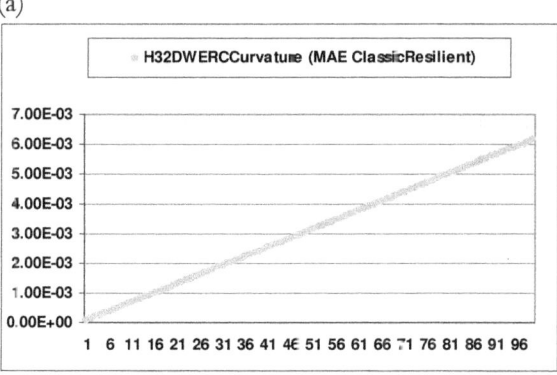

(c)

Figure 9: Ratio of improvement of: the hybrid formula over the classic-curvature formula (a), the hybrid formula over the resilient-curvature formula (b); and the MAE of the hybrid interpolation formula (c). The value *'the_A_const'* is set equal to -0.7045 ('WE').

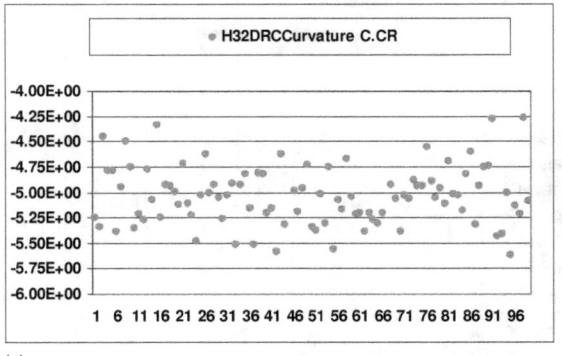

(a)

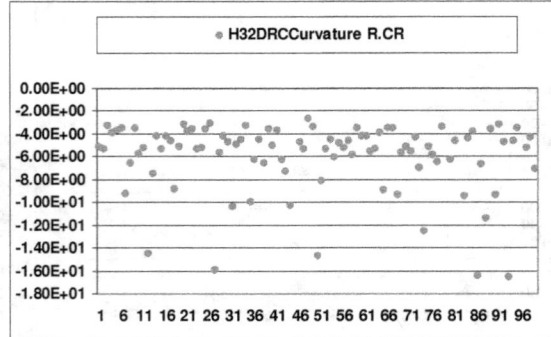

(b)

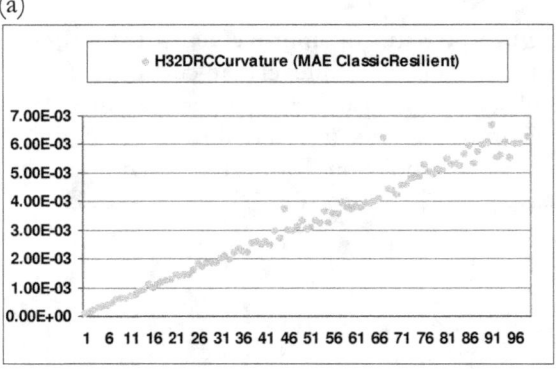

(c)

Figure 10: Ratio of improvement of: the hybrid formula over the classic-curvature formula (a), the hybrid formula over the resilient-curvature formula (b); and the MAE of the hybrid interpolation formula (c). The value *the_A_const* is set equal to 0.71411 ('With E').

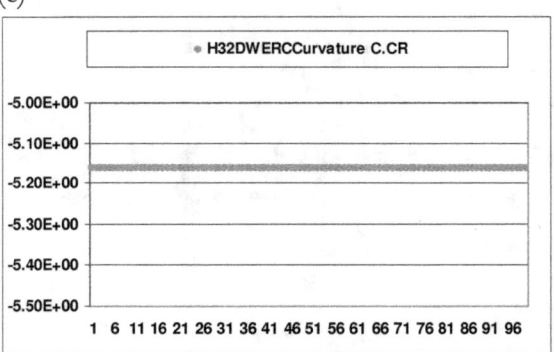

(a)

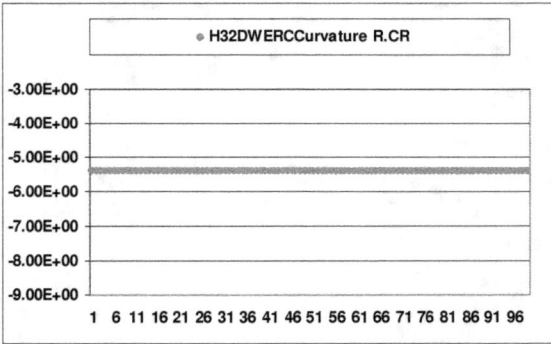

(b)

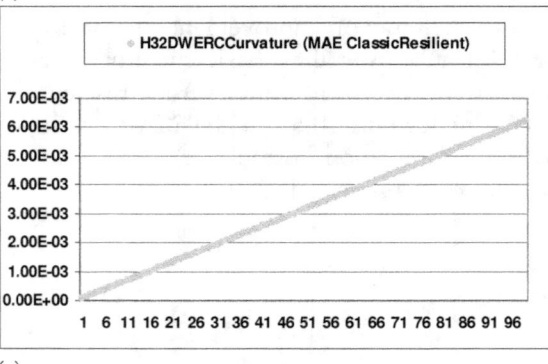

(c)

Figure 11: Ratio of improvement of: the hybrid formula over the classic-curvature formula (a), the hybrid formula over the resilient-curvature formula (b); and the MAE of the hybrid interpolation formula (c). The value *the_A_const* set equal to 0.71411 ('WE').

In figure 8 the ratio of improvement of the hybrid over the classic-curvature formula is between approximately 280 % and approximately 390 % in (a), and the ratio of improvement of the hybrid over the resilient-curvature formula is approximately 500 % in (b). The value of MAE reaches values of approximately 6.00 E-03 (see (c)). The transformation into the hybrid formulation is more beneficial to the resilient-curvature formula than it is to the classic-curvature formula.

Figure 9 shows results demonstrating the transformation into the hybrid formulation to be more beneficial to the classic-curvature formula (the ratio of improvement of the hybrid formula is approximately 333%, see (a)), than it is to the resilient-curvature formula (the ratio of improvement of the hybrid formula is approximately 388 %, see (b)); and it also shows the values of MAE of the hybrid formula with a behaviour of linear increase with increasing misplacements reaching approximately 6.00 E-03 (c).

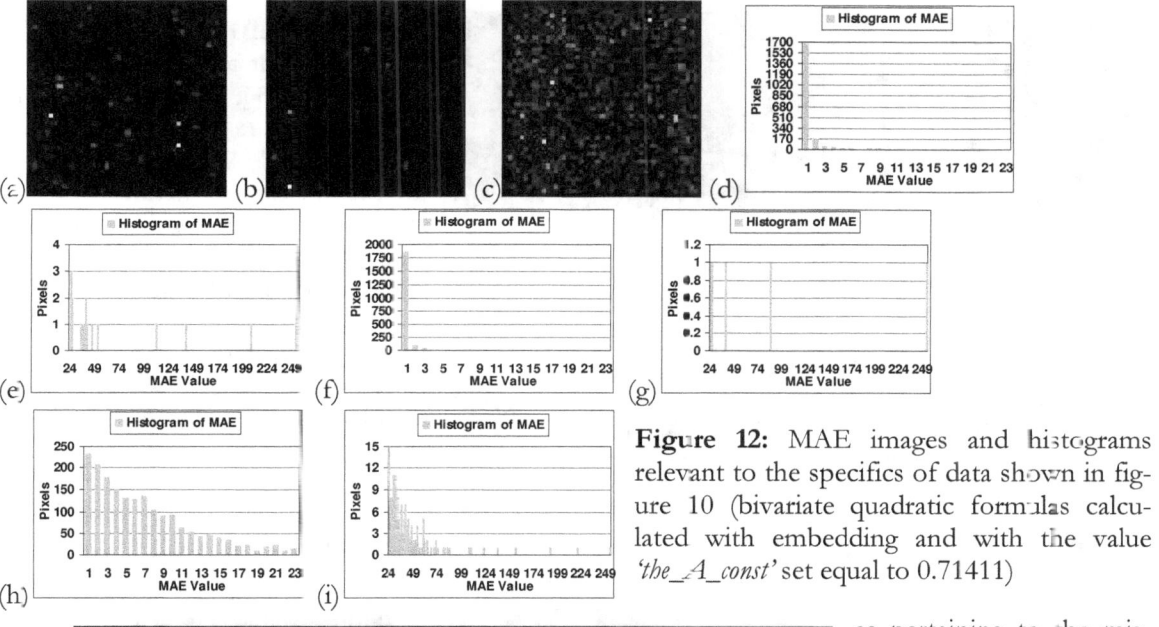

Figure 12: MAE images and histograms relevant to the specifics of data shown in figure 10 (bivariate quadratic formulas calculated with embedding and with the value 'the_A_const' set equal to 0.71411)

(H32D)				
Histogram Values			Image Values	
MAE at peak		MAE (average)	MAE	Formula
peak	interval			
1.672 E+03	[1]	7.941 E+00	1.273 E-02	Classic
1.857 E+03	[1]	7.941 E+00	1.074 E-02	Resilient
2.310 E+02	[1]	7.941 E+00	1.955 E-03	Hybrid

as pertaining to the misplacement (x, y) = (0.33, 0.33). Classic-curvature: MAE image in (a), histograms in (d) and (e). Resilient-curvature: an MAE image in (b), histograms in (f) and (g).

Classic-resilient hybrid: an MAE image in (c), histograms in (h) and (i). The table in (j) adds further information to the MAE images and histograms.

Figure 10 shows the hybrid formulation having a ratio of improvement over the classic-curvature formula of between approximately 425 % and approximately 550 % in (a), and a ratio of improvement over the resilient-curvature formula of between approximately 200 % and approximately 1,600 % in (b), and it also shows the MAE of the hybrid formula with a behaviour of quasi linear increase with increasing misplacements reaching values below approximately 7.00 E-03 (c).

Figure 11 shows a ratio of improvement of the hybrid formulation over the classic-curvature formula of between 510 % and 520 % in (a), a ratio of improvement of the hybrid formulation over the resilient-curvature formula of between 500 % and 600 % in (b), and it also shows that the MAE linearly increases with increasing misplacements up to the value of approximately 6.00 E-03 (c).

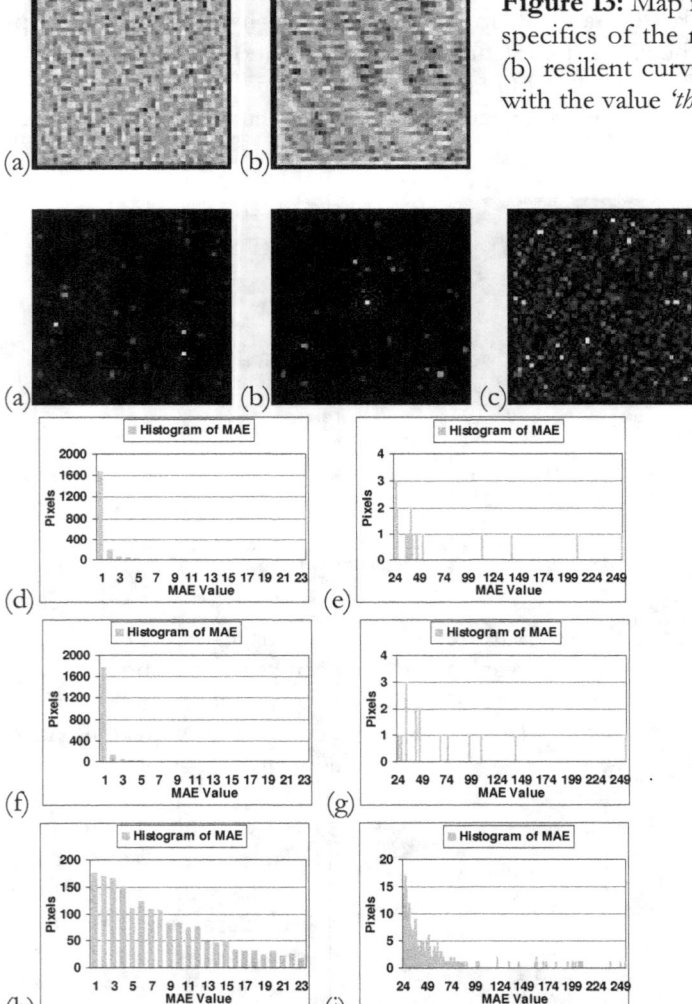

Figure 13: Map images of the curvature as relevant to the specifics of the results of figure 12: (a) classic curvature, (b) resilient curvature. Misplacement (x, y) = (0.33, 0.33) with the value *'the_A_const'* set equal to 0.71411.

Figure 14: MAE images and histograms relevant to the misplacement (x, y) = (0.33, 0.33) as per data shown in figure 11 (bivariate quadratic formulae, WE, *'the_A_const'* = 0.71411). Classic-curvature: the MAE image in (a), histograms in (d) and (e). Resilient-curvature: the MAE image in (b), histograms in (f) and (g). Classic-resilient hybrid: the MAE image in (c), histograms in (h) and (i). The table shown in (j) adds further information to the MAE images and histograms.

The cases presented in figures 3, 5, 7 and 9 are the ones, for the majority of the misplacements, with the benefits made to the classic-curvature and to the resilient curvature formulae which are most similar to each other in terms of the transformation into the hybrid formulation, and this shows that the performance of the two formulae are quite similar. Nevertheless it is still beneficial to both of them to be transformed into the hybrid formula.

(H32DWE)				
Histogram Values			Image Values	
MAE at peak		MAE (average)	MAE	Formula
peak	interval			
1.672 E+03	[1]	7.941 E+00	1.273 E-02	Classic
1.768 E+03	[1]	7.941 E+00	1.324 E-02	Resilient
1.750 E+02	[1]	7.941 E+00	2.067 E-03	Hybrid

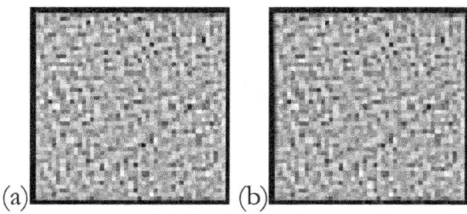

(a) (b)

Figure 15: Map images of the curvature as relevant to the specifics of the results of figure 14 ('WE'): (a) classic curvature, (b) resilient curvature. The two curvature maps are more similar to each other than the two maps seen in figure 13.

In figure 12 we analyze the case of figure 10 which is relevant to the misplacement $(x, y) = (0.33, 0.33)$ insofar as the MAE images which are residual after interpolation are concerned. Histograms were built of the images seen in (a), (b) and (c) and they are shown as: (d), (e) (classic-curvature formula), (f), (g) (resilient-curvature formula), and (h), (i) (classic-resilient-curvature hybrid formula). Each histogram of the MAE image is divided into two parts and in each of the first parts, the peak value is not displayed, so as to increase the visibility; instead it is placed in table (j). To be specific, the values of the MAE peaks are: 1.672 E+03, 1.857 E+03 and 2.310 E+02, classic, resilient and hybrid formulae respectively.

It is worth noting that even though the MAE image deriving from the hybrid formula (see (c)) is the one with the most widespread scattered grey pixels of the three, the MAE is the lowest of the three (1.955 E-03), as can be seen in table (j). It is also worth noting the results in the histograms in (d), (f) and (h) where the pixel count in the interval [1, 23] is 2008, 2020 and 1840 respectively (lowest in (h), the hybrid formula) and also worth looking at the histograms in (e), (g), (i), where the pixel count in the interval [24, 255] is 17, 5 and 185 respectively. Overall, since the MAE of classic, resilient and hybrid formulae are: 1.273 E-02, 1.074 E-02 and 1.955 E-03 respectively, the lowest being the one of the hybrid, we can summarize by saying that the information provided in figure 12 reminds us of the fact that the MAE image of the hybrid formula seen in (c) is the one with minimal residual error. What is argued is that although the wide spread of grey pixels is wider in (a) than it is in (b), the peak values of the histograms are 1.672 E+03 and 1.857 E+03 as is observable in (j) of figure 12.

In figure 13 we can observe maps of the curvature: from the classic-curvature formula in (a), and from the resilient-curvature formula in (b). The image in (b) is brighter than the image in (a) showing that the resilient-curvature in such a specific case has higher values than the classic-curvature.

On the basis of the curvature maps seen so far in figure 2(g) and figure 2(h) and in figures 13 and 15, we can suggest that the curvature tends to approximate the details of the image and this is clearly observable in the two curvature maps of figure 2, and to a lesser extent in figures 13 and 15, where perhaps because of the highly unstructured image (see figure 5 of chapter 4) it is not possible to make such an observation. Perhaps studying images with an intermediate level of details between those of figure 1 and figure 5 of chapter 4 and showing their curvature maps obtained through classic-curvature and resilient-curvature formulae would allow us to investigate further the idea that the curvature tends to approximate the details of the image.

The data of figure 14 are calculated with the bivariate quadratic formula without embedding. An observation can be made this time in relationship to the MAE images shown in (a) and (b), classic-curvature and resilient-curvature formulae respectively. The pixel count in the interval [1, 23] is 2008 (d), 2010 (f) and 1770 (h) When looking at the histograms in (e), (g) and (i), classic-curvature, resilient-curvature and classic-resilient hybrid respectively, we may observe that the pixel count in the interval [24, 255] is 17 (e), 15 (g) and 255 (i). Furthermore the values of the MAE relevant to the images shown in (a) and (b) are 1.273 E-02 and 1.324 E-02 respectively, thus the smaller residual between (a) and (b) is located in (a).

BIVARIATE CUBIC POLYNOMIALS

In this paragraph the experiments are relevant to the bivariate cubic polynomials calculated with and without embedding; to be specific, the investigation intends to ascertain how beneficial it would be to transform either of the classic-curvature or the resilient-curvature signal reconstruction formulae into the classic-resilient-curvature hybrid signal reconstruction formula. The notation employed in figures 16 and 17 is found in Tables I and II of the INTRODUCTION section. Data consists of the image shown in figure 5 of chapter 4.

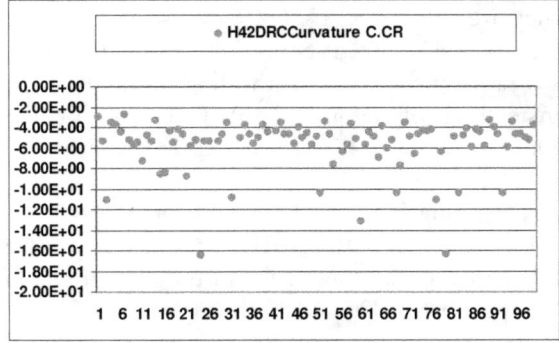

(a)

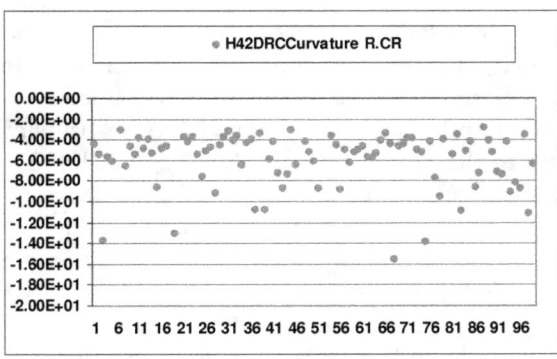

(b)

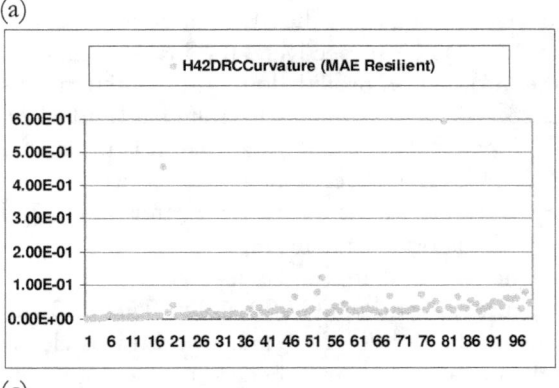

(c)

Figure 16: Graphs showing the ratio of improvement of the hybrid formula: over the classic-curvature formula (a), over the resilient-curvature formula (b), and the MAE of the hybrid formula (c). Polynomials were calculated with embedding.

In figures 16 and 17, as well as the graphs presented during the study of the test image with the bivariate polynomial formulae, we can discern how beneficial the pixel-by-pixel inclusion of the classic-curvature or the resilient-curvature signal reconstruction formulae into the hybrid signal reconstruction is.

The data in figure 16 are relevant to the polynomials calculated with embedding. The hybrid formula leads to an improvement over both classic-curvature and resilient formulae in between approximately 200 % and approximately 1,600 % (see figures 16(a) and 16(b)). The general behaviour of both classic-curvature and resilient curvature formulae is that of achieving a gain, so as to be complementary to each other, and this is immediately visible in the graphs of the MAE of the hybrid formula, which reaches values up to approximately 0.1 (see figure 16(c)).

In figure 17, the experimental sessions are those of the bivariate cubic polynomials calculated without embedding ('WE'). The general behaviour of the hybrid formula is consistent with data previously presented in that either the classic-curvature or the resilient-curvature formulae benefit from their transformation into the hybrid formula. The ratio of improvement of the classic-curvature formula is generally different from that seen in figure 16, it is between approximately 500 % and approximately 2,000 % (see interval (x, y) = [(0.01, 0.01), (0.30, 0.30)] in figure 17(a)) and between approximately 500 % and approximately 3,000 % (see the range (x, y) = [(0.31, 0.31), (0.99, 0.99) in figure 17(b)]).

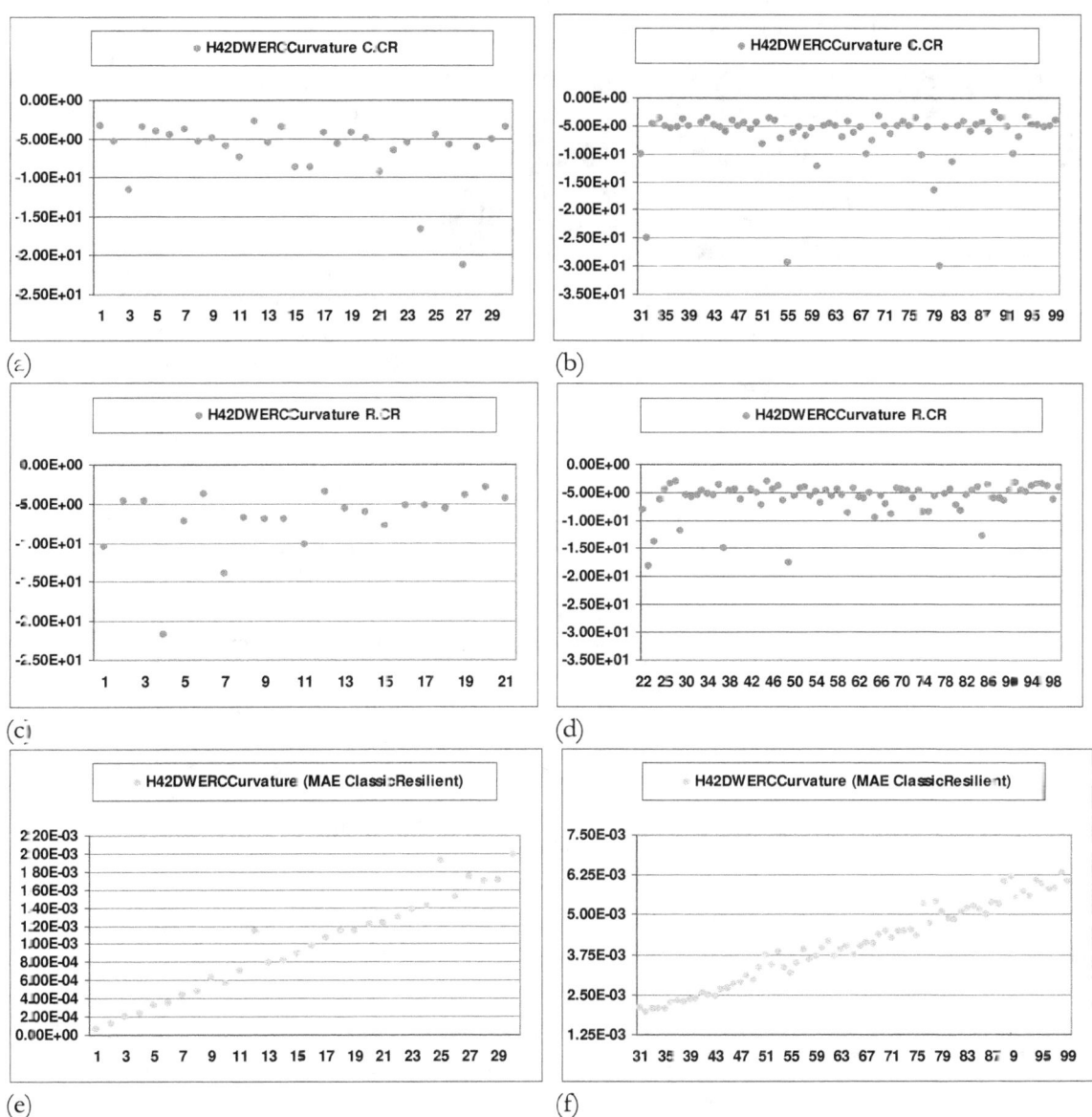

Figure 17: The ratio of improvement of the hybrid formula over the classic-curvature formula is shown in (a) and (b), while (c) and (d) show the ratio of improvement of the hybrid formula over the resilient-curvature formula and (e) and (f) display the graph of the Mean Absolute Error (MAE) of the hybrid formula. Polynomials were calculated without embedding.

Visual inspection indicates that the resilient-curvature formula is generally not so different from the classic-curvature formula in the interval $(x, y) = [(0.22, 0.22), (0.99, 0.99)]$ (see figures 17(b), 17(d)). The MAE of the hybrid formula increase up to the value of approximately 2.00 E-03 (see interval $(x, y) = [(0.01, 0.01), (0.30, 0.30)]$ in figure 17(e)) and continues to increase up to the value of approximately 6.25 E-03 (see interval $(x, y) = [(0.31, 0.31), (0.99, 0.99)$ in figure 17(f)]). These observation are made on the basis of the combined information of figure 17 in (a), (b), (d) (e) and (f).

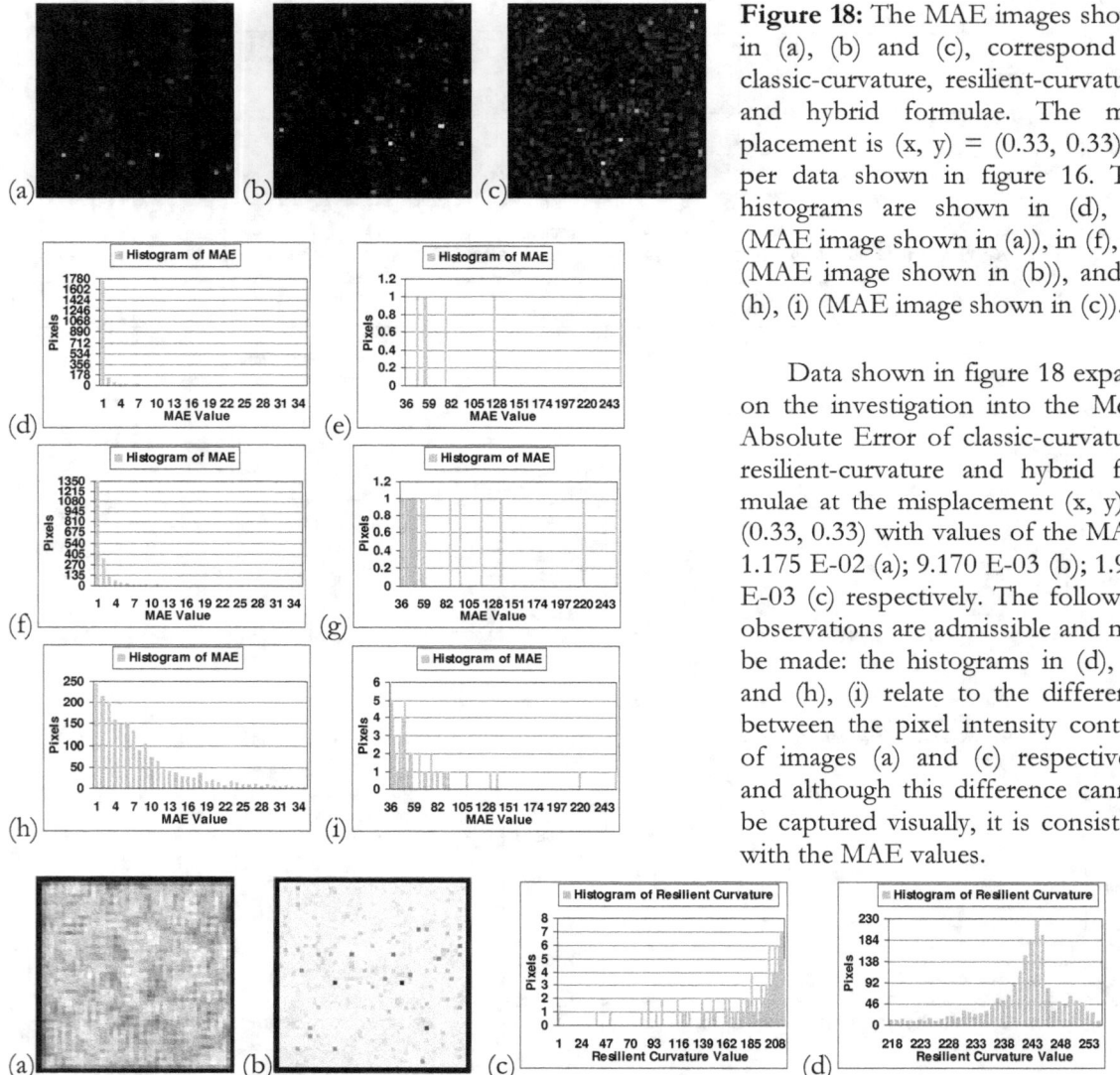

Figure 18: The MAE images shown in (a), (b) and (c), correspond to classic-curvature, resilient-curvature, and hybrid formulae. The misplacement is $(x, y) = (0.33, 0.33)$ as per data shown in figure 16. The histograms are shown in (d), (e) (MAE image shown in (a)), in (f), (g) (MAE image shown in (b)), and in (h), (i) (MAE image shown in (c)).

Data shown in figure 18 expand on the investigation into the Mean Absolute Error of classic-curvature, resilient-curvature and hybrid formulae at the misplacement $(x, y) = (0.33, 0.33)$ with values of the MAE: 1.175 E-02 (a); 9.170 E-03 (b); 1.960 E-03 (c) respectively. The following observations are admissible and may be made: the histograms in (d), (e) and (h), (i) relate to the difference between the pixel intensity content of images (a) and (c) respectively, and although this difference cannot be captured visually, it is consistent with the MAE values.

Figure 19: Map of the classic curvature in (a) and the resilient curvature in (b) accompanied by the histograms of the image in (b), which are shown in (c) and (d). Data are relevant to figure 16; misplacement is $(x, y) = (0.33, 0.33)$.

The wide spread in dark pixels is higher in (a) than it is in (c); the pixel count in the interval [1, 35] is 2018 (d), 2009 (f), 1964 (h) (see 2018 versus 1964), and the widespread of grey pixels is higher in (c)

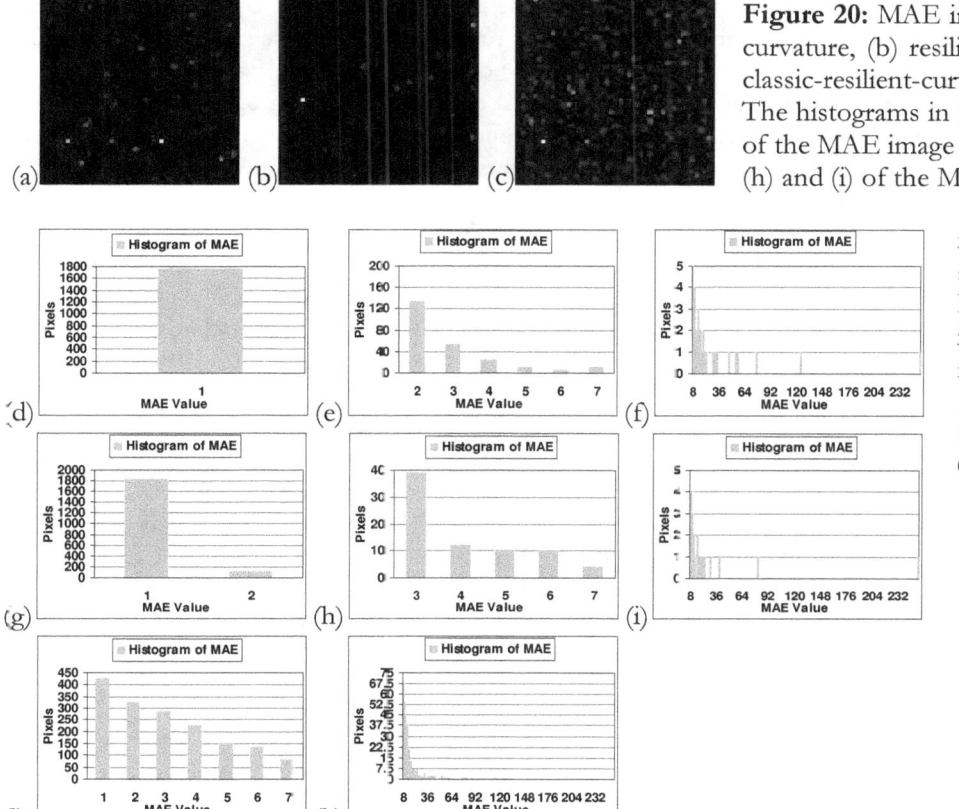

Figure 20: MAE images: (a) classic-curvature, (b) resilient-curvature, (c) classic-resilient-curvature hybrid. The histograms in (d), (e) and (f) are of the MAE image seen in (a), in (g), (h) and (i) of the MAE image seen in (b), and in (j) and (k) of the MAE image seen in (c). Data are relevant to figure 17 (misplacement (x, y) = (0.33, 0.33)).

than it is in (a); the pixel count in the interval [36, 255] is 7 (e), 16 (g), 61 (i) (see 61 versus 7). When compared to the histograms in (d) and (h), the histogram in (f) shows differences which relate to the difference existing between the MAE of the image in (b) and the MAE of the images in (a) and (c). Also, the comparison of the histogram in (g) with the histogram in (i) tells us about the difference of the MAE between the image in (b) and the image in (c), referring specifically to the wider spread of pixels in grey seen in (c) when compared to (b), (see 61 versus 16 in the interval [36, 255] as a case point).

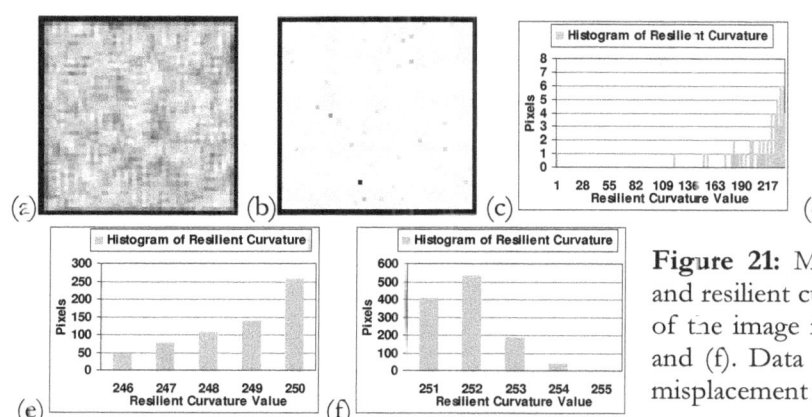

Figure 21: Map of classic curvature in (a) and resilient curvature in (b). The histograms of the image in (b) are shown in (c) (d), (e) and (f). Data are relevant to figure 17. The misplacement is (x, y) = (0.33, 0.33).

The curvature image maps shown in figure 19 outline in the specific case of $(x, y) = (0.33, 0.33)$ the lack of great variability in the numerical values of the resilient curvature (see the images in figures 19(a) and 19(b)) which, instead, is observable in the classic curvature shown in figure 19(a). The histograms in figures 19(c) and 19(d) are relevant to the resilient curvature shown in figure 19(b).

Values of MAE in the images in figures 20(a), 20(b) and 20(c) are: 1.175 E-02; 1.145 E-02; 2.080 E-03. Making a visual comparison, we can observe that the histograms in (i) and (k) tell us about the difference between the MAE images in (b) and in (c) in relationship to the wider spread of brighter (grey) pixels in (c) than in (b), which, when counted, were 23 (i) and 409 (k) respectively, in the interval [8, 255]. This information is additional to that obtained counting the pixels of the images in (c) and in (b) in the interval [1, 7] (they are 2,000 in (g), (h) and 1,616 in (j) respectively), and helps to give us an explanation to the difference between the numerical values of the MAE in (c) and in (b).

The maps of classic curvature and resilient curvature seen in figures 21(a) and 21(b) show a noticeable difference, which was not observable when the test image was studied with the bivariate quadratic polynomials (see figures 13 and 15). The difference consists in the fact that the image of the resilient curvature seen in (b) appears flatter in magnitude than that of the classic curvature seen in (a), and it tends towards high values of the curvature, as is shown in the histograms in (d), (e) and in (f). This case is somehow in contrast with what we can see in figure 19(b) where the large majority of the values of the resilient curvature are seen in figure 19(d) (see for instance the histogram value 226 at the location 244), whereas in the histograms seen in figures 21(e) and 21(f) the curvature values shift towards magnitude values higher than 226 (see range [250, 252]). The basic difference in formulation between figures 19 and 21 is that the cubic polynomials are calculated with and without embedding respectively.

BIVARIATE CUBIC LAGRANGE POLYNOMIALS

This section concludes the chapter through the extension of the study of the test image to the bivariate cubic Lagrange polynomials. Although, is not the point made through this paragraph it is worth mention that because of their math formulation, naturally, the hybrid formulae are constructed in such a way to avoid the arbitration induced through any transfer functions of the choice in order to process the convolution; and they are also able to derive one transfer function from the math form of the interpolator using the concept of curvature, as explained in chapter 4. The transfer function obtained from the math form of the interpolator is thus customized to the interpolator. The notation employed in figures 22 and 23 is found in Tables I and II.

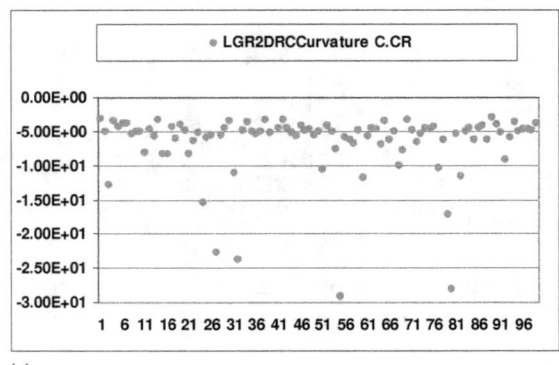

(a)

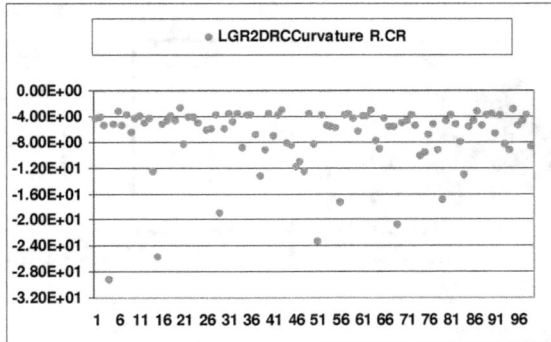

(b)

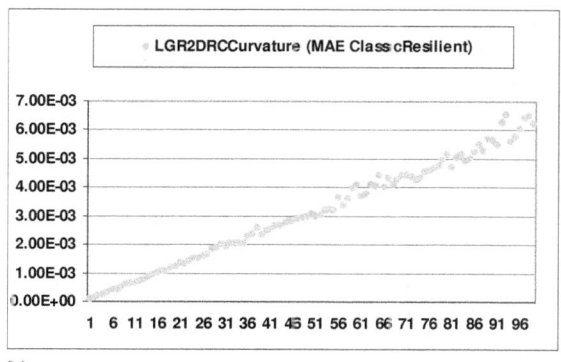

(a)

Figure 22: Graphs of the ratio of improvement of: the hybrid formula over the classic-curvature formula (a); the hybrid formula over the resilient-curvature formula (b); the MAE of the hybrid formula (c).

In figure 22, polynomials were calculated with embedding. The behaviour of the classic-curvature formula shows that the hybrid formula obtains a ratio of improvement which is, in the vast majority of the cases between approximately 500 % and approximately 3,000 % (see figure 22 in (a)). In figure 22, the behaviour of the resilient-curvature formula, on the other hand, favours the ratio of improvement of the hybrid to the extent that in the vast majority of the cases the ratio of improvement is between approximately 400 % and approximately 3,000 %, (see (b)), and these observations are relevant to the entire range of misplacements shown in the graphs. In graph (c) of figure 22 we can see that the behaviour of the MAE of the hybrid formula is such that it displays linear increase with increasing misplacement so as to reach the value of approximately 7.00 E-03 at $(x, y) = (0.99, 0.99)$, which is the farthest location, from the origin of the pixel used in the simulations.

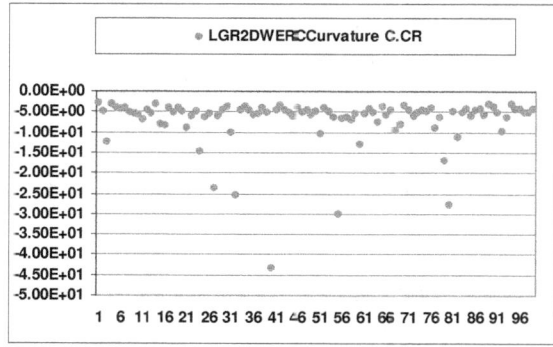

(a)

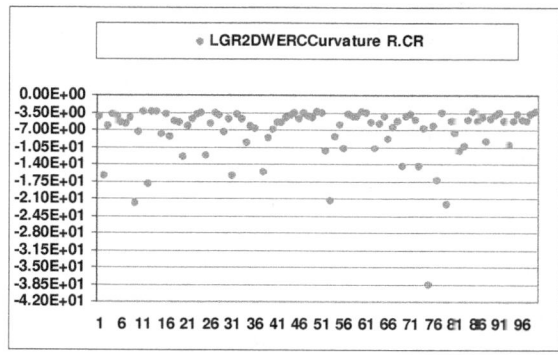

(b)

Figure 23: Ratio of improvement of: (a) the hybrid formula over the classic-curvature formula and in: (b) of the hybrid formula over the resilient-curvature formula. The MAE of the hybrid formula is shown in (c).

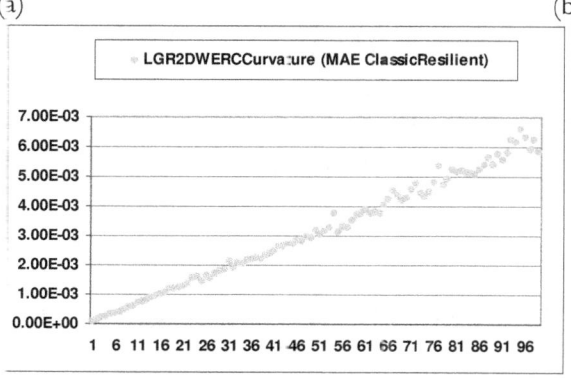

(c)

This section follows the same framework previously outlined in the chapter and thus ascertains which one of the two formulations, namely classic-curvature and resilient-curvature, offer the most benefit to the transformation into the hybrid formula. The behaviour of both classic-curvature and resilient curvature formulae as far as concerns to the ratio of improvement of the hybrid formula is observable in figure 23 (polynomials calculated without embedding) and is not dissimilar to that already seen in figure 22. The hybrid formula achieves an improvement over the classic-curvature formula, which ranges between approximately 500 % and approximately 1,500 % in the vast majority of the misplacements (see figure 23(a)), whereas the same hybrid formula reaches values of the ratio of improvement over the resilient-curvature formula, with a range between approximately 350 % and approximately 1,750 % in the vast majority of the misplacements (see figure 23(b)), thus providing an immediate justification to its being chosen on a pixel-by-pixel basis and at the given misplacement, without any generalizations whatsoever. The MAE of the hybrid formula is located in (c) and shows similarities with that seen in figure 22(c).

The value of the MAE found in images (a), (b) and (c) of figure 24 is: 1.175 E-02, 9.396 E-03, 2.032 E-03. Comparison of the histograms in (d), (e) (MAE image in (a)) with the histograms in (f), (g) (MAE image in (b)), can add information to clarify the difference arising through visual comparison of the two images in (a) and in (b).

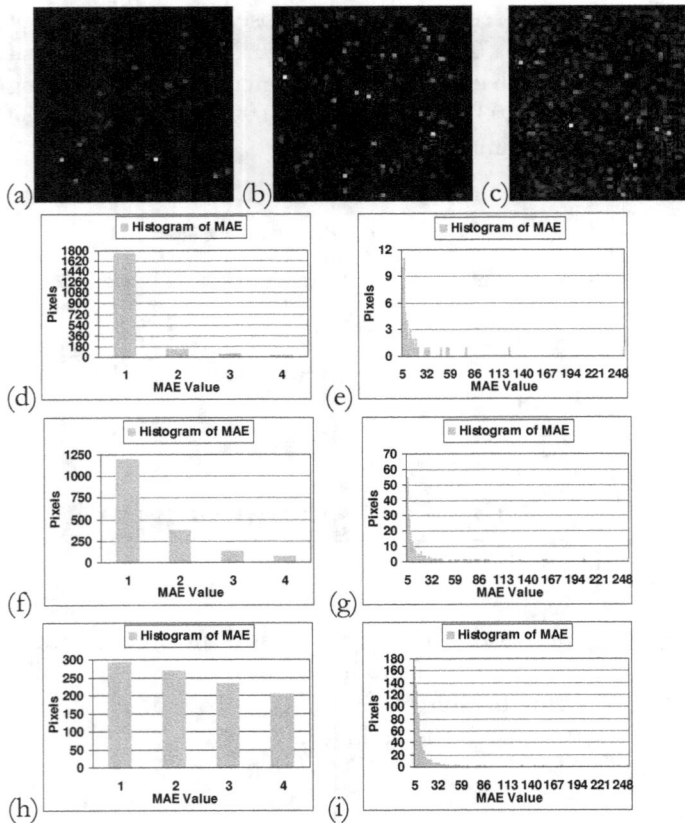

(a) (b) (c)

(d) (e)

(f) (g)

(h) (i)

Figure 24: The MAE images shown in (a), (b) and (c) refer to classic-curvature, resilient-curvature and classic-resilient-curvature hybrid signal reconstruction respectively. The histograms of the image in (a) are shown in (d) and (e), those of the image in (b) are shown in (f) and (g), and those of the image in (c) are shown in (h) and (i). The misplacement is (x, y) = (0.33, 0.33). Data are relevant to the graphs of figure 22.

We may list two aspects in particular to be observed here, namely: (i) the pixel count at position 1 in the histograms in (d) and in (f), which are 1747 and 1198 respectively. Also: (ii) the pixel count in images (a) and (b) as shown in their histograms (to be specific in the intervals [1, 4], [5, 255]), which show 1,957 in (d), 68 in (e) (MAE image in (a)); and 1,758 in (f), 267 in (g) (MAE image in (b)).

What tells us the most about the difference of the MAE 1.175 E-02 in (a) and 9.396 E-03 in (b), is the pixel count in the interval [1, 4], which is 1,957 versus 1,758, (a) and (b) respectively; along with the pixel count in the interval [5, 255], which is 68 versus 267.

Similarly, to add further information to the visual inspection of images (a) and (c), what tells us most about the values of the MAE: 1.175 E-02 and 2.032 E-03 respectively in (a) and (c) is the comparison between the histograms in (d), (e) and the histograms in (h), (i); these latter having a consistent number of pixels with the MAE different from that of the pixels of the histograms in (d) and in (e) (compare (h) versus (d) and (i) versus (e)). The pixel count in the intervals [1, 4], [5, 255] is 1,957; 68 in histograms (d), (e) (MAE image in (a)), and 996; 1029 in histograms (h), (i) (MAE image in (c)). With reference to the values of MAE: 1.175 E-02 and 2.032 E-03 respectively in (a) and (c), the difference existing between 68 (e) and 1029 (i) assumes a clear significance.

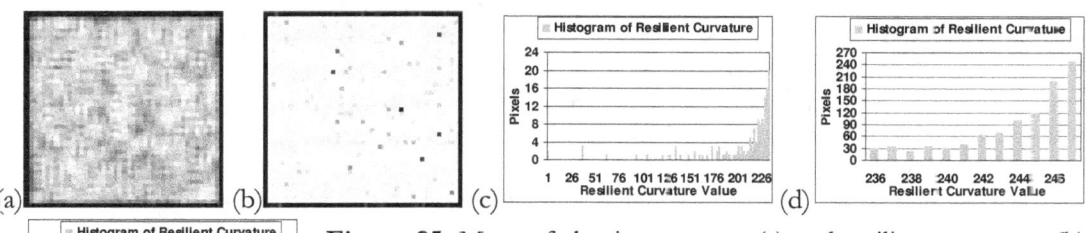

(a) (b) (c) (d)

Figure 25: Maps of classic curvature (a) and resilient curvature (b). Histograms of the image in (b) are shown in (c), (d) and (e). Likewise figure 24; the misplacement is (x, y) = (0.33, 0.33) and data are relevant to the graphs of figure 22.

(e)

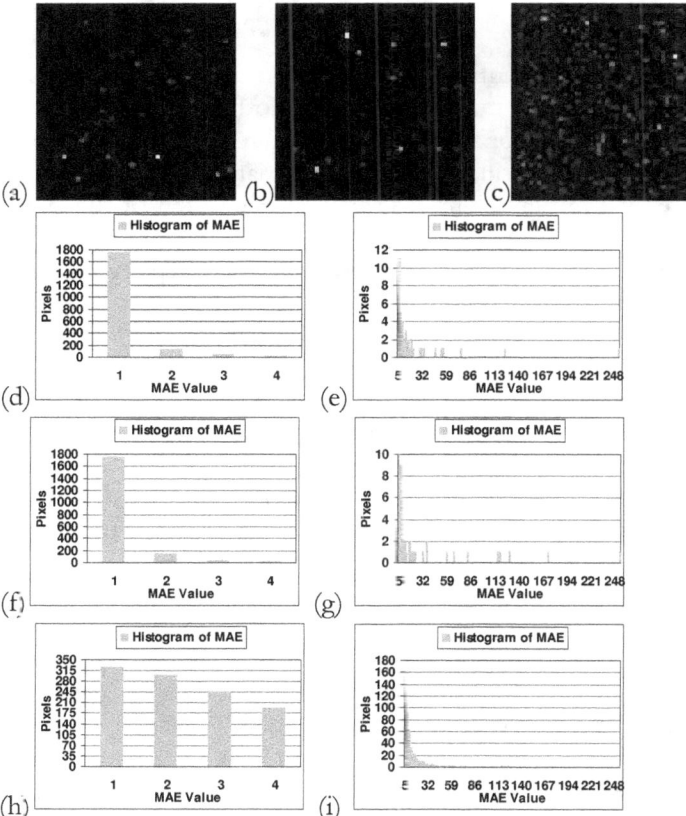

(a) (b) (c)

(d) (e)

(f) (g)

(h) (i)

Figure 26: MAE images obtained with the classic-curvature formula, resilient-curvature formula and classic-resilient-curvature hybrid formula are shown in (a), (b) and (c) respectively. The histograms of the MAE in (a) are shown in (d) and (e), those of the MAE in (b) are shown in (f) and (g), and those of the MAE in (c) are shown in (h) and (i). The misplacement is (x, y) = (0.33, 0.33). Data are relevant to the graphs of figure 23.

The behaviour of the resilient curvature map image seen in figure 25(b) is similar to the behaviour of the image seen in figure 21(b) (bivariate cubic polynomials). The map in figure 25(b) appears flatter than the classic-curvature map image seen in figure 25(a), and it tends to have values of the curvature that shift towards a high magnitude; and this

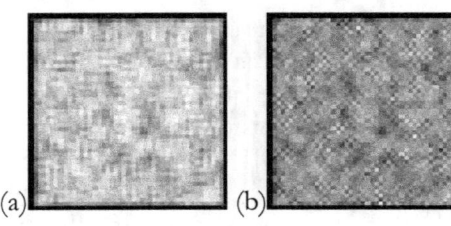

Figure 27: Map of classic-curvature in (a) and map of resilient-curvature in (b). The misplacement is (x, y) = (0.33, 0.33), and data are relevant to the graphs of figure 23.

(a)   (b)

tendency is also seen in the histograms in figures 25(c), 25(d) and 25(e). In figure 26 the MAE of the images seen in (a), (b) and (c) is: 1.175 E-02, 1.225 E-02, 2.068 E-03 respectively. To further the analysis beyond the visual inspection of (a) and (b), it is possible, when looking at the histograms in (d), (e) (relevant to the map of the MAE seen in (a)), to compare the magnitude of the MAE located in the histograms in (f), (g) (relevant to the MAE map seen in (b)).

It was calculated that in the intervals [1, 4] and [5, 255] the pixel count is 1,960, 68 in (d) and in (e), and is 1,960, 68 in (f) and in (g). Thus, what makes most of the difference between (a) and (b) is the distribution of pixels located in the histograms (e) and (g). As far as regards the MAE of (a) compared to the MAE of (c), comparison of the histogram in (e) versus the histogram in (i), provides us with the opportunity to observe the multitude of pixels in (e) versus those pixels located in (i). To be specific, the pixel count in the interval [5, 255] in (e) and (i) is 68 and 963 respectively. The maps of figure 27 are not alike; in fact classic curvature (a) and resilient curvature (b) differ in their numerical values. Similarly, we have observed in figure 13 the difference in numerical values of the two types of curvature within the context of the study of bivariate quadratic formulae calculated with embedding.

SUMMARY

To conclude this chapter, we may draw the reader's attention to the fact that the paragraphs above illustrate the methodology to be adopted to characterize mathematically the form of a novel class of interpolation functions to employ in signal reconstruction. These functions are the classic-resilient-curvature hybrid formulae and they have been presented in three forms: bivariate quadratic, bivariate cubic, and bivariate cubic Lagrange; with each form given in two variants: calculation of the signal with embedding and without embedding.

It would be advantageous to select each case, which means each pixel and each misplacement and treat it as a separate case without any similarity whatsoever; although it would be possible to take such similarities into consideration, instead we have chosen, for the sake of convenience and efficiency, to exclude one formula or the other (classic-curvature or resilient-curvature) giving our preference to the determinist choice which leads to the formulation of the hybrid signal reconstruction. This combines the better of the two formulae: the classic-curvature formula and the resilient-curvature formula, and does that on a pixel-by-pixel basis and at the given misplacement. The hybrid signal reconstruction makes it clear that a novel class of interpolation functions has been developed and it has been denoted classic-resilient-curvature hybrid formula: HYB (x, y) = f (0, 0) + PIC (x, y), where PIC (x, y) is the smaller value obtained from equations (1) and (2).

REFERENCES

Ciulla, C. (2009). *Improved Signal and Image Interpolation in Biomedical Applications: The Case of Magnetic Resonance Imaging (MRI)* – Medical Information Science Reference - IGI Global Publisher, Hershey, PA, U.S.A.

De Boor, C. (1978). *A practical guide to splines. Applied mathematical sciences.* New York, NY: Springer-

Verlag.

Hecklin, G., Nurnberger, G., Schumaker, L. L. & Zeilfelder, F. (2008) *A local Lagrange interpolation method based on C¹ cubic splines on freudenthal partitions.* Mathematics of Computation, 77(262), 1017-1036.

Jonic, S., Thevenaz, P., Zheng, G., Nolte, L. P., & Unser, M. (2006). *An optimized spline-based registration of a 3D CT to a set of C-arm images.* International Journal of Biomedical Imaging, 2006(Article ID 47197), 1-12.

Keys, R. G. (1981). *Cubic convolution interpolation for digital image processing.* IEEE Transactions on Acoustics, Speech, and Signal Processing, 29(6), 1153-1160.

Ledesma-Carbayo, M. J., Kybic, J., Desco, M., Santos, A., Suhling, M., Hunziker, P. & Unser, M. (2005). *Spatio-temporal nonrigid registration for ultrasound cardiac motion estimation.* IEEE Transactions on Medical Imaging, 24(9), 1113-1126.

Meijering, E. H. W., Zuiderveld, K. J. & Viergever, M. A. (1999). *Image reconstruction by convolution with symmetrical piecewise n-th order polynomial kernels.* IEEE Transactions on Image Processing, 8(2), 192-201.

Precioso, F., Barlaud, M., Blu, T. & Unser M. (2003). *Smoothing B-spline active contour for fast and robust image and video segmentation.* Proceedings of the International Conference on Image Processing, Spain, 1, 137-140.

Precioso, F., Barlaud, M., Blu, T. & Unser M. (2005). *Robust real-time segmentation of images and videos using a smooth-spline snake-based algorithm.* IEEE Transactions on Image Processing, 14(7), 910-924.

Schoenberg, I. J. (1946a). *Contributions to the problem of approximation of equidistant data by analytic functions. Part A. On the problem of smoothing or graduation. A first class of analytic approximation formulae.* Quarterly of Applied Mathematics, 4, 45-99.

Schoenberg, I. J. (1946b). *Contributions to the problem of approximation of equidistant data by analytic functions. Part A. On the problem of osculatory interpolation. A second class of analytic approximation formulae.* Quarterly of Applied Mathematics, 4, 112-141.

Schoenberg, I. J. (1969). *Cardinal interpolation and spline functions.* Journal of Approximation Theory, 2, 167-206.

Schoenberg, I. J. (1971). *On equidistant cubic spline interpolation.* Bulletin of the American Mathematical Society, 77(6), 1039-1044.

Sorzano, C. O. S., Blagov, M. S., Thevenaz, P., Myasnikova, E. M., Samsonova, M. G. & Unser M. (2006). *Algorithm for spline-based elastic registration in application to confocal images of gene expression.* Pattern Recognition and Image Analysis, 16(1), 93-96.

Unser, M., Aldroubi, A. & Eden, M. (1993a). *B-spline signal processing: Part I – theory.* IEEE Transactions on Signal Processing, 41(2), 821-833.

Unser, M., Aldroubi, A. & Eden, M. (1993b). *B-spline signal processing: Part II – efficient design and applications.* IEEE Transactions on Signal Processing, 41(2), 834-848.

Unser, M., Thevenaz, P. & Yaroslavsky, L. (1995). *Convolution-based interpolation for fast, high-quality rotation of images.* IEEE Transactions on Image Processing, 4(10), 1371-1381.

Unser, M. (1999). Splines: *A perfect fit for signal and image processing.* IEEE Signal Processing Magazine, 16(6), 22-38.

Unser, M. & Blu, T. (2005). *Cardinal exponential splines: Part I - theory and filtering algorithms.* IEEE Transactions on Signal Processing, 53(4), 1425-1438.

Unser, M. (2005). *Cardinal exponential splines: Part II - think analog, act digital.* IEEE Transactions on Signal Processing, 53(4), 1439-1449.

Unser, M. & Blu, T. (2006). *Generalized smoothing splines and the optimal discretization of the Wiener filter.* IEEE Transactions on Signal Processing, 53(6), 2146-2159.

CHAPTER 9

CALCULATION OF THE CURVATURE OF THE SIGNAL RESILIENT TO INTERPOLATION: TRIVARIATE POLYNOMIALS WITH AND WITHOUT EMBEDDING

INTRODUCTION

Let us recall what is the final purpose end aim of the exploratory research at this stage, namely to device three classes of interpolation functions called: clasic-curvature, resilient-curvature, classic-resilient-curvature hybrid formulae. This chapter deals with the three dimensional cases of the parametric quadratic polynomial, the cubic and the cubic Lagrange polynomials.

To calculate the total curvature of the three polynomials in their classic form the sum of the second order partial derivatives is computed including all the variates x, y, z and covariates xy, xz, yx, yz, zx, zy. The classic parametric quadratic polynomial requires that the total curvature is provided with: $[(\partial^2 (h_3(x, y, z)) / \partial x^2) + (\partial^2 (h_3(x, y, z)) / \partial y^2) + (\partial^2 (h_3(x, y, z)) / \partial z^2) + (\partial^2 (h_3(x, y, z)) / \partial x \partial y) + (\partial^2 (h_3(x, y, z)) / \partial y \partial x) + (\partial^2 (h_3(x, y, z)) / \partial x \partial z) + (\partial^2 (h_3(x, y, z)) / \partial z \partial x) + (\partial^2 (h_3(x, y, z)) / \partial y \partial z) + (\partial^2 (h_3(x, y, z)) / \partial z \partial y)].$

The classic cubic polynomial has its total curvature provided with: $[(\partial^2 (h_4(x, y, z)) / \partial x^2) + (\partial^2 (h_4(x, y, z)) / \partial y^2) + (\partial^2 (h_4(x, y, z)) / \partial z^2) + (\partial^2 (h_4(x, y, z)) / \partial x \partial y) + (\partial^2 (h_4(x, y, z)) / \partial y \partial x) + (\partial^2 (h_4(x, y, z)) / \partial x \partial z) + (\partial^2 (h_4(x, y, z)) / \partial z \partial x) + (\partial^2 (h_4(x, y, z)) / \partial y \partial z) + (\partial^2 (h_4(x, y, z)) / \partial z \partial y)].$

The classic cubic Lagrange polynomial has its total curvature in the sum: $[(\partial^2 (LGR_3(x, y, z)) / \partial x^2) + (\partial^2 (LGR_3(x, y, z)) / \partial y^2) + (\partial^2 (LGR_3(x, y, z)) / \partial z^2) + (\partial^2 (LGR_3(x, y, z)) / \partial x \partial y) + (\partial^2 (LGR_3(x, y, z)) / \partial y \partial x) + (\partial^2 (LGR_3(x, y, z)) / \partial x \partial z) + (\partial^2 (LGR_3(x, y, z)) / \partial z \partial x) + (\partial^2 (LGR_3(x, y, z)) / \partial y \partial z) + (\partial^2 (LGR_3(x, y, z)) / \partial z \partial y)].$ Considering that the forms with embedding: $h_3(x, y, z) = f(0, 0, 0) + \omega_1 * [-2a(x + y + z)^2 + 1/2(a+1)] + \omega_2 * [a(x + y + z)^2 - (2a + 1/2)(x + y + z) + 3/4(a+1)]$, $h_4(x, y, z) = f(0, 0, 0) + \omega_1 * [1/2(x + y + z)^3 - (x + y + z)^2 + 2/3] + \omega_2 * [-1/6(x + y + z)^3 + (x + y + z)^2 - 2(x + y + z) + 4/3]$, $LGR_3(x, y, z) = f(0, 0, 0) + \omega_1 * [(1/2)(x + y + z)^3 - (x + y + z)^2 - 1/2(x + y + z) + 1] + \omega_2 * [-(1/6)(x + y + z)^3 + (x + y + z)^2 - (11/6)(x + y + z) + 1]$, and without embedding: $h_3(x, y, z) = \omega_1 * [-2a(x + y + z)^2 + 1/2(a+1)] + \omega_2 * [a(x + y + z)^2 - (2a + 1/2)(x + y + z) + 3/4(a+1)]$, $h_4(x, y, z) = \omega_1 * [1/2(x + y + z)^3 - (x + y + z)^2 + 2/3] + \omega_2 * [-1/6(x + y + z)^3 + (x + y + z)^2 - 2(x + y + z) + 4/3]$, $LGR_3(x, y, z) = \omega_1 * [(1/2)(x + y + z)^3 - (x + y + z)^2 - 1/2(x + y + z) + 1] + \omega_2 * [-(1/6)(x + y + z)^3 + (x + y + z)^2 - (11/6)(x + y + z) + 1]$, differ only because of the term $f(0, 0, 0)$, which is the original signal value, the two total curvatures: classic with and without embedding are exactly the same.

To calculate the total curvature of the signal resilient to interpolation for the three interpolation functions: $h_3(x, y)$, $h_4(x, y)$, $LGR_3(x, y)$, with and without embedding, the starting equation is:
$R(x, y, z) = f(0, 0, 0) + \varrho(x, y, z)$, where $f(0, 0, 0)$ is the original signal and $\varrho(x, y, z)$ is the signal resilient to the interpolation functions herein treated. The total curvature is given in either case through the formula: $[(\partial^2 (\varrho(x, y, z)) / \partial x^2) + (\partial^2 (\varrho(x, y, z)) / \partial y^2) + (\partial^2 (\varrho(x, y, z)) / \partial z^2) + (\partial^2 (\varrho(x, y, z)) / \partial y \partial x) + (\partial^2 (\varrho(x, y, z)) / \partial x \partial y) + (\partial^2 (\varrho(x, y, z)) / \partial y \partial z) + (\partial^2 (\varrho(x, y, z)) / \partial z \partial y) + (\partial^2 (\varrho(x, y, z)) / \partial x \partial z) + (\partial^2 (\varrho(x, y, z)) / \partial z \partial x)].$

The formulation of the signal resilient to interpolation with embedding is different from the formulation without embedding, therefore the two total curvatures will differ in their math forms.

TRIVARIATE CUBIC LAGRANGE CALCULATED WITH EMBEDDING

Let us recall equations (29), (28), (24) and (25) of chapter 3 part III and number them here as (1), (2), (3) and (4):

$$\Pi^{(1)} = (x^2yz/2 + xy^2z/2 + xyz^2/2) \tag{1}$$

$$\Pi^{(2)} = (x^3yz/3 + xy^3z/3 + xz^3y/3 + x^2y^2z/2 + x^2yz^2/2 + xy^2z^2/2) \tag{2}$$

$$\Pi^{(3)} = (x^4yz/4 + x^2y^3z/6 + x^2yz^3/6 + x^3y^2z/3 + 1/3\ x^3yz^2 + x^2y^2z^2/4 + x^3y^2z/6 + xy^4z/4 +$$
$$xy^2z^3/6 + 1/3\ x^2y^3z + x^2y^2z^2/4 + xy^3z^2/3 + x^3yz^2/6 + xy^3z^2/6 + xyz^4/4 + x^2y^2z^2/4 + 1/3\ x^2yz^3$$
$$+ 1/3\ xy^2z^3) \tag{3}$$

$$\Pi^{(4)} = (x^5zy/5 + x^3y^3z/9 + x^3yz^3/9 + x^4y^2z/4 + x^4z^2y/4 + x^3y^2z^2/6 + xy^5z/5 + x^3y^3z/9 +$$
$$xy^3z^3/9 + x^2y^4z + x^2z^2y^3/6 + xy^4z^2/4 + x^3yz^3/9 + xz^3y^3/9 + xz^5y/5 + 1/6\ x^2y^2z^3 + 1/4\ x^2yz^4 +$$
$$1/4\ xy^2z^4 + x^4y^2z/4 + 1/4\ x^2y^4z + 1/6\ x^2y^2z^3 + 4/9\ x^3y^3z + x^3y^2z^2/3 + 1/3\ x^2y^3z^2 + x^4z^2y/4 +$$
$$x^2y^3z^2/6 + x^2yz^4 + x^3y^2z^2/3 + 4/9\ x^3yz^3 + 4/12\ x^2y^2z^3 + x^3y^2z^2/6 + xy^4z^2/4 + 1/4\ xy^2z^4 + 1/3$$
$$x^2y^3z^2 + 4/12\ x^2y^2z^3 + 4/9\ xy^3z^3) \tag{4}$$

From equations (1) through (4) it follows that:

$$\Pi^{(1)} = x\ y\ z * (x/2 + y/2 + z/2) \tag{5}$$

$$\Pi^{(2)} = x\ y\ z * (x^2/3 + y^2/3 + z^2/3 + xy/2 + xz/2 + yz/2) \tag{6}$$

$$\Pi^{(3)} = x\ y\ z * (x^3/4 + xy^2/6 + xz^2/6 + x^2y/3 + (1/3)\ x^2z + xyz/4 + x^2y/6 + y^3/4 + yz^2/6 +$$
$$(1/3)\ xy^2 + xyz/4 + y^2z/3 + x^2z/6 + y^2z/6 + z^3/4 + xyz/4 - (1/3)\ xz^2 + (1/3)\ yz^2) \tag{7}$$

$$\Pi^{(4)} = x\ y\ z * (x^4/5 + x^2y^2/9 + x^2z^2/9 + x^3y/4 + x^3z/4 + x^2yz/6 + y^4/5 + x^2y^2/9 + y^2z^2/9 + xy^3$$
$$+ xy^2z/6 + y^3z/4 + x^2z^2/9 + z^2y^2/9 + z^4/5 + 1/6\ xyz^2 + 1/4\ xz^3 + 1/4\ yz^3 + x^3y/4 + 1/4\ xy^3 -$$
$$1/6\ xyz^2 + 4/9\ x^2y^2 + x^2yz/3 + 1/3\ xy^2z + x^3z/4 + xy^2z/6 + xz^3 + x^2yz/3 + 4/9\ x^2z^2 + 4/12$$
$$xyz^2 + x^2yz/6 + y^3z/4 + 1/4\ yz^3 + 1/3\ xy^2z + 4/12\ xyz^2 + 4/9\ y^2z^2) \tag{8}$$

In view of equations (5) through (8) let us posit:

$$\Pi^{(1)} = x\,y\,z * \Lambda_{(1)} \tag{9}$$

$$\Pi^{(2)} = x\,y\,z * \Lambda_{(2)} \tag{10}$$

$$\Pi^{(3)} = x\,y\,z * \Lambda_{(3)} \tag{11}$$

$$\Pi^{(4)} = x\,y\,z * \Lambda_{(4)} \tag{12}$$

Calculation of First and Second Order Partial Derivatives of $\Lambda_{(1)}$, $\Lambda_{(2)}$, $\Lambda_{(3)}$, $\Lambda_{(4)}$

$$\Lambda_{(1)} = (x/2 + y/2 + z/2) \tag{13}$$

$$(\partial\,(\Lambda_{(1)})\,/\partial x) = (1/2);\; (\partial\,(\Lambda_{(1)})\,/\partial y) = (1/2);\; (\partial\,(\Lambda_{(1)})\,/\partial z) = (1/2) \tag{14}$$

$$(\partial^2\,(\Lambda_{(1)})\,/\partial x^2) = 0;\; (\partial^2\,(\Lambda_{(1)})\,/\partial y^2) = 0;\; (\partial^2\,(\Lambda_{(1)})\,/\partial z^2) = 0 \tag{15}$$

$$(\partial^2\,(\Lambda_{(1)})\,/\partial x\partial y) = 0;\; (\partial^2\,(\Lambda_{(1)})\,/\partial y\partial x) = 0 \tag{16}$$

$$(\partial^2\,(\Lambda_{(1)})\,/\partial x\partial z) = 0;\; (\partial^2\,(\Lambda_{(1)})\,/\partial z\partial x) = 0 \tag{17}$$

$$(\partial^2\,(\Lambda_{(1)})\,/\partial y\partial z) = 0;\; (\partial^2\,(\Lambda_{(1)})\,/\partial z\partial y) = 0 \tag{18}$$

$$\Lambda_{(2)} = (x^2/3 + y^2/3 + z^2/3 + xy/2 + xz/2 + yz/2) \tag{19}$$

$$(\partial\,(\Lambda_{(2)})\,/\partial x) = (\,(2/3)\,x + y/2 + z/2\,) \tag{20}$$

$$(\partial\,(\Lambda_{(2)})\,/\partial y) = (\,(2/3)\,y + x/2 + z/2\,) \tag{21}$$

$$(\partial\,(\Lambda_{(2)})\,/\partial z) = (\,(2/3)\,z + x/2 + y/2\,) \tag{22}$$

$$(\partial^2\,(\Lambda_{(2)})\,/\partial x^2) = (2/3);\; (\partial^2\,(\Lambda_{(2)})\,/\partial y^2) = (2/3);\; (\partial^2\,(\Lambda_{(2)})\,/\partial z^2) = (2/3) \tag{23}$$

$$(\partial^2\,(\Lambda_{(2)})\,/\partial x\partial y) = (\partial^2\,(\Lambda_{(2)})\,/\partial y\partial x) = (\partial^2\,(\Lambda_{(2)})\,/\partial x\partial z) = (\partial^2\,(\Lambda_{(2)})\,/\partial z\partial x) =$$

$$(\partial^2\,(\Lambda_{(2)})\,/\partial y\partial z) = (\partial^2\,(\Lambda_{(2)})\,/\partial z\partial y) = (1/2) \tag{24}$$

$$\Lambda_{(3)} = (x^3/4 + xy^2/6 + xz^2/6 + x^2y/3 + (1/3)\,x^2z + xyz/4 + x^2y/6 + y^3/4 + yz^2/6 + (1/3)\,xy^2 +$$

$$xyz/4 + y^2z/3 + x^2z/6 + y^2z/6 + z^3/4 + xyz/4 + (1/3)\,xz^2 + (1/3)\,yz^2) \tag{25}$$

$$(\partial\,(\Lambda_{(3)})\,/\partial x) = [\,(3/4)\,x^2 + y^2/6 + z^2/6 + (2/3)\,xy + (2/3)\,xz + yz/4 + (1/3)\,xy + (1/3)\,y^2 +$$

$$yz/4 + (1/3)\,xz + yz/4 + (1/3)\,z^2\,] \tag{26}$$

$$(\partial^2\,(\Lambda_{(3)})\,/\partial x^2) = [\,(3/2)\,x + (2/3)\,y + (2/3)\,z + (1/3)\,y + (1/3)\,z\,] = [\,(3/2)\,x + y + z\,] \tag{27}$$

$$(\partial\,(\Lambda_{(3)})\,/\partial y) = [\,(1/3)\,xy + x^2/3 + xz/4 + x^2/6 + (3/4)\,y^2 + z^2/6 + (2/3)\,xy + xz/4 + (2/3)\,yz$$

$$+ (1/3)\ yz + xz/4 + (1/3)\ z^2] \tag{28}$$

$$(\partial^2 (\Lambda_{(3)}) / \partial y^2) = [(1/3)\ x + (3/2)\ y + (2/3)\ x + (2/3)\ z + (1/3)\ z] = [(3/2)\ y + x + z] \tag{29}$$

$$(\partial (\Lambda_{(3)}) / \partial z) = [(1/3)\ xz + (1/3)\ x^2 + xy/4 + (1/3)\ yz + xy/4 + y^2/3 + x^2/6 + y^2/6 + (3/4)\ z^2$$

$$+ xy/4 + (2/3)\ xz + (2/3)\ yz] \tag{30}$$

$$(\partial^2 (\Lambda_{(3)}) / \partial z^2) = [(1/3)\ x + (1/3)\ y + (3/2)\ z + (2/3)\ x + (2/3)\ y] = [(3/2)\ z + x - y] \tag{31}$$

$$(\partial^2 (\Lambda_{(3)}) / \partial x \partial y) = [(1/3)\ y + (2/3)\ x + z/4 + (1/3)\ x + (2/3)\ y + z/4 + z/4] =$$

$$[(3/4)\ z + x + y] \tag{32}$$

$$(\partial^2 (\Lambda_{(3)}) / \partial x \partial z) = [(1/3)\ z + (2/3)\ x + y/4 + y/4 + (1/3)\ x + y/4 + (2/3)\ z] =$$

$$[(3/4)\ y + x + z] \tag{33}$$

$$(\partial^2 (\Lambda_{(3)}) / \partial y \partial x) = [(1/3)\ y + (2/3)\ x + z/4 + (1/3)\ x + (2/3)\ y + z/4 + z/4] =$$

$$[(3/4)\ z + x + y] \tag{34}$$

$$(\partial^2 (\Lambda_{(3)}) / \partial y \partial z) = [x/4 + (1/3)\ z - x/4 + (2/3)\ y + (1/3)\ y + x/4 + (2/3)\ z] =$$

$$[(3/4)\ x + y + z] \tag{35}$$

$$(\partial^2 (\Lambda_{(3)}) / \partial z \partial x) = [(1/3)\ z + (2/3)\ x + y/4 + y/4 + (1/3)\ x + y/4 + (2/3)\ z] =$$

$$[(3/4)\ y + x + z] \tag{36}$$

$$(\partial^2 (\Lambda_{(3)}) / \partial z \partial y) = [x/4 + (1/3)\ z + x/4 + (2/3)\ y + (1/3)\ y + x/4 + (2/3)\ z] =$$

$$[(3/4)\ x + y + z] \tag{37}$$

$$\Lambda_{(4)} = (x^4/5 + x^2y^2/9 + x^2z^2/9 + x^3y/4 + x^3z/4 + x^2yz/6 + y^4/5 + x^2y^2/9 + y^2z^2/9 + xy^3 +$$

$$xy^2z/6 + y^3z/4 + x^2z^2/9 + z^2y^2/9 + z^4/5 + 1/6\ xyz^2 + 1/4\ xz^3 + 1/4\ yz^3 + x^3y/4 + 1/4\ xy^3 +$$

$$1/6\ xyz^2 + 4/9\ x^2y^2 + x^2yz/3 + 1/3\ xy^2z + x^3z/4 + xy^2z/6 + xz^3 + x^2yz/3 + 4/9\ x^2z^2 + 4/12$$

$$xyz^2 + x^2yz/6 + y^3z/4 + 1/4\ yz^3 + 1/3\ xy^2z + 4/12\ xyz^2 + 4/9\ y^2z^2) = (x^4/5 + (6/9)\ x^2y^2 +$$

$$(6/9)\ x^2z^2 + (1/2)\ x^3y + (1/2)\ x^3z + x^2yz + y^4/5 + (6/9)\ y^2z^2 + (5/4)\ xy^3 + xy^2z + (1/2)\ y^3z +$$

$$z^4/5 + xyz^2 + (5/4)\ xz^3 + (1/2)\ yz^3) \tag{38}$$

$$(\partial\,(\Lambda_{(4)})\,/\partial x) = [\,(4/5)\,x^3 + (12/9)\,xy^2 + (12/9)\,xz^2 + (3/2)\,x^2y + (3/2)\,x^2z + 2\,xyz + (5/4)\,y^3$$

$$+\,y^2z + yz^2 + (5/4)\,z^3\,] \tag{39}$$

$$(\partial^2\,(\Lambda_{(4)})\,/\partial x^2) = [\,(12/5)\,x^2 + (12/9)\,y^2 + (12/9)\,z^2 + (6/2)\,xy + (6/2)\,xz + 2\,yz\,] \tag{40}$$

$$(\partial\,(\Lambda_{(4)})\,/\partial y) = [\,(12/9)\,x^2y + x^3/2 + x^2z + (4/5)\,y^3 + (12/9)\,yz^2 + (15/4)\,xy^2 + 2\,xyz + (3/2)$$

$$y^2z + xz^2 + (1/2)\,z^3\,] \tag{41}$$

$$(\partial^2\,(\Lambda_{(4)})\,/\partial y^2) = [\,(12/9)\,x^2 + (12/5)\,y^2 + (12/9)\,z^2 + (30/4)\,xy + 2\,xz + (6/2)\,yz\,] \tag{42}$$

$$(\partial\,(\Lambda_{(4)})\,/\partial z) = [\,(12/9)\,zx^2 + x^3/2 + x^2y + (12/9)\,y^2z + xy^2 + (1/2)\,y^3 + (4/5)\,z^3 + 2\,xyz +$$

$$(15/4)\,xz^2 + (3/2)\,yz^2\,] \tag{43}$$

$$(\partial^2\,(\Lambda_{(4)})\,/\partial z^2) = [\,(12/9)\,x^2 + (12/9)\,y^2 + (12/5)\,z^2 + 2\,xy + (30/4)\,xz + (6/2)\,yz\,] \tag{44}$$

$$(\partial^2\,(\Lambda_{(4)})\,/\partial x\partial y) = [\,(24/9)\,xy + (3/2)\,x^2 + 2\,xz + (15/4)\,y^2 + 2\,yz + z^2\,] \tag{45}$$

$$(\partial^2\,(\Lambda_{(4)})\,/\partial x\partial z) = [\,(24/9)\,xz + (3/2)\,x^2 + 2\,xy + y^2 + 2\,yz + (15/4)\,z^2\,] \tag{46}$$

$$(\partial^2\,(\Lambda_{(4)})\,/\partial y\partial x) = [\,(24/9)\,xy + (3/2)\,x^2 + 2\,xz + (15/4)\,y^2 + 2\,yz + z^2\,] \tag{47}$$

$$(\partial^2\,(\Lambda_{(4)})\,/\partial y\partial z) = [\,x^2 + (24/9)\,yz + 2\,xy + (3/2)\,y^2 + 2\,xz + (3/2)\,z^2\,] \tag{48}$$

$$(\partial^2\,(\Lambda_{(4)})\,/\partial z\partial x) = [\,(24/9)\,zx + (3/2)\,x^2 + 2\,xy + y^2 + 2\,yz + (15/4)\,z^2\,] \tag{49}$$

$$(\partial^2\,(\Lambda_{(4)})\,/\partial z\partial y) = [\,x^2 + (24/9)\,yz + 2\,xy + (3/2)\,y^2 + 2\,xz + (3/2)\,z^2\,] \tag{50}$$

Formulation of $\varrho\,(x,\,y,\,z)$

Equation (77) in chapter 2 part IV is numbered here as (51):

$$\Lambda_L = 9 * \{\,[\,(3/2)\,\omega_1^2 - \omega_1\,\omega_2 + (1/6)\,\omega_2^2\,]\,\Pi^{(4)} + [-\,4\,\omega_1^2 + (16/3)\,\omega_1\,\omega_2 - (8/6)\,\omega_2^2\,]\,\Pi^{(3)} +$$

$$[\,1/2\,\omega_1^2 - (54/6)\,\omega_1\,\omega_2 + (23/6)\,\omega_2^2\,]\,\Pi^{(2)} + [\,4\,\omega_1^2 + (28/6)\,\omega_1\,\omega_2 - (28/6)\,\omega_2^2\,]\,\Pi^{(1)} +$$

$$[\,-\,2\,\omega_1^2 + 2\,\omega_2^2\,]\,xyz\,\} \tag{51}$$

Let us posit:

$$\Lambda_{L1} = [\,(3/2)\,\omega_1^2 - \omega_1\,\omega_2 + (1/6)\,\omega_2^2\,] \tag{52}$$

$$\Lambda_{L2} = [-\,4\,\omega_1^2 + (16/3)\,\omega_1\,\omega_2 - (8/6)\,\omega_2^2\,] \tag{53}$$

$$\Lambda_{L3} = [\,1/2\,\omega_1^2 - (54/6)\,\omega_1\,\omega_2 + (23/6)\,\omega_2^2\,] \tag{54}$$

$$\Lambda_{L4} = [\, 4\, \omega_1^2 + (28/6)\, \omega \cdot \omega_2 - (28/6)\, \omega_2^2 \,] \tag{55}$$

$$\Lambda_{L5} = [\, -2\, \omega_1^2 + 2\, \omega_2^2 \,] \tag{56}$$

Equation (78) of chapter 2 part IV is numbered here as (57):

$$f\,(0, 0, 0) = \Lambda_L\, /\, \{\, 9\, xyz * \{\, -2\, \omega_1 + 2\, \omega_2 \,\} - 9 * \{\, \omega_1 * [\, 3\, \Pi^{(1)} - 2\, xyz\,] + \omega_2 * [\, -\Pi^{(1)} +$$

$$2\, xyz\,]\,\}\,\} \tag{57}$$

The formulation R $(x, y, z) = f\,(0, 0, 0) + \varrho\,(x, y, z)$, requires:

$$\varrho\,(x, y, z) = \Lambda_L\, /\, \{\, 9\, xyz * \{\, -2\, \omega_1 + 2\, \omega_2 \,\} - 9 * \{\, \omega_1 * [\, 3\, \Pi^{(1)} - 2\, xyz\,] + \omega_2 * [\, -\Pi^{(1)} + 2\, xyz\,]\,\}\,\}$$

Thus in view of equations (9) through (12) and equations (52) through (56), equation (51) becomes:

$$\Lambda_L = 9 * \{\, \Lambda_{L1} * x\, y\, z * \Lambda_{(4)} + \Lambda_{L2} * x\, y\, z * \Lambda_{(3)} + \Lambda_{L3} * x\, y\, z * \Lambda_{(2)} + \Lambda_{L4} * x\, y\, z * \Lambda_{(1)} +$$

$$xyz * \Lambda_{L5} \,\} = 9 * xyz \,\{\, \Lambda_{L1} * \Lambda_{(4)} + \Lambda_{L2} * \Lambda_{(3)} + \Lambda_{L3} * \Lambda_{(2)} + \Lambda_{L4} * \Lambda_{(1)} + \Lambda_{L5} \,\} \tag{58}$$

And:
$$\varrho\,(x, y, z) = \Lambda_L\, /\, \{\, 9\, xyz * \{\, -2\, \omega_1 + 2\, \omega_2 \,\} - 9 * \{\, \omega_1 * [\, 3\, \Pi^{(1)} - 2\, xyz\,] + \omega_2 * [\, -\Pi^{(1)} +$$

$$2\, xyz\,]\,\}\,\} = \Lambda_L\, /\, \{\, 9\, xyz * \{\, -2\, \omega_1 + 2\, \omega_2 \,\} - 9 * \{\, \omega_1 * [\, 3\, x\, y\, z * \Lambda_{(1)} - 2\, xyz\,] + \omega_2 * [$$

$$-x\, y\, z * \Lambda_{(1)} + 2\, xyz\,]\,\}\,\} = \Lambda_L\, /\, \{\, 9\, xyz * \{\, (-2\, \omega_1 + 2\, \omega_2\,) - \omega_1\,(3\, \Lambda_{(1)} - 2\,) + \omega_2\,(2 -$$

$$\Lambda_{(1)}\,)\,\}\,\} = 9 * xyz\,\{\, \Lambda_{L1} * \Lambda_{(4)} + \Lambda_{L2} * \Lambda_{(3)} + \Lambda_{L3} * \Lambda_{(2)} + \Lambda_{L4} * \Lambda_{(1)} + \Lambda_{L5} \,\}\, /\, \{\, 9\, xyz * \{$$

$$(-2\, \omega_1 + 2\, \omega_2\,) - \omega_1\,(3\, \Lambda_{(1)} - 2\,) + \omega_2\,(2 - \Lambda_{(1)}\,)\,\}\,\} = \{\, \Lambda_{L1} * \Lambda_{(4)} + \Lambda_{L2} * \Lambda_{(3)} + \Lambda_{L3} *$$

$$\Lambda_{(2)} + \Lambda_{L4} * \Lambda_{(1)} + \Lambda_{L5} \,\}\, /\, \{\, (-2\, \omega_1 + 2\, \omega_2\,) - \omega_1\,(3\, \Lambda_{(1)} - 2\,) + \omega_2\,(2 - \Lambda_{(1)}\,)\,\} \tag{59}$$

Let us posit:

$$\Lambda_{LOO} = \{\, (-2\, \omega_1 + 2\, \omega_2\,) - \omega_1\,(3\, \Lambda_{(1)} - 2\,) + \omega_2\,(2 - \Lambda_{(1)}\,)\,\} \tag{60}$$

It follows that:

$$(\partial\,(\Lambda_{LOO})\, /\, \partial x) = -3\, \omega_1\,(\partial\,(\Lambda_{(1)})\, /\, \partial x) - \omega_2\,(\partial\,(\Lambda_{(1)})\, /\, \partial x) = -(\partial\,(\Lambda_{(1)})\, /\, \partial x)\,[\omega_2 + 3\, \omega_1] \tag{61}$$

$$(\partial^2\,(\Lambda_{LOO})\, /\, \partial x^2) = -(\partial^2\,(\Lambda_{(1)})\, /\, \partial x^2)\,[\omega_2 + 3\, \omega_1] \tag{62}$$

$$(\partial\,(\Lambda_{LOO})\, /\, \partial y) = -(\partial\,(\Lambda_{(1)})\, /\, \partial y)\,[\omega_2 + 3\, \omega_1] \tag{63}$$

$$(\partial^2\,(\Lambda_{LOO})\, /\, \partial y^2) = -(\partial^2\,(\Lambda_{(1)})\, /\, \partial y^2)\,[\omega_2 + 3\, \omega_1] \tag{64}$$

$$(\partial\,(\Lambda_{LOO})\, /\, \partial z) = -(\partial\,(\Lambda_{(1)})\, /\, \partial z)\,[\omega_2 + 3\, \omega_1] \tag{65}$$

SIGNAL RESILIENT TO INTERPOLATION

$$(\partial^2 (\Lambda_{LOO}) / \partial z^2) = - (\partial^2 (\Lambda_{(1)}) / \partial z^2) [\omega_2 + 3 \, \omega_1] \tag{66}$$

$$(\partial^2 (\Lambda_{LOO}) / \partial x \partial y) = - (\partial^2 (\Lambda_{(1)}) / \partial x \partial y) [\omega_2 + 3 \, \omega_1] \tag{67}$$

$$(\partial^2 (\Lambda_{LOO}) / \partial y \partial x) = - (\partial^2 (\Lambda_{(1)}) / \partial y \partial x) [\omega_2 + 3 \, \omega_1] \tag{68}$$

$$(\partial^2 (\Lambda_{LOO}) / \partial x \partial z) = - (\partial^2 (\Lambda_{(1)}) / \partial x \partial z) [\omega_2 + 3 \, \omega_1] \tag{69}$$

$$(\partial^2 (\Lambda_{LOO}) / \partial z \partial x) = - (\partial^2 (\Lambda_{(1)}) / \partial z \partial x) [\omega_2 + 3 \, \omega_1] \tag{70}$$

$$(\partial^2 (\Lambda_{LOO}) / \partial y \partial z) = - (\partial^2 (\Lambda_{(1)}) / \partial y \partial z) [\omega_2 + 3 \, \omega_1] \tag{71}$$

$$(\partial^2 (\Lambda_{LOO}) / \partial z \partial y) = - (\partial^2 (\Lambda_{(1)}) / \partial z \partial y) [\omega_2 + 3 \, \omega_1] \tag{72}$$

In view of equations (59) and (60):

$$\varrho \, (x, y, z) = \{ \Lambda_{L1} * \Lambda_{(4)} + \Lambda_{L2} * \Lambda_{(3)} + \Lambda_{L3} * \Lambda_{(2)} + \Lambda_{L4} * \Lambda_{(1)} + \Lambda_{L5}\} / \{ \Lambda_{LOO} \} \tag{73}$$

Calculation of First and Second Order Partial Derivatives of ϱ (x, y, z)

$$(\partial (\varrho \, (x, y, z)) / \partial x) = \{ \{ \Lambda_{L1} * (\partial (\Lambda_{(4)}) / \partial x) + \Lambda_{L2} * (\partial (\Lambda_{(3)}) / \partial x) + \Lambda_{L3} * (\partial (\Lambda_{(2)}) / \partial x) +$$

$$\Lambda_{L4} * (\partial (\Lambda_{(1)}) / \partial x) \} * \{ \Lambda_{LOO} \} - \{ \Lambda_{L1} * \Lambda_{(4)} + \Lambda_{L2} * \Lambda_{(3)} + \Lambda_{L3} * \Lambda_{(2)} + \Lambda_{L4} * \Lambda_{(1)} + \Lambda_{L5}$$

$$\} * (\partial (\Lambda_{LOO}) / \partial x) \} / [\Lambda_{LOO}]^2 \tag{74}$$

Let us posit:

$$N_L = \{ \Lambda_{L1} * \Lambda_{(4)} + \Lambda_{L2} * \Lambda_{(3)} + \Lambda_{L3} * \Lambda_{(2)} + \Lambda_{L4} * \Lambda_{(1)} + \Lambda_{L5} \} \tag{75}$$

$$(\partial (N_L) / \partial x) = \{\Lambda_{L1} * (\partial (\Lambda_{(4)}) / \partial x) + \Lambda_{L2} * (\partial (\Lambda_{(3)}) / \partial x) + \Lambda_{L3} * (\partial (\Lambda_{(2)}) / \partial x) + \Lambda_{L4} * (\partial$$

$$(\Lambda_{(1)}) / \partial x)\} \tag{76}$$

$$(\partial^2 (N_L) / \partial x^2) = \{ \Lambda_{L1} * (\partial^2 (\Lambda_{(4)}) / \partial x^2) + \Lambda_{L2} * (\partial^2 (\Lambda_{(3)}) / \partial x^2) + \Lambda_{L3} * (\partial^2 (\Lambda_{(2)}) / \partial x^2) + \Lambda_{L4}$$

$$* (\partial^2 (\Lambda_{(1)}) / \partial x^2) \} \tag{77}$$

Thus:

$$(\partial (\varrho \, (x, y, z)) / \partial x) = \{ (\partial (N_L) / \partial x) * \Lambda_{LOO} - N_L * (\partial (\Lambda_{LOO}) / \partial x) \} / [\Lambda_{LOO}]^2 \tag{78}$$

$$(\partial^2 (\varrho \, (x, y, z)) / \partial x^2) = \{ \{ (\partial^2 (N_L) / \partial x^2) * [\Lambda_{LOO}] + (\partial (N_L) / \partial x) * (\partial (\Lambda_{LOO}) / \partial x) - (\partial$$

$$(N_L) / \partial x) * (\partial (\Lambda_{LOO}) / \partial x) - N_L * (\partial^2 (\Lambda_{LOO}) / \partial x^2) \} * [\Lambda_{LOO}]^2 - \{ (\partial (N_L) / \partial x) * \Lambda_{LOO} -$$

$$N_L * (\partial (\Lambda_{LOO}) / \partial x) \} * \{ 2 * \Lambda_{LOO} * (\partial (\Lambda_{LOO}) / \partial x) \} \} / [\Lambda_{LOO}]^4 \tag{79}$$

Let us posit:

$$(\partial (N_L) / \partial y) = \{\Lambda_{L1} * (\partial (\Lambda_{(4)}) / \partial y) + \Lambda_{L2} * (\partial (\Lambda_{(3)}) / \partial y) + \Lambda_{L3} * (\partial (\Lambda_{(2)}) / \partial y) + \Lambda_{L4} * (\partial (\Lambda_{(1)}) / \partial y)\}$$
(80)

$$(\partial^2 (N_L) / \partial y^2) = \{ \Lambda_{L1} * (\partial^2 (\Lambda_{(4)}) / \partial y^2) + \Lambda_{L2} * (\partial^2 (\Lambda_{(3)}) / \partial y^2) + \Lambda_{L3} * (\partial^2 (\Lambda_{(2)}) / \partial y^2) + \Lambda_{L4}$$

$$* (\partial^2 (\Lambda_{(1)}) / \partial y^2) \}$$
(81)

$$(\partial (N_L) / \partial z) = \{\Lambda_{L1} * (\partial (\Lambda_{(4)}) / \partial z) + \Lambda_{L2} * (\partial (\Lambda_{(3)}) / \partial z) + \Lambda_{L3} * (\partial (\Lambda_{(2)}) / \partial z) + \Lambda_{L4} * (\partial$$

$$(\Lambda_{(1)}) / \partial z)\}$$
(82)

$$(\partial^2 (N_L) / \partial z^2) = \{ \Lambda_{L1} * (\partial^2 (\Lambda_{(4)}) / \partial z^2) + \Lambda_{L2} * (\partial^2 (\Lambda_{(3)}) / \partial z^2) + \Lambda_{L3} * (\partial^2 (\Lambda_{(2)}) / \partial z^2) + \Lambda_{L4}$$

$$* (\partial^2 (\Lambda_{(1)}) / \partial z^2) \}$$
(83)

$$(\partial^2 (N_L) / \partial x \partial y) = \{\Lambda_{L1} * (\partial^2 (\Lambda_{(4)}) / \partial x \partial y) + \Lambda_{L2} * (\partial^2 (\Lambda_{(3)}) / \partial x \partial y) + \Lambda_{L3} * (\partial^2 (\Lambda_{(2)})$$

$$/ \partial x \partial y) + \Lambda_{L4} * (\partial^2 (\Lambda_{(1)}) / \partial x \partial y) \}$$
(84)

$$(\partial^2 (N_L) / \partial y \partial x) = \{\Lambda_{L1} * (\partial^2 (\Lambda_{(4)}) / \partial y \partial x) + \Lambda_{L2} * (\partial^2 (\Lambda_{(3)}) / \partial y \partial x) + \Lambda_{L3} * (\partial^2 (\Lambda_{(2)})$$

$$/ \partial y \partial x) + \Lambda_{L4} * (\partial^2 (\Lambda_{(1)}) / \partial y \partial x) \}$$
(85)

$$(\partial^2 (N_L) / \partial x \partial z) = \{\Lambda_{L1} * (\partial^2 (\Lambda_{(4)}) / \partial x \partial z) + \Lambda_{L2} * (\partial^2 (\Lambda_{(3)}) / \partial x \partial z) + \Lambda_{L3} * (\partial^2 (\Lambda_{(2)}) / \partial x \partial z)$$

$$+ \Lambda_{L4} * (\partial^2 (\Lambda_{(1)}) / \partial x \partial z) \}$$
(86)

$$(\partial^2 (N_L) / \partial z \partial x) = \{\Lambda_{L1} * (\partial^2 (\Lambda_{(4)}) / \partial z \partial x) + \Lambda_{L2} * (\partial^2 (\Lambda_{(3)}) / \partial z \partial x) + \Lambda_{L3} * (\partial^2 (\Lambda_{(2)}) / \partial z \partial x)$$

$$+ \Lambda_{L4} * (\partial^2 (\Lambda_{(1)}) / \partial z \partial x) \}$$
(87)

$$(\partial^2 (N_L) / \partial y \partial z) = \{\Lambda_{L1} * (\partial^2 (\Lambda_{(4)}) / \partial y \partial z) + \Lambda_{L2} * (\partial^2 (\Lambda_{(3)}) / \partial y \partial z) + \Lambda_{L3} * (\partial^2 (\Lambda_{(2)}) / \partial y \partial z)$$

$$+ \Lambda_{L4} * (\partial^2 (\Lambda_{(1)}) / \partial y \partial z) \}$$
(88)

$$(\partial^2 (N_L) / \partial z \partial y) = \{\Lambda_{L1} * (\partial^2 (\Lambda_{(4)}) / \partial z \partial y) + \Lambda_{L2} * (\partial^2 (\Lambda_{(3)}) / \partial z \partial y) + \Lambda_{L3} * (\partial^2 (\Lambda_{(2)}) / \partial z \partial y)$$

$$+ \Lambda_{L4} * (\partial^2 (\Lambda_{(1)}) / \partial z \partial y) \}$$
(89)

It follows that:

$$(\partial (\varrho (x, y, z)) / \partial y) = \{ (\partial (N_L) / \partial y) * \Lambda_{LOO} - N_L * (\partial (\Lambda_{LOO}) / \partial y) \} / [\Lambda_{LOO}]^2$$
(90)

$$(\partial^2 (\varrho (x, y, z)) / \partial y^2) = \{ \{ (\partial^2 (N_L) / \partial y^2) * [\Lambda_{LOO}] + (\partial (N_L) / \partial y) * (\partial (\Lambda_{LOO}) / \partial y) - (\partial$$

$$(N_L) / \partial y) * (\partial (\Lambda_{LOO}) / \partial y) - N_L * (\partial^2 (\Lambda_{LOO}) / \partial y^2) \} * [\Lambda_{LOO}]^2 - \{ (\partial (N_L) / \partial y) * \Lambda_{LOO} -$$

$$N_L * (\partial (\Lambda_{LOO}) / \partial y) \} * \{ 2 * \Lambda_{LOO} * (\partial (\Lambda_{LOO}) / \partial y) \} \} / [\Lambda_{LOO}]^4 \tag{91}$$

$$(\partial (\varrho (x, y, z)) / \partial z) = \{ (\partial (N_L) / \partial z) * \Lambda_{LOO} - N_L * (\partial (\Lambda_{LOO}) / \partial z) \} / [\Lambda_{LOO}]^2 \tag{92}$$

$$(\partial^2 (\varrho (x, y, z)) / \partial z^2) = \{ \{ (\partial^2 (N_L) / \partial z^2) * [\Lambda_{LOO}] + (\partial (N_L) / \partial z) * (\partial (\Lambda_{LOO}) / \partial z) - (\partial$$

$$(N_L) / \partial z) * (\partial (\Lambda_{LOO}) / \partial z) - N_L * (\partial^2 (\Lambda_{LOO}) / \partial z^2) \} * [\Lambda_{LOO}]^2 - \{ (\partial (N_L) / \partial z) * \Lambda_{LOO} -$$

$$N_L * (\partial (\Lambda_{LOO}) / \partial z) \} * \{ 2 * \Lambda_{LOO} * (\partial (\Lambda_{LOO}) / \partial z) \} \} / [\Lambda_{LOO}]^4 \tag{93}$$

$$(\partial^2 (\varrho (x, y, z)) / \partial x \partial y) = \{ \{ (\partial (\partial (N_L) / \partial x) / \partial y) * [\Lambda_{LOO}] - (\partial (N_L) / \partial y) * (\partial (\Lambda_{LOO}) / \partial x)$$

$$+ (\partial (N_L) / \partial x) * (\partial (\Lambda_{LOO}) / \partial y) - N_L * (\partial^2 (\Lambda_{LOO}) / \partial x \partial y) \} * [\Lambda_{LOO}]^2 - \{ (\partial (N_L) / \partial x) *$$

$$\Lambda_{LOO} - N_L * (\partial (\Lambda_{LOO}) / \partial x) \} * \{ 2 * \Lambda_{LOO} * (\partial (\Lambda_{LOO}) / \partial y) \} \} / [\Lambda_{LOO}]^4 \tag{94}$$

$$(\partial^2 (\varrho (x, y, z)) / \partial y \partial x) = \{ \{ (\partial (\partial (N_L) / \partial y) / \partial x) * [\Lambda_{LOO}] - (\partial (N_L) / \partial x) * (\partial (\Lambda_{LOO}) / \partial y)$$

$$+ (\partial (N_L) / \partial y) * (\partial (\Lambda_{LOO}) / \partial x) - N_L * (\partial^2 (\Lambda_{LOO}) / \partial y \partial x) \} * [\Lambda_{LOO}]^2 - \{ (\partial (N_L) / \partial y) *$$

$$\Lambda_{LOO} - N_L * (\partial (\Lambda_{LOO}) / \partial y) \} * \{ 2 * \Lambda_{LOO} * (\partial (\Lambda_{LOO}) / \partial x) \} \} / [\Lambda_{LOO}]^4 \tag{95}$$

$$(\partial^2 (\varrho (x, y, z)) / \partial z \partial x) = \{ \{ (\partial (\partial (N_L) / \partial z) / \partial x) * [\Lambda_{LOO}] - (\partial (N_L) / \partial x) * (\partial (\Lambda_{LOO}) / \partial z)$$

$$+ (\partial (N_L) / \partial z) * (\partial (\Lambda_{LOO}) / \partial x) - N_L * (\partial^2 (\Lambda_{LOO}) / \partial z \partial x) \} * [\Lambda_{LOO}]^2 - \{ (\partial (N_L) / \partial z) *$$

$$\Lambda_{LOO} - N_L * (\partial (\Lambda_{LOO}) / \partial z) \} * \{ 2 * \Lambda_{LOO} * (\partial (\Lambda_{LOO}) / \partial x) \} \} / [\Lambda_{LOO}]^4 \tag{96}$$

$$(\partial^2 (\varrho (x, y, z)) / \partial x \partial z) = \{ \{ (\partial (\partial (N_L) / \partial x) / \partial z) * [\Lambda_{LOO}] - (\partial (N_L) / \partial z) * (\partial (\Lambda_{LOO}) / \partial x)$$

$$+ (\partial (N_L) / \partial x) * (\partial (\Lambda_{LOO}) / \partial z) - N_L * (\partial^2 (\Lambda_{LOO}) / \partial x \partial z) \} * [\Lambda_{LOO}]^2 - \{ (\partial (N_L) / \partial x) *$$

$$\Lambda_{LOO} - N_L * (\partial (\Lambda_{LOO}) / \partial x) \} * \{ 2 * \Lambda_{LOO} * (\partial (\Lambda_{LOO}) / \partial z) \} \} / [\Lambda_{LOO}]^4 \tag{97}$$

$$(\partial^2 (\varrho (x, y, z)) / \partial y \partial z) = \{ \{ (\partial (\partial (N_L) / \partial y) / \partial z) * [\Lambda_{LOO}] - (\partial (N_L) / \partial z) * (\partial (\Lambda_{LOO}) / \partial y)$$

$$+ (\partial (N_L) / \partial y) * (\partial (\Lambda_{LOO}) / \partial z) - N_L * (\partial^2 (\Lambda_{LOO}) / \partial y \partial z) \} * [\Lambda_{LOO}]^2 - \{ (\partial (N_L) / \partial y) *$$

$$\Lambda_{LOO} - N_L * (\partial (\Lambda_{LOO}) / \partial y) \} * \{ 2 * \Lambda_{LOO} * (\partial (\Lambda_{LOO}) / \partial z) \} \} / [\Lambda_{LOO}]^4 \tag{98}$$

$$(\partial^2 (\varrho (x, y, z)) / \partial z \partial y) = \{ \{ (\partial (\partial (N_L) / \partial z) / \partial y) * [\Lambda_{LOO}] - (\partial (N_L) / \partial y) * (\partial (\Lambda_{LOO}) / \partial z)$$

$$+ (\partial (N_L) / \partial z) * (\partial (\Lambda_{LOO}) / \partial y) - N_L * (\partial^2 (\Lambda_{LOO}) / \partial z \partial y) \} * [\Lambda_{LOO}]^2 - \{ (\partial (N_L) / \partial z) *$$

$$\Lambda_{LOO} - N_L * (\partial (\Lambda_{LOO}) / \partial z) \} * \{ 2 * \Lambda_{LOO} * (\partial (\Lambda_{LOO}) / \partial y) \} \} / [\Lambda_{LOO}]^4 \tag{99}$$

Calculation of the Total Curvature of the Resilient Signal

$$(\partial^2 (R((x, y, z), f)) / \partial x^2) = (\partial^2 (\varrho (x, y, z)) / \partial x^2) \tag{100}$$

$$(\partial^2 (R((x, y, z), f)) / \partial y^2) = (\partial^2 (\varrho (x, y, z)) / \partial y^2) \tag{101}$$

$$(\partial^2 (R((x, y, z), f)) / \partial z^2) = (\partial^2 (\varrho (x, y, z)) / \partial z^2) \tag{102}$$

$$(\partial (\partial (R((x, y, z), f)) / \partial y \partial x) = (\partial^2 (\varrho (x, y, z)) / \partial y \partial x) \tag{103}$$

$$(\partial (\partial (R((x, y, z), f)) / \partial x \partial y) = (\partial^2 (\varrho (x, y, z)) / \partial x \partial y) \tag{104}$$

$$(\partial (\partial (R((x, y, z), f)) / \partial y) \partial z) = (\partial^2 (\varrho (x, y, z)) / \partial y \partial z) \tag{105}$$

$$(\partial (\partial (R((x, y, z), f)) / \partial z) \partial y) = (\partial^2 (\varrho (x, y, z)) / \partial z \partial y) \tag{106}$$

$$(\partial (\partial (R((x, y, z), f)) / \partial x) \partial z) = (\partial^2 (\varrho (x, y, z)) / \partial x \partial z) \tag{107}$$

$$(\partial (\partial (R((x, y, z), f)) / \partial z) \partial x) = (\partial^2 (\varrho (x, y, z)) / \partial z \partial x) \tag{108}$$

The total curvature $C(R((x, y, z), f))$ is:

$$C(R((x, y, z), f)) = (\partial^2 (R((x, y, z), f)) / \partial x^2) + (\partial^2 (R((x, y, z), f)) / \partial y^2) + (\partial^2 (R((x, y,$$

$$z), f)) / \partial z^2) + (\partial (\partial (R((x, y, z), f)) / \partial y) \partial x) + (\partial (\partial (R((x, y, z), f)) / \partial x) \partial y) + (\partial (\partial (R((x,$$

$$y, z), f)) / \partial y) \partial z) + (\partial (\partial (R((x, y, z), f)) / \partial z) \partial y) + (\partial (\partial (R((x, y, z), f)) / \partial z) \partial x) + (\partial (\partial ($$

$$R((x, y, z), f)) / \partial x) \partial z) \tag{109}$$

TRIVARIATE CUBIC LAGRANGE CALCULATED WITHOUT EMBEDDING

Equation (20) of chapter 3 part IV is reported here:

$$f(0, 0, 0) = \Lambda_L / \{ 9 xyz * \{ - 2 \omega_1 + 2 \omega_2 \} \} \tag{110}$$

$$\Lambda_L = 9 * \{ [(3/2) \omega_1^2 - \omega_1 \omega_2 + (1/6) \omega_2^2] \Pi^{(4)} + [- 4 \omega_1^2 + (16/3) \omega_1 \omega_2 - (8/6) \omega_2^2] \Pi^{(3)}$$

$$+ [1/2 \omega_1^2 - (54/6) \omega_1 \omega_2 + (23/6) \omega_2^2] \Pi^{(2)} + [4 \omega_1^2 + (28/6) \omega_1 \omega_2 - (28/6) \omega_2^2] \Pi^{(1)} +$$

$$[- 2 \omega_1^2 + 2 \omega_2^2] xyz \} \tag{111}$$

SIGNAL RESILIENT TO INTERPOLATION

In view of equations (9) through (12) and equations (52) through (56), let us posit:

$\Lambda_{L6} = \{ -2\,\omega_1 + 2\,\omega_2 \}$ so that equation (110) is:

$$f(0,0,0) = 9 * xyz \{ \Lambda_{L1} * \Lambda_{(4)} + \Lambda_{L2} * \Lambda_{(3)} + \Lambda_{L3} * \Lambda_{(2)} + \Lambda_{L4} * \Lambda_{(1)} + \Lambda_{L5} \} / \{ 9\,xyz * \Lambda_{L6} \} \quad (112)$$

The formulation $R(x, y, z) = f(0, 0, 0) + \varrho(x, y, z)$ requires that from equation (112):

$$\varrho(x, y, z) = 9 * xyz \{ \Lambda_{L1} * \Lambda_{(4)} + \Lambda_{L2} * \Lambda_{(3)} + \Lambda_{L3} * \Lambda_{(2)} + \Lambda_{L4} * \Lambda_{(1)} + \Lambda_{L5} \} / \{ 9\,xyz *$$

$$\Lambda_{L6} \} = \{ \Lambda_{L6} \}^{-1} * \{ \Lambda_{L1} * \Lambda_{(4)} + \Lambda_{L2} * \Lambda_{(3)} + \Lambda_{L3} * \Lambda_{(2)} + \Lambda_{L4} * \Lambda_{(1)} + \Lambda_{L5} \} \quad (113)$$

Given that Λ_{L6} is constant the calculation of the total curvature of the resilient signal is:

$$C(R((x, y, z), f)) = \{ \Lambda_{L6} \}^{-1} * \{ \Lambda_{L1} * \{ (\partial^2 (\Lambda_{(4)}) / \partial x^2) + (\partial^2 (\Lambda_{(4)}) / \partial y^2) + (\partial^2 (\Lambda_{(4)}) / \partial z^2) +$$

$$(\partial^2 (\Lambda_{(4)}) / \partial x \partial y) + (\partial^2 (\Lambda_{(4)}) / \partial y \partial x) + (\partial^2 (\Lambda_{(4)}) / \partial x \partial z) + (\partial^2 (\Lambda_{(4)}) / \partial z \partial x) + (\partial^2 (\Lambda_{(4)}) / \partial y \partial z)$$

$$+ (\partial^2 (\Lambda_{(4)}) / \partial z \partial y) \} + \Lambda_{L2} * \{ (\partial^2 (\Lambda_{(3)}) / \partial x^2) + (\partial^2 (\Lambda_{(3)}) / \partial y^2) + (\partial^2 (\Lambda_{(3)}) / \partial z^2) + (\partial^2 (\Lambda_{(3)})$$

$$/ \partial x \partial y) + (\partial^2 (\Lambda_{(3)}) / \partial y \partial x) + (\partial^2 (\Lambda_{(3)}) / \partial x \partial z) + (\partial^2 (\Lambda_{(3)}) / \partial z \partial x) + (\partial^2 (\Lambda_{(3)}) / \partial y \partial z) + (\partial^2$$

$$(\Lambda_{(3)}) / \partial z \partial y) \} + \Lambda_{L3} * \{ (\partial^2 (\Lambda_{(2)}) / \partial x^2) + (\partial^2 (\Lambda_{(2)}) / \partial y^2) + (\partial^2 (\Lambda_{(2)}) / \partial z^2) + (\partial^2 (\Lambda_{(2)})$$

$$/ \partial x \partial y) + (\partial^2 (\Lambda_{(2)}) / \partial y \partial x) + (\partial^2 (\Lambda_{(2)}) / \partial x \partial z) + (\partial^2 (\Lambda_{(2)}) / \partial z \partial x) + (\partial^2 (\Lambda_{(2)}) / \partial y \partial z) + (\partial^2$$

$$(\Lambda_{(2)}) / \partial z \partial y) \} + \Lambda_{L4} * \{ (\partial^2 (\Lambda_{(1)}) / \partial x^2) + (\partial^2 (\Lambda_{(1)}) / \partial y^2) + (\partial^2 (\Lambda_{(1)}) / \partial z^2) + (\partial^2 (\Lambda_{(1)}))$$

$$/ \partial x \partial y) + (\partial^2 (\Lambda_{(1)}) / \partial y \partial x) + (\partial^2 (\Lambda_{(1)}) / \partial x \partial z) + (\partial^2 (\Lambda_{(1)}) / \partial z \partial x) + (\partial^2 (\Lambda_{(1)}) / \partial y \partial z) +$$

$$(\partial^2 (\Lambda_{(1)}) / \partial z \partial y) \} \} \quad (114)$$

where the second order partial derivatives of $\Lambda_{(1)}$, $\Lambda_{(2)}$, $\Lambda_{(3)}$ and $\Lambda_{(4)}$ are calculated in the math developments of equations (13) through (50).

TRIVARIATE CUBIC B-SPLINE CALCULATED WITH EMBEDDING

Equations (55), (56), (57) and (58) of chapter 2 part III are reported in the section: "TRIVARIATE CUBIC LAGRANGE CALCULATED WITH EMBEDDING" and numbered as equations (1), (2), (3) and (4). Equation (63) in chapter 2 part III gives the math form of the signal resilient to the cubic B-Spline interpolation function as:

$$f(0, 0, 0) = \{ [3/2\,\omega_1^2 - 1/2\,\omega_1\omega_2 - 3/6\,\omega_2\omega_1 + 1/6\,\omega_2^2] \Pi^{(4)} + [-4\,\omega_1^2 + 2\,\omega_1\omega_2 + 4/3\,\omega_2\omega_1$$

$$- 4/3\,\omega_2^2] \Pi^{(3)} + [2\,\omega_1^2 - 2\,\omega_1\omega_2 - 8\,\omega_2\omega_1 + 4\,\omega_2^2] \Pi^{(2)} + [2\,\omega_1^2 - 2/3\,\omega_1\omega_2 + 8\,\omega_2\omega_1 -$$

$$16/3\,\omega_2^2] \Pi^{(1)} + [-4/3\,\omega_1^2 + 4/3\,\omega_1\omega_2 - 8/3\,\omega_2\omega_1 + 8/3\,\omega_2^2] xyz \} / \{ xyz * [-2\,\omega_1 + 2\,\omega_2$$

$$] - \{ \omega_1 * [3 \, \Pi^{(1)} - 2 \, xyz] + \omega_2 * [- \Pi^{(1)} + 2 \, xyz] \} \} \tag{115}$$

In view of equations (9) through (12), considering the position: $\Lambda_{L6} = \{ - 2 \, \omega_1 + 2 \, \omega_2 \}$, and positing:

$$\Delta_{L1} = [3/2 \, \omega_1^2 - 1/2 \, \omega_1 \omega_2 - 3/6 \, \omega_2 \omega_1 + 1/6 \, \omega_2^2] \tag{116}$$

$$\Delta_{L2} = [- 4 \, \omega_1^2 + 2 \, \omega_1 \, \omega_2 + 4/3 \, \omega_2 \, \omega_1 - 4/3 \, \omega_2^2] \tag{117}$$

$$\Delta_{L3} = [2 \, \omega_1^2 - 2 \, \omega_1 \, \omega_2 - 8 \, \omega_2 \, \omega_1 + 4 \, \omega_2^2] \tag{118}$$

$$\Delta_{L4} = [2 \, \omega_1^2 - 2/3 \, \omega_1 \, \omega_2 + 8 \, \omega_2 \, \omega_1 - 16/3 \, \omega_2^2] \tag{119}$$

$$\Delta_{L5} = [- 4/3 \, \omega_1^2 + 4/3 \, \omega_1 \, \omega_2 - 8/3 \, \omega_2 \, \omega_1 + 8/3 \, \omega_2^2] \tag{120}$$

Also, considering that from equation (60):

$$\Lambda_{LOO} = \{ (- 2 \, \omega_1 + 2 \, \omega_2) - \omega_1 (3 \, \Lambda_{(1)} - 2) + \omega_2 (2 - \Lambda_{(1)}) \} =$$

$$\{ \Lambda_{L6} - \omega_1 * (3 \, \Lambda_{(1)} - 2) + \omega_2 * (2 - \Lambda_{(1)}) \}, \text{ equation (115) can be written as:}$$

$$f (0, 0, 0) = \{ \Delta_{L1} * \Lambda_{(4)} + \Delta_{L2} * \Lambda_{(3)} + \Delta_{L3} * \Lambda_{(2)} + \Delta_{L4} * \Lambda_{(1)} + \Delta_{L5} \} / \{ \Lambda_{L6} - \{ \omega_1 * [3 \, \Lambda_{(1)} $$

$$- 2] + \omega_2 * [- \Lambda_{(1)} + 2] \} \} = \{ \Delta_{L1} * \Lambda_{(4)} + \Delta_{L2} * \Lambda_{(3)} + \Delta_{L3} * \Lambda_{(2)} + \Delta_{L4} * \Lambda_{(1)} + \Delta_{L5} \} / \{$$

$$\Lambda_{LOO} \} \tag{121}$$

The formulation $R (x, y, z) = f (0, 0, 0) + \varrho (x, y, z)$ requires that:

$$\varrho (x, y, z) = \{ \Delta_{L1} * \Lambda_{(4)} + \Delta_{L2} * \Lambda_{(3)} + \Delta_{L3} * \Lambda_{(2)} + \Delta_{L4} * \Lambda_{(1)} + \Delta_{L5} \} / \{ \Lambda_{LOO} \} \tag{122}$$

Let us posit:

$$\Omega_L = \{ \Delta_{L1} * \Lambda_{(4)} + \Delta_{L2} * \Lambda_{(3)} + \Delta_{L3} * \Lambda_{(2)} + \Delta_{L4} * \Lambda_{(1)} + \Delta_{L5} \} \tag{123}$$

Keeping in mind that Δ_{L5} is constant, the calculation of the first and second order partial derivatives of Ω_L furnishes the following set of equations:

$$(\partial (\Omega_L) / \partial x) = \{ \Delta_{L1} * (\partial (\Lambda_{(4)}) / \partial x) + \Delta_{L2} * (\partial (\Lambda_{(3)}) / \partial x) + \Delta_{L3} * (\partial (\Lambda_{(2)}) / \partial x) + \Delta_{L4} * (\partial$$

$$(\Lambda_{(1)}) / \partial x) \} \tag{124}$$

$$(\partial^2 (\Omega_L) / \partial x^2) = \{ \Delta_{L1} * (\partial^2 (\Lambda_{(4)}) / \partial x^2) + \Delta_{L2} * (\partial^2 (\Lambda_{(3)}) / \partial x^2) + \Delta_{L3} * (\partial^2 (\Lambda_{(2)}) / \partial x^2) + \Delta_{L4}$$

$$* (\partial^2 (\Lambda_{(1)}) / \partial x^2) \} \tag{125}$$

$$(\partial (\Omega_L) / \partial y) = \{ \Delta_{L1} * (\partial (\Lambda_{(4)}) / \partial y) + \Delta_{L2} * (\partial (\Lambda_{(3)}) / \partial y) + \Delta_{L3} * (\partial (\Lambda_{(2)}) / \partial y) + \Delta_{L4} * (\partial$$

$$(\Lambda_{(1)}) / \partial y) \} \tag{126}$$

$$(\partial^2 (\Omega_L) / \partial y^2) = \{ \Delta_{L1} * (\partial^2 (\Lambda_{(4)}) / \partial y^2) + \Delta_{L2} * (\partial^2 (\Lambda_{(3)}) / \partial y^2) + \Delta_{L3} * (\partial^2 (\Lambda_{(2)}) / \partial y^2) + \Delta_{L4}$$

$$* (\partial^2 (\Lambda_{(1)}) / \partial y^2) \} \tag{127}$$

$$(\partial (\Omega_L) / \partial z) = \{ \Delta_{L1} * (\partial (\Lambda_{(4)}) / \partial z) + \Delta_{L2} * (\partial (\Lambda_{(3)}) / \partial z) + \Delta_{L3} * (\partial (\Lambda_{(2)}) / \partial z) + \Delta_{L4} * (\partial$$

$$(\Lambda_{(1)}) / \partial z) \} \tag{128}$$

$$(\partial^2 (\Omega_L) / \partial z^2) = \{ \Delta_{L1} * (\partial^2 (\Lambda_{(4)}) / \partial z^2) + \Delta_{L2} * (\partial^2 (\Lambda_{(3)}) / \partial z^2) + \Delta_{L3} * (\partial^2 (\Lambda_{(2)}) / \partial z^2) + \Delta_{L4}$$

$$* (\partial^2 (\Lambda_{(1)}) / \partial z^2) \} \tag{129}$$

$$(\partial^2 (\Omega_L) / \partial x \partial y) = \{ \Delta_{L1} * (\partial^2 (\Lambda_{(4)}) / \partial x \partial y) + \Delta_{L2} * (\partial^2 (\Lambda_{(3)}) / \partial x \partial y) + \Delta_{L3} * (\partial^2 (\Lambda_{(2)}) / \partial x \partial y)$$

$$+ \Delta_{L4} * (\partial^2 (\Lambda_{(1)}) / \partial x \partial y) \} \tag{130}$$

$$(\partial^2 (\Omega_L) / \partial y \partial x) = \{ \Delta_{L1} * (\partial^2 (\Lambda_{(4)}) / \partial y \partial x) + \Delta_{L2} * (\partial^2 (\Lambda_{(3)}) / \partial y \partial x) + \Delta_{L3} * (\partial^2 (\Lambda_{(2)}) / \partial y \partial x)$$

$$+ \Delta_{L4} * (\partial^2 (\Lambda_{(1)}) / \partial y \partial x) \} \tag{131}$$

$$(\partial^2 (\Omega_L) / \partial x \partial z) = \{ \Delta_{L1} * (\partial^2 (\Lambda_{(4)}) / \partial x \partial z) + \Delta_{L2} * (\partial^2 (\Lambda_{(3)}) / \partial x \partial z) + \Delta_{L3} * (\partial^2 (\Lambda_{(2)}) / \partial x \partial z)$$

$$+ \Delta_{L4} * (\partial^2 (\Lambda_{(1)}) / \partial x \partial z) \} \tag{132}$$

$$(\partial^2 (\Omega_L) / \partial z \partial x) = \{ \Delta_{L1} * (\partial^2 (\Lambda_{(4)}) / \partial z \partial x) + \Delta_{L2} * (\partial^2 (\Lambda_{(3)}) / \partial z \partial x) + \Delta_{L3} * (\partial^2 (\Lambda_{(2)}) / \partial z \partial x)$$

$$+ \Delta_{L4} * (\partial^2 (\Lambda_{(1)}) / \partial z \partial x) \} \tag{133}$$

$$(\partial^2 (\Omega_L) / \partial y \partial z) = \{ \Delta_{L1} * (\partial^2 (\Lambda_{(4)}) / \partial y \partial z) + \Delta_{L2} * (\partial^2 (\Lambda_{(3)}) / \partial y \partial z) + \Delta_{L3} * (\partial^2 (\Lambda_{(2)}) / \partial y \partial z)$$

$$+ \Delta_{L4} * (\partial^2 (\Lambda_{(1)}) / \partial y \partial z) \} \tag{134}$$

$$(\partial^2 (\Omega_L) / \partial z \partial y) = \{ \Delta_{L1} * (\partial^2 (\Lambda_{(4)}) / \partial z \partial y) + \Delta_{L2} * (\partial^2 (\Lambda_{(3)}) / \partial z \partial y) + \Delta_{L3} * (\partial^2 (\Lambda_{(2)}) / \partial z \partial y)$$

$$+ \Delta_{L4} * (\partial^2 (\Lambda_{(1)}) / \partial z \partial y) \} \tag{135}$$

It follows that:

$$(\partial (\varrho (x, y, z)) / \partial x) = \{ (\partial (\Omega_L) / \partial x) * \Lambda_{LOO} - \Omega_L * (\partial (\Lambda_{LOO}) / \partial x) \} / [\Lambda_{LOO}]^2 \tag{136}$$

$$(\partial^2 (\varrho (x, y, z)) / \partial x^2) = \{ \{ (\partial^2 (\Omega_L) / \partial x^2) * [\Lambda_{LOO}] + (\partial (\Omega_L) / \partial x) * (\partial (\Lambda_{LOO}) / \partial x) - (\partial$$

$$(\Omega_L) / \partial x) * (\partial (\Lambda_{LOO}) / \partial x) - \Omega_L * (\partial^2 (\Lambda_{LOO}) / \partial x^2) \} * [\Lambda_{LOO}]^2 - \{ (\partial (\Omega_L) / \partial x) * \Lambda_{LOO} -$$

$$\Omega_L * (\partial (\Lambda_{LOO}) / \partial x) \} * \{ 2 * \Lambda_{LOO} * (\partial (\Lambda_{LOO}) / \partial x) \} \} / [\Lambda_{LOO}]^4 \tag{137}$$

$$(\partial (\varrho (x, y, z)) / \partial y) = \{ (\partial (\Omega_L) / \partial y) * \Lambda_{LOO} - \Omega_L * (\partial (\Lambda_{LOO}) / \partial y) \} / [\Lambda_{LOO}]^2 \tag{138}$$

$$(\partial^2 (\varrho\,(x, y, z))\,/\partial y^2) = \{\,\{\,(\partial^2\,(\Omega_L)\,/\partial y^2)*[\Lambda_{LOO}] + (\partial\,(\Omega_L)\,/\partial y)*(\partial\,(\Lambda_{LOO})\,/\partial y) - (\partial\,(\Omega_L)\,/\partial y)*(\partial\,(\Lambda_{LOO})\,/\partial y) - \Omega_L*(\partial^2\,(\Lambda_{LOO})\,/\partial y^2)\,\}*[\Lambda_{LOO}]^2 - \{\,(\partial\,(\Omega_L)\,/\partial y)*\Lambda_{LOO} - \Omega_L*(\partial\,(\Lambda_{LOO})\,/\partial y)\,\}*\{\,2*\Lambda_{LOO}*(\partial\,(\Lambda_{LOO})\,/\partial y)\,\}\,\}\,/\,[\Lambda_{LOO}]^4 \tag{139}$$

$$(\partial\,(\varrho\,(x, y, z))\,/\partial z) = \{\,(\partial\,(\Omega_L)\,/\partial z)*\Lambda_{LOO} - \Omega_L*(\partial\,(\Lambda_{LOO})\,/\partial z)\,\}\,/\,[\Lambda_{LOO}]^2 \tag{140}$$

$$(\partial^2 (\varrho\,(x, y, z))\,/\partial z^2) = \{\,\{\,(\partial^2\,(\Omega_L)\,/\partial z^2)*[\Lambda_{LOO}] + (\partial\,(\Omega_L)\,/\partial z)*(\partial\,(\Lambda_{LOO})\,/\partial z) - (\partial\,(\Omega_L)\,/\partial z)*(\partial\,(\Lambda_{LOO})\,/\partial z) - \Omega_L*(\partial^2\,(\Lambda_{LOO})\,/\partial z^2)\,\}*[\Lambda_{LOO}]^2 - \{\,(\partial\,(\Omega_L)\,/\partial z)*\Lambda_{LOO} - \Omega_L*(\partial\,(\Lambda_{LOO})\,/\partial z)\,\}*\{\,2*\Lambda_{LOO}*(\partial\,(\Lambda_{LOO})\,/\partial z)\,\}\,\}\,/\,[\Lambda_{LOO}]^4 \tag{141}$$

$$(\partial^2 (\varrho\,(x, y, z))\,/\partial x\partial y) = \{\,\{\,(\partial\,(\partial\,(\Omega_L)\,/\partial x)\,/\partial y)*[\Lambda_{LOO}] - (\partial\,(\Omega_L)\,/\partial y)*(\partial\,(\Lambda_{LOO})\,/\partial x) + (\partial\,(\Omega_L)\,/\partial x)*(\partial\,(\Lambda_{LOO})\,/\partial y) - \Omega_L*(\partial^2\,(\Lambda_{LOO})\,/\partial x\partial y)\,\}*[\Lambda_{LOO}]^2 - \{\,(\partial\,(\Omega_L)\,/\partial x)*\Lambda_{LOO} - \Omega_L*(\partial\,(\Lambda_{LOO})\,/\partial x)\,\}*\{\,2*\Lambda_{LOO}*(\partial\,(\Lambda_{LOO})\,/\partial y)\,\}\,\}\,/\,[\Lambda_{LOO}]^4 \tag{142}$$

$$(\partial^2 (\varrho\,(x, y, z))\,/\partial y\partial x) = \{\,\{\,(\partial\,(\partial\,(\Omega_L)\,/\partial y)\,/\partial x)*[\Lambda_{LOO}] - (\partial\,(\Omega_L)\,/\partial x)*(\partial\,(\Lambda_{LOO})\,/\partial y) + (\partial\,(\Omega_L)\,/\partial y)*(\partial\,(\Lambda_{LOO})\,/\partial x) - \Omega_L*(\partial^2\,(\Lambda_{LOO})\,/\partial y\partial x)\,\}*[\Lambda_{LOO}]^2 - \{\,(\partial\,(\Omega_L)\,/\partial y)*\Lambda_{LOO} - \Omega_L*(\partial\,(\Lambda_{LOO})\,/\partial y)\,\}*\{\,2*\Lambda_{LOO}*(\partial\,(\Lambda_{LOO})\,/\partial x)\,\}\,\}\,/\,[\Lambda_{LOO}]^4 \tag{143}$$

$$(\partial^2 (\varrho\,(x, y, z))\,/\partial z\partial x) = \{\,\{\,(\partial\,(\partial\,(\Omega_L)\,/\partial z)\,/\partial x)*[\Lambda_{LOO}] - (\partial\,(\Omega_L)\,/\partial x)*(\partial\,(\Lambda_{LOO})\,/\partial z) + (\partial\,(\Omega_L)\,/\partial z)*(\partial\,(\Lambda_{LOO})\,/\partial x) - \Omega_L*(\partial^2\,(\Lambda_{LOO})\,/\partial z\partial x)\,\}*[\Lambda_{LOO}]^2 - \{\,(\partial\,(\Omega_L)\,/\partial z)*\Lambda_{LOO} - \Omega_L*(\partial\,(\Lambda_{LOO})\,/\partial z)\,\}*\{\,2*\Lambda_{LOO}*(\partial\,(\Lambda_{LOO})\,/\partial x)\,\}\,\}\,/\,[\Lambda_{LOO}]^4 \tag{144}$$

$$(\partial^2 (\varrho\,(x, y, z))\,/\partial x\partial z) = \{\,\{\,(\partial\,(\partial\,(\Omega_L)\,/\partial x)\,/\partial z)*[\Lambda_{LOO}] - (\partial\,(\Omega_L)\,/\partial z)*(\partial\,(\Lambda_{LOO})\,/\partial x) + (\partial\,(\Omega_L)\,/\partial x)*(\partial\,(\Lambda_{LOO})\,/\partial z) - \Omega_L*(\partial^2\,(\Lambda_{LOO})\,/\partial x\partial z)\,\}*[\Lambda_{LOO}]^2 - \{\,(\partial\,(\Omega_L)\,/\partial x)*\Lambda_{LOO} - \Omega_L*(\partial\,(\Lambda_{LOO})\,/\partial x)\,\}*\{\,2*\Lambda_{LOO}*(\partial\,(\Lambda_{LOO})\,/\partial z)\,\}\,\}\,/\,[\Lambda_{LOO}]^4 \tag{145}$$

$$(\partial^2 (\varrho\,(x, y, z))\,/\partial y\partial z) = \{\,\{\,(\partial\,(\partial\,(\Omega_L)\,/\partial y)\,/\partial z)*[\Lambda_{LOO}] - (\partial\,(\Omega_L)\,/\partial z)*(\partial\,(\Lambda_{LOO})\,/\partial y) + (\partial\,(\Omega_L)\,/\partial y)*(\partial\,(\Lambda_{LOO})\,/\partial z) - \Omega_L*(\partial^2\,(\Lambda_{LOO})\,/\partial y\partial z)\,\}*[\Lambda_{LOO}]^2 - \{\,(\partial\,(\Omega_L)\,/\partial y)*$$

$$\Lambda_{\text{LOO}} - \Omega_L * (\partial\,(\Lambda_{\text{LOO}})\,/\partial y)\,\}\,*\,\{\,2\,*\,\Lambda_{\text{LOO}}\,*\,(\partial\,(\Lambda_{\text{LOO}})\,/\partial z)\,\}\,\}\,/\,[\,\Lambda_{\text{LOO}}\,]^4 \tag{146}$$

$$(\partial^2\,(\varrho\,(x,\,y,\,z))\,/\partial z \partial y) = \{\,\{\,(\partial\,(\partial\,(\Omega_L)\,/\partial z)\,/\partial y)\,*\,[\,\Lambda_{\text{LOO}}\,] - (\partial\,(\Omega_L)\,/\partial y)\,*\,(\partial\,(\Lambda_{\text{LOO}})\,/\partial z)$$

$$+ (\partial\,(\Omega_L)\,/\partial z)\,*\,(\partial\,(\Lambda_{\text{LOO}})\,/\partial y) - \Omega_L\,*\,(\partial^2\,(\Lambda_{\text{LOO}})\,/\partial z \partial y)\,\}\,*\,[\,\Lambda_{\text{LOO}}\,]^2 - \{\,(\partial\,(\Omega_L)\,/\partial z)\,*$$

$$\Lambda_{\text{LOO}} - \Omega_L\,*\,(\partial\,(\Lambda_{\text{LOO}})\,/\partial z)\,\}\,*\,\{\,2\,*\,\Lambda_{\text{LOO}}\,*\,(\partial\,(\Lambda_{\text{LOO}})\,/\partial y)\,\}\,\}\,/\,[\,\Lambda_{\text{LOO}}\,]^4 \tag{147}$$

Using the knowledge resulting from equations (100) through (108), the total curvature $C(\,R((x,\,y,\,z),\,f)\,)$ is:

$$C(\,R((x,\,y,\,z),\,f)\,) = (\partial^2\,(\,R((x,\,y,\,z),\,f)\,)\,/\partial x^2) + (\partial^2\,(\,R((x,\,y,\,z),\,f)\,)\,/\partial y^2) + (\partial^2\,(\,R((x,\,y,$$

$$z),\,f)\,)\,/\partial z^2) + (\partial\,(\partial\,(\,R((x,\,y,\,z),\,f)\,)\,/\partial y)\partial x) + (\partial\,(\partial\,(\,R((x,\,y,\,z),\,f)\,)\,/\partial x)\partial y) + (\partial\,(\partial\,(\,R((x,$$

$$y,\,z),\,f)\,)\,/\partial y)\partial z) + (\partial\,(\partial\,(\,R((x,\,y,\,z),\,f)\,)\,/\partial z)\partial y) + (\partial\,(\partial\,(\,R((x,\,y,\,z),\,f)\,)\,/\partial z)\partial x) + (\partial\,(\partial\,($$

$$R((x,\,y,\,z),\,f)\,)\,/\partial x)\partial z) \tag{148}$$

TRIVARIATE CUBIC B-SPLINE CALCULATED WITHOUT EMBEDDING

Equations (29), (28), (24) and (25) of chapter 3 part III are reported in the section: "TRIVARIATE CUBIC LAGRANGE CALCULATED WITH EMBEDDING" and numbered in this chapter as equations (1), (2), (3) and (4), and using the positions:

$$\Pi^{(1)} = x\,y\,z\,*\,\Lambda_{(1)},\,\Pi^{(2)} = x\,y\,z\,*\,\Lambda_{(2)}\,,\,\Pi^{(3)} = x\,y\,z\,*\,\Lambda_{(3)},\,\Pi^{(4)} = x\,y\,z\,*\,\Lambda_{(4)}.$$

The signal resilient to the cubic trivariate B-Spline interpolation function is given in equation (35) in chapter 3 part III in the following math form:

$$f\,(0,\,0,\,0) = \Lambda_\gamma\,/\,\{\,xyz\,*\,[-2\,\omega_1 + 2\,\omega_2]\,\} \tag{149}$$

$$\Lambda_\gamma = \{\,[3/2\,\omega_1^2 - 1/2\,\omega_1\omega_2 - 3/6\,\omega_2\,\omega_1 + 1/6\,\omega_2^2]\,\Pi^{(4)} + [-4\,\omega_1^2 + 2\,\omega_1\,\omega_2 + 4/3\,\omega_2\,\omega_1 - 4/3$$

$$\omega_2^2]\,\Pi^{(3)} + [\,2\,\omega_1^2 - 2\,\omega_1\,\omega_2 - 8\,\omega_2\,\omega_1 + 4\,\omega_2^2]\,\Pi^{(2)} + [2\,\omega_1^2 - 2/3\,\omega_1\,\omega_2 + 8\,\omega_2\,\omega_1 - 16/3$$

$$\omega_2^2\,]\,\Pi^{(1)} + [-4/3\,\omega_1^2 + 4/3\,\omega_1\,\omega_2 - 8/3\,\omega_2\,\omega_1 + 8/3\,\omega_2^2]\,xyz\,\} \tag{150}$$

In view of equations (9), (10), (11) and (12), is true that:

$$\Pi^{(1)} = x\,y\,z\,*\,\Lambda_{(1)}\,;\,\Pi^{(2)} = x\,y\,z\,*\,\Lambda_{(2)}\,;\,\Pi^{(3)} = x\,y\,z\,*\,\Lambda_{(3)}\,;\,\Pi^{(4)} = x\,y\,z\,*\,\Lambda_{(4)}.$$

Thus equation (150) becomes:

$$\Lambda_\gamma = \{\,\{\,[3/2\,\omega_1^2 - 1/2\,\omega_1\omega_2 - 3/6\,\omega_2\,\omega_1 + 1/6\,\omega_2^2]\,\Lambda_{(4)} + [-4\,\omega_1^2 + 2\,\omega_1\,\omega_2 + 4/3\,\omega_2\,\omega_1 - 4/3$$

$$\omega_2^2]\,\Lambda_{(3)} + [\,2\,\omega_1^2 - 2\,\omega_1\,\omega_2 - 8\,\omega_2\,\omega_1 + 4\,\omega_2^2]\,\Lambda_{(2)} + [2\,\omega_1^2 - 2/3\,\omega_1\,\omega_2 + 8\,\omega_2\,\omega_1 - 16/3\,\omega_2^2$$

$$] \Lambda_{(1)} + [- 4/3 \omega_1{}^2 + 4/3 \omega_1 \omega_2 - 8/3 \omega_2 \omega_1 + 8/3 \omega_2{}^2] \} * xyz \} = \{ \Lambda_\tau * xyz \} \tag{151}$$

Where:

$$\Lambda_\tau = \{ [3/2 \omega_1{}^2 - 1/2 \omega_1 \omega_2 - 3/6 \omega_2 \omega_1 + 1/6 \omega_2{}^2] \Lambda_{(4)} + [- 4 \omega_1{}^2 + 2 \omega_1 \omega_2 + 4/3 \omega_2 \omega_1 - 4/3$$

$$\omega_2{}^2] \Lambda_{(3)} + [2 \omega_1{}^2 - 2 \omega_1 \omega_2 - 8 \omega_2 \omega_1 + 4 \omega_2{}^2] \Lambda_{(2)} + [2 \omega_1{}^2 - 2/3 \omega_1 \omega_2 + 8 \omega_2 \omega_1 - 16/3 \omega_2{}^2$$

$$] \Lambda_{(1)} + [- 4/3 \omega_1{}^2 + 4/3 \omega_1 \omega_2 - 8/3 \omega_2 \omega_1 + 8/3 \omega_2{}^2] \} \tag{152}$$

Equation (149) is now written as:

$$f (0, 0, 0) = \Lambda_\tau / \{ [- 2 \omega_1 + 2 \omega_2] \} \tag{153}$$

From equations (116), (117), (118), (119) and (120), and the position: $\Lambda_{L6} = \{ - 2 \omega_1 + 2 \omega_2 \}$ equation (153) is written as:

$$f (0, 0, 0) = \{ \Lambda_{L6} \}^{-1} * \{ \Delta_{I1} * \Lambda_{(4)} + \Delta_{L2} * \Lambda_{(3)} + \Delta_{L3} * \Lambda_{(2)} + \Delta_{L4} * \Lambda_{(1)} + \Delta_{L5} \} \tag{154}$$

The formulation $R (x, y, z) = f (0, 0, 0) + \varrho (x, y, z)$ requires that:

$$\varrho (x, y, z) = \{ \Lambda_{L6} \}^{-1} * \{ \Delta_{I1} * \Lambda_{(4)} + \Delta_{L2} * \Lambda_{(3)} + \Delta_{L3} * \Lambda_{(2)} + \Delta_{L4} * \Lambda_{(1)} + \Delta_{L5} \} \tag{155}$$

Calculation of the Total Curvature of the Resilient Signal

The total curvature is calculated just as in equation (114) with the exception that are found the terms $\Delta_{L1}, \Delta_{L2}, \Delta_{L3}, \Delta_{L4},$ of equations (116) through (119) are found in place of the terms $\Lambda_{L1}, \Lambda_{L2}, \Lambda_{L3}, \Lambda_{L4}$. It is worth noting that

$\Delta_{L5} = [- 4/3 \omega_1{}^2 + 4/3 \omega_1 \omega_2 - 8/3 \omega_2 \omega_1 + 8/3 \omega_2{}^2]$ of equation (120) is constant and thus its derivatives are zero.

$$C(R((x, y, z), f)) = \{ \Lambda_{L6} \}^{-1} * \{ \Delta_{L1} * \{ (\partial^2 (\Lambda_{(4)}) / \partial x^2) + (\partial^2 (\Lambda_{(4)}) / \partial y^2) + (\partial^2 (\Lambda_{(4)}) / \partial z^2) +$$

$$(\partial^2 (\Lambda_{(4)}) / \partial x \partial y) + (\partial^2 (\Lambda_{(4)}) / \partial y \partial x) + (\partial^2 (\Lambda_{(4)}) / \partial x \partial z) + (\partial^2 (\Lambda_{(4)}) / \partial z \partial x) + (\partial^2 (\Lambda_{(4)}) / \partial y \partial z)$$

$$+ (\partial^2 (\Lambda_{(4)}) / \partial z \partial y) \} + \Delta_{L2} * \{ (\partial^2 (\Lambda_{(3)}) / \partial x^2) + (\partial^2 (\Lambda_{(3)}) / \partial y^2) + (\partial^2 (\Lambda_{(3)}) / \partial z^2) + (\partial^2 (\Lambda_{(3)})$$

$$/ \partial x \partial y) + (\partial^2 (\Lambda_{(3)}) / \partial y \partial x) + (\partial^2 (\Lambda_{(3)}) / \partial x \partial z) + (\partial^2 (\Lambda_{(3)}) / \partial z \partial x) + (\partial^2 (\Lambda_{(3)}) / \partial y \partial z) + (\partial^2$$

$$(\Lambda_{(3)}) / \partial z \partial y) \} + \Delta_{L3} * \{ (\partial^2 (\Lambda_{(2)}) / \partial x^2) + (\partial^2 (\Lambda_{(2)}) / \partial y^2) + (\partial^2 (\Lambda_{(2)}) / \partial z^2) + (\partial^2 (\Lambda_{(2)})$$

$$/ \partial x \partial y) + (\partial^2 (\Lambda_{(2)}) / \partial y \partial x) + (\partial^2 (\Lambda_{(2)}) / \partial x \partial z) + (\partial^2 (\Lambda_{(2)}) / \partial z \partial x) + (\partial^2 (\Lambda_{(2)}) / \partial y \partial z) + (\partial^2$$

$$(\Lambda_{(2)}) / \partial z \partial y) \} + \Delta_{L4} * \{ (\partial^2 (\Lambda_{(1)}) / \partial x^2) + (\partial^2 (\Lambda_{(1)}) / \partial y^2) + (\partial^2 (\Lambda_{(1)}) / \partial z^2) + (\partial^2 (\Lambda_{(1)})$$

$$/ \partial x \partial y) + (\partial^2 (\Lambda_{(1)}) / \partial y \partial x) + (\partial^2 (\Lambda_{(1)}) / \partial x \partial z) + (\partial^2 (\Lambda_{(1)}) / \partial z \partial x) + (\partial^2 (\Lambda_{(1)}) / \partial y \partial z) +$$

$$(\partial^2 (\Lambda_{(1)}) / \partial z \partial y) \} \} \tag{156}$$

The second order partial derivatives of $\Lambda_{(1)}$, $\Lambda_{(2)}$, $\Lambda_{(3)}$ and $\Lambda_{(4)}$ were given in equations (13) through (50).

TRIVARIATE QUADRATIC B-SPLINE CALCULATED WITH EMBEDDING

The signal resilient to the trivariate quadratic B-Spline interpolation function of equation (4) in chapter 2 part II is given in equation (35) and reported here:

$$f(1, 0, 0) = - \{ \varphi_1 * \Xi_1 + \varphi_2 * \Xi_2 \} / \{ \Xi_1 + \Xi_2 \} \tag{157}$$

Where it is posited:

$$\Xi_1 = [- 2a (x^3yz/3 + xy^3z/3 + xz^3y/3 + x^2y^2z/2 + x^2yz^2/2 + xy^2z^2/2) + 1/2 (a+1) xyz] \tag{158}$$

$$\Xi_2 = [a (x^3yz/3 + xy^3z/3 + xz^3y/3 + x^2y^2z/2 + x^2yz^2/2 + xy^2z^2/2) - (2a + 1/2) (x^2yz/2 +$$

$$xy^2z/2 + xyz^2/2) + 3/4 (a+1) xyz] \tag{159}$$

$$\varphi_1 = [f(-1, 0, 1) + f(-1, 0, 0) + f(1, 0, -1) + f(-1, -1, 1) + f(-1, -1, 0) + f(1, -1, 0) + f(1,$$

$$-1, -1) + f(-1, 1, 1) + f(-1, 1, 0) + f(1, 1, 0) + f(1, 1, -1)] \tag{160}$$

$$\varphi_2 = [f(-1, 0, 1) + f(0, 0, 1) + f(1, 0, 1) + f(-1, 0, 0) + f(-1, 0, -1) + f(0, 0, -1) + f(1, 0,$$

$$-1) + f(-1, -1, 1) + f(0, -1, 1) + f(1, -1, 1) + f(-1, -1, 0) + f(0, -1, 0) + f(1, -1, 0) + f(-1,$$

$$-1, -1) + f(0, -1, -1) + f(1, -1, -1) + f(-1, 1, 1) + f(0, 1, 1) + f(1, 1, 1) + f(-1, 1, 0) + f(0,$$

$$1, 0) + f(1, 1, 0) + f(-1, 1, -1) + f(0, 1, -1) + f(1, 1, -1)] \tag{161}$$

Equation (158) can be written as:

$$\Xi_1 = [- 2a (x^3yz/3 + xy^3z/3 + xz^3y/3 + x^2y^2z/2 + x^2yz^2/2 + xy^2z^2/2) + 1/2 (a+1) xyz] = xyz$$

$$[-2a (x^2/3 + y^2/3 + z^2/3 + xy/2 + xz/2 + yz/2) + 1/2 (a+1)] \tag{162}$$

Equation (159) can be written as:

$$\Xi_2 = [a (x^3yz/3 + xy^3z/3 + xz^3y/3 + x^2y^2z/2 + x^2yz^2/2 + xy^2z^2/2) - (2a + 1/2) (x^2yz/2 +$$

$$xy^2z/2 + xyz^2/2) + 3/4 (a+1) xyz] = xyz [a (x^2/3 + y^2/3 + z^2/3 + xy/2 + xz/2 + yz/2) - (2a$$

$$+ 1/2) (x/2 + y/2 + z/2) + 3/4 (a+1)] \tag{163}$$

Let us posit the term that equations (162) and (163) have in common:

$$\Xi_C = (x^2/3 + y^2/3 + z^2/3 + xy/2 + xz/2 + yz/2) \tag{164}$$

To obtain:

$$\Xi_1 = xyz \, [\, -2a \, (x^2/3 + y^2/3 + z^2/3 + xy/2 + xz/2 + yz/2) + 1/2 \, (a+1) \,] =$$

$$xyz \, [\, -2a \; \Xi_C + 1/2 \, (a+1) \,] \tag{165}$$

$$\Xi_2 = xyz \, [\, a \, (x^2/3 + y^2/3 + z^2/3 + xy/2 + xz/2 + yz/2) - (2a + 1/2) \, (x/2 + y/2 + z/2) + 3/4$$

$$(a+1) \,] = xyz \, [\, a \, \Xi_C - (2a + 1/2) \, (x/2 + y/2 + z/2) + 3/4 \, (a+1) \,] \tag{166}$$

Now let us write equation (157) as:

$$f \, (1, 0, 0) = - \, \{ \, \varphi_1 * \Xi_1 + \varphi_2 * \Xi_2 \, \} \, / \, \{ \, \Xi_1 + \Xi_2 \, \} = - \, xyz \, \{ \, \varphi_1 * [\, -2a \; \Xi_C + 1/2 \, (a+1) \,] + \varphi_2$$

$$* \, [\, a \, \Xi_C - (2a + 1/2) \, (x/2 + y/2 + z/2) + 3/4 \, (a+1) \,] \, \} \, / \, \{ \, xyz \, \{ \, [\, -2a \; \Xi_C + 1/2 \, (a+1) \,] + [\, a$$

$$\Xi_C - (2a + 1/2) \, (x/2 + y/2 + z/2) + 3/4 \, (a+1) \,] \, \} \, \} =$$

$$- \, \{ \, \varphi_1 * [\, -2a \; \Xi_C + 1/2 \, (a+1) \,] + \varphi_2 * [\, a \, \Xi_C - (2a + 1/2) \, (x/2 + y/2 + z/2) + 3/4 \, (a+1) \,] \, \} \, /$$

$$\{ \, [\, -2a \; \Xi_C + 1/2 \, (a+1) \,] + [\, a \, \Xi_C - (2a + 1/2) \, (x/2 + y/2 + z/2) + 3/4 \, (a+1) \,] \, \} \tag{167}$$

Let us posit:

$$\Xi_N = - \, \{ \, \varphi_1 * [\, -2a \; \Xi_C + 1/2 \, (a+1) \,] + \varphi_2 * [\, a \, \Xi_C - (2a + 1/2) \, (x/2 + y/2 + z/2) + 3/4 \, (a+1) \,] \, \} \tag{168}$$

$$\Xi_D = \{ \, [\, -2a \; \Xi_C + 1/2 \, (a+1) \,] + [\, a \, \Xi_C - (2a + 1/2) \, (x/2 + y/2 + z/2) + 3/4 \, (a+1) \,] \, \} \tag{169}$$

Calculation of the second order partial derivatives of Ξ_N and Ξ_D

$$(\partial \, (\Xi_N) \, / \partial x) = \varphi_1 * 2a \, (\partial \, (\Xi_C) \, / \partial x) - \varphi_2 * a \, (\partial \, (\Xi_C) \, / \partial x) + \varphi_2 * (2a + 1/2) \, (1/2) =$$

$$(\, 2\varphi_1 - \varphi_2) * a \, (\partial \, (\Xi_C) \, / \partial x) + \varphi_2 * (2a + 1/2) \, (1/2) \tag{170}$$

$$(\partial \, (\Xi_N) \, / \partial y) = \varphi_1 * 2a \, (\partial \, (\Xi_C) \, / \partial y) - \varphi_2 * a \, (\partial \, (\Xi_C) \, / \partial y) + \varphi_2 * (2a + 1/2) \, (1/2) =$$

$$(\, 2\varphi_1 - \varphi_2) * a \, (\partial \, (\Xi_C) \, / \partial y) + \varphi_2 * (2a + 1/2) \, (1/2) \tag{171}$$

$$(\partial \, (\Xi_N) \, / \partial z) = \varphi_1 * 2a \, (\partial \, (\Xi_C) \, / \partial z) - \varphi_2 * a \, (\partial \, (\Xi_C) \, / \partial z) + \varphi_2 * (2a + 1/2) \, (1/2) =$$

$$(\, 2\varphi_1 - \varphi_2) * a \, (\partial \, (\Xi_C) \, / \partial z) + \varphi_2 * (2a + 1/2) \, (1/2) \tag{172}$$

$$(\partial^2 \, (\Xi_N) \, / \partial x^2) = \varphi_1 * 2a \, (\partial^2 \, (\Xi_C) \, / \partial x^2) - \varphi_2 * a \, (\partial^2 \, (\Xi_C) \, / \partial x^2) = (\, 2\varphi_1 - \varphi_2) * a \, (\partial^2 \, (\Xi_C) \, / \partial x^2) \tag{173}$$

$$(\partial^2\,(\Xi_N)\,/\partial y^2) = \varphi_1 * 2a\,(\partial^2\,(\Xi_C)\,/\partial y^2) - \varphi_2 * a\,(\partial^2\,(\Xi_C)\,/\partial y^2) = (\,2\varphi_1 - \varphi_2\,) * a\,(\partial^2\,(\Xi_C)\,/\partial y^2) \tag{174}$$

$$(\partial^2\,(\Xi_N)\,/\partial z^2) = \varphi_1 * 2a\,(\partial^2\,(\Xi_C)\,/\partial z^2) - \varphi_2 * a\,(\partial^2\,(\Xi_C)\,/\partial z^2) = (\,2\varphi_1 - \varphi_2\,) * a\,(\partial^2\,(\Xi_C)\,/\partial z^2) \tag{175}$$

$$(\partial\,(\partial\,(\Xi_N)\,/\partial x)\partial y) = \varphi_1 * 2a\,(\partial\,(\partial\,(\Xi_C)\,/\partial x)\partial y) - \varphi_2 * a\,(\partial\,(\partial\,(\Xi_C)\,/\partial x)\partial y) =$$
$$(\,2\varphi_1 - \varphi_2\,) * a\,(\partial\,(\partial\,(\Xi_C)\,/\partial x)\partial y) \tag{176}$$

$$(\partial\,(\partial\,(\Xi_N)\,/\partial y)\partial x) = \varphi_1 * 2a\,(\partial\,(\partial\,(\Xi_C)\,/\partial y)\partial x) - \varphi_2 * a\,(\partial\,(\partial\,(\Xi_C)\,/\partial y)\partial x) =$$
$$(\,2\varphi_1 - \varphi_2\,) * a\,(\partial\,(\partial\,(\Xi_C)\,/\partial y)\partial x) \tag{177}$$

$$(\partial\,(\partial\,(\Xi_N)\,/\partial x)\partial z) = \varphi_1 * 2a\,(\partial\,(\partial\,(\Xi_C)\,/\partial x)\partial z) - \varphi_2 * a\,(\partial\,(\partial\,(\Xi_C)\,/\partial x)\partial z) =$$
$$(\,2\varphi_1 - \varphi_2\,) * a\,(\partial\,(\partial\,(\Xi_C)\,/\partial x)\partial z) \tag{178}$$

$$(\partial\,(\partial\,(\Xi_N)\,/\partial z)\partial x) = \varphi_1 * 2a\,(\partial\,(\partial\,(\Xi_C)\,/\partial z)\partial x) - \varphi_2 * a\,(\partial\,(\partial\,(\Xi_C)\,/\partial z)\partial x) =$$
$$(\,2\varphi_1 - \varphi_2\,) * a\,(\partial\,(\partial\,(\Xi_C)\,/\partial z)\partial x) \tag{179}$$

$$(\partial\,(\partial\,(\Xi_N)\,/\partial y)\partial z) = \varphi_1 * 2a\,(\partial\,(\partial\,(\Xi_C)\,/\partial y)\partial z) - \varphi_2 * a\,(\partial\,(\partial\,(\Xi_C)\,/\partial y)\partial z) =$$
$$(\,2\varphi_1 - \varphi_2\,) * a\,(\partial\,(\partial\,(\Xi_C)\,/\partial y)\partial z) \tag{180}$$

$$(\partial\,(\partial\,(\Xi_N)\,/\partial z)\partial y) = \varphi_1 * 2a\,(\partial\,(\partial\,(\Xi_C)\,/\partial z)\partial y) - \varphi_2 * a\,(\partial\,(\partial\,(\Xi_C)\,/\partial z)\partial y) =$$
$$(\,2\varphi_1 - \varphi_2\,) * a\,(\partial\,(\partial\,(\Xi_C)\,/\partial z)\partial y) \tag{181}$$

$$(\partial\,(\Xi_D)\,/\partial x) = -2a\,(\partial\,(\Xi_C)\,/\partial x) + a\,(\partial\,(\Xi_C)\,/\partial x) - (2a + 1/2)\,(1/2) =$$
$$- a\,(\partial\,(\Xi_C)\,/\partial x) - (2a + 1/2)\,(1/2) \tag{182}$$

$$(\partial\,(\Xi_D)\,/\partial y) = -2a\,(\partial\,(\Xi_C)\,/\partial y) + a\,(\partial\,(\Xi_C)\,/\partial y) - (2a + 1/2)\,(1/2) =$$
$$- a\,(\partial\,(\Xi_C)\,/\partial y) - (2a + 1/2)\,(1/2) \tag{183}$$

$$(\partial\,(\Xi_D)\,/\partial z) = -2a\,(\partial\,(\Xi_C)\,/\partial z) + a\,(\partial\,(\Xi_C)\,/\partial z) - (2a + 1/2)\,(1/2) =$$
$$- a\,(\partial\,(\Xi_C)\,/\partial z) - (2a + 1/2)\,(1/2) \tag{184}$$

$$(\partial^2\,(\Xi_D)\,/\partial x^2) = -2a\,(\partial^2\,(\Xi_C)\,/\partial x^2) + a\,(\partial^2\,(\Xi_C)\,/\partial x^2) = - a\,(\partial^2\,(\Xi_C)\,/\partial x^2) \tag{185}$$

$$(\partial^2\,(\Xi_D)\,/\partial y^2) = -2a\,(\partial^2\,(\Xi_C)\,/\partial y^2) + a\,(\partial^2\,(\Xi_C)\,/\partial y^2) = - a\,(\partial^2\,(\Xi_C)\,/\partial y^2) \tag{186}$$

$$(\partial^2\,(\Xi_D)\,/\partial z^2) = -2a\,(\partial^2\,(\Xi_C)\,/\partial z^2) + a\,(\partial^2\,(\Xi_C)\,/\partial z^2) = - a\,(\partial^2\,(\Xi_C)\,/\partial z^2) \tag{187}$$

$$(\partial\,(\partial\,(\Xi_D)\,/\partial x)\partial y) = -2a\,(\partial\,(\partial\,(\Xi_C)\,/\partial x)\partial y) + a\,(\partial\,(\partial\,(\Xi_C)\,/\partial x)\partial y) = - a\,(\partial\,(\partial\,(\Xi_C)\,/\partial x)\partial y) \tag{188}$$

$$(\partial \, (\partial \, (\Xi_D) \, / \partial y) \partial x) = -2a \, (\partial \, (\partial \, (\Xi_C) \, / \partial y) \partial x) + a \, (\partial \, (\partial \, (\Xi_C) \, / \partial y) \partial x) = - a \, (\partial \, (\partial \, (\Xi_C) \, / \partial y) \partial x) \tag{189}$$

$$(\partial \, (\partial \, (\Xi_D) \, / \partial x) \partial z) = -2a \, (\partial \, (\partial \, (\Xi_C) \, / \partial x) \partial z) + a \, (\partial \, (\partial \, (\Xi_C) \, / \partial x) \partial z) = - a \, (\partial \, (\partial \, (\Xi_C) \, / \partial x) \partial z) \tag{190}$$

$$(\partial \, (\partial \, (\Xi_D) \, / \partial z) \partial x) = -2a \, (\partial \, (\partial \, (\Xi_C) \, / \partial z) \partial x) + a \, (\partial \, (\partial \, (\Xi_C) \, / \partial z) \partial x) = - a \, (\partial \, (\partial \, (\Xi_C) \, / \partial z) \partial x) \tag{191}$$

$$(\partial \, (\partial \, (\Xi_D) \, / \partial y) \partial z) = -2a \, (\partial \, (\partial \, (\Xi_C) \, / \partial y) \partial z) + a \, (\partial \, (\partial \, (\Xi_C) \, / \partial y) \partial z) = - a \, (\partial \, (\partial \, (\Xi_C) \, / \partial y) \partial z) \tag{192}$$

$$(\partial \, (\partial \, (\Xi_D) \, / \partial z) \partial y) = -2a \, (\partial \, (\partial \, (\Xi_C) \, / \partial z) \partial y) + a \, (\partial \, (\partial \, (\Xi_C) \, / \partial z) \partial y) = - a \, (\partial \, (\partial \, (\Xi_C) \, / \partial z) \partial y) \tag{193}$$

From equation (164):

$\Xi_C = (x^2/3 + y^2/3 + z^2/3 + xy/2 + xz/2 + yz/2)$, it follows that the calculation of first and second order partial derivatives is:

$$(\partial \, (\Xi_C) \, / \partial x) = [\, (2/3)x + (1/2) \, (y + z) \,] \tag{194}$$

$$(\partial \, (\Xi_C) \, / \partial y) = [\, (2/3)y + (1/2) \, (x - z) \,] \tag{195}$$

$$(\partial \, (\Xi_C) \, / \partial z) = [\, (2/3)z + (1/2) \, (x - y) \,] \tag{196}$$

$$(\partial^2 \, (\Xi_C) \, / \partial x^2) = (\partial^2 \, (\Xi_C) \, / \partial y^2) = (\partial^2 \, (\Xi_C) \, / \partial z^2) = (2/3) \tag{197}$$

$$(\partial \, (\partial \, (\Xi_C) \, / \partial x) \partial y) = (\partial \, (\partial \, (\Xi_C) \, / \partial y) \partial x) = (\partial \, (\partial \, (\Xi_C) \, / \partial x) \partial z) = (\partial \, (\partial \, (\Xi_C) \, / \partial z) \partial x) =$$

$$(\partial \, (\partial \, (\Xi_C) \, / \partial y) \partial z) = (\partial \, (\partial \, (\Xi_C) \, / \partial z) \partial y) = (1/2) \tag{198}$$

Calculation of the the second order partial derivatives of $\varrho \, (x, y, z)$

The formulation R $(x, y, z) = f \, (0, 0, 0) + \varrho \, (x, y, z)$ requires with the use of equations (167), (168), (169) that:

$$\varrho \, (x, y, z) = (\, \Xi_N \, / \, \Xi_D \,) \tag{199}$$

It follows that:

$$(\partial \, (\varrho \, (x, y, z)) \, / \partial x) = \{ \, [(\partial \, (\Xi_N) \, / \partial x) * \Xi_D - (\partial \, (\Xi_D) \, / \partial x) * \Xi_N \,] \, / \, [\, \Xi_D \,]^2 \, \} \tag{200}$$

$$(\partial \, (\varrho \, (x, y, z)) \, / \partial y) = \{ \, [(\partial \, (\Xi_N) \, / \partial y) * \Xi_D - (\partial \, (\Xi_D) \, / \partial y) * \Xi_N \,] \, / \, [\, \Xi_D \,]^2 \, \} \tag{201}$$

$$(\partial \, (\varrho \, (x, y, z)) \, / \partial z) = \{ \, [(\partial \, (\Xi_N) \, / \partial z) * \Xi_D - (\partial \, (\Xi_D) \, / \partial z) * \Xi_N \,] \, / \, [\, \Xi_D \,]^2 \, \} \tag{202}$$

$$(\partial^2 \, (\varrho \, (x, y, z)) \, / \partial x^2) = \{ \, \{ \, [(\partial^2 \, (\Xi_N) \, / \partial x^2) * \Xi_D + (\partial \, (\Xi_N) \, / \partial x) * (\partial \, (\Xi_D) \, / \partial x) - (\partial^2 \, (\Xi_D) \, / \partial x^2) *$$

$$\Xi_N - (\partial \, (\Xi_D) \, / \partial x) * (\partial \, (\Xi_N) \, / \partial x) \,] \, \} * \{ \, [\, \Xi_D \,]^2 \, \} - [(\partial \, (\Xi_N) \, / \partial x) * \Xi_D - (\partial \, (\Xi_D) \, / \partial x) * \Xi_N \,] *$$

$$\{ 2 * [\Xi_D] * (\partial (\Xi_D) / \partial x) \} \} / \{ [\Xi_D]^4 \} \tag{203}$$

$$(\partial^2 (\varrho (x, y, z)) / \partial y^2) = \{ \{ [(\partial^2 (\Xi_N) / \partial y^2) * \Xi_D + (\partial (\Xi_N) / \partial y) * (\partial (\Xi_D) / \partial y) - (\partial^2 (\Xi_D) / \partial y^2) *$$

$$\Xi_N - (\partial (\Xi_D) / \partial y) * (\partial (\Xi_N) / \partial y)] \} * \{ [\Xi_D]^2 \} - [(\partial (\Xi_N) / \partial y) * \Xi_D - (\partial (\Xi_D) / \partial y) * \Xi_N] *$$

$$\{ 2 * [\Xi_D] * (\partial (\Xi_D) / \partial y) \} \} / \{ [\Xi_D]^4 \} \tag{204}$$

$$(\partial^2 (\varrho (x, y, z)) / \partial z^2) = \{ \{ [(\partial^2 (\Xi_N) / \partial z^2) * \Xi_D + (\partial (\Xi_N) / \partial z) * (\partial (\Xi_D) / \partial z) - (\partial^2 (\Xi_D) / \partial z^2) *$$

$$\Xi_N - (\partial (\Xi_D) / \partial z) * (\partial (\Xi_N) / \partial z)] \} * \{ [\Xi_D]^2 \} - [(\partial (\Xi_N) / \partial z) * \Xi_D - (\partial (\Xi_D) / \partial z) * \Xi_N] *$$

$$\{ 2 * [\Xi_D] * (\partial (\Xi_D) / \partial z) \} \} / \{ [\Xi_D]^4 \} \tag{205}$$

$$(\partial (\partial (\varrho (x, y, z)) / \partial x) \partial y) = \{ \{ [(\partial (\partial (\Xi_N) / \partial x) \partial y) * \Xi_D - (\partial (\partial (\Xi_D) / \partial x) \partial y) * \Xi_N + (\partial (\Xi_N)$$

$$/ \partial x) * (\partial (\Xi_D) / \partial y) - (\partial (\Xi_D) / \partial x) * (\partial (\Xi_N) / \partial y)] \} * \{ [\Xi_D]^2 \} - [(\partial (\Xi_N) / \partial x) * \Xi_D - (\partial (\Xi_D) / \partial x) *$$

$$\Xi_N] * \{ 2 * [\Xi_D] * (\partial (\Xi_D) / \partial y) \} \} / \{ [\Xi_D]^4 \} \tag{206}$$

$$(\partial (\partial (\varrho (x, y, z)) / \partial y) \partial x) = \{ \{ [(\partial (\partial (\Xi_N) / \partial y) \partial x) * \Xi_D - (\partial (\partial (\Xi_D) / \partial y) \partial x) * \Xi_N + (\partial (\Xi_N)$$

$$/ \partial y) * (\partial (\Xi_D) / \partial x) - (\partial (\Xi_D) / \partial y) * (\partial (\Xi_N) / \partial x)] \} * \{ [\Xi_D]^2 \} - [(\partial (\Xi_N) / \partial y) * \Xi_D - (\partial (\Xi_D) / \partial y) *$$

$$\Xi_N] * \{ 2 * [\Xi_D] * (\partial (\Xi_D) / \partial x) \} \} / \{ [\Xi_D]^4 \} \tag{207}$$

$$(\partial (\partial (\varrho (x, y, z)) / \partial x) \partial z) = \{ \{ [(\partial (\partial (\Xi_N) / \partial x) \partial z) * \Xi_D - (\partial (\partial (\Xi_D) / \partial x) \partial z) * \Xi_N + (\partial (\Xi_N)$$

$$/ \partial x) * (\partial (\Xi_D) / \partial z) - (\partial (\Xi_D) / \partial x) * (\partial (\Xi_N) / \partial z)] \} * \{ [\Xi_D]^2 \} - [(\partial (\Xi_N) / \partial x) * \Xi_D - (\partial (\Xi_D) / \partial x) *$$

$$\Xi_N] * \{ 2 * [\Xi_D] * (\partial (\Xi_D) / \partial z) \} \} / \{ [\Xi_D]^4 \} \tag{208}$$

$$(\partial (\partial (\varrho (x, y, z)) / \partial z) \partial x) = \{ \{ [(\partial (\partial (\Xi_N) / \partial z) \partial x) * \Xi_D - (\partial (\partial (\Xi_D) / \partial z) \partial x) * \Xi_N + (\partial (\Xi_N)$$

$$/ \partial z) * (\partial (\Xi_D) / \partial x) - (\partial (\Xi_D) / \partial z) * (\partial (\Xi_N) / \partial x)] \} * \{ [\Xi_D]^2 \} - [(\partial (\Xi_N) / \partial z) * \Xi_D - (\partial (\Xi_D) / \partial z) *$$

$$\Xi_N] * \{ 2 * [\Xi_D] * (\partial (\Xi_D) / \partial x) \} \} / \{ [\Xi_D]^4 \} \tag{209}$$

$$(\partial (\partial (\varrho (x, y, z)) / \partial y) \partial z) = \{ \{ [(\partial (\partial (\Xi_N) / \partial y) \partial z) * \Xi_D - (\partial (\partial (\Xi_D) / \partial y) \partial z) * \Xi_N + (\partial (\Xi_N)$$

$$/ \partial y) * (\partial (\Xi_D) / \partial z) - (\partial (\Xi_D) / \partial y) * (\partial (\Xi_N) / \partial z)] \} * \{ [\Xi_D]^2 \} - [(\partial (\Xi_N) / \partial y) * \Xi_D - (\partial (\Xi_D) / \partial y) *$$

$$\Xi_N] * \{ 2 * [\Xi_D] * (\partial (\Xi_D) / \partial z) \} \} / \{ [\Xi_D]^4 \} \tag{210}$$

$$(\partial \, (\partial \, (\varrho \, (x, y, z)) \, / \partial z) \partial y) = \{ \, \{ \, [(\partial \, (\partial \, (\Xi_N) \, / \partial z) \partial y) * \Xi_D - (\partial \, (\partial \, (\Xi_D) \, / \partial z) \partial y) * \Xi_N + (\partial \, (\Xi_N)$$

$$/ \partial z) * (\partial \, (\Xi_D) \, / \partial y) - (\partial \, (\Xi_D) \, / \partial z) * (\partial \, (\Xi_N) \, / \partial y) \,] \, \} * \{ \, [\, \Xi_D \,]^2 \, \} - [(\partial \, (\Xi_N) \, / \partial z) * \Xi_D - (\partial \, (\Xi_D) \, / \partial z) *$$

$$\Xi_N \,] * \{ \, 2 * [\, \Xi_D \,] * (\partial \, (\Xi_D) \, / \partial y) \, \} \, \} / \{ \, [\, \Xi_D \,]^4 \, \} \tag{211}$$

Calculation of the Total Curvature of the Resilient Signal

On the basis of the knowledge provided in equations (100) through (108), the total curvature $C(R((x, y, z), f))$ is obtained adding up equations (203) through (211) as follows:

$$C(R((x, y, z), f)) = (\partial^2 \, (R((x, y, z), f)) \, / \partial x^2) + (\partial^2 \, (R((x, y, z), f)) \, / \partial y^2) + (\partial^2 \, (R((x, y,$$

$$z), f)) \, / \partial z^2) + (\partial \, (\partial \, (R((x, y, z), f)) \, / \partial y) \partial x) + (\partial \, (\partial \, (R((x, y, z), f)) \, / \partial x) \partial y) + (\partial \, (\partial \, (R((x,$$

$$y, z), f)) \, / \partial y) \partial z) + (\partial \, (\partial \, (R((x, y, z), f)) \, / \partial z) \partial y) + (\partial \, (\partial \, (R((x, y, z), f)) \, / \partial z) \partial x) + (\partial \, (\partial \, ($$

$$R((x, y, z), f)) \, / \partial x) \partial z) \tag{212}$$

TRIVARIATE QUADRATIC B-SPLINE CALCULATED WITHOUT EMBEDDING

Equation (28) in chapter 3 part II furnishes the value of the signal resilient to the trivariate quadratic B-Spline interpolation function of equation (4) (chapter 3 part II) as:

$$f \, (0, 0, 0) = \{ \, \omega_1 * [- 2a \, (x^2/3 + y^2/3 + z^2/3 + xy/2 + xz/2 + yz/2) + 1/2 \, (a+1) \,] + \omega_2 * [a$$

$$(x^2/3 + y^2/3 + z^2/3 + xy/2 + xz/2 + yz/2) - (2a + 1/2) \, (x/2 + y/2 + z/2) + 3/4 \, (a+1) \,] \, \} \tag{213}$$

Placing equation (164), which posits

$$\Xi_C = (x^2/3 + y^2/3 + z^2/3 + xy/2 + xz/2 + yz/2)$$ into equation (213) furnishes:

$$f \, (0, 0, 0) = \{ \, \omega_1 * [- 2a \, \Xi_C + 1/2 \, (a+1) \,] + \omega_2 * [a \, \Xi_C - (2a + 1/2) \, (x/2 + y/2 + z/2) + 3/4$$

$$(a+1) \,] \, \} \tag{214}$$

Calculation of the the second order partial derivatives of $\varrho \, (x, y, z)$

The formulation $R \, (x, y, z) = f \, (0, 0, 0) + \varrho \, (x, y, z)$ requires:

$$\varrho \, (x, y, z) = \{ \, \omega_1 * [- 2a \, \Xi_C + 1/2 \, (a+1) \,] + \omega_2 * [a \, \Xi_C - (2a + 1/2) \, (x/2 + y/2 + z/2) + 3/4$$

$$(a+1) \,] \, \} \tag{215}$$

$$(\partial \, (\varrho \, (x, y, z)) \, / \partial x) = [- 2a \, \omega_1 * (\partial \, (\Xi_C) \, / \partial x)] + \omega_2 * [a \, (\partial \, (\Xi_C) \, / \partial x) - (2a + 1/2) \, (1/2) \,] =$$

$$(\partial \, (\Xi_C) \, / \partial x) * [- 2a \, \omega_1 + a \, \omega_2] - \omega_2 \, (2a + 1/2) \, (1/2) \tag{216}$$

$$(\partial\,(\varrho\,(x,\,y,\,z))\,/\partial y) = [\text{-}\,2a\,\omega_1 * (\partial\,(\Xi_C)\,/\partial y)] + \omega_2 * [\,a\,(\partial\,(\Xi_C)\,/\partial y) \text{-} (2a + 1/2)\,(1/2)\,] =$$

$$(\partial\,(\Xi_C)\,/\partial y) * [\text{-}\,2a\,\omega_1 + a\,\omega_2] \text{-} \omega_2\,(2a + 1/2)\,(1/2) \tag{217}$$

$$(\partial\,(\varrho\,(x,\,y,\,z))\,/\partial z) = [\text{-}\,2a\,\omega_1 * (\partial\,(\Xi_C)\,/\partial z)] + \omega_2 * [\,a\,(\partial\,(\Xi_C)\,/\partial z) \text{-} (2a + 1/2)\,(1/2)\,] =$$

$$(\partial\,(\Xi_C)\,/\partial z) * [\text{-}\,2a\,\omega_1 + a\,\omega_2] \text{-} \omega_2\,(2a + 1/2)\,(1/2) \tag{218}$$

$$(\partial^2\,(\varrho\,(x,\,y,\,z))\,/\partial x^2) = (\partial^2\,(\Xi_C)\,/\partial x^2) * [\text{-}\,2a\,\omega_1 + a\,\omega_2] \tag{219}$$

$$(\partial^2\,(\varrho\,(x,\,y,\,z))\,/\partial y^2) = (\partial^2\,(\Xi_C)\,/\partial y^2) * [\text{-}\,2a\,\omega_1 + a\,\omega_2] \tag{220}$$

$$(\partial^2\,(\varrho\,(x,\,y,\,z))\,/\partial z^2) = (\partial^2\,(\Xi_C)\,/\partial z^2) * [\text{-}\,2a\,\omega_1 + a\,\omega_2] \tag{221}$$

$$(\partial\,(\partial\,(\varrho\,(x,\,y,\,z))\,/\partial x)\partial y) = (\partial\,(\partial\,(\Xi_C)\,/\partial x)\partial y) * [\text{-}\,2a\,\omega_1 + a\,\omega_2] \tag{222}$$

$$(\partial\,(\partial\,(\varrho\,(x,\,y,\,z))\,/\partial y)\partial x) = (\partial\,(\partial\,(\Xi_C)\,/\partial y)\partial x) * [\text{-}\,2a\,\omega_1 + a\,\omega_2] \tag{223}$$

$$(\partial\,(\partial\,(\varrho\,(x,\,y,\,z))\,/\partial x)\partial z) = (\partial\,(\partial\,(\Xi_C)\,/\partial x)\partial z) * [\text{-}\,2a\,\omega_1 + a\,\omega_2] \tag{224}$$

$$(\partial\,(\partial\,(\varrho\,(x,\,y,\,z))\,/\partial z)\partial x) = (\partial\,(\partial\,(\Xi_C)\,/\partial z)\partial x) * [\text{-}\,2a\,\omega_1 + a\,\omega_2] \tag{225}$$

$$(\partial\,(\partial\,(\varrho\,(x,\,y,\,z))\,/\partial y)\partial z) = (\partial\,(\partial\,(\Xi_C)\,/\partial y)\partial z) * [\text{-}\,2a\,\omega_1 + a\,\omega_2] \tag{226}$$

$$(\partial\,(\partial\,(\varrho\,(x,\,y,\,z))\,/\partial z)\partial y) = (\partial\,(\partial\,(\Xi_C)\,/\partial z)\partial y) * [\text{-}\,2a\,\omega_1 + a\,\omega_2] \tag{227}$$

Calculation of the Total Curvature of the Resilient Signal

On the basis of the knowledge provided in equations (100) through (108), the total curvature C(R((x, y, z), f)) is obtained adding up equations (219) through (227) as follows:

$$C(\,R((x,\,y,\,z),\,f)\,) = (\partial^2\,(\,R((x,\,y,\,z),\,f)\,)\,/\partial x^2) + (\partial^2\,(\,R((x,\,y,\,z),\,f)\,)\,/\partial y^2) + (\partial^2\,(\,R((x,\,y,\,z),\,f)\,)\,/\partial z^2) + (\partial\,(\partial\,(\,R((x,\,y,\,z),\,f)\,)\,/\partial y)\partial x) + (\partial\,(\partial\,(\,R((x,\,y,\,z),\,f)\,)\,/\partial x)\partial y) + (\partial\,(\partial\,(\,R((x,\,y,\,z),\,f)\,)\,/\partial y)\partial z) + (\partial\,(\partial\,(\,R((x,\,y,\,z),\,f)\,)\,/\partial z)\partial y) + (\partial\,(\partial\,(\,R((x,\,y,\,z),\,f)\,)\,/\partial z)\partial x) + (\partial\,(\partial\,(\,R((x,\,y,\,z),\,f)\,)\,/\partial x)\partial z) \tag{228}$$

Where the numerical values of:

$(\partial^2\,(\Xi_C)\,/\partial x^2)$, $(\partial^2\,(\Xi_C)\,/\partial y^2)$, $(\partial^2\,(\Xi_C)\,/\partial z^2)$, $(\partial\,(\partial\,(\Xi_C)\,/\partial x)\partial y)$, $(\partial\,(\partial\,(\Xi_C)\,/\partial y)\partial x)$, $(\partial\,(\partial\,(\Xi_C)\,/\partial x)\partial z)$, $(\partial\,(\partial\,(\Xi_C)\,/\partial z)\partial x)$, $(\partial\,(\partial\,(\Xi_C)\,/\partial y)\partial z)$, $(\partial\,(\partial\,(\Xi_C)\,/\partial z)\partial y)$ are given in equations (197), (198).

SUMMARY

The math developments in this chapter calculate the total curvature of three trivariate interpolation functions, with and without embedding. The outcome of these math deductions will be seen in chapter 10 where the results will be presented along with the total curvature maps to substantiate the meaning and relevance of the total curvature in the calculation of the interpolation functions with the custom transfer function induced through the new concept of Pixel Intensity Correction (PIC). PIC(x, y, z) = Shift * tan(total curvature). In each case the interpolation function with a customized transfer function will have the following simplified math forms: I (x, y, z) = f (0, 0, 0) + PIC (x, y, z), classic-curvature formula; R (x, y, z) = f (0, 0, 0) + PIC (x, y, z), resilient-curvature formula, and HYB (x, y, z) = I (x, y, z) if I (x, y, z) < R (x, y, z) or HYB (x, y, z) = R (x, y, z) if I (x, y, z) ≥ R (x, y, z), classic-resilient-curvature hybrid formula.

REFERENCES

Ciulla, C. (2009). *Improved Signal and Image Interpolation in Biomedical Applications: The Case of Magnetic Resonance Imaging (MRI)* – Medical Information Science Reference - IGI Global Publisher, Hershey, PA, U.S.A.

CHAPTER 10

CLASSIC-RESILIENT-CURVATURE HYBRID INTERPOLATION: TRIVARIATE POLYNOMIALS WITH AND WITHOUT EMBEDDING

INTRODUCTION

This chapter intends to expand on the concept of classic-resilient-curvature hybrid interpolation to reconstruct signals modelled through trivariate functions. The functions used to provide the presentation herein are in the variables X, Y and Z and are of the second and third order. The math of these functions is illustrated in chapter 2 parts II through IV, (i) with embedding: trivariate quadratic B-Spline, trivariate cubic B-Spline, trivariate cubic Lagrange formulae; and in chapter 3 parts II through IV (ii) without embedding ('WE'): trivariate quadratic B-Spline, trivariate cubic B-Spline, trivariate cubic Lagrange formulae.

The investigation of bivariate polynomials has illustrated that is possible, through the use of the curvature of the model interpolation function, to devise the math form of the transfer function of the interpolator. Thus the arbitration of the transfer function of the choice is removed in both classic and resilient interpolation.

Given that on a pixel-by-pixel basis it is not necessarily true that the classic-curvature formula outperforms the resilient-curvature formula, and nor the opposite, it is possible as well as auspicious to merge the two types of signal reconstruction formulae into the classic-resilient-curvature hybrid interpolation HYB(x, y, z), which is devised to be classic-curvature or resilient-curvature depending on which one of the two paradigms reveals the lower interpolation error. The results presented here serve to support such a novel methodology in signal reconstruction in 3D.

As an advantage, the hybrid formulation HYB(x, y, z) presents lower Mean Absolute Error (MAE) than either classic-curvature or resilient-curvature formulae; we may also draw the reader's attention to the real possibility of deriving transfer functions directly from the math form of second and third degree order interpolators through the use of the second order partial derivatives of the signal, which are added up across the variate (x, y, z) and covariates (xy, xz, yx, yz, zx, zy) to calculate what is called the total curvature, as shown in chapter 9.

The presentation of the results is similar to the one given in chapter 8; the math forms of both classic-curvature and resilient-curvature formulae and, in addition, the derivation of the transfer functions pertinent to the trivariate interpolation formulae are given in chapter 9.

This chapter is divided in two main sections, one quantitative conducted using chaotic image data (see figure 10) and one qualitative conducted mainly using T2 MRI data. In the quantitative section the graphs of the MAE are presented to support the argument in favour of classic-resilient-curvature hybrid signal reconstruction. In the qualitative section, images of the curvature are presented and their nature investigated so that relationships given through the math derivation between the image and the total curvature of the image shall be ascertained.

Work on Lagrange interpolation functions is to be found in the following texts and papers: Dumpster et al. (1999), Fuchs and Delyon (2002), Hecklin et al. (2008), Kubayi and Lubinsky (2004), Marvasti (1990), Radzyner, and Bason (1972), Waldron (1997), Xie and Zhou (2001), Ye (2003). We hold it to be proper to mention these works although the aims and scopes of this chapter are different, as shall be seen in the following sections.

TRIVARIATE POLYNOMIALS WITH AND WITHOUT EMBEDDING: QUANTITATIVE EVALUATION

In reading the graphs of figures 1 through 6 it is necessary to introduce to the reader the following notation:

Classic-curvature With E	Classic-curvature WE	Resilient-curvature With E	Resilient-curvature WE
Quadratic			
H33DRCCurvature C.CR	H33DWERCCurvature C.CR	H33DRCCurvature R.CR	H33DWERCCurvature R.CR
Cubic			
H43DRCCurvature C.CR	H43DWERCCurvature C.CR	H43DRCCurvature R.CR	H43DWERCCurvature R.CR
Cubic Lagrange			
LGR3DRCCurvature C.CR	LGR3DWERCCurvature C.CR	LGR3DRCCurvature R.CR	LGR3DWERCCurvature R.CR

Table I: The headings found in the graphs shown in figures 1 through 6, and employed in indicating the ratio of improvement of the hybrid interpolation formula over the classic-curvature (C.CR) or over the resilient-curvature interpolation formulae (R.CR).

With E	WE
H33DRCCurvature (MAE ClassicResilient)	H33DWERCCurvature (MAE ClassicResilient)
H43DRCCurvature (MAE ClassicResilient)	H43DWERCCurvature (MAE ClassicResilient)
LGR3DRCCurvature (MAE ClassicResilient)	LGR3DWERCCurvature (MAE ClassicResilient)

Table II: The headings of the graphs shown in figures 1 through 6 to indicate the notation used to identify the MAE of the hybrid signal reconstruction throughout the three types of interpolation formulae: trivariate quadratic, trivariate cubic and trivariate cubic Lagrange, with embedding ('With E') and without embedding ('WE').

The headings seen in Tables I and II are of the graphs of figures 1 through 6, calculated with embedding ('With E') and without embedding ('WE'). This notation is employed in presenting the results of the trivariate quadratic (H3), the trivariate cubic (H4) and the trivariate cubic Lagrange (LGR) formulae to identify the MAE of the classic-curvature, the resilient-curvature and hybrid formulae and the ratio of improvement of the hybrid formula in the various cases.

In either of the cases (i) classic-curvature: I (x, y, z) = f (0, 0, 0) + PIC (x, y, z), and (ii) resilient-curvature: R (x, y, z) = f (0, 0, 0) + PIC (x, y, z) it is possible to obtain an improvement in signal reconstruction when comparing the aforementioned two interpolation formulae to the Hybrid signal reconstruction, that is to say equation (1) or (2), depending on which one of the two has the lower Mean Absolute Error (MAE) on a pixel-by-pixel basis. The Pixel-Intensity-Correction (PIC) acts as the transfer function obtained from the math form of the interpolation function.

$$I (x, y, z) = f (0, 0, 0) + PIC (x, y, z) \qquad (1)$$

$$R (x, y, z) = f (0, 0, 0) + PIC (x, y, z) \qquad (2)$$

SIGNAL RESILIENT TO INTERPOLATION

The interpolators calculated with embedding are: (i) the trivariate formula seen in equations (1), (2), (3) in chapter 2 part II (quadratic B-Spline), (ii) the trivariate cubic B-Spline seen in equation (36) in chapter 2 part III, and (iii) the trivariate cubic Lagrange seen in equation (64) in chapter 2 part IV. Positions made through equations (1) and (2) of part II of chapter 2 hold true also for equation (36) in part III of chapter 2 and equation (64) in part IV of chapter 2.

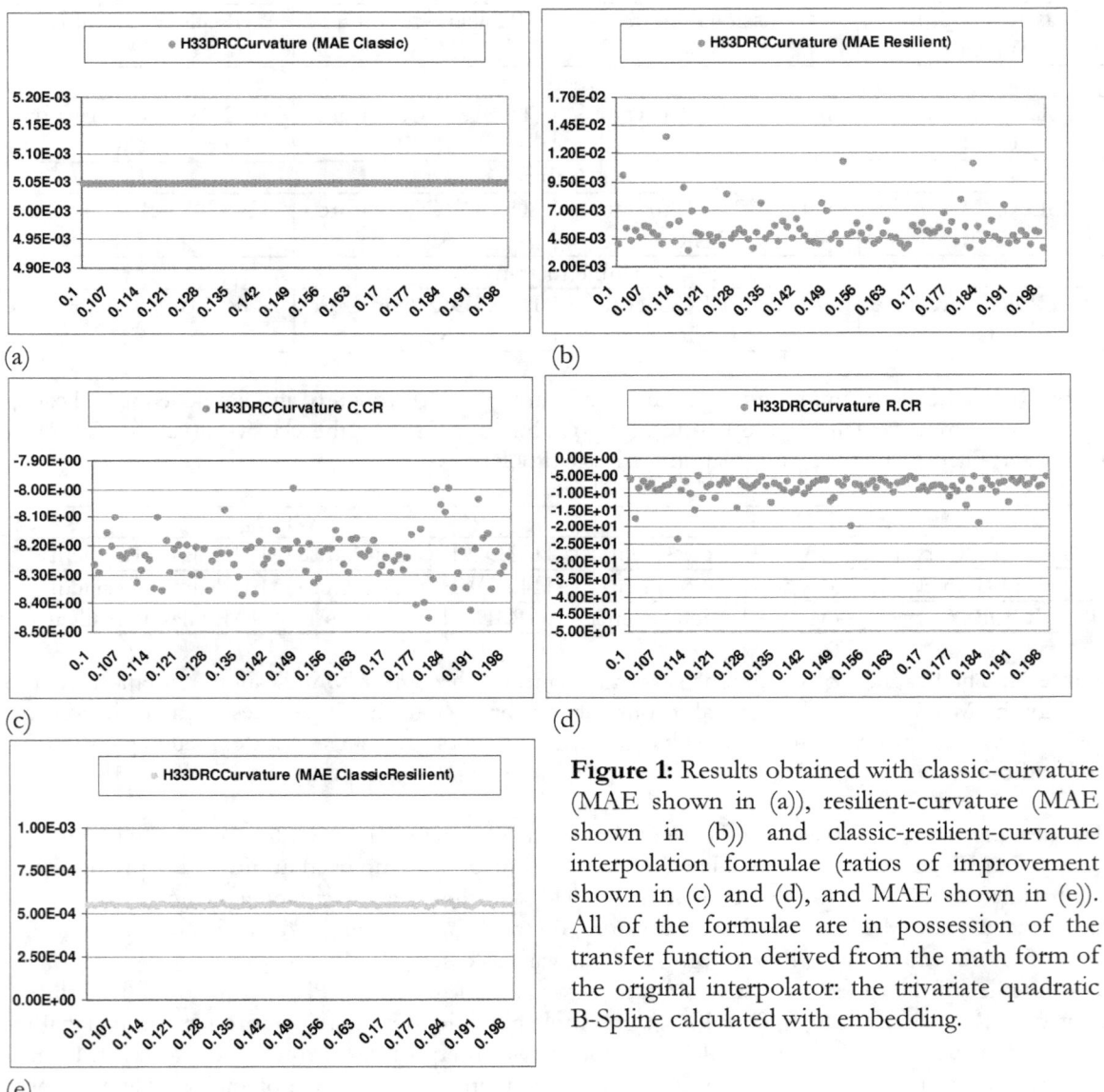

(a) (b) (c) (d) (e)

Figure 1: Results obtained with classic-curvature (MAE shown in (a)), resilient-curvature (MAE shown in (b)) and classic-resilient-curvature interpolation formulae (ratios of improvement shown in (c) and (d), and MAE shown in (e)). All of the formulae are in possession of the transfer function derived from the math form of the original interpolator: the trivariate quadratic B-Spline calculated with embedding.

Those interpolators calculated without embedding are located in: (i) equations (1), (2), (3) in chapter 3 part II (quadratic B-Spline); (ii) equations (1), (2) and (3) in chapter 3 part III (trivariate cubic B-Spline); and (iii) equations (1), (2) and (3) in chapter 3 part IV (trivariate cubic Lagrange). From the aforementioned math forms, the transfer function PIC(x, y, z) is derived and added up as it is the fraction of pixel intensity correction to the value of f (0, 0, 0), and this is indicated in equations (1) and (2) of this chapter.

The trivariate formula seen in equations (1), (2) and (3) in chapter 2 part II was employed to derive the results herein reported of the quadratic signal reconstruction obtained through classic-curvature, resilient-curvature and classic-resilient-curvature hybrid trivariate quadratic formulae.

Results shown in figure 1 are relevant to classic-curvature, resilient-curvature interpolation formulae indicated in equations (1), (2) of this chapter, which were derived from the math form of the original interpolator: the trivariate quadratic B-Spline calculated with embedding. Results of figure 1 are also relevant to the classic-resilient-curvature hybrid interpolation formula which is, on a pixel-by-pixel basis, either equation (1) or equation (2), depending on which one between (1) and (2) releases the lowest Mean Absolute Error (MAE).

In (a) we find the MAE of the classic-curvature, in (b) we find the MAE of the resilient-curvature, in (c) we find the ratio of improvement of the classic-resilient-curvature hybrid formula over the classic-curvature formula, in (d) we find the ratio of improvement of the classic-resilient-curvature hybrid formula over the resilient-curvature formula, and lastly in (e) we find the MAE of the classic-resilient-curvature hybrid formula.

In the figures below (from figure 1 to figure 6), the rigid transformation applied is a rotation of progressive steps of 0.001 E+00 deg, about each of the three rotational angles defined in the three dimensional space, of the array [0.1 E+00, 0.1 E+00, 0.1 E+00]. Each rotation (about X, Y and Z axis of the image volume coordinate system) ranges into the closed interval [0.1 E+00 deg, 0.198 E+00 deg]. This means that each step is made of the rigid rotation of 0.001 E+00 deg about X, Y and Z axis (three simultaneous rotations) of the array [0.1 E+00, 0.1 E+00, 0.1 E+00]. Each rotation made at each step adds on to the previous rigid rotation. The abscissa in the graphs shows the progression of the rotations (each of the three angles starts from the value of 0.1 E+00 deg).

The MAE in figure 1(a) is quasi constant: 5.05 E-03; and is attributable to the math form of the second order partial derivatives of the bivariate quadratic classic formula as seen in equation (19) of chapter 2 part II and recalled below:

Since, $(\partial^2 (h_3(x, y, z)) /\partial x^2) = (\partial^2 (h_3(x, y, z)) /\partial y^2) = (\partial^2 (h_3(x, y, z)) /\partial z^2) = (\partial^2 (h_3(x,$

$y, z)) /\partial x\partial y) = (\partial^2 (h_3(x, y, z)) /\partial y\partial x) = (\partial^2 (h_3(x, y, z)) /\partial x\partial z) = (\partial^2 (h_3(x, y, z))$

$/\partial z\partial x) = (\partial^2 (h_3(x, y, z)) /\partial y\partial z) = (\partial^2 (h_3(x, y, z)) /\partial z\partial y) = \{ - 4a \omega_1 + 2a \omega_2 \}$ \hfill (3)

The total curvature which is the sum of the nine second order partial derivatives in (3), is constant and independent from the misplacement and with the use of equation (2) is responsible for the error. In this case the classic-curvature formula becomes effective for the purposes of the signal reconstruction. The value of 'a' was set at *the_A_const'* = 0.745.

In figure 1(b) the MAE of the resilient-curvature formula ranges between values of 2.0 E-03 and 1.45 E-02, showing a beneficial effect in signal reconstruction. The formulation of the total curvature of the trivariate quadratic B-Spline calculated with embedding is shown in the calculations reported in chapter 9 (see C(R((x, y, z), f)) in equation (212)).

In figure 1(c) we highlight the improvement obtained through the use of the hybrid formula over the classic-curvature formula with values of the ratio of improvement mostly between approximately 790 % and approximately 850 %.

In figure 1(d) the improvement of the hybrid formula over the resilient-curvature formula is approximately between 500 % and 1,000 % where the majority of the scattered points in the graph are located.

In figure 1(e) it is worth noting the value of the MAE of the hybrid formula which is stationary (quasi constant), between values of 5.00 E-04 and 7.50 E-04.

Similarly as we have seen already with signal reconstruction performed with embedding with the trivariate quadratic B-Spline, the MAE of the formula that uses the classic-curvature is low enough to make these specific paradigms (classic-curvature and resilient-curvature) effective on the trivariate quadratic B-Spline calculated without embedding (see figure 2).

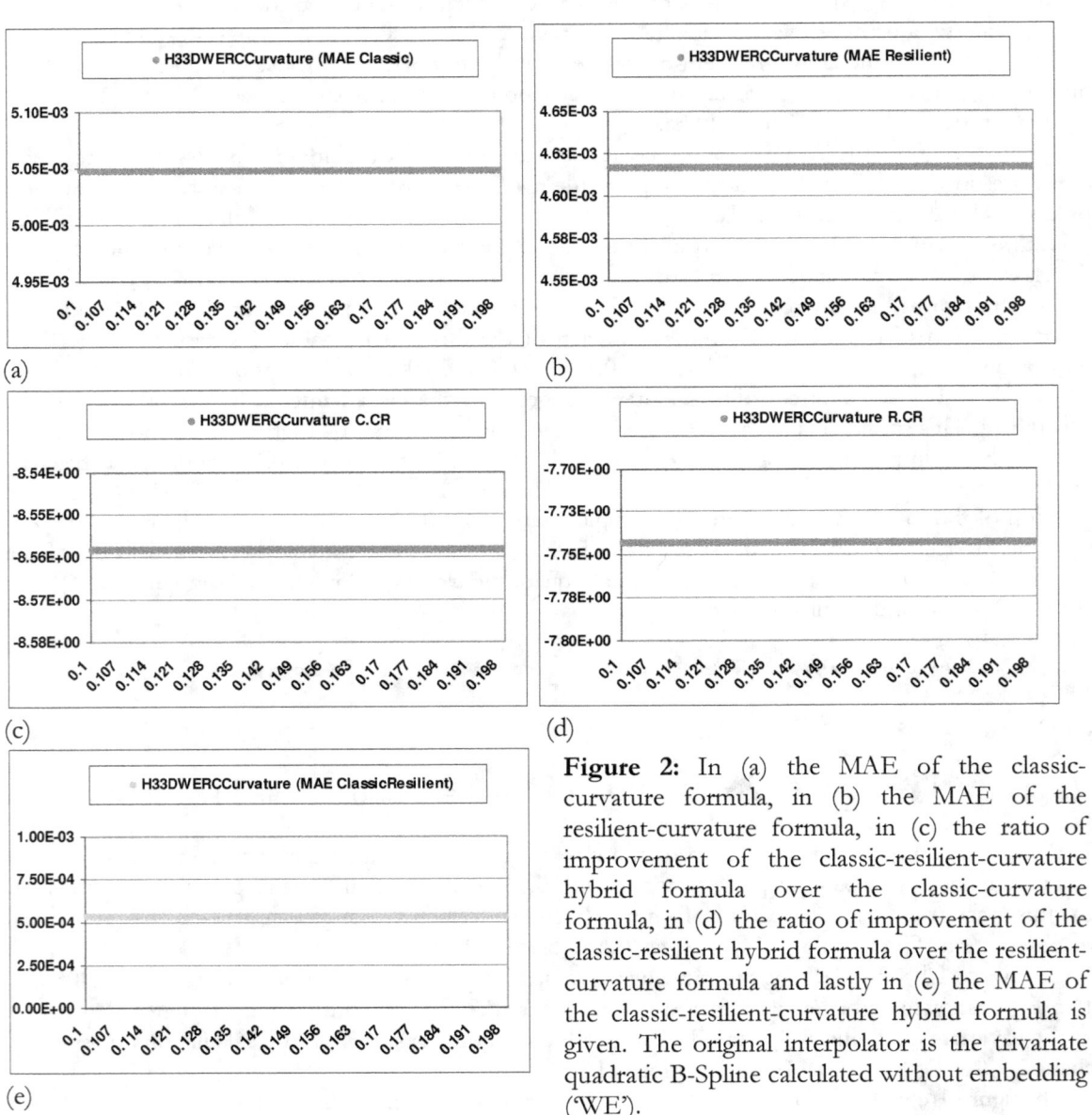

(a)

(b)

(c)

(d)

(e)

Figure 2: In (a) the MAE of the classic-curvature formula, in (b) the MAE of the resilient-curvature formula, in (c) the ratio of improvement of the classic-resilient-curvature hybrid formula over the classic-curvature formula, in (d) the ratio of improvement of the classic-resilient hybrid formula over the resilient-curvature formula and lastly in (e) the MAE of the classic-resilient-curvature hybrid formula is given. The original interpolator is the trivariate quadratic B-Spline calculated without embedding ('WE').

On the same note, the results of the ratio of improvement of the hybrid formula on classic-curvature and resilient-curvature formulae indicate the MAE to be approximately 856 % (classic-curvature, see (c)) and the MAE at approximately 775 % (resilient-curvature, see (d)) even given the small values of error as visible in figures 2(a) and 2(b), 5.05 E-03 and approximately 4.63 E-03 respectively.

Results shown in figure 2 were obtained with classic-curvature, resilient-curvature and classic-resilient-curvature (hybrid) interpolation formulae with transfer functions derived from the math form of the original interpolator: the trivariate quadratic B-Spline calculated without embedding ('WE'). The same notation employed in figure 1 is adopted in figures 2 through 6 regarding the rotation angles and their range.

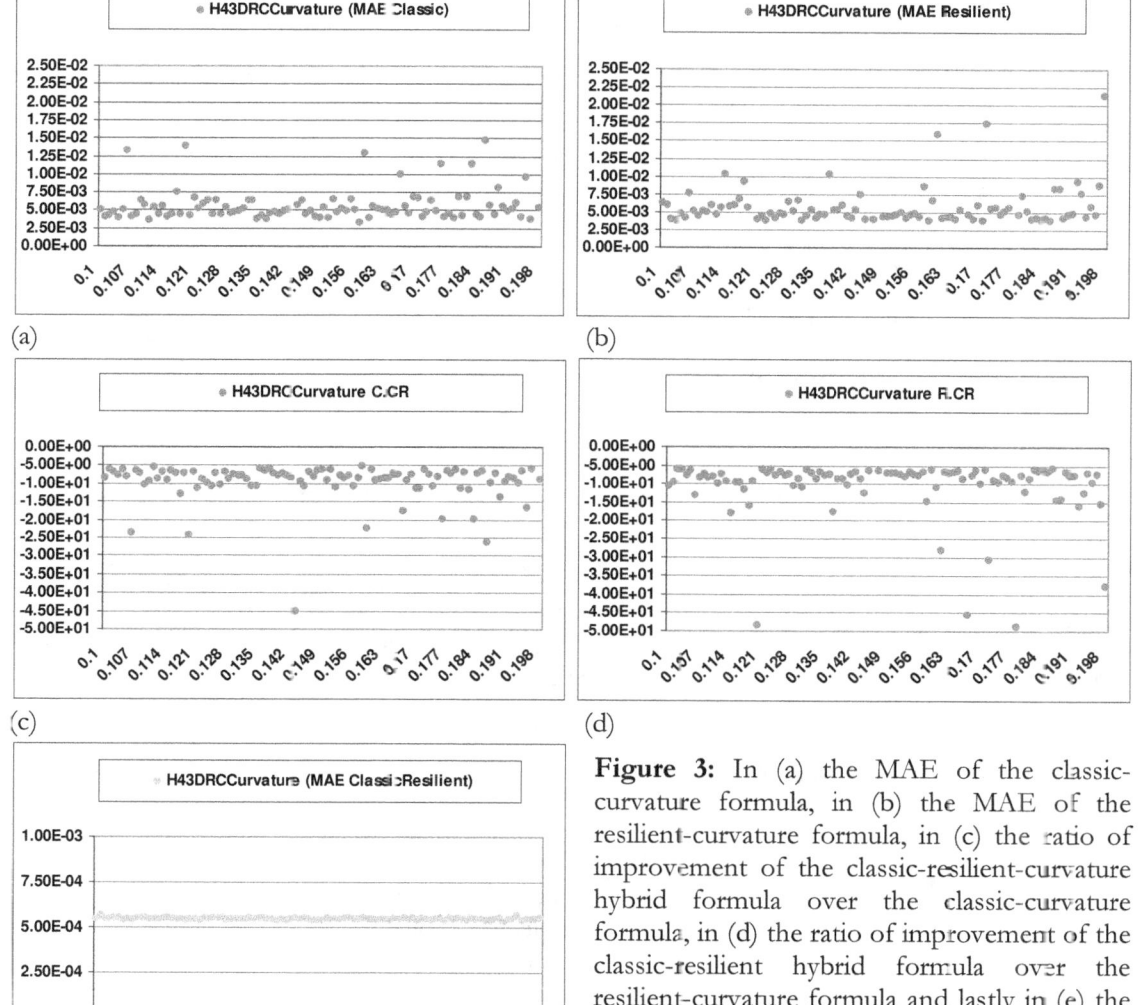

(a)

(b)

(c)

(d)

(e)

Figure 3: In (a) the MAE of the classic-curvature formula, in (b) the MAE of the resilient-curvature formula, in (c) the ratio of improvement of the classic-resilient-curvature hybrid formula over the classic-curvature formula, in (d) the ratio of improvement of the classic-resilient hybrid formula over the resilient-curvature formula and lastly in (e) the MAE of the classic-resilient-curvature hybrid formula is given. The original interpolator is the trivariate cubic B-Spline calculated with embedding.

It is worth recalling that the math form from which we calculate the total curvature of the trivariate quadratic B-Spline (classic-curvature formula without embedding) is shown in equation (19) in chapter 3 part II. As can be seen from equation (19) in chapter 3 part II the curvature is constant and dependent on the pixel intensity values and the 'a' constant only (the value of 'a' was set at 'the_A_const' = 0.745), thus permitting us to derive the custom transfer function, which still does allow reduction of the interpolation error.

To summarize, we can say that also the trivariate quadratic B-Spline without embedding does offer the option of obtaining math forms of classic-curvature and resilient-curvature formulae, which are effective in reaching their goal, which is that of lowering the interpolation error. The classic-curvature and resilient-curvature quadratic formulae calculated with embedding and without embedding (see figures 1 and 2) are in possession of the transfer function derived from the math form of the original interpolator and can lower the interpolation error.

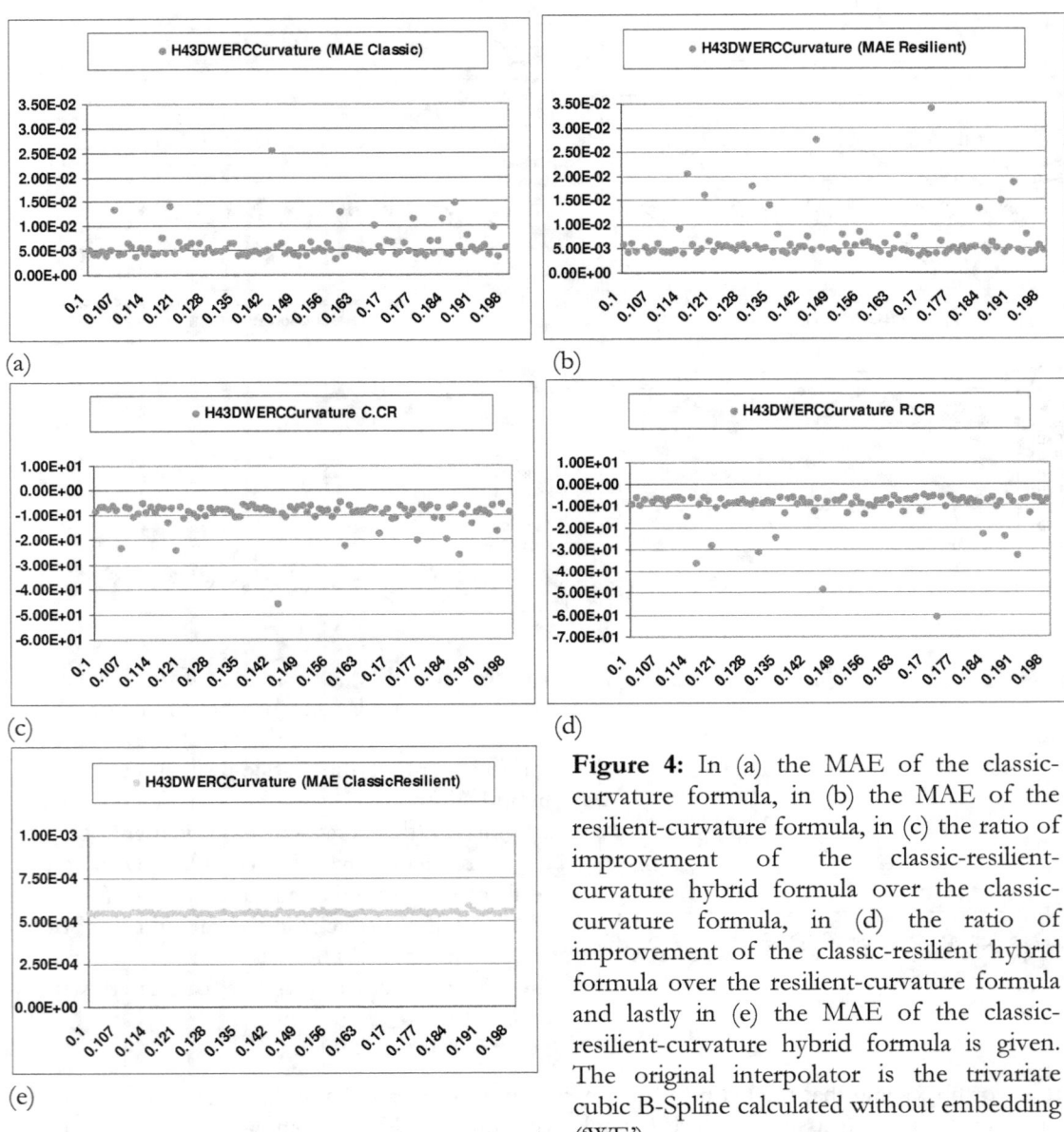

Figure 4: In (a) the MAE of the classic-curvature formula, in (b) the MAE of the resilient-curvature formula, in (c) the ratio of improvement of the classic-resilient-curvature hybrid formula over the classic-curvature formula, in (d) the ratio of improvement of the classic-resilient hybrid formula over the resilient-curvature formula and lastly in (e) the MAE of the classic-resilient-curvature hybrid formula is given. The original interpolator is the trivariate cubic B-Spline calculated without embedding ('WE').

In figure 3 the results refer to the trivariate cubic formulae calculated with embedding. In figure 3(a) the classic-curvature signal reconstruction permits MAE values between approximately 1.00 E-02 and approximately 2.50 E-03, similarly in figure 3(b) the values of the MAE are relevant to the resilient-curvature formula and range between approximately 2.50 E-03 and approximately 1.00 E-02, leaving room to improve such paradigms with the hybrid formula having a ratio of improvement between approximately 500 % and approximately 1,500 % in the case of the classic-curvature (see (c)) and also between approximately 500 % and approximately 1,500 % in the case of the resilient-curvature paradigm (see (d)). In figure 3(e) we can see the MAE of the hybrid formula with values of approximately 5.00 E-04 across the full range of rotations reported in the abscissa of the graph.

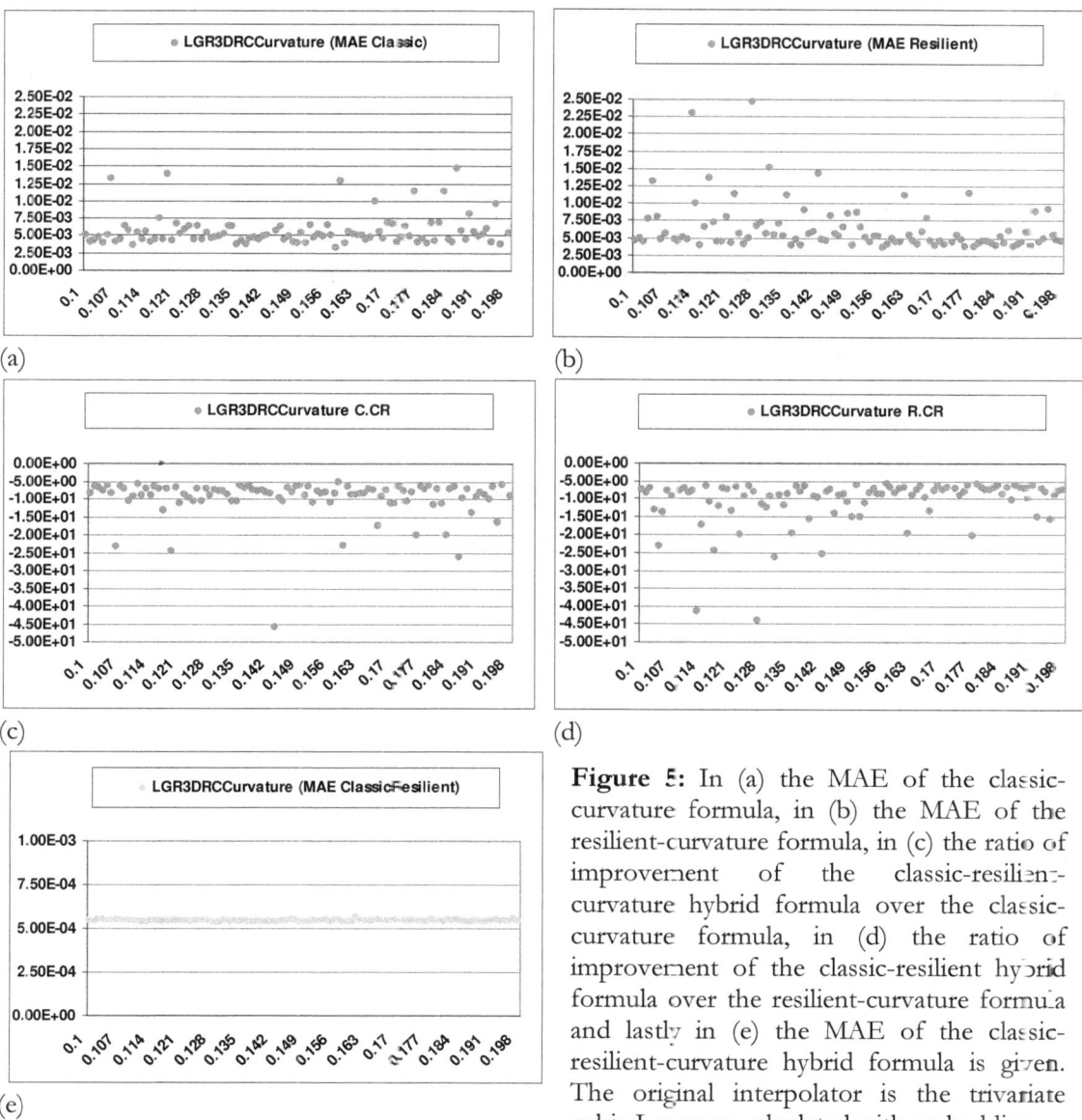

(a)

(b)

(c)

(d)

(e)

Figure 5: In (a) the MAE of the classic-curvature formula, in (b) the MAE of the resilient-curvature formula, in (c) the ratio of improvement of the classic-resilient-curvature hybrid formula over the classic-curvature formula, in (d) the ratio of improvement of the classic-resilient hybrid formula over the resilient-curvature formula and lastly in (e) the MAE of the classic-resilient-curvature hybrid formula is given. The original interpolator is the trivariate cubic Lagrange calculated with embedding.

SIGNAL RESILIENT TO INTERPOLATION

In figure 4 the results of the trivariate cubic B-Spline are those calculated without embedding. In (a) we can see the MAE of the classic-curvature formula with values between approximately 5.00 E-03 and approximately 1.50 E-02, and with scattered points that reach approximately 2.50 E-02 from the centre of 5.00 E-03. In (b) the resilient-curvature formula shows MAE values centred around 5.00 E-03 and with scattered points that reach approximately 3.50 E-02 from the centre of 5.00 E-03. The ratio of improvement of the hybrid formula over the classic-curvature formula indicate in figure 4(c) values of approximately 1,000 % with a few scattered points located between approximately 1,000 % and approximately 5,000 %.

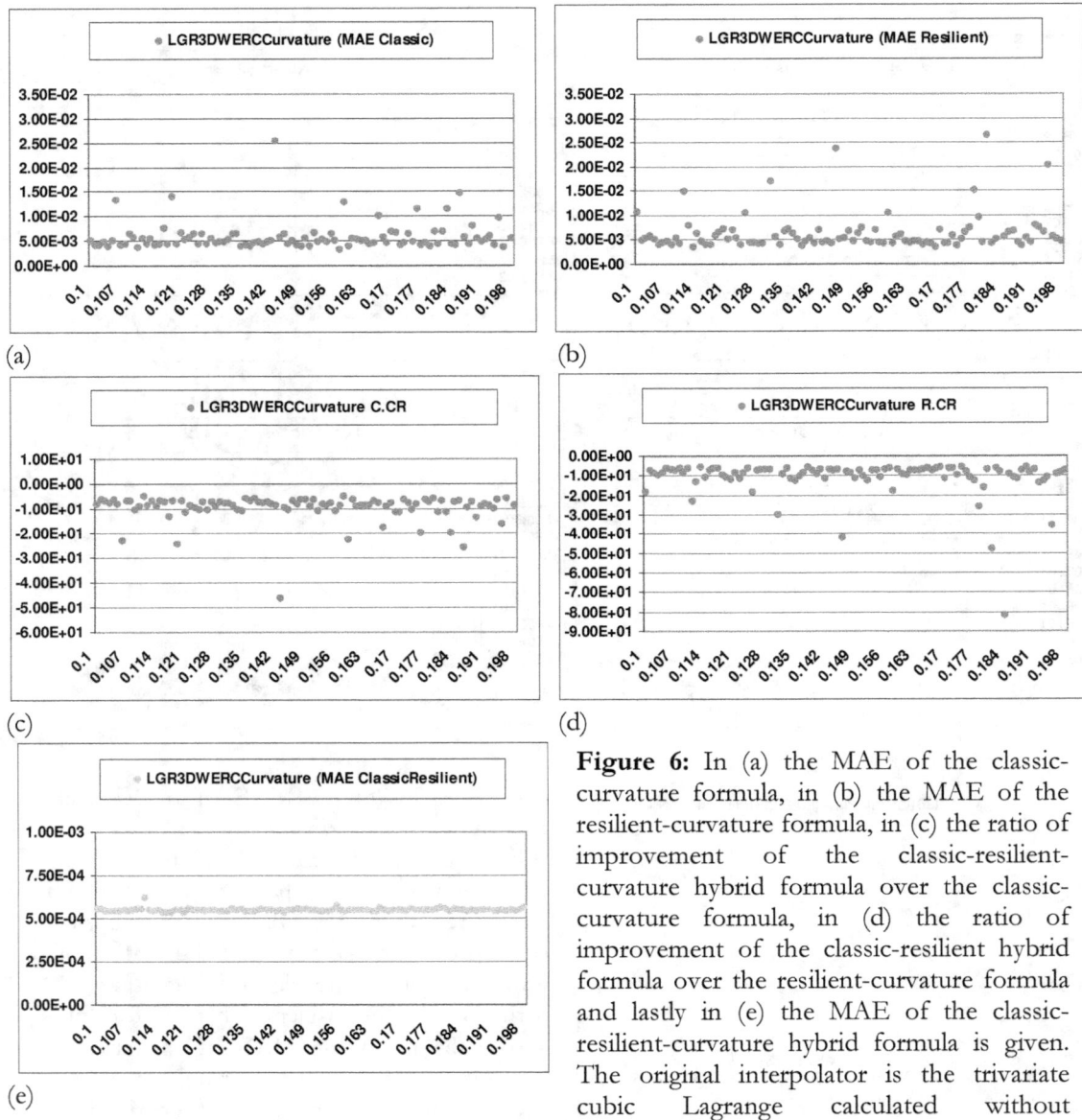

(a)

(b)

(c)

(d)

(e)

Figure 6: In (a) the MAE of the classic-curvature formula, in (b) the MAE of the resilient-curvature formula, in (c) the ratio of improvement of the classic-resilient-curvature hybrid formula over the classic-curvature formula, in (d) the ratio of improvement of the classic-resilient hybrid formula over the resilient-curvature formula and lastly in (e) the MAE of the classic-resilient-curvature hybrid formula is given. The original interpolator is the trivariate cubic Lagrange calculated without embedding ('WE').

The improvement of the resilient-curvature formula seen in (d) shows values centred at approximately 1,000 % with a few scattered points located in between approximately 1,000 % and approximately 6,000 %. Finally, figure 4(e) shows the graph of the MAE of the hybrid formula with values approximately above the quasi constant behaviour of 5.00 E-04. In the set of experiments shown in figure 4, both of the classic-curvature and the resilient-curvature formulae benefit from the hybrid formula. In figure 5 we can observe the behaviour of the results that refer to the trivariate cubic Lagrange formulae calculated with embedding; to be specific, in (a) the MAE of the classic-curvature formula which is centred around the quasi constant value of 5.00 E-03 and a discernible scattering of points whose value is between approximately 1.50 E-02 and approximately 2.50 E-03 are shown. In (b) we can see the behaviour of the resilient-curvature signal reconstruction formula with points of the MAE seen in the graph between approximately 2.50 E-02 and approximately 2.50 E-03, centred on the same quasi constant value of approximately 5.00 E-03. In (c) and (d) we can see the values of the ratio of improvement of the hybrid formula over the classic-curvature and the resilient-curvature formulae respectively, with a behaviour which is similar in that the improvement is approximately between the quasi constant values of 500 % and 1,000 %, with some scattering which brings the improvement right up to the value of approximately 4,500 %. This means that the hybrid formula can outperform either of the two formulae: classic-curvature and resilient curvature (either of the two benefits from the hybrid formula in lowering the interpolation error). In figure 5(e) we can observe the MAE of the hybrid formula which is approximately 5.00 E-04 across the full span of rigid rotations. The rigid rotation applied is of 0.001 E+00 deg about X, Y and Z axis (three simultaneous rotations) of the array [0.1 E+00, 0.1 E+00, 0.1 E+00].

In figure 6 the presentation of the results refers to the trivariate cubic Lagrange formulae calculated without embedding. In (a) we can see the graph of the MAE of the classic-curvature formula which is centred on approximately 5.00 E-03; in (b) we can see the graph of the MAE of the resilient-curvature formula which is centred on approximately the same value of 5.00 E-03; some scattering is manifest in both of (a) and (b) up to the value of 2.50 E+02. In figures 6(c) and figure 6(d) we can see the ratio of improvement of the hybrid formula (the so called classic-resilient-curvature signal reconstruction formula) over the classic-curvature and the resilient-curvature formulae respectively, with values approximately centred at 1,000 % in both graphs (c) and (d). Similarly to what is seen in figure 5(e), figure 6(e) also shows the MAE of the hybrid formula to be a quasi constant value of approximately 5.00 E-04 across the full span of rigid rotations, confirming the benefit obtained from the use of the hybrid formula in favour of the interpolation error of both classic-curvature and resilient-curvature formulae.

In equations (59) and (113) of chapter 9 we give the math form of the fraction ϱ (x, y, z) of pixel intensity correction used to calculate the total curvature C(R((x, y, z), f)) of the two forms of trivariate Lagrange polynomials: with and without embedding respectively. In equations (122) and (155) we give the math form of ϱ (x, y, z) obtained through the math process employed to calculate the total curvature C(R((x, y, z), f)) of the two forms of trivariate cubic B-Spline polynomials: with and without embedding. Finally in equations (199) and (215) we give the ϱ (x, y, z) of pixel intensity correction used to calculate the total curvature of the two forms of trivariate quadratic B-Splines: with and without embedding. The reader may appreciate that quite a remarkable difference exists between equations (199), (215) and equations (122), (155) and equations (59), (113), and what immediately comes to our attention is that in equations (199) and (215), the ϱ (x, y, z) is not entirely dependent on convolutions of pixel intensity with misplacement because of the presence of the *the_A_const*, whereas equations (122), (155) and equations (59), (113) are not dependent on any parametric value, and this is to be taken into account while looking at the behaviour of the MAE observed in the case of the two forms of trivariate quadratic B-Splines, with and without embedding, as seen in figures 1 and 2.

TOTAL CURVATURE: UNIT ANALYSIS

The unit analysis is relevant to the processing done to input the *total curvature* of the classic signal and/or the Signal Resilient to Interpolation (SRI) in the function tangent so as to calculate the Pixel Intensity Correction (PIC) in the following customary and consistent form for all the interpolation formulae treated in this book. PIC = Shift * tan (*total curvature*), where the shift is calculated with the Pythagorean Theorem and the *total curvature* is calculated in the way we have seen in the math developments presented in several preceding chapters. We may immediately acknowledge that the unit of the *total curvature* is $[mm * I]^k$ where k is the exponent, mm is millimeters and I is the unit of pixel intensity. The unit $[mm * I]$ is located in the convolution of misplacement with pixel intensity (see interpolation formulae) and also in: the derivatives of the interpolation formula; intensity-curvature terms (before and after interpolation); and certainly also in the *total curvature*.

To derive k, the exponent, let us undertake the following generalized form of math deduction process. In the polynomial (interpolation formula) let n be the maximum exponent. Thus the polynomial is of order n.

At the first step of the math deduction process, the two intensity-curvature terms are calculated, the intensity-curvature term before interpolation called E_O and the intensity-curvature term after interpolation called E_{IN}.

Let us recall that E_O requires the integration of the second order derivative of the interpolation function calculated at the grid point (for instance x = 0 in 1D), whereas in the calculation of E_{IN} the second order derivative is calculated at the intra-node location (for instance x $\neq$ 0 in 1D).

In fact, the integrand of E_O and E_{IN} is made of the product between two terms (such product is called: *Intensity-curvature multiplication*, see section titled 'DEFINITION OF TECHNICAL TERMS'). In the case of E_O the two terms are: the numerical value of the signal (constant, not a variable) and the second order derivative term (order t = (n-2)). For instance, if n = 2 (which means that the interpolation formula is quadratic), the second order derivative term has order t = (n-2) = 0. In the case of E_{IN} the two terms are: the interpolation formula (order n) and the second order derivative term (order t = (n-2)). Because of the product between two terms, the integrand of E_O and E_{IN} has order (n-2) and (n+t) respectively, and this happens in 1D, 2D and 3D. Therefore, before the calculation of the integral in either 1D, 2D or 3D, the product between the two terms (called intensity-curvature multiplication, and which is the integrand of E_O and E_{IN}), the order of the product is (n-2) (in the case of E_O; the constant (value of the signal) does not change the order of the product), or the order of the product is (n+t) (in the case of E_{IN}).

To summarize, before the integral, the integrand of E_O has order (n-2) and E_{IN} has order (n+t), with t = 0, 1 for quadratic and cubic interpolation functions respectively.

Calculation of E_O and E_{IN} requires that the integrals in 1D, 2D or 3D are calculated. Therefore at the second step, both of E_O and E_{IN} undergo single (1D), double (2D) or triple (3D) integration. In general, after the integration, the order of E_O and E_{IN} is: (n-2+m) and (n+t+m) respectively; where m is the dimensionality (m = 1 in 1D, m = 2 in 2D and m = 3 in 3D).
Specifically, after the integration, the order of E_O is (n-2+1) in 1D, (n-2+2) in 2D, and (n-2+3) in 3D; and the order of E_{IN} is (n+t+1) in 1D, (n+t+2) in 2D, and (n+t+3) in 3D. When the interpolation formula is quadratic, t = (n-2) = 0. When the interpolation formula is cubic, t = (n-2) = 1.

In relationship to both quadratic and cubic interpolation formulae, the integration makes the order of E_O to be (n-2+m), and the order of E_{IN} to be (n+t+m). The higher is the maximum exponent of the interpolation formula (e. g. quadratic, cubic, etc.), the higher (t+m) becomes because t = (n-2). To have a readily observable example, see the calculations in equations (30) through (49) in chapter 11, in the paragraph titled 'EQUATION OF THE INTENSITY CURVATURE TERMS: RESILIENT FORMULATION'.

At this point, as far as regards the calculation of the *total curvature* (see definitions of: *Total curvature of an interpolation function* and *Total curvature of the Signal Resilient to Interpolation* in the section titled 'DEFINITION OF TECHNICAL TERMS') of either the classic signal or the SRI, the procedure requires the calculation of all of the second order derivatives. The number of second order derivatives is 4 in 2D and is 9 in 3D. Also, in some cases, the SRI is a ratio between a numerator and a denominator.

As seen in the formulae of the *total curvature* in the preceding chapters, because of the calculation of the second order derivatives of the interpolation formula with order n, in either 1D, 2D or 3D, the order of the *total curvature* may change from (n-2) to (n-2+r) depending on r. What determines r is the calculation of the *total curvature* of the classic signal or the SRI (depending on which case is considered), which entails that r depends on: (i) the fact as to if the *total curvature* is calculated of the classic signal (r ≥ 0) or the SRI (r ≥ 0); (ii) the fact as to if the SRI is a ratio (r ≥ 0) or not (r ≥ 0); and (iii) the exponent of the variables which are not derived because they are kept constant (because are partial in relationship to the variable used to calculate the second order derivatives of which the *total curvature* comprises of). For both of the cases (either classic signal or SRI) (iii) is true and r ≥ 0, because while calculating first and second order partial derivatives (for instance in the case of an interpolation formula with two independent variables: x and y) with respect to the variable x, the variable y is kept constant because is partial in relationship to the variable x.

Finally, we can generalize by saying that: (i) the order of the polynomial interpolation function is n; and (ii) after the aforementioned process, the order of the polynomial form of the *total curvature* of the classic signal is (n-2+r), and the order of the polynomial form of the *total curvature* of the SRI is (n-2+r) with r ≥ 0. Thus, the unit of the *total curvature* will be [mm * I] raised to the power of (n-2−r) (in the case of the classic signal) and (n-2+r), (in the case of the SRI), which is: $[mm * I]^k$; where k = (n-2) or k = (n-2+r).

RESULTS WITH THE TRIVARIATE QUADRATIC POLYNOMIALS

The value of the MAE furnished through 3D trivariate quadratic signal reconstruction using the curvature of the classic interpolation formula is comparable that one MAE of the two other 3D cubic paradigms of signal reconstruction using the curvature (trivariate cubic and trivariate cubic Lagrange).

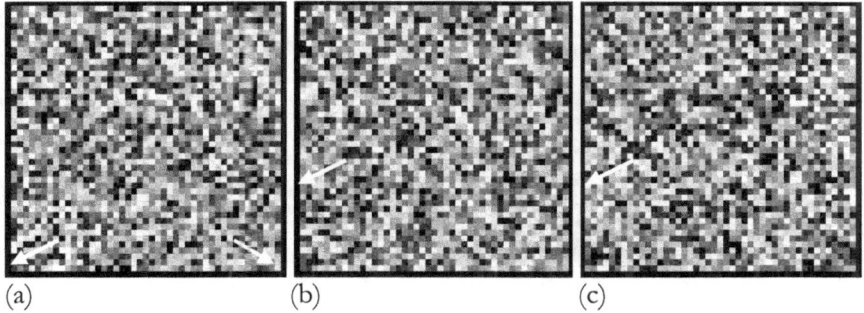

(a) (b) (c)

Figure 7: Original signal images, Bottom − Top View in (a) which is the third slice from the bottom; Left − Right View in (b) which is the third slice from the left; Front − Back View in (c) which is the third slice from the front.

This is readily observable when looking at figures 1(e), 2(e), 3(e), 4(e), 5(e) and 6(e), which show the value of MAE at approximately 5.00 E-04.

SIGNAL RESILIENT TO INTERPOLATION

Initially the original signal is contained in a discretized box of 47x47x47 pixels. The original signal is generated with the zero outer most stripe (spatial location (x, y, z) = (1, 1, 1) and spatial location (x, y, z) = (47, 1, 1), see white arrows in figure 7(a)). Before the original signal seen in figure 7 is used to calculate the curvature images (both classic and resilient), computer processing is performed on it, specifically, lines of 47 pixels (herein defined as lines) are zero filled in each slice of the volume. The scope of the computer processing is that one of zero filling the signal. More specifically, the signal is zero filled with additional $N_x = 4$ lines along the X direction in the bottom - top view, the signal is zero filled with additional $N_y = 4$ lines in the Y direction in the left – right view, and also the signal is zero filled with additional $N_z = 4$ lines along the Z direction in the front – back view. The result of the computer processing is that of placing the original non-zero valued signal inside a discretized box of 41x41x41 pixels, therefore the original signal is zero filled inside the 3D volume of 47x47x47 pixels. The same computer processing is performed after the calculation of both classic and resilient curvature images, and in such way the 3D volume containing the curvature images is cropped from the initial size of 47x47x47 pixels down to the size of 41x41x41 pixels.

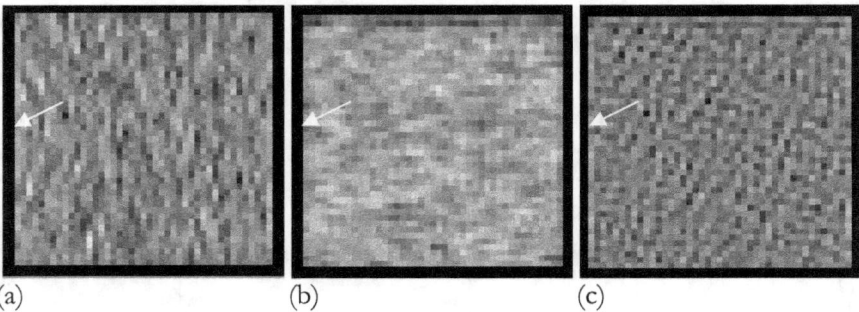

(a)   (b)   (c)

Figure 8: Classic Curvature Images, Bottom – Top View in (a) which is the third slice from the bottom; Left – Right View in (b) which is the third slice from the left; Front – Back View in (c) which is the third slice from the front. The three angles of the rigid rotation were 0.1 deg, and the three misplacements X, Y and Z were 0.1.

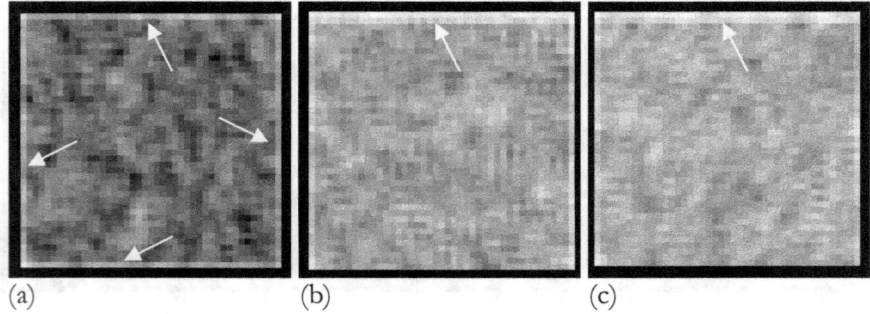

(a)   (b)   (c)

Figure 9: Resilient Curvature Images, Bottom – Top View in (a) which is the third slice from the bottom; Left – Right View in (b) which is the third slice from the left; Front – Back View in (c) which is the third slice from the front. The three angles of the rigid rotation were 0.1 deg, and the three misplacements X, Y and Z were 0.1.

The aforementioned explanation is in order to understand as to why figure 7 displays the signal in three composite views of 46x46x46 non-zero pixels and also as to why figures 8 and 9 display the curvature images in three composite views of 41x41x41 pixels.

During the calculation of the curvature images seen in figures 8 and 9, processing is done on neighbourhoods of 3x3x3 pixels and the first neighbourhood which is processed has its center located at the spatial position $(x, y, z) = (3, 3, 3)$.

The first processed neighbourhood has thus spatial extent along the X direction with center at the location $(x, y, z) = (3, 3, 3)$, and extends from $(x, y, z) = (2, 3, 3)$ to $(x, y, z) = (4, 3, 3)$. Similarly, the first processed neighbourhood has thus spatial extent along the Y direction with center at the location $(x, y, z) = (3, 3, 3)$, and extends from $(x, y, z) = (3, 2, 3)$ to $(x, y, z) = (3, 4, 3)$. And, the first processed neighbourhood has thus spatial extent along the Z direction with center at the location $(x, y, z) = (3, 3, 3)$, and extends from $(x, y, z) = (3, 3, 2)$ to $(x, y, z) = (3, 3, 4)$.

Consequentially, the curvature image has high curvature stripes which are determined because of the gradient existing across the pixel intensity values at the aforementioned (x, y, z) locations and the complete processing across the image, and such high curvature stripes are seen in figure 9(a) (indicated by the arrows) at the outer most edges of the bottom - top view of the volume (XY plane).

An additional consequence is visible in the curvature image seen in figure 9(b) (see arrow) where the high curvature stripes (2) are visible at the left edge of the left – right view of the volume (YZ plane) and are determined because of the gradient existing across the pixel intensity values along the Z direction.

Another consequence is visible in the curvature image seen in figure 9(c) (see arrow) where the high curvature stripes (2) are visible at the top edge of the front – back view of the volume (XZ plane) and are determined because of the gradient existing across the pixel intensity values along the Z direction.

The processing of the curvature images (both classic and resilient) includes the original signal pixels inside the closed interval [3, 45] of the 47x47 matrix relevant to bottom – top view of the volume (XY plane), also of the 47x47 matrix relevant to front – back view of the volume (XZ plane), and also of the 47x47 matrix relevant to the left – right view of the volume (YZ plane). The processing is performed through an algorithm using three loops, the outer-most running along the Z direction, the inner running along the X direction, and the inner-most running along the Y direction. Practically, this entails that during processing the center of the neighbourhood of f (0, 0, 0) moves along the Y axis of those pixels located in the left – right views of the volume (YZ planes), from the location $(x, y, z) = (3, 3, 3)$ to the location $(x, y, z) = (3, 45, 3)$.

This implies that three curvature stripes are determined through the gradient existing across the pixels located along the Z direction: one stripe located in the bottom - top views of the volume (XY plane) and is seen in figure 9(a) (in one slice, indicated by the bottom arrow); and two more stripes located in the left – right views of the volume (YZ planes) and they are seen in figure 9(b) (in one slice, see the arrow) at the top of the curvature image, and they also seen in figure 10 (as slices) in slice 44 and slice 45. Another implication is that two more curvature stripes are determined through the gradient existing across the pixels located along the Z direction: in the front – back views of the volume (XZ planes) and they are seen in figure 9(c) (in one slice, see the arrow) at the top of the curvature image, and they are also seen in figure 10 (as slices) in slice 44 and slice 45. The aforementioned curvature stripes are well pronounced in the resilient curvature images as shown in figure 9.

.

QUALITATIVE EVALUATION: TRIVARIATE LAGRANGE CURVATURE CALCULATED WITH EMBEDDING

This section investigates the nature of the curvature images obtained through both classic-curvature and resilient-curvature formulae derived from the trivariate Lagrange polynomial calculated with embedding. In the investigation, the following assumption is made: *curvature images reproduce to a large extent the shape and details of the data structure in the original image volume.* In order to ascertain and to make assessments on such an hypothesis we may conduct the following experimentation.

The image volume is processed with the aforementioned image reconstruction paradigm (trivariate Lagrange polynomial) and the curvature images are calculated and they are shown in figure 10. The next step is that of calculating the difference image at each slice of the volumes (see rows numbered from 3 through 45 in figure 11). The difference images calculated are: (i) the original signal minus the classic-curvature (shown in the central column in figure 11), and (ii) the original signal minus the resilient-curvature (shown in the right most column in figure 11).

Clearly, any deviations from zero constitute a difference between volumes: original image volume versus classic-curvature image volume, and original image volume versus resilient-curvature image volume, and such a difference is in contrast with the working hypothesis. In order to increase the information content provided to the reader with the curvature images (classic and resilient) shown in figure 11, the histogram of the difference between the resilient curvature volume and the classic curvature volume is calculated and shown in figure 12 and is called the 'Histogram of the Difference (Resilient Curvature - Classic Curvature)'.

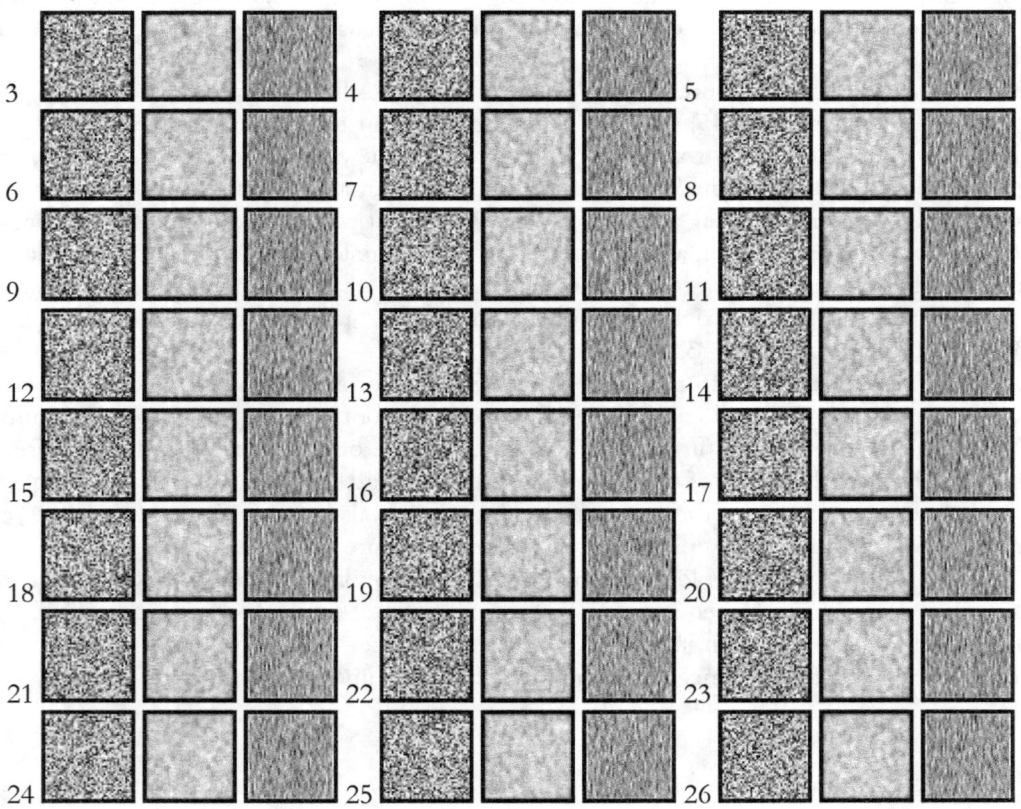

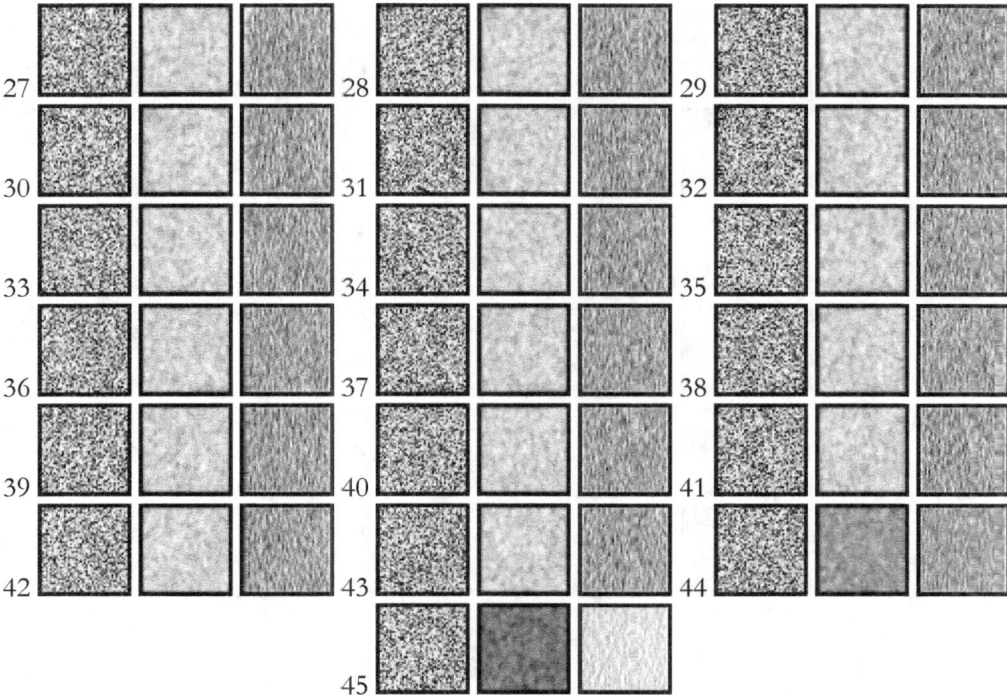

Figure 10: The image volume shown in the left-most column in 2D XY plane slices, from bottom (third slice (3)) to top (forty-fifth slice (45)); 2D views of corresponding curvature images: classic curvature to the right of the image volume (central column) and resilient curvature in the right-most column of the picture.

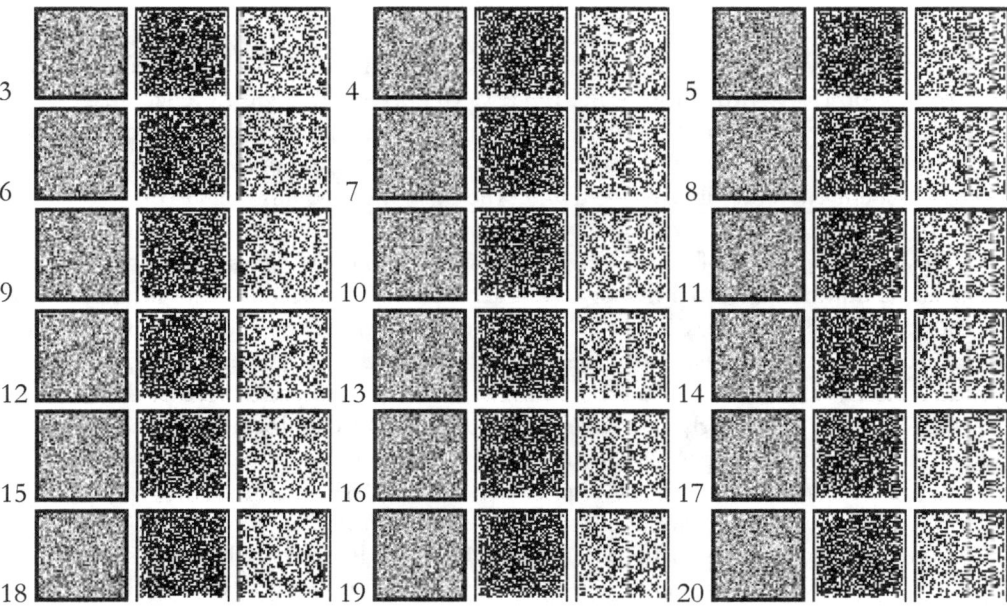

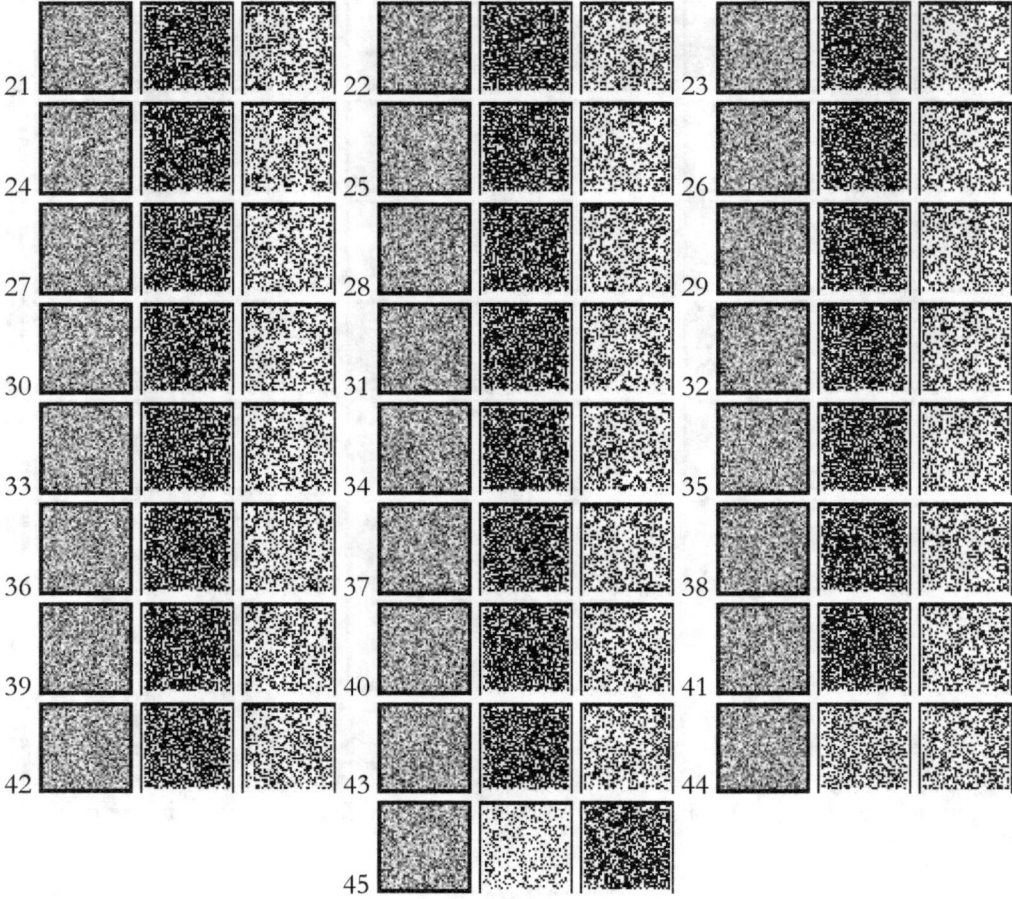

Figure 11: The image volume shown in the left-most column in 2D XY plane slices, from bottom (third slice (3)) to top (forty-fifth slice (45)); 2D views of corresponding difference images: (i) the difference between the original image (shown in left-most column) and the classic-curvature image is shown in central column and (ii) the difference between the original image (shown in left-most column) and the resilient-curvature image is shown in the right-most column of the picture.

Figure 10 shows the slices of the original signal volume and the slices of the two curvature volumes; each slice is numbered from 3 through 45. Each volume is made of 47 slices and slices 1 and 47 are null as per data structure. Slices 1, 2, 46, 47 are not part of the classic-curvature volume nor of the resilient-curvature volume, because of the neighbourhood of the interpolation formulae which starts at slice 3 and ends at slice 45. In the curvature images, slice 3 is comprehensive of data of the original volume in slice 2, 3 and 4. Similarly slice 45 is comprehensive of data located in slice 44, 45 and 46 in the original volume. This concept was illustrated in the figures 7, 8 and 9 of the preceding section along with information pertinent to the interpretation of the MAE results of the hybrid formulae obtained from the math forms of trivariate quadratic, cubic and cubic Lagrange polynomials.

When evaluating the pictures in figures 10 and 11 the reader may wish to consider that the base of scaling of the image volume is the numerical interval [0, 32767] and the base of scaling of the curvature volume is [0, 32757]. Figure 11 shows difference images: original signal minus classic curvature (centre column) and original signal minus resilient-curvature (right-most column). Each slice of the volumes is numbered from slice 3 through slice 45.

In figure 12, positive difference values are indicative of dissimilarity between the resilient curvature image volume and the classic curvature image volume. The values in figure 12 are all positive confirming what can be seen in figure 11 where is observable that the magnitude of the numerical values of the resilient curvature images is higher than the magnitude of the of the numerical values of the classic curvature images. Figure 12 clearly shows that in this particular experimental session the resilient curvature has numerical values different from the numerical values of the classic curvature, and does not provide evidence in support to the hypothesis that classic and resilient curvature images reproduce to a large extent the data structure of the original image volume (which is shown in figures 10 and 11).

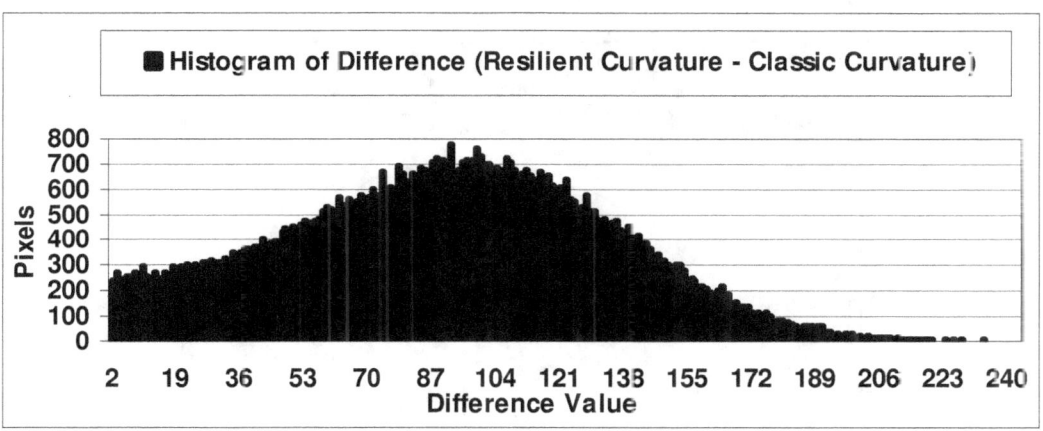

Figure 12: 'Histogram of the Difference (Resilient Curvature - Classic Curvature)' obtained subtracting the following two terms: (i) resilient-curvature image volume and (ii) classic-curvature image volume. Data at the 'Difference Value' 1 were omitted to increase visibility. The number of pixels is 11865.

SIMILARITY BETWEEN IMAGE DATA AND CURVATURE IMAGES' MAPS

Naturally, more investigation is needed to further sustain the aforementioned hypothesis which reads as: *the curvature images reproduce to a large extent the shape and details of the data structure in the original image volume*. In the following paragraph slices of a T2 Magnetic Resonance Imaging (MRI) volume (Courtesy of: R. S. Swenson, www.Dartmouth.edu/~rswenson/Atlas) were processed with the three types of interpolation formulae (quadratic, cubic and cubic Lagrange) in their two variants with and without embedding, which give rise to the classic-curvature, resilient-curvature, and classic-resilient-curvature hybrid signal reconstruction formulae.

This paragraph shows that both classic-curvature and resilient-curvature formulations are able to calculate maps of curvature images that closely resemble features and details of the image data. Figures 13 through 24 show data in support of such a statement, with calculations made with the classic-curvature and resilient-curvature signal reconstruction respectively.

Each figure is organized in three rows, in the top row we show the original T2 MRI, in the second and third row from the top we show the resulting curvature maps obtained with the given polynomial, without embedding and with embedding respectively.

Original

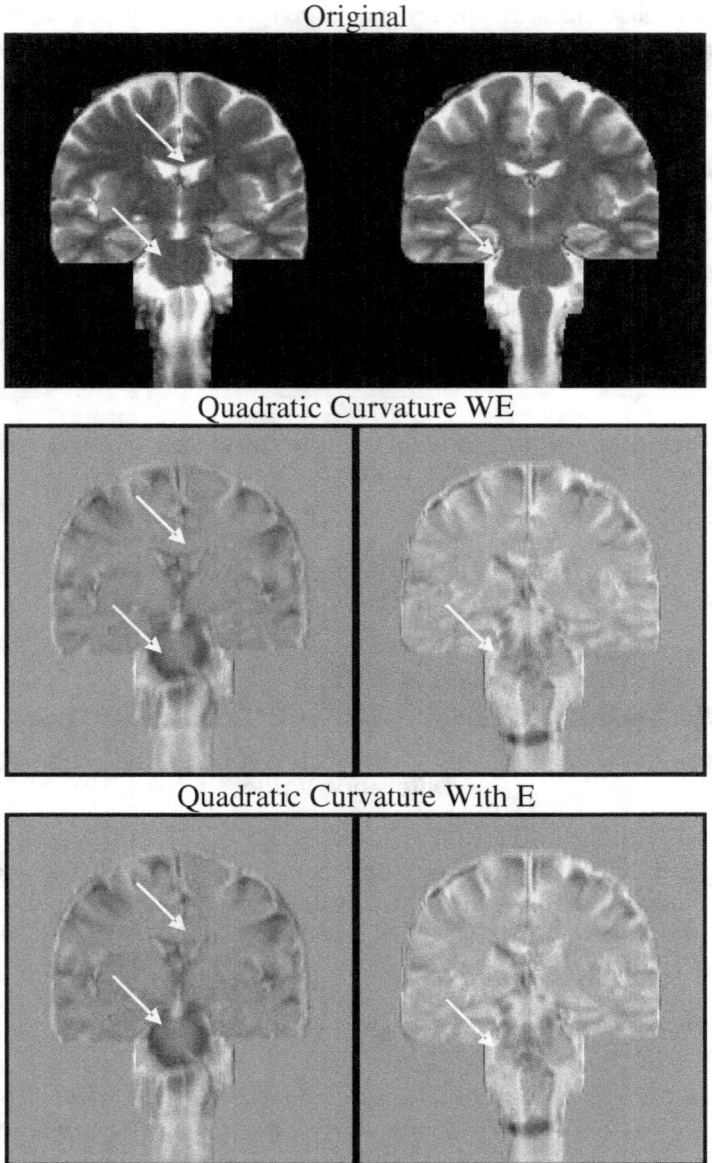

Figure 13: In the top row we have the original T2 MRI data, in the middle and lower rows respectively we have the classic curvature maps obtained with quadratic polynomials without embedding ('WE') and with embedding ('With E').

In figure 13, in the images at the left, there are two arrows pointing to the ventricles and the thalamus of the human brain, whose purpose is that of highlighting the similarity in the features of the curvature maps with the original T2 MRI at the top row. Such similarity is more pronounced in the curvature images located at the left than it is in the curvature images located at the right of the figure. In these three pictures, the cortical structures are also well mapped in the curvature images located at the left. In figure 13, in the images at the right, which follow on from each other from front to rear, the thalamic structures are better mapped than the ventricles are (see arrows).

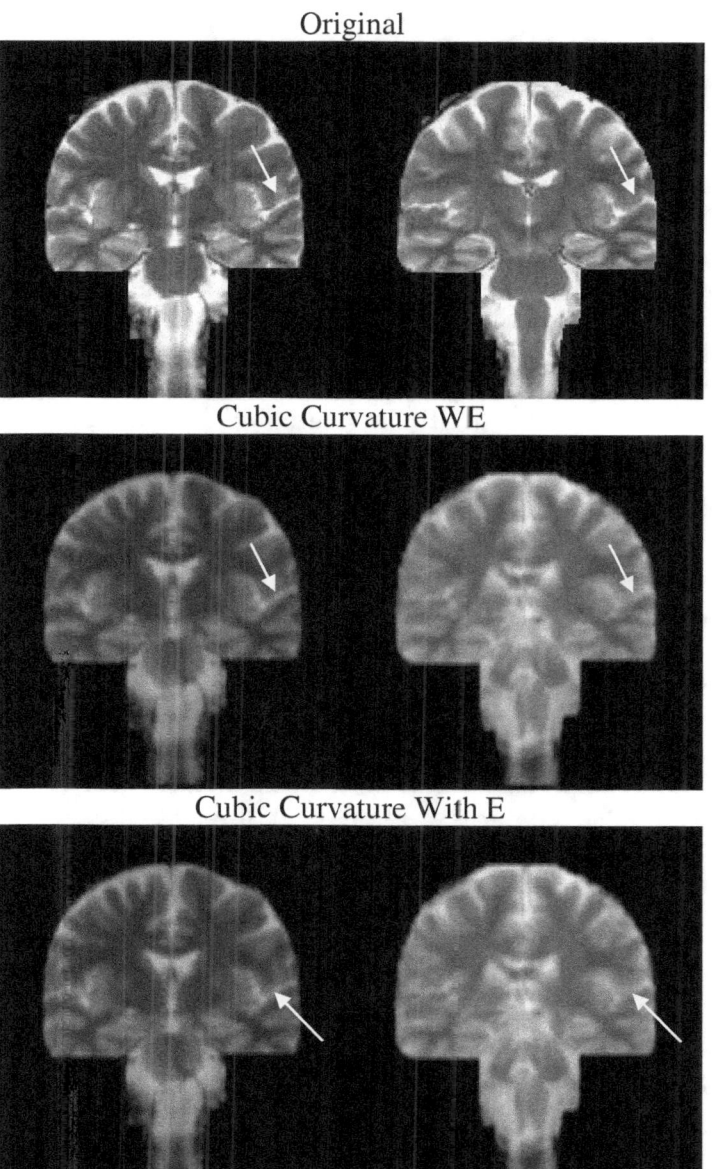

Original

Cubic Curvature WE

Cubic Curvature With E

Figure 14: The top row shows the original T2 MRI data, the middle and lower rows show respectively classic curvature maps obtained with cubic polynomials, without embedding ('WE') and with embedding ('With E').

Ventricles and thalamic structures are well mapped in the curvature images, also when the signal reconstruction is conducted with cubic polynomials, and this is shown in the left of figure 14. The reader should not misinterpret the images in the middle and lower row as those being the result of signal reconstruction.

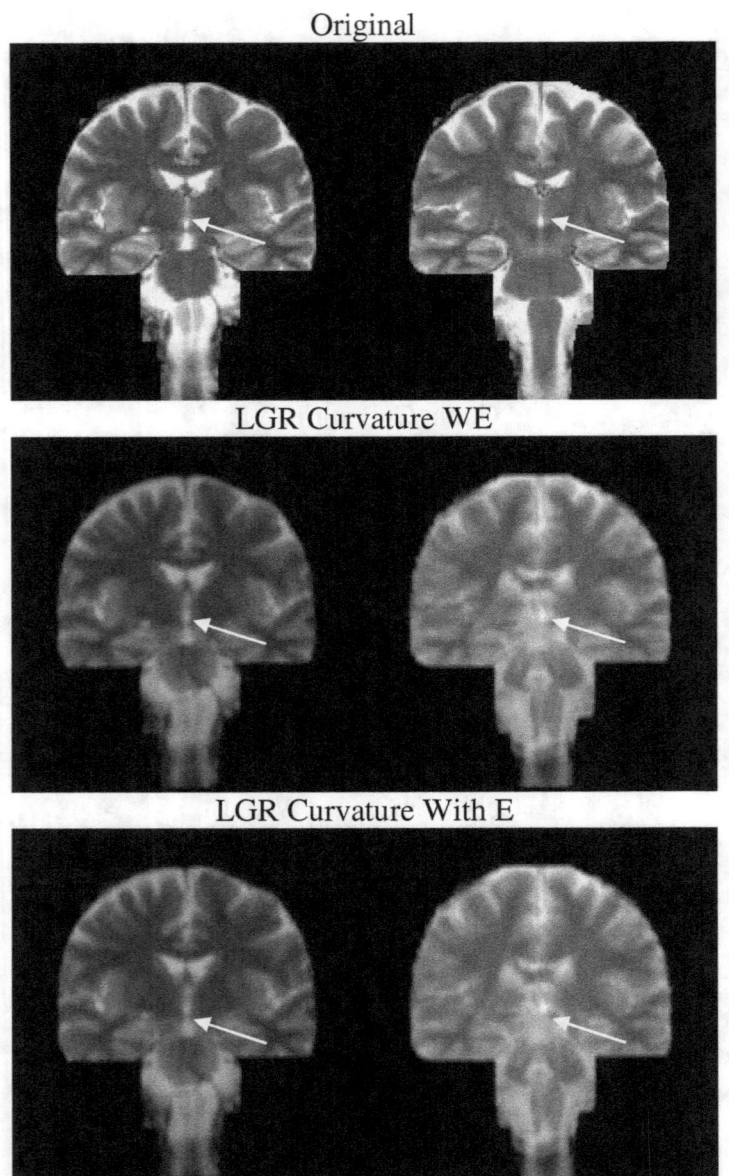

Figure 15: The Original T2 MRI data and the classic curvature maps obtained with cubic Lagrange polynomials without embedding ('WE') and with embedding ('With E'), are shown in the top, middle and lower rows respectively.

The images in the middle and lower row are curvature images and as such they determine the math form of the transfer function and they function as an intermediate step towards the determination of signal interpolation. The arrows in figure 14 point to other features of the images highlighting the similarity of details (shapes) between the images at the top (original) and the curvature images without and with embedding. Such details are more pronounced in the curvature images located at the left than they are in the curvature images located at the left of the figure.

Original

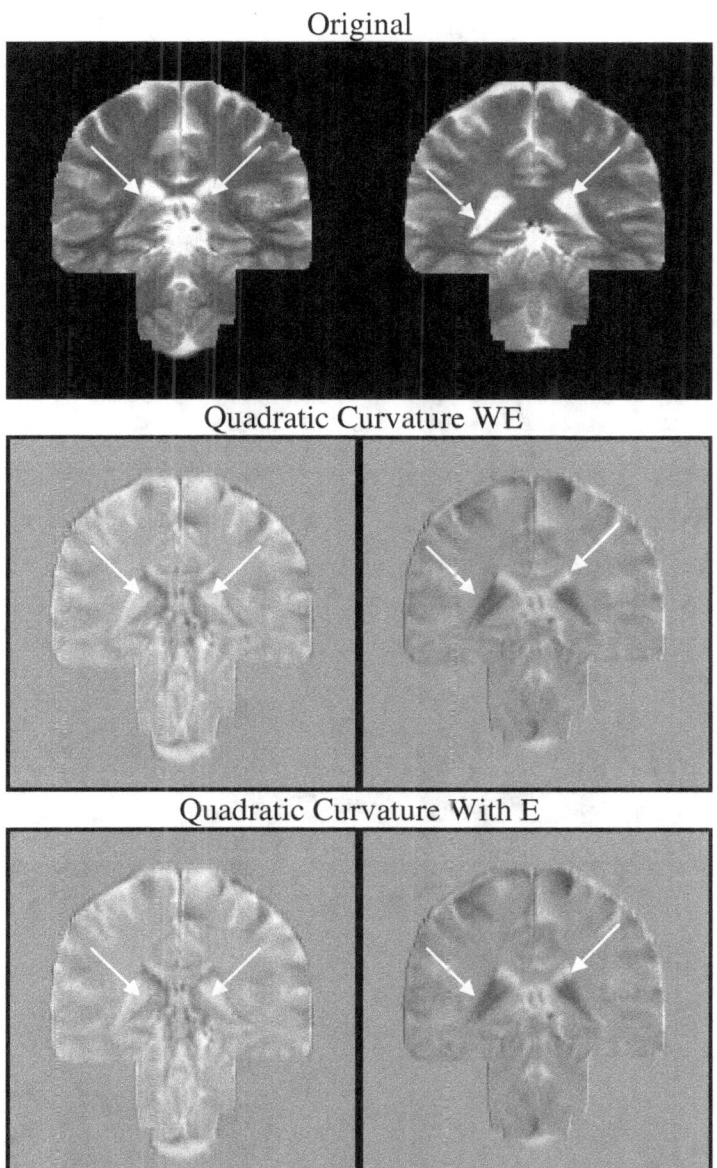

Quadratic Curvature WE

Quadratic Curvature With E

Figure 16: The Original T2 MRI data and the classic curvature maps obtained with quadratic polynomials without embedding ('WE') and with embedding ('With E') are shown in the top, middle and lower rows respectively.

The arrows in figure 15 show another aspect of the capability of the curvature images to reproduce through their maps the features of the original T2 MRI. In this case the focus is on the brain regions located below the ventricles and above the thalamus (see the white arrows). Throughout figures 13, 14 and 15 we can also clearly see that the curvature images placed in the right column are slightly less exhaustive in showing similarity of details with the original T2 when their degree of mapping is compared to the images located in the left column.

Original

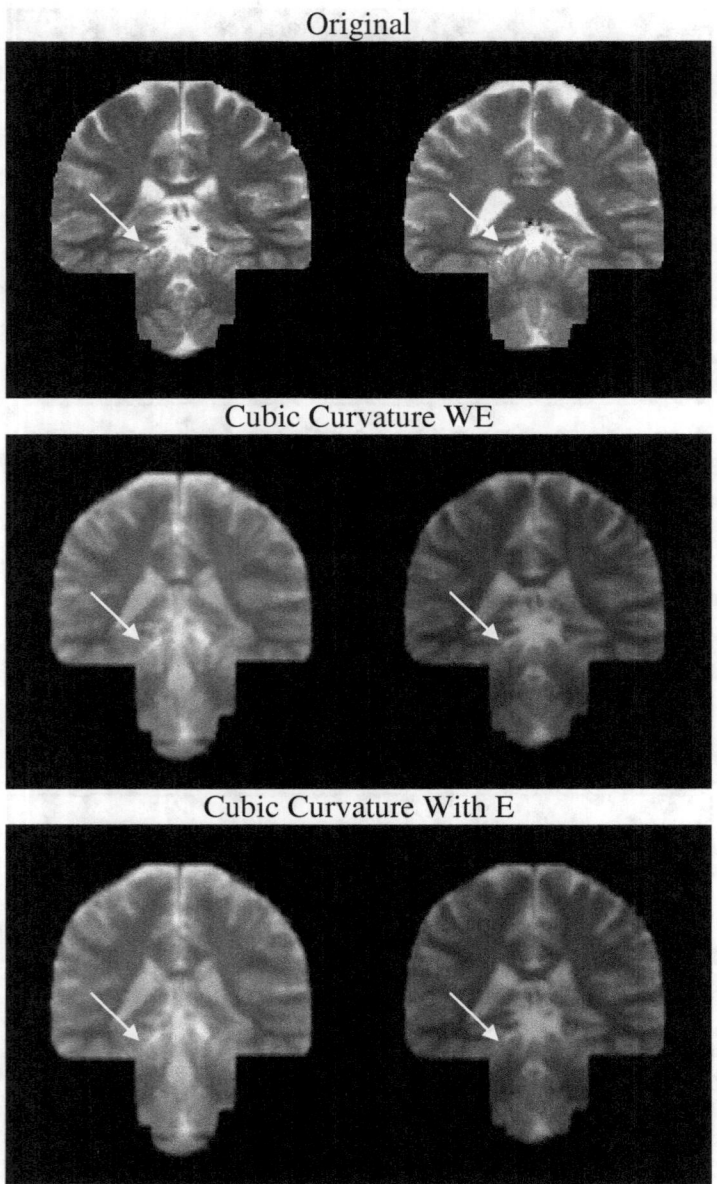

Cubic Curvature WE

Cubic Curvature With E

Figure 17: The Original T2 MRI data and classic curvature maps obtained with polynomials without embedding ('WE') and with embedding ('With E'). These images were calculated with the cubic formulations.

Nevertheless these data give far more support to the statement that the curvature maps reproduce more details and shapes of the original images, than the synthetic images shown earlier in this chapter.

The data shown in figure 16 are different from the T2 MRI shown in figures 13 through 15. Again the effort made is in finding similarities between the original images and the classic curvature image maps. Looking at the arrows, the brain structures that most closely resemble features of the original image are the right and left ventricles of the human brain.

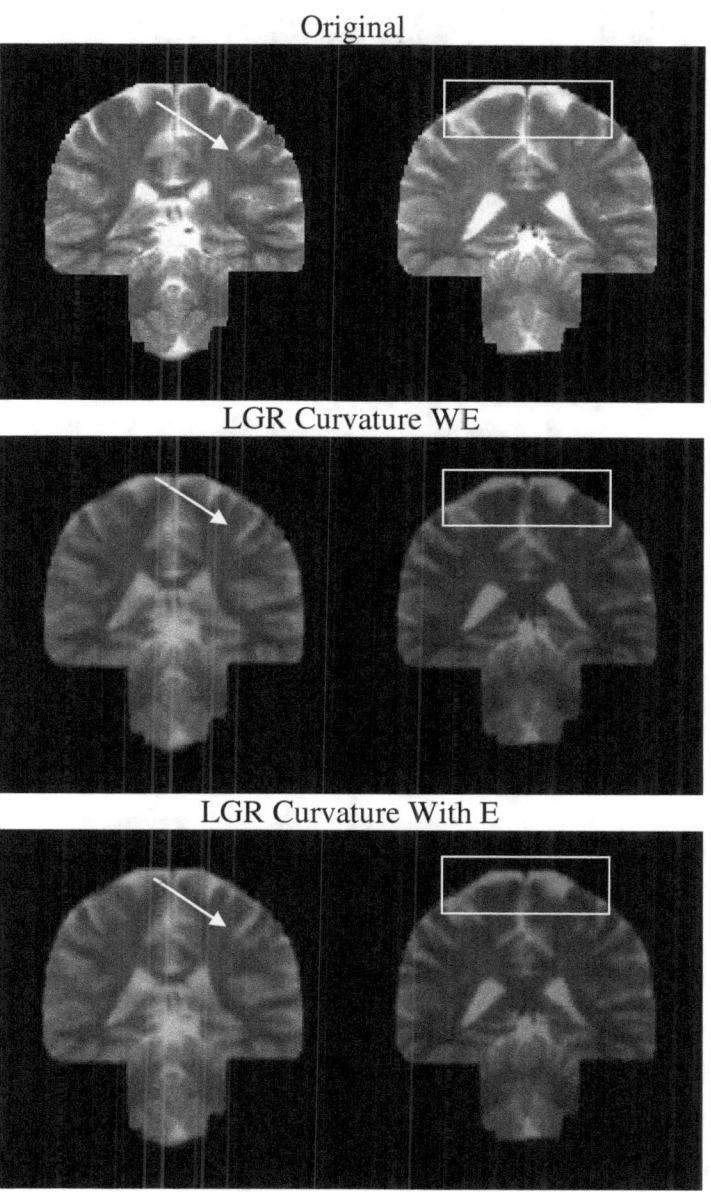

Figure 18: The Original T2 MRI data and classic curvature maps obtained with polynomials without embedding ('WE') and with embedding ('With E'). These images were calculated with the cubic Lagrange formulations.

An interesting feature that correlates the brain images located at the top row to the curvature maps seen in the lower two rows from the top is what seen in the luminosity of the ventricles, bright in the image located at the upper most row at the right, and dark in the image located in the upper most row at the left.

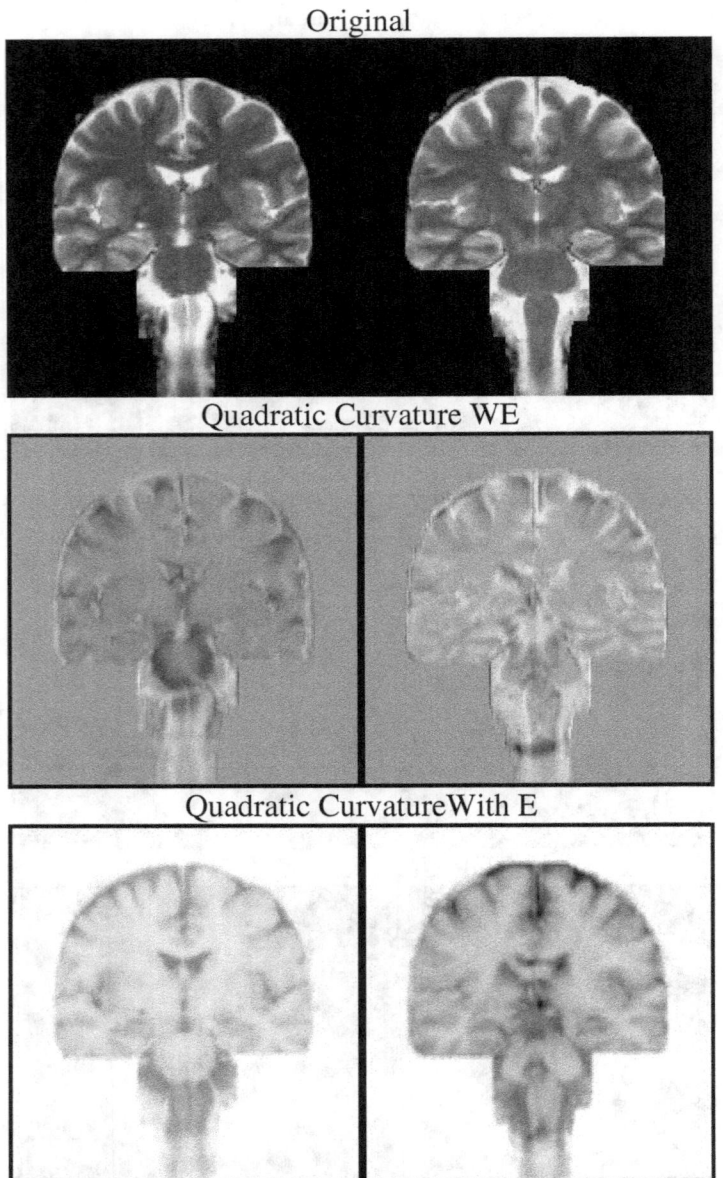

Figure 19: Original T2 MRI data and resilient curvature maps obtained with quadratic polynomials without embedding ('WE') and with embedding ('With E').

The curvature images reproduce the inverse of such a difference, dark seen at the image at the right and bright seen at the image at the left. Such behavior is in support to the manifest change in image gradient shown in the images at the top most row (see arrows), and the manifest change in image gradient, correlates well with the change in curvature discernible in the curvature images shown in the two rows below the top row. The aforementioned behavior is confirmed through the maps that have been calculated by the computer program, as shown clearly in these pictures.

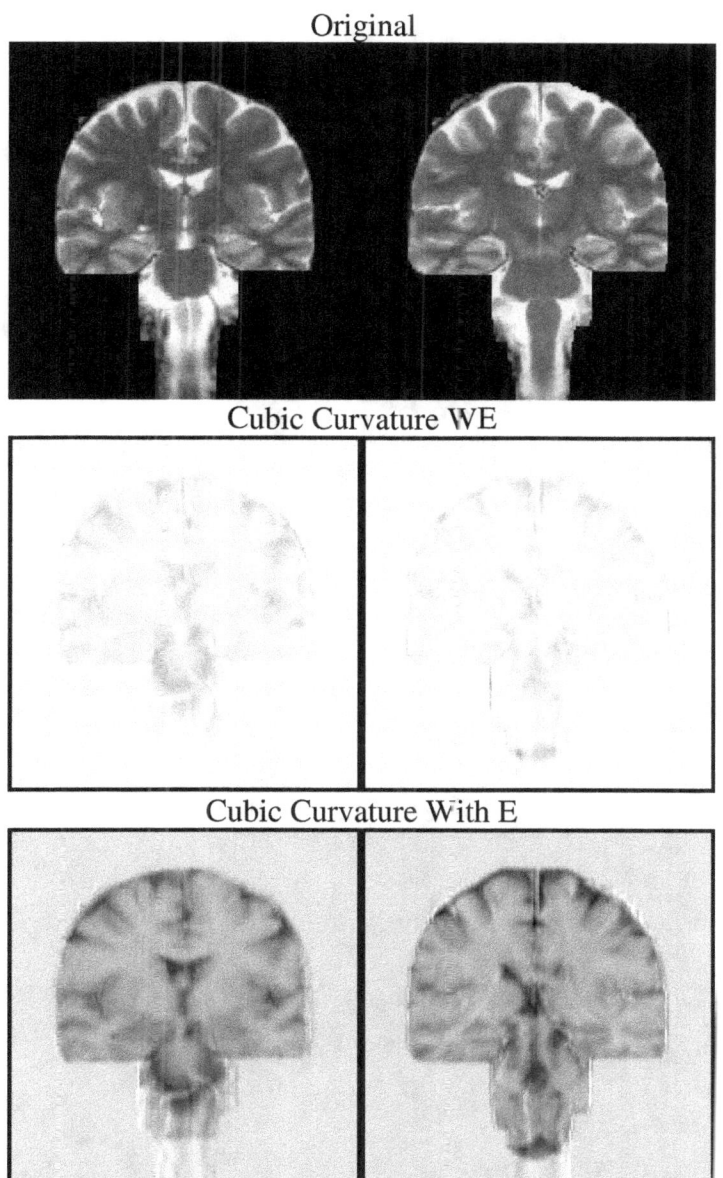

Figure 20: The Original T2 MRI data and resilient curvature maps obtained with cubic polynomials without embedding ('WE') and with embedding ('With E').

In figures 16 and 17, the arrows point to the brain structure, where similarity is seen in the high level of detail the original images and the classic curvature images have in common, (leaving aside the ventricles which were the object of attention in figure 15). The images in figures 16 and 17 were obtained with quadratic and cubic formulations respectively. It is worth noting that the value of the curvature images is that of being the preliminary step in the determination of the Pixel Intensity Correction (PIC).

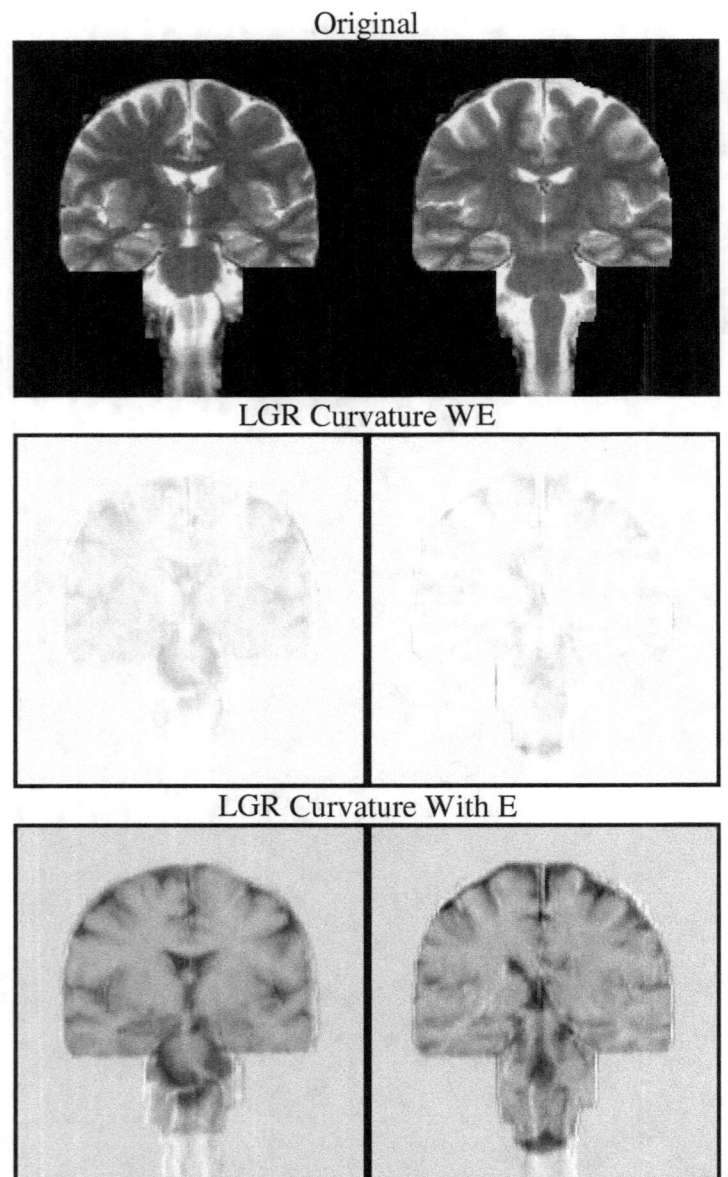

Figure 21: The Original T2 MRI data and resilient curvature maps obtained with cubic Lagrange polynomials without embedding ('WE') and with embedding ('With E').

The curvature image sets the basis of signal reconstruction. The math for deriving these images has been fully developed in the previous chapters of the book. The interpolated image admits a relationship with the curvature images because of the Pixel-Intensity-Correction (PIC) (acting as the transfer function, see equations (1) and (2)), and the relationship is that the curvature aids the calculation of the PIC and the PIC determines re-sampling so that interpolation takes place. The curvature image is thus an intermediate step in the overall process of the signal reconstruction. Two more details of similarity between the original T2 images and the curvature maps are highlighted in figure 18 and they are shown in the column at the left by means of the arrow and in the column at the right by means of the rectangle.

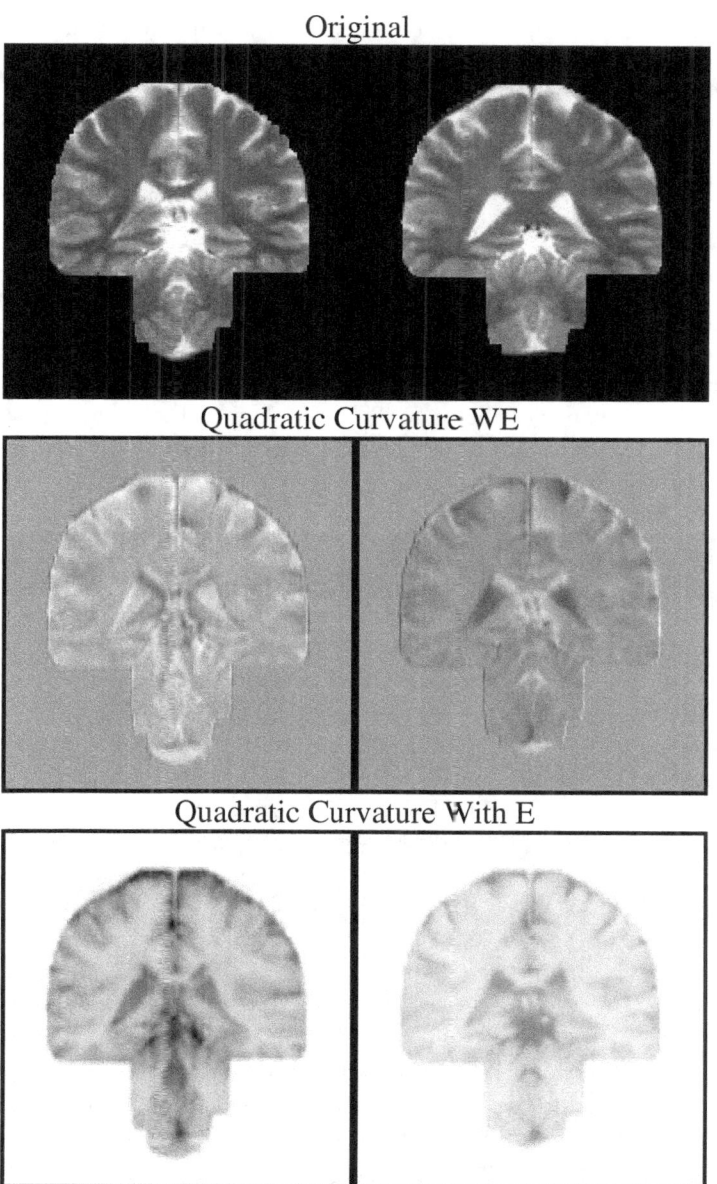

Figure 22: Resilient curvature maps obtained with quadratic polynomials without embedding ('WE') and with embedding ('With E') are compared to the original T2 MRI data shown in the top row.

Curvature maps of the cubic and cubic Lagrange formulae show similar behaviour in figures 14 and 17 (cubic) and in figures 15 and 18 (cubic Lagrange), whereas the curvature images obtained with quadratic polynomials show different behaviour, as is observable in figures 13 and 16.

A general observation which might be immediately made by the observer when looking at the resilient curvature images is that there is a clear difference in the grey level between those images calculated without embedding and those calculated with embedding. The difference is relevant to two aspects of the math development. One aspect of the math developments makes one difference between: (i) the classic curvature, which is immediately obtainable through the second order partial derivatives of the interpolation formula; and (ii) the resilient curvature, which is the total curvature of the resilient signal. The total curvature of the resilient signal is calculated through the second order partial derivatives of the resilient signal.

Another aspect of the math developments makes another difference which is in the calculation of the total curvature between the two formulations without and with embedding. Thus, the difference consists in the math act on the pixel to be reconstructed, which is, with embedding ('With E') or without embedding ('WE'). The math formulation with embedding differs from the math formulation without embedding. The difference exists in the calculation of both classic signal and resilient signal, and thus also in the total curvature of both classic signal and resilient signal. The difference is directly related to the appearance of the images seen in figures 19 through 24.

In figure 19 (quadratic polynomials) one remarkable feature of the resilient curvature images calculated with embedding is its capability of depicting the cortical features quite clearly and to discriminate between the white and grey matter of the brain. When looking at the bottom row in figure 19, the remarkable feature is more clearly visible in the resilient curvature image located at the left than it is in the resilient curvature image located at the right.

(x, y, z)	MAE Classic	MAE Resilient	MAE ClassicResilient	Ratio C.CR	Ratio R.CR
		(H33DRCCurvature)			
(0.048, 0.049, 0.051)	0.0001	0.0009	4.45 E-05	-3.3832	-19.6418
		(H33DWERCCurvature)			
(0.048, 0.049, 0.051)	0.0002	0.0002	0.0002	-8.2 E-14	-2.7 E-14
		(H43DRCCurvature)			
(0.048, 0.049, 0.051)	0.0013	0.0003	4.23 E-05	-31.3563	-8.0798
		(H43DWERCCurvature)			
(0.048, 0.049, 0.051)	0.0013	0.0001	5.45 E-05	-24.136	-1.2438
		(LGR3DRCCurvature)			
(0.048, 0.049, 0.051)	0.0013	0.0007	4.34 E-05	-30.5318	-17.185
		(LGR3DWERCCurvature)			
(0.048, 0.049, 0.051)	0.0013	0.0001	4.81 E-05	-27.5004	-1.2461

Table III. Along the rows from left to right, we give the 3D misplacement (x, y, z), the mean absolute error of the classic-curvature, of the resilient-curvature and of the classic-resilient curvature formulae, the ratio of improvement of the hybrid formula over the classic-curvature formula (Ratio C.CR) and over the resilient-curvature formula (Ratio R.CR). Each of the entries is relevant to the interpolation paradigms employed: quadratic, cubic, cubic Lagrange, all of the three with embedding ('With E') and without embedding ('WE').

As far as figures 20 (cubic polynomials) and 21 (cubic Lagrange polynomials) are concerned, the features of the images in grey are generally more discernible than the features of the curvature images in white and they also show higher sensitivity to the original image features. Moreover, the area inside the cortex of the human brain presents two clear and distinctive regions of curvature in the images calculated with embedding, whereas such distinction in curvature is less discernible in the images calculated without embedding. The thalamic structures are the ones that are most well-defined in the resilient curvature images, along with the outer parts of the cortical region, where a neat change in curvature is observable, especially in the images calculated with embedding.

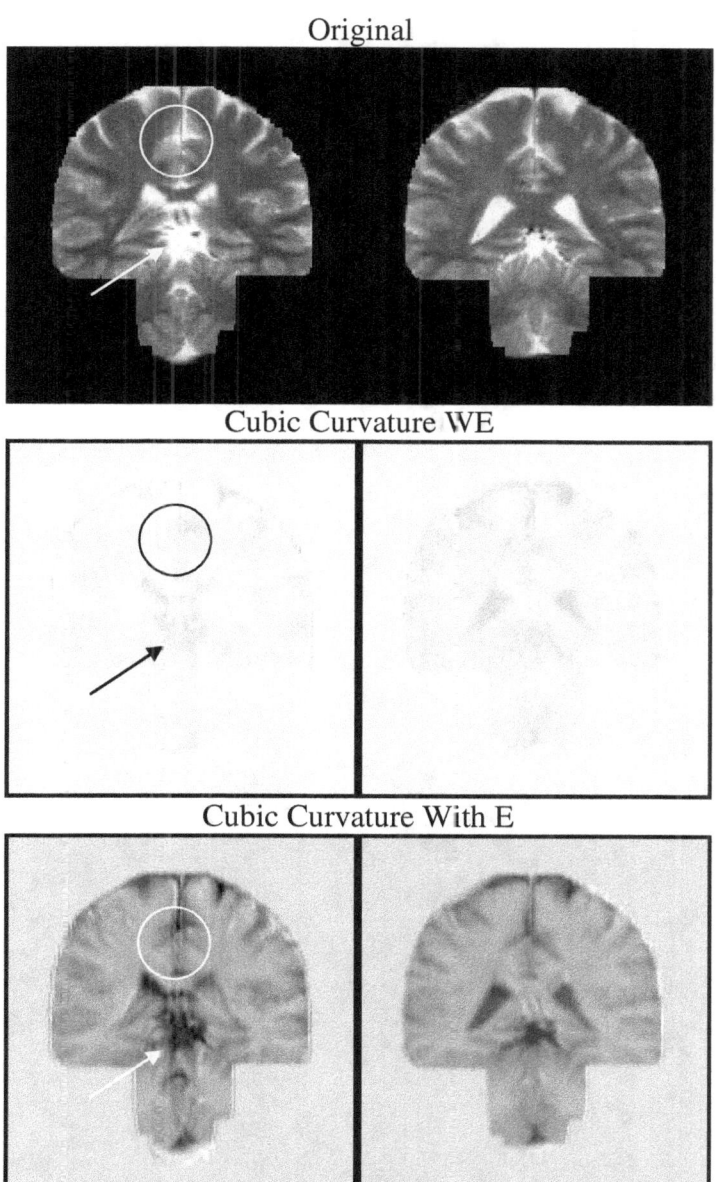

Figure 23: Resilient curvature maps obtained with cubic polynomials without embedding ('WE') and with embedding ('With E') are compared to the original T2 MRI data shown in the top row.

SIGNAL RESILIENT TO INTERPOLATION

In figure 21, as much as in the rest of the figures showing both classic and resilient curvature images, it is important to note that the region outside the brain which in its original form is filled with zeros, shows a constant and persistent value of curvature (arc tang of the angle subtended by the horizontal with the tangent to the first order derivative curve of the interpolation function). It could not be otherwise, since signal fluctuations are absent in neighborhoods filled with the same value of pixel intensity.

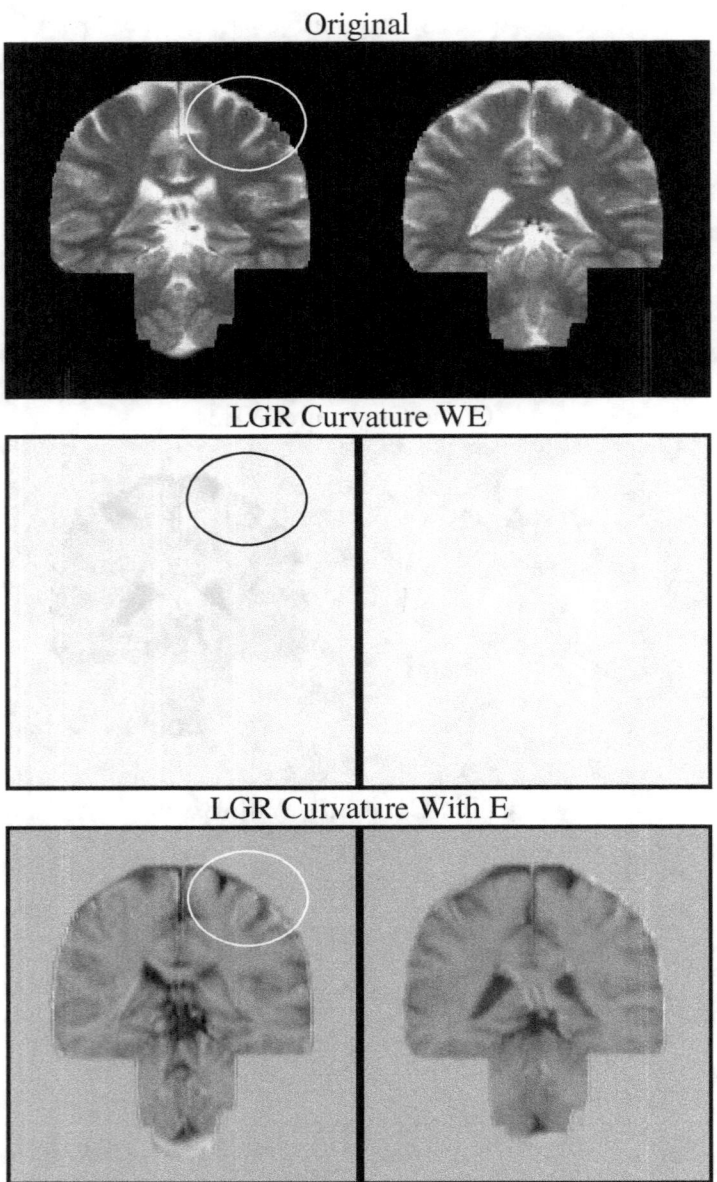

Figure 24: Resilient curvature maps obtained with cubic Lagrange polynomials without embedding ('WE') and with embedding ('With E') are compared to the original T2 MRI data shown in the top row.

The borderline between the outside of the brain (seen in flat black in the original images) and the brain is also quite well-defined. The frame in black visible in the resilient curvature images shown in figures 19 through 24 was artificially created to enhance the appearance of the curvature images and this does not make any difference because the zero filled regions extends as far as the last two horizontal rows and the two vertical columns of pixels of the image. The original image volumes were processed with a rigid transformation involving the value of the three rotation angles and the three misplacements along x, y and z coordinates. The values of the angles were all 0.01 deg and those of the misplacements were 0.05 along x, y and z, to obtain a 3D shift calculated as $(x, y, z) = (0.048, 0.049, 0.051)$, as indicated in Table III.

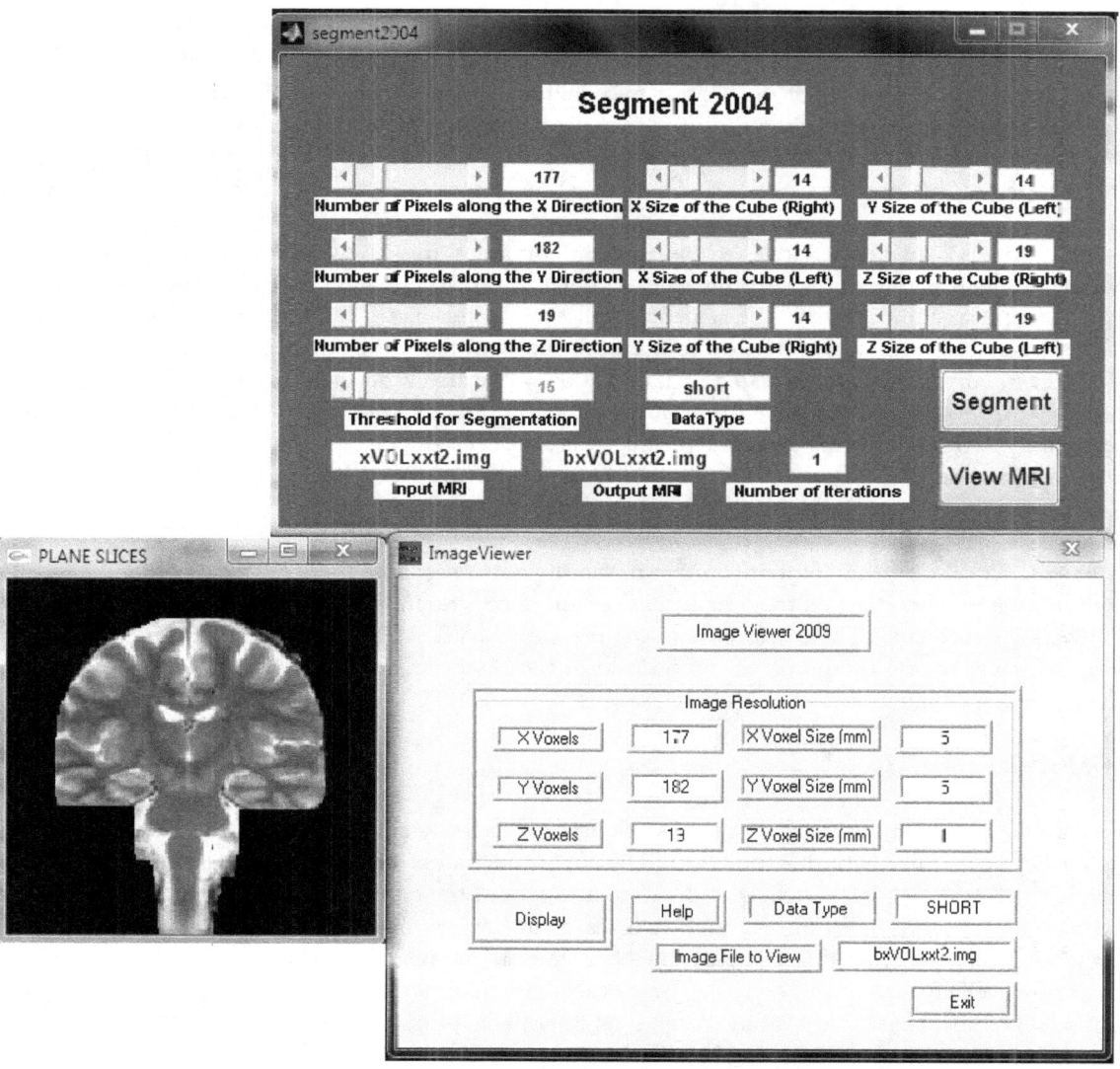

Figure 25: The computer program employed to segment the brain slices of the T2 MRI. © Carlo Ciulla.

The change of curvature inside the ventricles, between the classic curvature image at the left and the classic curvature image at the right (see in figure 22), is also seen in figures 16, 17 and 18, between the resilient curvature images at the left and the resilient curvature images at the right.

We can also observe in Table III that the major improvements in using the hybrid formula were to the benefit of the resilient-curvature formula when quadratic polynomials with embedding (H33DRCCurvature) were involved (compare the values of -3.3832 with -19.6418), thus the classic-curvature formula with embedding outperformed the resilient-curvature formula with embedding when quadratic polynomials were employed. In the experiments involving the quadratic parametric polynomials, the value of the 'the_A_const' was set to 5.4554.

On the other hand, the remaining rows of Table III indicate the superiority of the resilient-curvature formula over the classic-curvature formula (see Ratio R.CR < Ratio C.CR) and this makes the hybrid formula gain less from the classic formula than from the resilient formula, as can be seen in the values of Ratio C.CR and Ratio R.CR in the last two columns at the far right of the table. In figure 23 the following observation is in order: the circular cluster of brain structures seen inside the white circle (in the original image at the left) above the ventricles presents a noteworthy level of detail and this is seen and mapped in the same way in the resilient curvature images (see two further circles in the curvature images).

In figure 23, inside the circles in the resilient curvature images, we can also see the junction between the brain hemispheres and this level of detail is remarkable. We may also notice the different way of mapping the curvature; the structure indicated by means of the arrow is seen as bright in the resilient curvature map calculated without embedding and the same structure is also seen as dark in the resilient curvature map calculated with embedding (see arrows in the three images at the left).

To corroborate the fact that the resilient curvature images with a higher level of grey furnish a better map in terms of being able to reproduce details of the original images more faithfully, figure 24 shows three locations indicated through the ellipses in the images in the left column, where the cortex is more clearly distinguishable in the map calculated with embedding ('With E') (see inside the ellipses).

Figure 25 shows the computer program employed to segment the T2 MRI volume, which was built in house and connected to a graphical user interface written in Visual C++ ® to visualize the result of the processing. The interface was written with MATLAB®; the program segments the skull from the brain tissue by means of a threshold in the three directions x, y and z, using a user input given number of iterations.

SUMMARY

To summarize, we can say that (i) it is possible to devise classic-curvature, resilient-curvature and classic-resilient-curvature hybrid formulae in 3D also; (ii) the nature of the curvature images reproduce to a wide extent the shape and details of the data structure of the original image volume. The level of detail in the curvature image, however, increases greatly when the original image has a well-defined shape and behaviour, so that the curvature becomes a true map of the features of the image. This concept is well correlated and consistent with the mathematical meaning of the curvature. It is important to remember that the curvature image is an intermediate stage and not the final stage of the signal reconstruction process; to be more specific it aids the calculation of the customized transfer function derived from the math form of the interpolator called as the Pixel-Intensity-Correction (PIC). From the PIC the interpolation process can then be finalized and the re-sampled image can be calculated. In the first instance, therefore, this chapter adds confirmation to the conceptualization and practice of the Pixel Intensity Correction (PIC) as a useful device to reduce the interpolation error; in the second instance it also confirms that it is possible to reduce even further the interpolation error when the sig-

ral reconstruction is performed on a pixel-by-pixel basis through the classic-resilient-curvature hybrid interpolation formula, and finally it ascertains the nature of the curvature images and establishes that their behaviour is in accordance with that of the tangent to the first order derivative of the original signal images.

REFERENCES

Ciulla, C. (2009). *Improved Signal and Image Interpolation in Biomedical Applications: The Case of Magnetic Resonance Imaging (MRI)* – Medical Information Science Reference - IGI Global Publisher. Hershey, PA, U.S.A.

Dumpster, A.G. & Murphy, N.P. (1999). *Lagrange Interpolator Filters and Binomial Windows*. Sign Proc 76, 81-91.

Fuchs, J.J. & Delyon, B. (2002). *Min-Max Interpolators and Lagrange Interpolation Formula*. ISCAS IEEE Int Symp, 4 IV-429 - IV 432.

Hecklin, G., Nurnberger, G., Schumaker, L. L. & Zeilfelder, F. (2008) *A local Lagrange interpolation method based on C^1 cubic splines on freudenthal partitions*. Mathematics of Computation, 77(262), 1017-1036.

Kubayi, D. G. & Lubinsky, D.S. (2004). *A Hilbert Transform Representation of the Error in Lagrange Interpolation*. J Approx Theory 129, 94-100.

Marvasti, F. A. (1990). *Extension of Lagrange Interpolation to 2-D Nonuniform Samples in Polar Coordinates*. IEEE Trans Circ Syst 37, 567-568.

Radzyner, R. & Bason, P.T. (1972). *An Error Bound for Lagrange Interpolation of Low-Pass Functions*. IEEE Trans Inf Theory 18, 569-671.

Waldron, S. (1997). *A Multivariate Form of Hardy's Inequality and L_p-error Bounds for Multivariate Lagrange Interpolation Schemes*. SIAM J Math Anal 28, 233-258.

Xie, T. & Zhou, X. (2001). *A Modification of Lagrange Interpolation*. Acta Math Hungar 92, 285-297.

Ye, Z. (2003). *Linear Phase Lagrange Interpolation Filter Using Odd number of Basepoints*. Proc IEEE ICASSP; VI-237- VI-239.

CHAPTER 11

RESILIENT INTERPOLATION: AN INVESTIGATION ON THE DIMENSIONALITY OF THE POLYNOMIALS, THE MATHEMATICS

INTRODUCTION

Following the scheme of math deduction of chapter 9, in this chapter we go further, looking at three different interpolation functions which cover the three dimensions, which means that their math form is similar, except for the dimensionality.

The interpolation functions are all calculated with embedding and are defined as: $g_4(x) = f(0) + \alpha_2 * a [x^3 + 1/2 x^2 + 1/4 x + 1] + \alpha_3 * a [x^2 + 2x + 1]$, in 1D; $g_4(x, y) = f(0, 0) + \alpha_2 * a [(x + y)^3 + 1/2 (x + y)^2 + 1/4 (x + y) + 1] + \alpha_3 * a [(x + y)^2 + 2 (x + y) + 1]$, in 2D; and $g_4(x, y, z) = f(0, 0, 0) + \omega_1 * a [(x + y + z)^3 + 1/2 (x + y + z)^2 + 1/4 (x + y + z) + 1] + \omega_2 * a [(x + y + z)^2 + 2 (x + y + z) + 1]$, in 3D.

The calculation of the total curvature is carried out in this chapter for both classic signal and resilient signal, following the same methodology as the preceding chapters. The total curvature of the classic signal is given through the sum of second order parial derivatives in all variates and covariates. $[(\partial^2 (g_4(x)) /\partial x^2)]$ in 1D; $[(\partial^2 (g_4(x, y)) /\partial x^2) + (\partial^2 (g_4(x, y)) /\partial y^2) + (\partial^2 (g_4(x, y)) /\partial x \partial y) + (\partial^2 (g_4(x, y)) /\partial y \partial x)]$ in 2D; $[(\partial^2 (g_4(x, y, z)) /\partial x^2) + (\partial^2 (g_4(x, y, z)) /\partial y^2) + (\partial^2 (g_4(x, y, z)) /\partial z^2) + (\partial^2 (g_4(x, y, z)) /\partial x \partial y) + (\partial^2 (g_4(x, y, z)) /\partial y \partial x) + (\partial^2 (g_4(x, y, z)) /\partial x \partial z) + (\partial^2 (g_4(x, y, z)) /\partial y \partial z) + (\partial^2 (g_4(x, y, z)) /\partial z \partial x) + (\partial^2 (g_4(x, y, z)) /\partial z \partial y)]$ in 3D.

The calculation of the signals resilient to interpolation is: $R(x) = f(0) + \varrho(x)$, in 1D; $R(x, y) = f(0, 0) + \varrho(x, y)$, in 2D, and $R(x, y, z) = f(0, 0, 0) + \varrho(x, y, z)$, in 3D.

The total curvature is calculated as: $[(\partial^2 (\varrho(x)) /\partial x^2)]$ in 1D; $[(\partial^2 (\varrho(x, y)) /\partial x^2) + (\partial^2 (\varrho(x, y)) /\partial y^2) + (\partial^2 (\varrho(x, y)) /\partial x \partial y) + (\partial^2 (\varrho(x, y)) /\partial y \partial x)]$ in 2D; and $[(\partial^2 (\varrho(x, y, z)) /\partial x^2) + (\partial^2 (\varrho(x, y, z)) /\partial y^2) + (\partial^2 (\varrho(x, y, z)) /\partial z^2) + (\partial^2 (\varrho(x, y, z)) /\partial y \partial x) + (\partial^2 (\varrho(x, y, z)) /\partial x \partial y) + (\partial^2 (\varrho(x, y, z)) /\partial y \partial z) + (\partial^2 (\varrho(x, y, z)) /\partial z \partial y) + (\partial^2 (\varrho(x, y, z)) /\partial x \partial z) + (\partial^2 (\varrho(x, y, z)) /\partial z \partial x)]$ in 3D.

THE POLYNOMIAL FORMS CALCULATED WITH EMBEDDING

Let the polynomials be defined as follows:

$$g_4(x) = f(0) + \alpha_2 * a [x^3 + 1/2 x^2 + 1/4 x + 1] + \alpha_3 * a [x^2 + 2x + 1] \tag{1}$$

$$g_4(x, y) = f(0, 0) + \alpha_2 * a [(x + y)^3 + 1/2 (x + y)^2 + 1/4 (x + y) + 1] + \alpha_3 * a [(x + y)^2 + 2 (x + y) + 1] \tag{2}$$

Let us posit:

$$\alpha_2 = [f(1/2, 1/2) + f(-1/2, -1/2) + f(2/3, 2/3) + f(-2/3, -2/3) + f(-1, -1) + f(1, 1) + f(3/2, 3/2) + f(-3/2, -3/2)] \tag{3}$$

$$\alpha_3 = [f(1/2, 1/2) + f(-1/2, -1/2) + f(-1, -1) + f(1, 1)] \tag{4}$$

$g_4(x, y, z) = f(0, 0, 0) + \omega_1 * a[(x + y + z)^3 + 1/2(x + y + z)^2 + \frac{1}{4}(x + y + z) + 1] + \omega_2 *$

$a[(x + y + z)^2 + 2(x + y - z) + 1]$ (5)

Let us posit:

$\omega_1 = [f(-1, 0, 1) + f(-1, 0, 0) + f(1, 0, 0) + f(1, 0, -1) + f(-1, -1, 1) + f(-1, -1, 0) + f(1,$

$-1, 0) + f(1, -1, -1) + f(-1, 1, 1) + f(-1, 1, 0) + f(1, 1, 0) + f(1, 1, -1)]$ (6)

$\omega_2 = [f(-1, 0, 1) + f(0, 0, 1) + f(1, 0, 1) + f(-1, 0, 0) + f(1, 0, 0) + f(-1, 0, -1) + f(0, 0,$

$-1) + f(1, 0, -1) + f(-1, -1, 1) + f(0, -1, 1) + f(1, -1, 1) + f(-1, -1, 0) + f(0, -1, 0) + f(1,$

$-1, 0) + f(-1, -1, -1) + f(0, -1, -1) + f(1, -1, -1) + f(-1, 1, 1) + f(0, 1, 1) + f(1, 1, 1) +$

$f(-1, 1, 0) + f(0, 1, 0) + f(1, 1, 0) + f(-1, 1, -1) + f(0, 1, -1) + f(1, 1, -1)]$ (7)

Calculation of the Second Order Derivatives

$(\partial(g_4(x))/\partial x) = (\partial(f(0) + \alpha_2 * a[x^3 + 1/2 x^2 + \frac{1}{4} x + 1] + \alpha_3 * a[x^2 + 2x + 1])/\partial x) =$

$\alpha_2 * a[3x^2 + x + \frac{1}{4}] + \alpha_3 * a[2x + 2]$ (8)

$(\partial^2(g_4(x))/\partial x^2) = (\partial(\alpha_2 * a[3x^2 + x + \frac{1}{4}] + \alpha_3 * a[2x + 2])/\partial x) =$

$\alpha_2 * a[6x + 1] + 2\alpha_3 * a$ (9)

$(\partial(g_4(x, y))/\partial x) = (\partial(f(0, 0) + \alpha_2 * a[(x + y)^3 + 1/2(x + y)^2 + \frac{1}{4}(x + y) + 1] + \alpha_3* a[(x$

$+ y)^2 + 2(x + y) + 1])/\partial x) = \alpha_2 * a[3(x + y)^2 + (x + y) + \frac{1}{4}] + \alpha_3* a[2(x + y) + 2]$ (10)

$(\partial^2(g_4(x, y))/\partial x^2) = (\partial(\alpha_2 * a[3(x + y)^2 + (x + y) + \frac{1}{4}] + \alpha_3* a[2(x + y) + 2])/\partial x) =$

$\alpha_2 * a[6(x + y) + 1] + 2\alpha_3* a$ (11)

$(\partial(g_4(x, y))/\partial y) = (\partial(f(0, 0) + \alpha_2 * a[(x + y)^3 + 1/2(x + y)^2 + \frac{1}{4}(x + y) + 1] + \alpha_3* a[(x$

$+ y)^2 + 2(x + y) + 1])/\partial y = \alpha_2 * a[3(x + y)^2 + (x + y) + \frac{1}{4}] + \alpha_3* a[2(x + y) + 2]$ (12)

$(\partial^2(g_4(x, y))/\partial y^2) = (\partial(\alpha_2 * a[3(x + y)^2 + (x + y) + \frac{1}{4}] + \alpha_3* a[2(x + y) + 2])/\partial y) =$

$\alpha_2 * a[6(x + y) + 1] + 2\alpha_3* a$ (13)

SIGNAL RESILIENT TO INTERPOLATION

$$(\partial^2 (g_4(x, y)) / \partial x \partial y) = (\partial^2 (f (0, 0) + \alpha_2 * a [(x + y)^3 + 1/2 (x + y)^2 + \tfrac{1}{4} (x + y) + 1] + \alpha_3 *$$

$$a [(x + y)^2 + 2 (x + y) + 1]) / \partial x \partial y) = (\partial (\alpha_2 * a [3 (x + y)^2 + (x + y) + \tfrac{1}{4}] + \alpha_3 * a [2 (x$$

$$+ y) + 2]) / \partial y) = \alpha_2 * a [6 (x + y) + 1] + 2 \alpha_3 * a \tag{14}$$

$$(\partial^2 (g_4(x, y)) / \partial y \partial x) = (\partial^2 (f (0, 0) + \alpha_2 * a [(x + y)^3 + 1/2 (x + y)^2 + \tfrac{1}{4} (x + y) + 1] + \alpha_3 *$$

$$a [(x + y)^2 + 2 (x + y) + 1]) / \partial y \partial x) = (\partial (\alpha_2 * a [3 (x + y)^2 + (x + y) + \tfrac{1}{4}] + \alpha_3 * a [2 (x$$

$$+ y) + 2]) / \partial x) = \alpha_2 * a [6 (x + y) + 1] + 2 \alpha_3 * a \tag{14.a}$$

$$(\partial (g_4(x, y, z)) / \partial x) = (\partial (f (0, 0, 0) + \omega_1 * a [(x + y + z)^3 + 1/2 (x + y + z)^2 + \tfrac{1}{4} (x + y + z)$$

$$+ 1] + \omega_2 * a [(x + y + z)^2 + 2 (x + y + z) + 1]) / \partial x) = \omega_1 * a [3 (x + y + z)^2 + (x + y +$$

$$z) + \tfrac{1}{4}] + \omega_2 * a [2 (x + y + z) + 2] \tag{15}$$

$$(\partial^2 (g_4(x, y, z)) / \partial x^2) = (\partial (\omega_1 * a [3 (x + y + z)^2 + (x + y + z) + \tfrac{1}{4}] + \omega_2 * a [2 (x + y +$$

$$z) + 2]) / \partial x) = \omega_1 * a [6 (x + y + z) + 1] + 2 \omega_2 * a \tag{16}$$

$$(\partial (g_4(x, y, z)) / \partial y) = (\partial (f (0, 0, 0) + \omega_1 * a [(x + y + z)^3 + 1/2 (x + y + z)^2 + \tfrac{1}{4} (x + y + z)$$

$$+ 1] + \omega_2 * a [(x + y + z)^2 + 2 (x + y + z) + 1]) / \partial y) = \omega_1 * a [3 (x + y + z)^2 + (x + y +$$

$$z) + \tfrac{1}{4}] + \omega_2 * a [2 (x + y + z) + 2] \tag{17}$$

$$(\partial^2 (g_4(x, y, z)) / \partial y^2) = (\partial (\omega_1 * a [3 (x + y + z)^2 + (x + y + z) + \tfrac{1}{4}] + \omega_2 * a [2 (x + y +$$

$$z) + 2]) / \partial y) = \omega_1 * a [6 (x + y + z) + 1] + 2 \omega_2 * a \tag{18}$$

$$(\partial (g_4(x, y, z)) / \partial z) = (\partial (f (0, 0, 0) + \omega_1 * a [(x + y + z)^3 + 1/2 (x + y + z)^2 + \tfrac{1}{4} (x + y + z)$$

$$+ 1] + \omega_2 * a [(x + y + z)^2 + 2 (x + y + z) + 1]) / \partial z) = \omega_1 * a [3 (x + y + z)^2 + (x + y +$$

$$z) + \tfrac{1}{4}] + \omega_2 * a [2 (x + y + z) + 2] \tag{19}$$

$$(\partial^2 (g_4(x, y, z)) / \partial z^2) = (\partial (\omega_1 * a [3 (x + y + z)^2 + (x + y + z) + \tfrac{1}{4}] + \omega_2 * a [2 (x + y +$$

$$z) + 2]) / \partial z) = \omega_1 * a [6 (x + y + z) + 1] + 2 \omega_2 * a \tag{20}$$

$(\partial^2 (g_4(x, y, z)) / \partial x \partial y) = (\partial^2 (f (0, 0, 0) + \omega_1 * a [(x + y + z)^3 + 1/2 (x + y + z)^2 + \frac{1}{4} (x + y + z) + 1] + \omega_2 * a [(x + y + z)^2 + 2 (x + y + z) + 1]) / \partial x \partial y) = (\partial (\omega_1 * a [3 (x + y + z)^2 + (x + y + z) + \frac{1}{4}] + \omega_2 * a [2 (x + y + z) + 2]) / \partial y) =$

$\omega_1 * a [6 (x + y + z) + 1] + 2 \omega_2 * a \qquad (21)$

$(\partial^2 (g_4(x, y, z)) / \partial y \partial x) = (\partial^2 (f (0, 0, 0) + \omega_1 * a [(x + y + z)^3 + 1/2 (x + y + z)^2 + \frac{1}{4} (x + y + z) + 1] + \omega_2 * a [(x + y + z)^2 + 2 (x + y + z) + 1]) / \partial y \partial x) = (\partial (\omega_1 * a [3 (x + y + z)^2 + (x + y + z) + \frac{1}{4}] + \omega_2 * a [2 (x + y + z) + 2]) / \partial x) =$

$\omega_1 * a [6 (x + y + z) + 1] + 2 \omega_2 * a \qquad (22)$

$(\partial^2 (g_4(x, y, z)) / \partial x \partial z) = (\partial^2 (f (0, 0, 0) + \omega_1 * a [(x + y + z)^3 + 1/2 (x + y + z)^2 + \frac{1}{4} (x + y + z) + 1] + \omega_2 * a [(x + y + z)^2 + 2 (x + y + z) + 1]) / \partial x \partial z) = (\partial (\omega_1 * a [3 (x + y + z)^2 + (x + y + z) + \frac{1}{4}] + \omega_2 * a [2 (x + y + z) + 2]) / \partial z) =$

$\omega_1 * a [6 (x + y + z) + 1] + 2 \omega_2 * a \qquad (23)$

$(\partial^2 (g_4(x, y, z)) / \partial y \partial z) = (\partial^2 (f (0, 0, 0) + \omega_1 * a [(x + y + z)^3 + 1/2 (x + y + z)^2 + \frac{1}{4} (x + y + z) + 1] + \omega_2 * a [(x + y + z)^2 + 2 (x + y + z) + 1]) / \partial y \partial z) = (\partial (\omega_1 * a [3 (x + y + z)^2 + (x + y + z) + \frac{1}{4}] + \omega_2 * a [2 (x + y + z) + 2]) / \partial z) =$

$\omega_1 * a [6 (x + y + z) + 1] + 2 \omega_2 * a \qquad (24)$

$(\partial^2 (g_4(x, y, z)) / \partial z \partial x) = (\partial^2 (f (0, 0, 0) + \omega_1 * a [(x + y + z)^3 + 1/2 (x + y + z)^2 + \frac{1}{4} (x + y + z) + 1] + \omega_2 * a [(x + y + z)^2 + 2 (x + y + z) + 1]) / \partial z \partial x) = (\partial (\omega_1 * a [3 (x + y + z)^2 + (x + y + z) + \frac{1}{4}] + \omega_2 * a [2 (x + y + z) + 2]) / \partial x) =$

$\omega_1 * a [6 (x + y + z) + 1] + 2 \omega_2 * a \qquad (25)$

$(\partial^2 (g_4(x, y, z)) / \partial z \partial y) = (\partial^2 (f (0, 0, 0) + \omega_1 * a [(x + y + z)^3 + 1/2 (x + y + z)^2 + \frac{1}{4} (x + y + z) + 1] + \omega_2 * a [(x + y + z)^2 + 2 (x + y + z) + 1]) / \partial z \partial y) = (\partial (\omega_1 * a [3 (x + y + z)^2$

$$+ (x + y + z) + \tfrac{1}{4} \,] \; + \omega_2 * a \,[\, 2\,(x + y + z) + 2\,]\,)\, /\partial y) =$$

$$\omega_1 * a \,[\, 6\,(x + y + z) + 1\,] + 2\,\omega_2 * a \qquad (26)$$

CLASSIC-CURVATURE FORMULA: CALCULATION OF THE TOTAL CURVATURE

$$C(\, I((x), f)\,) = (\partial^2\,(\,g_4(x)\,)\, /\partial x^2) = \{\,\alpha_2 * a\,[6x + 1] + 2\,\alpha_3 * a\,\} \qquad (27)$$

$$C(\, I((x, y), f)\,) = (\partial^2\,(\,g_4(x, y)\,)\, /\partial x^2) + (\partial^2\,(\,g_4(x, y)\,)\, /\partial y^2) + (\partial^2\,(g_4(x, y))\, /\partial x\partial y) + (\partial^2$$

$$(g_4(x, y))\, /\partial y\partial x) = (\alpha_2 * a\,[6\,(x + y) + 1\,] + 2\,\alpha_3 * a)\; + (\alpha_2 * a\,[6\,(x + y) + 1\,] + 2\,\alpha_3 * a) +$$

$$(\alpha_2 * a\,[6\,(x + y) + 1\,] + 2\,\alpha_3 * a) + (\alpha_2 * a\,[6\,(x + y) + 1\,] + 2\,\alpha_3 * a)\; = 4 * \{\,\alpha_2 * a\,[6\,(x +$$

$$y) + 1\,] + 2\,\alpha_3 * a\,\} \qquad (28)$$

$$C(\, I((x, y, z), f)\,) = (\partial^2\,(g_4(x, y, z))\, /\partial x^2) + (\partial^2\,(g_4(x, y, z))\, /\partial y^2) + (\partial^2\,(g_4(x, y, z))\, /\partial z^2) +$$

$$(\partial^2\,(g_4(x, y, z))\, /\partial x\partial y) + (\partial^2\,(g_4(x, y, z))\, /\partial y\partial x) + (\partial^2\,(g_4(x, y, z))\, /\partial x\partial z) + (\partial^2\,(g_4(x, y,$$

$$z))\, /\partial y\partial z) + (\partial^2\,(g_4(x, y, z))\, /\partial z\partial x) + (\partial^2\,(g_4(x, y, z))\, /\partial z\partial y) = (\omega_1 * a\,[\, 6\,(x + y + z) + 1\,]$$

$$+ 2\,\omega_2 * a) + (\omega_1 * a\,[\, 6\,(x + y + z) + 1\,] + 2\,\omega_2 * a) + (\omega_1 * a\,[\, 6\,(x + y + z) + 1\,] + 2\,\omega_2$$

$$* a) + (\omega_1 * a\,[\, 6\,(x + y + z) + 1\,] + 2\,\omega_2 * a) + (\omega_1 * a\,[\, 6\,(x + y + z) + 1\,] + 2\,\omega_2 * a) +$$

$$(\omega_1 * a\,[\, 6\,(x + y + z) + 1\,] + 2\,\omega_2 * a) + (\omega_1 * a\,[\, 6\,(x + y + z) + 1\,] + 2\,\omega_2 * a) + (\omega_1 * a$$

$$[\, 6\,(x + y + z) + 1\,] + 2\,\omega_2 * a) + (\omega_1 * a\,[\, 6\,(x + y + z) + 1\,] + 2\,\omega_2 * a) =$$

$$9 * \{\,\omega_1 * a\,[\, 6\,(x + y + z) + 1\,] + 2\,\omega_2 * a\,\} \qquad (29)$$

Equations (27) (28) and (29) are the total curvature of the classic-curvature signal reconstruction formula in the unidimensional variate, the bivariate and trivariate cases respectively.

EQUATION OF THE INTENSITY-CURVATURE TERMS: RESILIENT FORMULATION

Calculation of the Intensity-Curvature Terms Before Interpolation

$$\{\,(\partial^2\,(g_4(x))\, /\partial x^2)\,\}\,(0) = \{\,\alpha_2 * a\,[6x + 1] + 2\,\alpha_3 * a\,\}\,(0) = \{\,\alpha_2 * a + 2\,\alpha_3 * a\,\} \qquad (30)$$

$$E_o = E_o\,(x) = \int_0^x f\,(0) * \{\,(\partial^2\,(g_4(x))\, /\partial x^2)\,\}\,(0)\,dx = \int_0^x f\,(0) * \{\,\alpha_2 * a + 2\,\alpha_3 * a\,\}\,dx = f\,(0)$$

$$* \{\,\alpha_2 * a + 2\,\alpha_3 * a\,\} * x \qquad (31)$$

$\{ (\partial^2 (g_4(x, y)) /\partial x^2) + (\partial^2 (g_4(x, y)) /\partial x \partial y) + (\partial^2 (g_4(x, y))/\partial y \partial x) + (\partial^2 (g_4(x, y)) /\partial y^2) \}$

$(0, 0) = 4 * \{ \alpha_2 * a [6 (x + y) + 1] + 2 \alpha_3 * a \} (0, 0) = 4 * \{ \alpha_2 * a + 2 \alpha_3 * a \}$ (32)

$E_o = E_o (x, y) = \int\limits_{0}^{x}\int\limits_{0}^{y} f (0, 0) * \{ (\partial^2 (g_4(x, y)) /\partial x^2) + (\partial^2 (g_4(x, y)) /\partial x \partial y) + (\partial^2 (g_4(x, y))$

$/\partial y \partial x) + (\partial^2 (g_4(x, y)) /\partial y^2) \} (0, 0) \, dx \, dy =$

$\int\limits_{0}^{x}\int\limits_{0}^{y} f (0, 0) * 4 * \{ \alpha_2 * a + 2 \alpha_3 * a \} \, dx \, dy = f (0, 0) * 4 * \{ \alpha_2 * a + 2 \alpha_3 * a \} * xy$ (33)

$\{ (\partial^2 (g_4(x, y, z)) /\partial x^2) + (\partial^2 (g_4(x, y, z)) /\partial y^2) + (\partial^2 (g_4(x, y, z)) /\partial z^2) + (\partial^2 (g_4(x, y, z))$

$/\partial x \partial y) + (\partial^2(g_4(x, y, z)) /\partial y \partial x) + (\partial^2 (g_4(x, y, z)) /\partial x \partial z) + (\partial^2 (g_4(x, y, z)) /\partial z \partial x) + (\partial^2$

$(g_4(x, y, z)) /\partial y \partial z) + (\partial^2 (g_4(x, y, z)) /\partial z \partial y)\} (0, 0, 0) = 9 * \{ \omega_1 * a [6 (x + y + z) + 1]$

$+ 2 \omega_2 * a \} (0, 0, 0) = 9 * \{ \omega_1 * a + 2 \omega_2 * a \}$ (34)

$E_o = E_o (x, y, z) = \int\limits_{0}^{x}\int\limits_{0}^{y}\int\limits_{0}^{z} f (0, 0, 0) * \{ (\partial^2 (g_4(x, y, z)) /\partial x^2) + (\partial^2 (g_4(x, y, z)) /\partial y^2) + (\partial^2$

$(g_4(x, y, z)) /\partial z^2) + (\partial^2 (g_4(x, y, z)) /\partial x \partial y) + (\partial^2 (g_4(x, y, z)) /\partial y \partial x) + (\partial^2 (g_4(x, y, z))$

$/\partial x \partial z) + (\partial^2 (g_4(x, y, z)) /\partial z \partial x) + (\partial^2 (g_4(x, y, z)) /\partial y \partial z) + (\partial^2 (g_4(x, y, z)) /\partial z \partial y)\} (0,$

$0, 0) \, dx \, dy \, dz =$

$\int\limits_{0}^{x}\int\limits_{0}^{y}\int\limits_{0}^{z} f (0, 0, 0) * 9 * \{ \omega_1 * a + 2 \omega_2 * a \} = f (0, 0, 0) * 9 * \{ \omega_1 * a + 2 \omega_2 * a \} * xyz$ (35)

Calculation of the Intensity-Curvature Terms After Interpolation

$E_{IN} = E_{IN} (x) = \int\limits_{0}^{x} g_4(x) * \{ (\partial^2 (g_4(x, y)) /\partial x^2) \} \, dx = \int\limits_{0}^{x} \{ f (0) + \alpha_2 * a [x^3 + 1/2 x^2 + 1/4 x + 1$

$] + \alpha_3 * a [x^2 + 2 x + 1] \} * \{ \alpha_2 * a [6x + 1] + 2 \alpha_3 * a \} \, dx = \int\limits_{0}^{x} f (0) * \{ \alpha_2 * a [6x + 1] +$

$2 \alpha_3 * a \} + \alpha_2 * a [x^3 + 1/2 x^2 + 1/4 x + 1] * \{ \alpha_2 * a [6x + 1] + 2 \alpha_3 * a \} + \alpha_3 * a [x^2 + 2 x$

$+ 1] * \{ \alpha_2 * a [6x + 1] + 2 \alpha_3 * a \} \, dx = f (0) * \{ \alpha_2 * a [3x^2 + x] + 2 \alpha_3 * a x\} + \int\limits_{0}^{x} (\alpha_2 a)^2$

$[6x^4 + 3x^3 + 3/2 x^2 + 6x + x^3 + 1/2 x^2 + 1/4 x + 1] + \alpha_2 * a [x^3 + 1/2 x^2 + 1/4 x + 1] * 2 \alpha_3 * a$

$+ \alpha_3\,\alpha_2\,a^2\,[\,6x^3 + 12\,x^2 + 6x + x^2 + 2\,x + 1\,] + \alpha_3\,a\,[\,x^2 + 2\,x + 1\,] * 2\,\alpha_3\,a \;\; dx = f\,(0) * \{\,\alpha_2$

$* a\,[3x^2 + x] + 2\,\alpha_3 * a\,x\,\} + (\alpha_2\,a)^2\,[6/5\,x^5 + \tfrac{3}{4}\,x^4 + 1/2\,x^3 + 3x^2 + x^4/4 + 1/6\,x^3 + 1/8\,x^2 +$

$x\,] + \alpha_2 * a\,[x^4/4 + 1/6\,x^3 + 1/8\,x^2 + x\,] * 2\,\alpha_3\,a + \alpha_3\,\alpha_2\,a^2\;[\,3/2\,x^4 + 4\,x^3 + 3x^2 + x^3/3 + x^2$

$+ x\,] + \alpha_3\,a\,[\,x^3/3 + x^2 + x\,] * 2\,\alpha_3\,a$ (36)

$E_{IN} = E_{IN}\,(x,\,y) = \displaystyle\int_0^x\!\!\int_0^y g_4(x,\,y) * \{\,(\partial^2\,(g_4(x,\,y))\,/\partial x^2) + (\partial^2\,(\,g_4(x,\,y)\,)\,)\,/\partial x\partial y) + (\partial^2\,(\,g_4(x,\,y)$

$)\,/\partial y\partial x) + (\partial^2\,(g_4(x,\,y))\,/\partial y^2)\,\}\;dx\,dy = \displaystyle\int_0^x\!\!\int_0^y \{\,f\,(0,\,0) + \alpha_2 * a\,[(x + y)^3 + 1/2\,(x + y)^2 + \tfrac{1}{4}\,(x$

$+ y) + 1\,] + \alpha_3 * a\,[\,(x + y)^2 + 2\,(x + y) + 1\,]\,\} * 4 * \{\,\alpha_2 * a\,[6\,(x + y) + 1\,] + 2\,\alpha_3 * a\,\}\;dx$

$dy = 4 * \displaystyle\int_0^x\!\!\int_0^y f\,(0,\,0)\,\{\,\alpha_2 * a\,[6\,(x + y) + 1\,] + 2\,\alpha_3 * a\,\} + \alpha_2 * a\,[\,(x + y)^3 + 1/2\,(x + y)^2 + \tfrac{1}{4}$

$(x + y) + 1\,]\; * \{\,\alpha_2 * a\,[6\,(x + y) + 1\,] + 2\,\alpha_3 * a\,\} + \alpha_3 * a\,[\,(x + y)^2 + 2\,(x + y) + 1\,] * \{\,\alpha_2$

$* a\,[6\,(x + y) + 1\,] + 2\,\alpha_3 * a\,\}\;dx\,dy =$

$4 * f\,(0,\,0)\,\{\,\alpha_2 * a\,[6\,(yx^2/2 + xy^2/2) + xy\,] + 2\,\alpha_3 * a\,xy\,\} + 4 * \displaystyle\int_0^x\!\!\int_0^y (\alpha_2\,a)^2\,[6(x + y)^4 +$

$6/2\,(x + y)^3 + 3/2\,(x + y)^2 + 6\,(x + y) + (x + y)^3 + 1/2\,(x + y)^2 + \tfrac{1}{4}\,(x + y) + 1\,] + 2\,\alpha_3\,\alpha_2\,a^2$

$* [\,(x + y)^3 + 1/2\,(x + y)^2 + \tfrac{1}{4}\,(x + y) + 1\,]\; + \alpha_3\,\alpha_2 * a^2\,[\,6(x + y)^3 + 12\,(x + y)^2 + 6\,(x + y)$

$+ (x + y)^2 + 2\,(x + y) + 1\,] + 2\,(\alpha_3\,a)^2\,[\,(x + y)^2 + 2\,(x + y) + 1\,]\;dx\,dy =$

$4 * f\,(0,\,0)\,\{\,\alpha_2 * a\,[6\,(yx^2/2 + xy^2/2) + xy\,] + 2\,\alpha_3 * a\,xy\,\} + 4 * \displaystyle\int_0^x\!\!\int_0^y [\,(\alpha_2\,a)^2\;6(x + y)^4\,] +$

$[\,4\,(\alpha_2\,a)^2 + 2\,\alpha_3\,\alpha_2\,a^2 + 6\,\alpha_3\,\alpha_2 * a^2\,]\,[\,(x + y)^3\,] + [\,2\,(\alpha_2\,a)^2 + \alpha_3\,\alpha_2\,a^2 + 13\,\alpha_3\,\alpha_2 * a^2 + 2$

$(\alpha_3\,a)^2\,]\,[\,(x + y)^2\,] + [\,25/4\,(\alpha_2\,a)^2 + 1/2\,\alpha_3\,\alpha_2\,a^2 + 8\,\alpha_3\,\alpha_2 * a^2 + 4\,(\alpha_3\,a)^2\,]\,[\,(x + y)\,] + [$

$(\alpha_2\,a)^2\; + 2\,\alpha_3\,\alpha_2\,a^2\; + \alpha_3\,\alpha_2 * a^2 + 2\,(\alpha_3\,a)^2\,]\;dx\,dy$ (37)

Let us calculate:

$\displaystyle\int_0^x\!\!\int_0^y (x + y)^4\,dx\,dy = \int_0^x\!\!\int_0^y (x + y)^2\,(x + y)^2\,dx\,dy = \int_0^x\!\!\int_0^y (x^2 + 2xy + y^2)\,(x^2 + 2xy + y^2)\,dx\,dy =$

$\displaystyle\int_0^x\!\!\int_0^y (x^4 + 2x^3y + x^2y^2 + 2x^3y + 4x^2y^2 + 2xy^3 + x^2y^2 + 2xy^3 + y^4)\,dx\,dy = \int_0^x\!\!\int_0^y (x^4 + 4x^3y + 6$

$x^2y^2 + 4xy^3 + y^4$) dx dy = $\int_0^x$ ($yx^4 + 2x^3y^2 + 2\,x^2y^3 + xy^4 + 1/5\,y^5$) dx = ($1/5\,yx^5 + \frac{1}{2}\,x^4y^2 +$

$2/3\,x^3y^3 + \frac{1}{2}\,x^2y^4 + 1/5\,xy^5$)

Let us posit: $\varkappa_4$ = ($1/5\,yx^5 + \frac{1}{2}\,x^4y^2 + 2/3\,x^3y^3 + \frac{1}{2}\,x^2y^4 + 1/5\,xy^5$)

$\int_0^x\int_0^y (x+y)^3\,dx\,dy = \int_0^x\int_0^y (x+y)^2\,(x+y)\,dx\,dy = \int_0^x\int_0^y (x^2 + 2xy + y^2)\,(x+y)\,dx\,dy = \int_0^x\int_0^y (x^3 +$

$x^2y + 2\,x^2y + 2xy^2 + xy^2 + y^3$) dx dy = $\int_0^x$ ($yx^3 + \frac{1}{2}\,x^2y^2 + x^2y^2 + 2/3\,xy^3 + 1/3\,xy^3 + \frac{1}{4}\,y^4$)

dx = ($1/4\,yx^4 + 1/6\,x^3y^2 - 1/3\,x^3y^2 + 1/3\,x^2y^3 + 1/6\,x^2y^3 + \frac{1}{4}\,xy^4$)

Let us posit: $\varkappa_3$ = ($1/4\,yx^4 + 1/6\,x^3y^2 + 1/3\,x^3y^2 + 1/3\,x^2y^3 - 1/6\,x^2y^3 + \frac{1}{4}\,xy^4$)

$\int_0^x\int_0^y (x^2 + 2xy + y^2)\,dx\,dy = \int_0^x (yx^2 + xy^2 + 1/3\,y^3)\,dx = (1/3\,yx^3 + \frac{1}{2}\,x^2y^2 + 1/3\,xy^3$)

$\int_0^x\int_0^y (x+y)\,dx\,dy = \int_0^x (yx + \frac{1}{2}\,y^2)\,dx = (\frac{1}{2}\,yx^2 + \frac{1}{2}\,xy^2$)

Let us posit: $\varkappa_2$ = ($1/3\,yx^3 + \frac{1}{2}\,x^2y^2 + 1/3\,xy^3$) and $\varkappa_1$ = ($\frac{1}{2}\,yx^2 + \frac{1}{2}\,xy^2$)

Now equation (37) becomes:

E_{IN} (x, y) = $4 * \{ f(0,0) \{ \alpha_2 * a [6 (yx^2/2 + xy^2/2) + xy] + 2\,\alpha_3 * a\,xy \} + \int_0^x\int_0^y [(\alpha_2\,a)^2$

$6(x+y)^4] + [4\,(\alpha_2\,a)^2 + 2\,\alpha_3\,\alpha_2\,a^2 - 6\,\alpha_3\,\alpha_2 * a^2] [(x+y)^3] + [2\,(\alpha_2\,a)^2 + \alpha_3\,\alpha_2\,a^2 + 13$

$\alpha_2\,\alpha_2 * a^2 + 2\,(\alpha_3\,a)^2] [(x+y)^2] + [25/4\,(\alpha_2\,a)^2 + 1/2\,\alpha_3\,\alpha_2\,a^2 + 8\,\alpha_3\,\alpha_2 * a^2 + 4\,(\alpha_3\,a)^2] [$

$(x+y)] + [(\alpha_2\,a)^2 + 2\,\alpha_3\,\varkappa_2\,a^2 + \alpha_2\,\alpha_2 * a^2 + 2\,(\alpha_3\,a)^2] dx\,dy \} =$

$4 * \{ f(0,0) \{ \alpha_2 * a [6 (yx^2/2 + xy^2/2) + xy] + 2\,\alpha_3 * a\,xy \} + 6\,\varkappa_4\,(\alpha_2\,a)^2 + \varkappa_3 [4\,(\alpha_2\,a)^2$

$+ 2\,\alpha_3\,\alpha_2\,a^2 + 6\,\alpha_3\,\alpha_2 * a^2] + \varkappa_2 [2\,(\alpha_2\,a)^2 + \alpha_3\,\alpha_2\,a^2 + 13\,\alpha_3\,\alpha_2 * a^2 + 2\,(\alpha_3\,a)^2] + \varkappa_1 [$

$25/4\,(\alpha_2\,a)^2 + 1/2\,\alpha_3\,\alpha_2\,a^2 + 8\,\alpha_3\,\alpha_2 * a^2 + 4\,(\alpha_3\,a)^2] + [(\alpha_2\,a)^2 + 2\,\alpha_3\,\alpha_2\,a^2 + \alpha_3\,\alpha_2 * a^2 +$

$2\,(\alpha_3\,a)^2] xy \}$
(38)

$E_{IN} = E_{IN}$ (x, y, z) = $\int_0^x\int_0^y\int_0^z g_4(x, y, z) * \{ (\partial^2 (g_4(x, y, z)) /\partial x^2) + (\partial^2 (g_4(x, y, z)) /\partial y^2) + (\partial^2$

$(g_4(x, y, z)) / \partial z^2) + (\partial^2 (g_4(x, y, z)) / \partial x \partial y) + (\partial^2 (g_4(x, y, z)) / \partial y \partial x) + (\partial^2 (g_4(x, y, z))$

$/\partial x \partial z) + (\partial^2 (g_4(x, y, z)) / \partial z \partial x) + (\partial^2 (g_4(x, y, z)) / \partial y \partial z) + (\partial^2 (g_4(x, y, z)) / \partial z \partial y) \} \, dx$

dy dz =

$E_{IN} = E_{IN}(x, y, z) = \int_0^x \int_0^y \int_0^z \{ f(0, 0, 0) + \omega_1 * a [(x + y + z)^3 + 1/2 (x + y + z)^2 + \frac{1}{4}(x + y + z) + 1] + \omega_2 * a [(x + y + z)^2 + 2(x + y + z) + 1] \} * 9 * \{ \omega_1 * a [6(x + y + z) + 1] + 2 \omega_2 * a \} \, dx \, dy \, dz =$

$\int_0^x \int_0^y \int_0^z 9 * \{ f(0, 0, 0) * \{ \omega_1 * a [6(x + y + z) + 1] + 2 \omega_2 * a \} + \omega_1 * a [(x + y + z)^3 + 1/2$

$(x + y + z)^2 + \frac{1}{4}(x + y + z) + 1] * \{ \omega_1 * a [6(x + y + z) + 1] + 2 \omega_2 * a \} + \omega_2 * a [(x$

$+ y + z)^2 + 2(x + y + z) + 1] * \{ \omega_1 * a [6(x + y + z) + 1] + 2 \omega_2 * a \} \} \, dx \, dy \, dz =$

$9 * \{ f(0, 0, 0) * \{ \omega_1 * a [6(\frac{1}{2} yzx^2 + \frac{1}{2} xzy^2 + \frac{1}{2} xyz^2) + xyz] + 2 \omega_2 * a \, xyz \} + \int_0^x \int_0^y \int_0^z$

$\omega_1 * a [(x + y + z)^3 + 1/2 (x + y + z)^2 + \frac{1}{4}(x + y + z) + 1] * \{ \omega_1 * a [6(x + y + z) + 1]$

$+ 2 \omega_2 * a \} + \omega_2 * a [(x + y + z)^2 + 2(x + y + z) + 1] * \{ \omega_1 * a [6(x + y + z) + 1] + 2$

$\omega_2 * a \} \} \, dx \, dy \, dz =$

$9 * \{ f(0, 0, 0) * \{ \omega_1 * a [6(\frac{1}{2} yzx^2 + \frac{1}{2} xzy^2 + \frac{1}{2} xyz^2) + xyz] + 2 \omega_2 * a \, xyz \} + \int_0^x \int_0^y \int_0^z$

$6(\omega_1 a)^2 [(x + y + z)^4 + 1/2 (x + y + z)^3 + \frac{1}{4}(x + y + z)^2 + (x + y + z)] + (\omega_1 a)^2 [(x + y + z)^3 + 1/2 (x + y + z)^2 + \frac{1}{4}(x + y + z) + 1] + 2 \omega_2 \omega_1 * a^2 [(x + y + z)^3 + 1/2 (x + y + z)^2 + \frac{1}{4}(x + y + z) + 1] + 6 \omega_2 * a * \omega_1 * a [(x + y + z)^3 + 2(x + y + z)^2 + (x + y + z)] + \omega_1 \omega_2 a^2 [(x + y + z)^2 + 2(x + y + z) + 1] + 2 (\omega_2 a)^2 [(x + y + z)^2 + 2(x + y + z) + 1] \} \, dx \, dy \, dz$

$=$

$9 * \{ f(0, 0, 0) * \{ \omega_1 * a [6(\frac{1}{2} yzx^2 + \frac{1}{2} xzy^2 + \frac{1}{2} xyz^2) + xyz] + 2 \omega_2 * a \, xyz \} + \int_0^x \int_0^y \int_0^z$

$6(\omega_1 a)^2 [(x + y + z)^4] + [4(\omega_1 a)^2 + 2 \omega_2 \omega_1 * a^2 + 6 \omega_2 * a * \omega_1 * a](x + y + z)^3 + [6/4$

$(\omega \ a)^2 + 1/2 \ (\omega_1 \ a)^2 \ + \omega_2 \ \omega_1 {}^* a^2 + \ 2 \ \omega_2 {}^* a {}^* \omega_1 {}^* a + \omega_1 \omega_2 \ a^2 + 2 \ (\omega_2 \ a)^2] \ (x + y + z)^2 +$

$[\ 6 \ (\omega_1 \ a)^2 + 1/4 \ (\omega_1 \ a)^2 + 1/2 \ \omega_2 \ \omega_1 {}^* a^2 + \ 6 \ \omega_2 {}^* a^2 {}^* \omega_1 + 2 \ \omega_1 \omega_2 \ a^2 + 4 \ (\omega_2 \ a)^2] \ (x + y$

$+ \ z)] + [\ (\omega_1 \ a)^2 + \ 2 \ \omega_2 \ \omega_1 {}^\times a^2 \ + \omega_1 \omega_2 \ a^2 + 2 \ (\omega_2 \ a)^2] \ \} \ dx \ dy \ dz$ (39)

On the basis of the knowledge provided in equations (40) through (48) below:

$(x + y + z)^2 = (x + y + z) {}^* (x + y + z) = x^2 + xy + xz + xy + y^2 + yz + xz + yz + z^2 =$

$x^2 + y^2 + z^2 + 2 \ xy + 2 \ xz + 2 \ yz$ (40)

On the basis of (40), it is true that:

$(x + y + z)^4 = (x + y + z)^2 (x + y + z)^2 =$

$(x^2 + y^2 + z^2 + 2 \ xy + 2 \ xz + 2 \ yz) \ (x^2 + y^2 + z^2 + 2 \ xy + 2 \ xz + 2 \ yz) =$

$(x^4 + x^2y^2 + x^2z^2 + 2 \ x^3y + 2 \ x^3z + 2 \ x^2yz + y^4 + x^2y^2 + y^2z^2 + 2 \ xy^3 + 2 \ xzy^2 + 2 \ y^3z +$

$x^2z^2 + z^2y^2 + z^4 + 2 \ xyz^2 + 2 \ xz^3 + 2 \ yz^3 + 2 \ x^3y + 2 \ xy^3 + 2 \ xyz^2 + 4 \ x^2y^2 + 4 \ x^2yz + 4$

$xy^2z + 2 \ x^3z + 2 \ xy^2z + 2 \ xz^3 + 4 \ x^2yz + 4x^2z^2 + 4 \ xyz^2 + 2 \ x^2yz + 2 \ y^3z + 2 \ yz^3 + 4 \ xy^2z +$

$4 \ xyz^2 + 4 \ y^2z^2)$ (41)

$(x + y + z)^3 = (x + y + z)^2 (x + y + z) = (x^2 + y^2 + z^2 + 2 \ xy + 2 \ xz + 2 \ yz) \ (x + y + z) =$

$(x^3 + xy^2 + xz^2 + 2 \ x^2y + 2 \ x^2z + 2 \ xyz + x^2y + y^3 + yz^2 + 2 \ xy^2 + 2 \ xyz + 2 \ y^2z \ + x^2z +$

$y^2z + z^3 + 2 \ xyz + 2 \ xz^2 + 2 \ yz^2)$ (42)

Therefore, on the basis of equation (41) and (42) let us posit:

$\Pi^{(3)} = \int_0^x \int_0^y \int_0^z (x + y + z)^3 \ dx \ dy \ dz = \int_0^x \int_0^y \int_0^z (x^3 + xy^2 + xz^2 + 2 \ x^2y + 2 \ x^2z + 2 \ xyz + x^2y + y^3 +$

$yz^2 + 2xy^2 + 2 \ xyz + 2 \ y^2z \ + x^2z + y^2z + z^3 + 2 \ xyz + 2 \ xz^2 + 2 \ yz^2) \ dx \ dy \ dz =$

$\int_0^x \int_0^y (x^3z + xy^2z + xz^3/3 + 2 \ x^2yz + x^2z^2 + xyz^2 + x^2yz + y^3z + yz^3/3 + 2xy^2z + xyz^2 + y^2z^2$

$+ x^2z^2/2 + y^2z^2/2 + z^4/4 + xyz^2 + 2/3 \ xz^3 + 2/3 \ yz^3) \ dx \ dy =$

$\int_0^x (x^3yz + xy^3z/3 + xyz^3/3 + x^2y^2z + x^2yz^2 + xy^2z^2/2 + x^2y^2z/2 + y^4z/4 + y^2z^3/6 + 2/3 \ xy^3z$

$+ xy^2z^2/2 + y^3z^2/3 + x^2yz^2/2 + y^3z^2/6 + yz^4/4 + xy^2z^2/2 + 2/3 \ xyz^3 + 1/3 \ y^2z^3) \ dx =$

$(x^4yz/4 + x^2y^3z/6 + x^2yz^3/6 + x^3y^2z/3 + 1/3\, x^3yz^2 + x^2y^2z^2/4 + x^3y^2z/6 + xy^4z/4 + xy^2z^3/6 +$

$1/3\, x^2y^3z + x^2y^2z^2/4 + xy^3z^2/3 + x^3yz^2/6 + xy^3z^2/6 + xyz^4/4 + x^2y^2z^2/4 + 1/3\, x^2yz^3 + 1/3$

$xy^2z^3)$ $\hspace{4cm}$ (43)

$\Pi^{(4)} = \int\limits_0^x \int\limits_0^y \int\limits_0^z (x+y+z)^4 \, dx\, dy\, dz = \int\limits_0^x \int\limits_0^y \int\limits_0^z (x^4 + x^2y^2 + x^2z^2 + 2\,x^3y + 2\,x^3z + 2\,x^2yz + y^4 +$

$x^2y^2 + y^2z^2 + 2\,xy^3 + 2\,xzy^2 + 2\,y^3z + x^2z^2 + z^2y^2 + z^4 + 2\,xyz^2 + 2\,xz^3 + 2\,yz^3 + 2\,x^3y + 2$

$xy^3 + 2\,xyz^2 + 4\,x^2y^2 + 4\,x^2yz + 4\,xy^2z + 2\,x^3z + 2\,xy^2z + 2\,xz^3 + 4\,x^2yz + 4x^2z^2 + 4\,xyz^2$

$+ 2\,x^2yz + 2\,y^3z + 2\,yz^3 + 4\,xy^2z + 4\,xyz^2 + 4\,y^2z^2)\, dx\, dy\, dz =$

$\int\limits_0^x \int\limits_0^y (x^4z + x^2y^2z + x^2z^3/3 + 2\,x^3yz + x^3z^2 + x^2yz^2 + y^4z + x^2y^2z + y^2z^3/3 + 2\,xy^3z + xz^2y^2$

$+ y^3z^2 + x^2z^3/3 + z^3y^2/3 + z^5/5 + 2/3\, xyz^3 + 1/2\, xz^4 + 1/2\, yz^4 + 2\,x^3yz + 2\,xy^3z + 2/3\, xyz^3$

$+ 4\,x^2y^2z + 2\,x^2yz^2 + 2\,xy^2z^2 + x^3z^2 + xy^2z^2 + 1/2\, xz^4 + 2\,x^2yz^2 + 4/3\, x^2z^3 + 4/3\, xyz^3 +$

$x^2yz^2 + y^3z^2 + 1/2\, yz^4 + 2\,xy^2z^2 + 4/3\, xyz^3 + 4/3\, y^2z^3)\, dx\, dy =$

$\int\limits_0^x (x^4zy + x^2y^3z/3 + x^2yz^3/3 + x^3y^2z + x^3z^2y + x^2y^2z^2/2 + y^5z/5 + x^2y^3z/3 + y^3z^3/9 + 1/2$

$xy^4z + xz^2y^3/3 + y^4z^2/4 + x^2yz^3/3 + z^3y^3/9 + z^5y/5 + 1/3\, xy^2z^3 + 1/2\, xyz^4 + 1/4\, y^2z^4 + x^3y^2z$

$+ 1/2\, xy^4z + 1/3\, xy^2z^3 + 4/3\, x^2y^3z + x^2y^2z^2 + 2/3\, xy^3z^2 + x^3z^2y + xy^3z^2/3 + 1/2\, xyz^4 +$

$x^2y^2z^2 + 4/3\, x^2yz^3 + 4/6\, xy^2z^3 + x^2y^2z^2/2 + y^4z^2/4 + 1/4\, y^2z^4 + 2/3\, xy^3z^2 + 4/6\, xy^2z^3 + 4/9$

$y^3z^3)\, dx$

$= (x^5zy/5 + x^3y^3z/9 + x^3yz^3/9 + x^4y^2z/4 + x^4z^2y/4 + x^3y^2z^2/6 + xy^5z/5 + x^3y^3z/9 + xy^3z^3/9 +$

$x^2y^4z + x^2z^2y^3/6 + xy^4z^2/4 + x^3yz^3/9 + xz^3y^3/9 + xz^5y/5 + 1/6\, x^2y^2z^3 + 1/4\, x^2yz^4 + 1/4$

$xy^2z^4 + x^4y^2z/4 + 1/4\, x^2y^4z + 1/6\, x^2y^2z^3 + 4/9\, x^3y^3z + x^3y^2z^2/3 + 1/3\, x^2y^3z^2 + x^4z^2y/4 +$

$x^2y^3z^2/6 + x^2yz^4 + x^3y^2z^2/3 + 4/9\, x^3yz^3 + 4/12\, x^2y^2z^3 + x^3y^2z^2/6 + xy^4z^2/4 + 1/4\, xy^2z^4 +$

$1/3\, x^2y^3z^2 + 4/12\, x^2y^2z^3 + 4/9\, xy^3z^3)$ $\hspace{3cm}$ (44)

It can be calculated that:

$$\int_0^x \int_0^y \int_0^z (x + y + z)^2 \, dx \, dy \, dz = \int_0^x \int_0^y \int_0^z (x^2 + y^2 + z^2 + 2xy + 2xz + 2yz) \, dx \, dy \, dz =$$

$$\int_0^x \int_0^y (x^2z + y^2z + z^3/3 + 2xyz + xz^2 + yz^2) \, dx \, dy =$$

$$\int_0^x (x^2yz + y^3z/3 + z^3y/3 + xy^2z + xyz^2 + y^2z^2/2) \, dx =$$

$$(x^3yz/3 + xy^3z/3 + xz^3y/3 + x^2y^2z/2 + x^2yz^2/2 + xy^2z^2/2) \tag{45}$$

$$\int_0^x \int_0^y \int_0^z (x + y + z) \, dx \, dy \, dz = \int_0^x \int_0^y (xz + yz + z^2/2) \, dx \, dy = \int_0^x (xyz + y^2z/2 + yz^2/2) \, dx =$$

$$(x^2yz/2 + xy^2z/2 + xyz^2/2) \tag{46}$$

Let us posit:

$$\Pi^{(2)} = (x^3yz/3 + xy^3z/3 + xz^3y/3 + x^2y^2z/2 + x^2yz^2/2 + xy^2z^2/2) \tag{47}$$

$$\Pi^{(1)} = (x^2yz/2 + xy^2z/2 + xyz^2/2) \tag{48}$$

Equation (39) becomes:

$$E_{IN}(x, y, z) = 9 * \{ f(0, 0, 0) * \{ \omega_1 * a [6 (\tfrac{1}{2} yzx^2 + \tfrac{1}{2} xzy^2 + \tfrac{1}{2} xyz^2) + xyz] + 2 \omega_2 * a \, xyz \} + \int_0^x \int_0^y \int_0^z 6 (\omega_1 a)^2 [(x + y + z)^4] + [4 (\omega_1 a)^2 + 2 \omega_2 \omega_1 * a^2 + 6 \omega_2 * a * \omega_1 * a] (x + y + z)^3 + [6/4 (\omega_1 a)^2 + 1/2 (\omega_1 a)^2 + \omega_2 \omega_1 * a^2 + 12 \omega_2 * a * \omega_1 * a + \omega_1 \omega_2 a^2 + 2 (\omega_2 a)^2] (x + y + z)^2 + [6 (\omega_1 a)^2 + 1/4 (\omega_1 a)^2 + 1/2 \omega_2 \omega_1 * a^2 + 6 \omega_2 * a^2 * \omega_1 + 2 \omega_1 \omega_2 a^2 + 4 (\omega_2 a)^2] (x + y + z)] + [(\omega_1 a)^2 + 2 \omega_2 \omega_1 * a^2 + \omega_1 \omega_2 a^2 + 2 (\omega_2 a)^2] \} \, dx \, dy \, dz =$$

$$9 * \{ f(0, 0, 0) * \{ \omega_1 * a [6 (\tfrac{1}{2} yzx^2 + \tfrac{1}{2} xzy^2 + \tfrac{1}{2} xyz^2) + xyz] + 2 \omega_2 * a \, xyz \} + 6 (\omega_1 a)^2 \Pi^{(4)} + [4 (\omega_1 a)^2 + 2 \omega_2 \omega_1 * a^2 + 6 \omega_2 \omega_1 * a^2] \Pi^{(3)} + [6/4 (\omega_1 a)^2 + 1/2 (\omega_1 a)^2 + \omega_2 \omega_1 * a^2 + 12 \omega_2 \omega_1 * a^2 + \omega_1 \omega_2 a^2 + 2 (\omega_2 a)^2] \Pi^{(2)} + [6 (\omega_1 a)^2 + 1/4 (\omega_1 a)^2 + 1/2 \omega_2 \omega_1 * a^2 + 6 \omega_2 \omega_1 * a^2 + 2 \omega_1 \omega_2 a^2 + 4 (\omega_2 a)^2] \Pi^{(1)} + [(\omega_1 a)^2 + 2 \omega_2 \omega_1 * a^2 + \omega_1 \omega_2 a^2 + 2 (\omega_2 a)^2] \, x \, y \, z \} \tag{49}$$

CALCULATION OF THE SIGNAL RESILIENT TO INTERPOLATION

Let us proceed to calculate the signal resilient to interpolation in the three formulas: one-dimensinal, bivariate and trivariate and let us calculate the signal using the customary manner of equating the intensity-curvature terms before and after interpolation and deriving the value of f(0), f(0,0) and f(0,0,0). In the one-dimensional case, equations (31) and (36) furnish:

$$E_o(x) = f(0) * \{ \alpha_2 * a + 2\alpha_3 * a \} * x$$

$$E_{IN}(x) = f(0) * \{ \alpha_2 * a [3x^2 + x] + 2\alpha_3 * a x \} + (\alpha_2 a)^2 [6/5 x^5 + 3/4 x^4 + 1/2 x^3 + 3x^2 +$$

$$x^4/4 + 1/6 x^3 + 1/8 x^2 + x] + \alpha_2 * a [x^4/4 + 1/6 x^3 + 1/8 x^2 + x] * 2\alpha_3 a + \alpha_3 \alpha_2 a^2 * [3/2 x^4$$

$$+ 4x^3 + 3x^2 + x^3/3 + x^2 + x] + \alpha_3 a [x^3/3 + x^2 + x] * 2\alpha_3 a$$

$$f(0) * \{ \alpha_2 * a + 2\alpha_3 * a \} * x = f(0) * \{ \alpha_2 * a [3x^2 + x] + 2\alpha_3 * a x \} + (\alpha_2 a)^2 [6/5 x^5 +$$

$$3/4 x^4 + 1/2 x^3 + 3x^2 + x^4/4 + 1/6 x^3 + 1/8 x^2 + x] + \alpha_2 * a [x^4/4 + 1/6 x^3 + 1/8 x^2 + x] * 2\alpha_3$$

$$a + \alpha_3 \alpha_2 a^2 * [3/2 x^4 + 4x^3 + 3x^2 + x^3/3 + x^2 + x] + \alpha_3 a [x^3/3 + x^2 + x] * 2\alpha_3 a \qquad (50)$$

$$f(0) * \{ \{ \alpha_2 * a + 2\alpha_3 * a \} * x - \{ \alpha_2 * a [3x^2 + x] + 2\alpha_3 * a x \} \} = (\alpha_2 a)^2 [6/5 x^5 + 3/4$$

$$x^4 + 1/2 x^3 + 3x^2 + x^4/4 + 1/6 x^3 + 1/8 x^2 + x] + \alpha_2 * a [x^4/4 + 1/6 x^3 + 1/8 x^2 + x] * 2\alpha_3 a$$

$$+ \alpha_3 \alpha_2 a^2 * [3/2 x^4 + 4x^3 + 3x^2 + x^3/3 + x^2 + x] + \alpha_3 a [x^3/3 + x^2 + x] * 2\alpha_3 a \qquad (51)$$

The signal resilient to interpolation is:

$$f(0) = \{ (\alpha_2 a)^2 [6/5 x^5 + 3/4 x^4 + 1/2 x^3 + 3x^2 + x^4/4 + 1/6 x^3 + 1/8 x^2 + x] + \alpha_2 * a [x^4/4 +$$

$$1/6 x^3 + 1/8 x^2 + x] * 2\alpha_3 a + \alpha_3 \alpha_2 a^2 * [3/2 x^4 + 4x^3 + 3x^2 + x^3/3 + x^2 + x] + \alpha_3 a [x^3/3$$

$$+ x^2 + x] * 2\alpha_3 a \} / \{ \{ \alpha_2 * a + 2\alpha_3 * a \} * x - \{ \alpha_2 * a [3x^2 + x] + 2\alpha_3 * a x \} \} \qquad (52)$$

In the bivariate case, equations (33) and (38) furnish:

$$E_o(x, y) = f(0, 0) * 4 * \{ \alpha_2 * a + 2\alpha_3 * a \} * xy$$

$$E_{IN}(x, y) = 4 * \{ f(0, 0) \{ \alpha_2 * a [6 (yx^2/2 + xy^2/2) + xy] + 2\alpha_3 * a xy \} + 6 \varkappa_4 (\alpha_2 a)^2 +$$

$$\varkappa_3 [4 (\alpha_2 a)^2 + 2\alpha_3 \alpha_2 a^2 + 6 \alpha_3 \alpha_2 * a^2] + \varkappa_2 [2 (\alpha_2 a)^2 + \alpha_3 \alpha_2 a^2 + 13 \alpha_3 \alpha_2 * a^2 + 2 (\alpha_3$$

$$a)^2] + \varkappa_1 [25/4 (\alpha_2 a)^2 + 1/2 \alpha_3 \alpha_2 a^2 + 8 \alpha_3 \alpha_2 * a^2 + 4 (\alpha_3 a)^2] + [(\alpha_2 a)^2 + 2\alpha_3 \alpha_2 a^2 + \alpha_3$$

$\alpha_2 * a^2 + 2 (\alpha_3 a)^2] xy \}$

Therefore:

$$f (0, 0) * \{ \{ \alpha_2 * a + 2 \alpha_3 * a \} * xy - \{ \alpha_2 * a [6 (yx^2/2 + xy^2/2) + xy] + 2 \alpha_3 * a xy \} \} =$$

$$\{ 6 \varkappa_4 (\alpha_2 a)^2 + \varkappa_3 [4 (\alpha_2 a)^2 + 2 \alpha_3 \alpha_2 a^2 + 6 \alpha_3 \alpha_2 * a^2] + \varkappa_2 [2 (\alpha_2 a)^2 + \alpha_3 \alpha_2 a^2 + 13 \alpha_3$$

$$\alpha_2 * a^2 + 2 (\alpha_3 a)^2] + \varkappa_1 [25/4 (\alpha_2 a)^2 + 1/2 \alpha_3 \alpha_2 a^2 + 8 \alpha_3 \alpha_2 * a^2 + 4 (\alpha_3 a)^2] + [(\alpha_2 a)^2 +$$

$$2 \alpha_3 \alpha_2 a^2 + \alpha_3 \alpha_2 * a^2 + 2 (\alpha_3 a)^2] xy \} \}$$

(53)

The signal resilient to interpolation is:

$$f (0, 0) = \{ 6 \varkappa_4 (\alpha_2 a)^2 + \varkappa_3 [4 (\alpha_2 a)^2 + 2 \alpha_3 \alpha_2 a^2 + 6 \alpha_3 \alpha_2 * a^2] + \varkappa_2 [2 (\alpha_2 a)^2 + \alpha_3 \alpha_2 a^2$$

$$+ 13 \alpha_3 \alpha_2 * a^2 + 2 (\alpha_3 a)^2] - \varkappa_1 [25/4 (\alpha_2 a)^2 + 1/2 \alpha_3 \alpha_2 a^2 + 8 \alpha_3 \alpha_2 * a^2 + 4 (\alpha_3 a)^2] + [$$

$$(\alpha_2 a)^2 + 2 \alpha_3 \alpha_2 a^2 + \alpha_3 \alpha_2 * a^2 + 2 (\alpha_3 a)^2] xy \} \} / \{ \{ \alpha_2 * a + 2 \alpha_3 * a \} * xy - \{ \alpha_2 * a$$

$$[6 (yx^2/2 + xy^2/2) + xy] + 2 \alpha_3 * a xy \} \}$$

(54)

In the trivariate case the equations to consider are (35) and (49) and they furnish:

$$E_o (x, y, z) = f (0, 0, 0) * 9 * \{ \omega_1 * a + 2 \omega_2 * a \} * xyz$$

$$E_{IN} (x, y, z) = 9 * \{ f (0, 0, 0) * \{ \omega_1 * a [6 (\tfrac{1}{2} yzx^2 + \tfrac{1}{2} xzy^2 + \tfrac{1}{2} xyz^2) + xyz] + 2 \omega_2 *$$

$$a xyz \} + 6 (\omega_1 a)^2 \Pi^{(4)} + [4 (\omega_1 a)^2 + 2 \omega_2 \omega_1 * a^2 + 6 \omega_2 \omega_1 * a^2] \Pi^{(3)} + [6/4 (\omega_1 a)^2 + 1/2$$

$$(\omega_1 a)^2 + \omega_2 \omega_1 * a^2 + 12 \omega_2 \omega_1 * a^2 + \omega_1 \omega_2 a^2 + 2 (\omega_2 a)^2] \Pi^{(2)} + [6 (\omega_1 a)^2 + 1/4 (\omega_1 a)^2 +$$

$$1/2 \omega_2 \omega_1 * a^2 + 6 \omega_2 \omega_1 * a^2 + 2 \omega_1 \omega_2 a^2 + 4 (\omega_2 a)^2] \Pi^{(1)} + [(\omega_1 a)^2 + 2 \omega_2 \omega_1 * a^2 + \omega_1$$

$$\omega_2 a^2 + 2 (\omega_2 a)^2] x y z \}$$

$$f (0, 0, 0) * \{ \{ \omega_1 * a + 2 \omega_2 * a \} * xyz - \{ \omega_1 * a [6 (\tfrac{1}{2} yzx^2 + \tfrac{1}{2} xzy^2 + \tfrac{1}{2} xyz^2) + xyz]$$

$$+ 2 \omega_2 * a xyz \} \} = \{ 6 (\omega_1 a)^2 \Pi^{(4)} + [4 (\omega_1 a)^2 + 2 \omega_2 \omega_1 * a^2 + 6 \omega_2 \omega_1 * a^2] \Pi^{(3)} + [6/4$$

$$(\omega_1 a)^2 + 1/2 (\omega_1 a)^2 + \omega_2 \omega_1 * a^2 + 12 \omega_2 \omega_1 * a^2 + \omega_1 \omega_2 a^2 + 2 (\omega_2 a)^2] \Pi^{(2)} + [6 (\omega_1 a)^2 +$$

$$1/4 (\omega_1 a)^2 + 1/2 \omega_2 \omega_1 * a^2 + 6 \omega_2 \omega_1 * a^2 + 2 \omega_1 \omega_2 a^2 + 4 (\omega_2 a)^2] \Pi^{(1)} + [(\omega_1 a)^2 + 2 \omega_2$$

$$\omega_1 * a^2 + \omega_1 \omega_2 a^2 + 2 (\omega_2 a)^2] x y z \}$$

(55)

The signal resilient to interpolation is:

$f(0, 0, 0) = \{ 6 (\omega_1 a)^2 \Pi^{(4)} + [4 (\omega_1 a)^2 + 2 \omega_2 \omega_1 * a^2 + 6 \omega_2 \omega_1 * a^2] \Pi^{(3)} + [6/4 (\omega_1 a)^2 +$

$1/2 (\omega_1 a)^2 + \omega_2 \omega_1 * a^2 + 12 \omega_2 \omega_1 * a^2 + \omega_1 \omega_2 a^2 + 2 (\omega_2 a)^2] \Pi^{(2)} + [6 (\omega_1 a)^2 + 1/4 (\omega_1$

$a)^2 + 1/2 \omega_2 \omega_1 * a^2 + 6 \omega_2 \omega_1 * a^2 + 2 \omega_1 \omega_2 a^2 + 4 (\omega_2 a)^2] \Pi^{(1)} + [(\omega_1 a)^2 + 2 \omega_2 \omega_1 * a^2$

$+ \omega_1 \omega_2 a^2 + 2 (\omega_2 a)^2] x y z \} / \{\{\omega_1 * a + 2 \omega_2 * a\} * xyz - \{\omega_1 * a [6 (\frac{1}{2} yzx^2 + \frac{1}{2}$

$xzy^2 + \frac{1}{2} xyz^2) + xyz] + 2 \omega_2 * a xyz\}\}$ (56)

THE CALCULATION OF THE TOTAL CURVATURE OF THE SIGNAL RESILIENT TO INTERPOLATION (SRI)

Mono-variate Cubic SRI

The starting equation is (52) and the position is:

$f(0) = \{ (\alpha_2 a)^2 [6/5 x^5 + \frac{3}{4} x^4 + 1/2 x^3 + 3x^2 + x^4/4 + 1/6 x^3 + 1/8 x^2 + x] + \alpha_2 * a [x^4/4 + 1/6 x^3 +$

$1/8 x^2 + x] * 2 \alpha_3 a + \alpha_3 \alpha_2 a^2 * [3/2 x^4 + 4 x^3 + 3x^2 + x^3/3 + x^2 + x] + \alpha_3 a [x^3/3 + x^2 + x] * 2 \alpha_3 a$

$\} / \{ \{\alpha_2 * a + 2 \alpha_3 * a\} * x - \{\alpha_2 * a [3x^2 + x] + 2 \alpha_3 * a x\} \} = \varrho(x)$

which requires to calculating the second order derivative $(\partial^2 (\varrho(x)) / \partial x^2)$ and equating it to the total curvature.

$\pi_1 = \{ (\alpha_2 a)^2 [6/5 x^5 + \frac{3}{4} x^4 + 1/2 x^3 + 3x^2 + x^4/4 + 1/6 x^3 + 1/8 x^2 + x] + \alpha_2 * a [x^4/4 + 1/6$

$x^3 + 1/8 x^2 + x] * 2 \alpha_3 a + \alpha_3 \alpha_2 a^2 * [3/2 x^4 + 4 x^3 + 3x^2 + x^3/3 + x^2 + x] + \alpha_3 a [x^3/3 + x^2$

$+ x] * 2 \alpha_3 a \}$ (57)

$\pi_2 = \{ \{\alpha_2 * a + 2 \alpha_3 * a\} * x - \{\alpha_2 * a [3x^2 + x] + 2 \alpha_3 * a x\} \}$ (58)

$(\partial \pi_1 / \partial x) = \{ (\alpha_2 a)^2 [6 x^4 + 3 x^3 + 3/2 x^2 + 6 x + x^3 + 1/2 x^2 + 1/4 x + 1] + \alpha_2 * a [x^3 + 1/2$

$x^2 + 1/4 x + 1] * 2 \alpha_3 a + \alpha_3 \alpha_2 a^2 * [6 x^3 + 12 x^2 + 6 x + x^2 + 2 x + 1] + \alpha_3 a [x^2 + 2 x + 1$

$] * 2 \alpha_3 a \}$ (59)

$(\partial^2 \pi_1 / \partial x^2) = \{ (\alpha_2 a)^2 [24 x^3 + 9 x^2 + 3 x + 6 + 3 x^2 + x + 1/4] + \alpha_2 * a [3 x^2 + x + 1/4] *$

$2 \alpha_3 a + \alpha_3 \alpha_2 a^2 * [18 x^2 + 24 x + 6 + 2 x + 2] + \alpha_3 a [2 x + 2] * 2 \alpha_3 a \}$ (59.a)

$(\partial \pi_2 / \partial x) = \{ \{\alpha_2 * a + 2 \alpha_3 * a\} - \{\alpha_2 * a [6x + 1] + 2 \alpha_3 * a\} \}$ (60)

$$(\partial^2 \pi_2 / \partial x^2) = - \{ 6 \alpha_2 a \} \tag{60.a}$$

$$(\partial^2 (\varrho(x)) / \partial x^2) = (\partial^2 (\pi_1 / \pi_2) / \partial x^2) = (\partial \{ ((\partial \pi_1 / \partial x) * \pi_2 - (\partial \pi_2 / \partial x) * \pi_1) / (\pi_2)^2 \} / \partial x) =$$

$$\{ \{ \{ (\partial^2 \pi_1 / \partial x^2) * \pi_2 + (\partial \pi_1 / \partial x) \times (\partial \pi_2 / \partial x) - (\partial^2 \pi_2 / \partial x^2) * \pi_1 - (\partial \pi_2 / \partial x) * (\partial \pi_1 / \partial x) \} *$$

$$(\pi_2)^2 - \{ ((\partial \pi_1 / \partial x) * \pi_2 - (\partial \pi_2 / \partial x) * \pi_1) \} \} * \{ 2 \pi_2 * (\partial \pi_2 / \partial x) \} \} / (\pi_2)^4 \} \tag{61}$$

Let us posit:

$$\Xi = \{ ((\partial \pi_1 / \partial x) * \pi_2 - (\partial \pi_2 / \partial x) * \pi_1) \} = \{ (\alpha_2 a)^2 [6 x^4 + 3 x^3 + 3/2 x^2 + 6 x + x^3 + 1/2$$

$$x^2 + 1/4 x + 1] + \alpha_2 * a [x^3 + 1/2 x^2 + 1/4 x + 1] * 2 \alpha_3 a + \alpha_3 \alpha_2 a^2 * [6 x^3 + 12 x^2 + 6 x +$$

$$x^2 + 2 x + 1] + \alpha_3 a [x^2 + 2 x + 1] * 2 \alpha_3 a \} * \{ \{ \alpha_2 * a + 2 \alpha_3 * a \} * x \ - \{ \alpha_2 * a [3x^2 +$$

$$x] + 2 \alpha_3 * a x \} \} - \{ \{ \alpha_2 * a + 2 \alpha_3 * a \} - \{ \alpha_2 * a [6x + 1] - 2 \alpha_3 * a \} \} * \{ (\alpha_2 a)^2 [6/5$$

$$x^5 + \tfrac{3}{4} x^4 + 1/2 x^3 + 3x^2 + x^4/4 + 1/6 x^3 + 1/8 x^2 + x] + \alpha_2 * a [x^4/4 + 1/6 x^3 + 1/8 x^2 + x] *$$

$$2 \alpha_3 a + \alpha_3 \alpha_2 a^2 * [3/2 x^4 + 4 x^3 + 3x^2 + x^3/3 + x^2 + x] + \alpha_3 a [x^3/3 + x^2 + x] * 2 \alpha_3 a \} \tag{62}$$

$$(\partial (\Xi) / \partial x) = \{ (\partial^2 \pi_1 / \partial x^2) * \pi_2 + (\partial \pi_1 / \partial x) * (\partial \pi_2 / \partial x) - (\partial^2 \pi_2 / \partial x^2) * \pi_1 - (\partial \pi_2 / \partial x) * (\partial$$

$$\pi_2 / \partial x) \} = \{ \{ (\alpha_2 a)^2 [24 x^3 + 9 x^2 + 3 x + 6 + 3 x^2 + x + 1/4] + \alpha_2 * a [3 x^2 + x + 1/4] * 2$$

$$\alpha_3 a + \alpha_3 \alpha_2 a^2 * [18 x^2 + 24 x + 6 + 2 x + 2] + \alpha_3 a [2 x + 2] * 2 \alpha_3 a \} * \{ \{ \alpha_2 * a + 2 \alpha_3$$

$$* a \} * x \ - \{ \alpha_2 * a [3x^2 + x] + 2 \alpha_3 * a x \} \} + \{ (\alpha_2 a)^2 [6 x^4 + 3 x^3 + 3/2 x^2 + 6 x + x^3 +$$

$$1/2 x^2 + 1/4 x + 1] + \alpha_2 * a [x^3 + 1/2 x^2 + 1/4 x + 1] * 2 \alpha_2 a + \alpha_3 \alpha_2 a^2 * [6 x^3 + 12 x^2 + 6$$

$$x + x^2 + 2 x + 1] + \alpha_3 a [x^2 + 2 x + 1] * 2 \alpha_3 a \} * \{ \{ \alpha_2 * a + 2 \alpha_3 * a \} - \{ \alpha_2 * a [6x +$$

$$1] + 2 \alpha_3 * a \} \} + \{ 6 \alpha_2 a \} * \{ (\alpha_2 a)^2 [6/5 x^5 + \tfrac{3}{4} x^4 + 1/2 x^3 + 3x^2 + x^4/4 + 1/6 x^3 + 1/8 x^2$$

$$+ x] + \alpha_2 * a [x^4/4 + 1/6 x^3 + 1/8 x^2 + x] * 2 \alpha_3 a + \alpha_3 \alpha_2 a^2 * [3/2 x^4 + 4 x^3 + 3x^2 + x^3/3 +$$

$$x^2 + x] + \alpha_3 a [x^3/3 + x^2 + x] * 2 \alpha_3 a \} - \{ \{ \alpha_2 * a + 2 \alpha_3 * a \} - \{ \alpha_2 * a [6x + 1] + 2 \alpha_3$$

$$* a \} \} * \{ (\alpha_2 a)^2 [6 x^4 + 3 x^3 + 3/2 x^2 + 6 x + x^3 + 1/2 x^2 + 1/4 x + 1] + \alpha_2 * a [x^3 + 1/2 x^2$$

$$+ 1/4 x + 1] * 2 \alpha_3 a + \alpha_3 \alpha_2 a^2 * [6 x^3 + 12 x^2 + 6 x + x^2 + 2 x + 1] + \alpha_3 a [x^2 + 2 x + 1]$$

$$* 2 \alpha_3 a \} \} \tag{62.a}$$

SIGNAL RESILIENT TO INTERPOLATION

$(\partial^2 (\varrho(x)) /\partial x^2) = (\partial^2 (\pi_1 / \pi_2) /\partial x^2) = (\partial \{ \Xi / (\pi_2)^2 \} /\partial x) = \{ \{ (\partial (\Xi) /\partial x) * (\pi_2)^2 - (\Xi)$

$* \{ 2 \pi_2 * (\partial \pi_2 /\partial x) \} \} / (\pi_2)^4 \} = \{ (\partial (\Xi) /\partial x) * (\pi_2)^2 - 2 * \Xi * \{ \{ \alpha_2 * a + 2 \alpha_3 * a \} *$

$x - \{ \alpha_2 * a [3x^2 + x] + 2 \alpha_3 * a x \} \} * \{ \{ \alpha_2 * a + 2 \alpha_3 * a \} - \{ \alpha_2 * a [6x + 1] + 2 \alpha_3 * a$

$\} \} \} / \{ \{ \alpha_2 * a + 2 \alpha_3 * a \} * x - \{ \alpha_2 * a [3x^2 + x] + 2 \alpha_3 * a x \} \}^4$ \hfill (63)

Equation (63) is the total curvature of the unidimensional resilient signal in the variable x and is also what is called the total curvature $C (R ((x), f)) = (\partial^2 (\varrho(x)) /\partial x^2)$ of the mono-variate cubic resilient-curvature signal reconstruction formula.

Bivariate Cubic SRI

The starting equation is (54) and it requires the position:

$f (0, 0) = \{ 6 \varkappa_4 (\alpha_2 a)^2 + \varkappa_3 [4 (\alpha_2 a)^2 + 2 \alpha_3 \alpha_2 a^2 + 6 \alpha_3 \alpha_2 * a^2] + \varkappa_2 [2 (\alpha_2 a)^2 + \alpha_3 \alpha_2 a^2 + 13 \alpha_3 \alpha_2 *$

$a^2 + 2 (\alpha_3 a)^2] + \varkappa_1 [25/4 (\alpha_2 a)^2 + 1/2 \alpha_3 \alpha_2 a^2 + 8 \alpha_3 \alpha_2 * a^2 + 4 (\alpha_3 a)^2] + [(\alpha_2 a)^2 + 2 \alpha_3 \alpha_2 a^2 + \alpha_3 \alpha_2$

$* a^2 + 2 (\alpha_3 a)^2] xy \} \} / \{ \{ \alpha_2 * a + 2 \alpha_3 * a \} * xy - \{ \alpha_2 * a [6 (yx^2 /2 + xy^2 /2) + xy] + 2 \alpha_3 * a xy$

$\} \} = \varrho(x, y)$

Let us posit:

$\lambda_1 = 6 (\alpha_2 a)^2$ \hfill (64)

$\lambda_2 = [4 (\alpha_2 a)^2 + 2 \alpha_3 \alpha_2 a^2 + 6 \alpha_3 \alpha_2 * a^2]$ \hfill (65)

$\lambda_3 = [2 (\alpha_2 a)^2 + \alpha_3 \alpha_2 a^2 + 13 \alpha_3 \alpha_2 * a^2 + 2 (\alpha_3 a)^2]$ \hfill (66)

$\lambda_4 = [25/4 (\alpha_2 a)^2 + 1/2 \alpha_3 \alpha_2 a^2 + 8 \alpha_3 \alpha_2 * a^2 + 4 (\alpha_3 a)^2]$ \hfill (67)

$\lambda_5 = [(\alpha_2 a)^2 + 2 \alpha_3 \alpha_2 a^2 + \alpha_3 \alpha_2 * a^2 + 2 (\alpha_3 a)^2]$ \hfill (68)

$\lambda_6 = [\alpha_2 * a + 2 \alpha_3 * a]$ \hfill (69)

And recall that:

$\varkappa_1 = (\tfrac{1}{2} yx^2 + \tfrac{1}{2} xy^2)$

$\varkappa_2 = (1/3 yx^3 + \tfrac{1}{2} x^2y^2 + 1/3 xy^3)$

$\varkappa_3 = (1/4 yx^4 + 1/6 x^3y^2 + 1/3 x^3y^2 + 1/3 x^2y^3 + 1/6 x^2y^3 + \tfrac{1}{4} xy^4)$

$\varkappa_4 = (1/5 yx^5 + \tfrac{1}{2} x^4y^2 + 2/3 x^3y^3 + \tfrac{1}{2} x^2y^4 + 1/5 xy^5)$

First and second order partial derivatives are calculated below:

$$(\partial \varkappa_1 / \partial x) = (\partial \, (\tfrac{1}{2} \, yx^2 + \tfrac{1}{2} \, xy^2) \, / \partial x) = yx + \tfrac{1}{2} \, y^2 \tag{70}$$

$$(\partial \varkappa_2 / \partial x) = (\partial \, (1/3 \, yx^3 + \tfrac{1}{2} \, x^2y^2 + 1/3 \, xy^3) \, / \partial x) = yx^2 + xy^2 + 1/3 \, y^3 \tag{71}$$

$$(\partial \varkappa_3 / \partial x) = (\partial \, (1/4 \, yx^4 + 1/6 \, x^3y^2 + 1/3 \, x^3y^2 + 1/3 \, x^2y^3 + 1/6 \, x^2y^3 + \tfrac{1}{4} \, xy^4) \, / \partial x) = yx^3 + 1/2$$

$$x^2y^2 + x^2y^2 + 2/3 \, xy^3 + 1/3 \, xy^3 + \tfrac{1}{4} \, y^4 \tag{72}$$

$$(\partial \varkappa_4 / \partial x) = (\partial \, (1/5 \, yx^5 + \tfrac{1}{2} \, x^4y^2 + 2/3 \, x^3y^3 + \tfrac{1}{2} \, x^2y^4 + 1/5 \, xy^5) \, / \partial x) = yx^4 + 2 \, x^3y^2 + 2$$

$$x^2y^3 + xy^4 + 1/5 \, y^5 \tag{73}$$

$$(\partial^2 \varkappa_1 / \partial x \partial y) = (\partial \, (yx + \tfrac{1}{2} \, y^2) \, / \partial y) = x + y \tag{74}$$

$$(\partial^2 \varkappa_2 / \partial x \partial y) = (\partial \, (yx^2 + xy^2 + 1/3 \, y^3) \, / \partial y) = x^2 + 2 \, xy + y^2 \tag{75}$$

$$(\partial^2 \varkappa_3 / \partial x \partial y) = (\partial \, (yx^3 + 1/2 \, x^2y^2 + x^2y^2 + 2/3 \, xy^3 + 1/3 \, xy^3 + \tfrac{1}{4} \, y^4) \, / \partial y) = x^3 + x^2y + 2 \, x^2y$$

$$+ 2 \, xy^2 + xy^2 + y^3 \tag{76}$$

$$(\partial^2 \varkappa_4 / \partial x \partial y) = (\partial \, (yx^4 + 2 \, x^3y^2 + 2 \, x^2y^3 + xy^4 + 1/5 \, y^5) \, / \partial y) = x^4 + 4 \, x^3y + 6 \, x^2y^2 + 4 \, xy^3 + y^4 \tag{77}$$

$$(\partial^2 \varkappa_1 / \partial x^2) = (\partial \, (yx + \tfrac{1}{2} \, y^2) \, / \partial x) = y \tag{78}$$

$$(\partial^2 \varkappa_2 / \partial x^2) = (\partial \, (yx^2 + xy^2 + 1/3 \, y^3) \, / \partial x) = 2yx + y^2 \tag{79}$$

$$(\partial^2 \varkappa_3 / \partial x^2) = (\partial \, (yx^3 + 1/2 \, x^2y^2 + x^2y^2 + 2/3 \, xy^3 + 1/3 \, xy^3 + \tfrac{1}{4} \, y^4) \, / \partial x) = 3yx^2 + xy^2 + 2xy^2$$

$$+ 2/3 \, y^3 + 1/3 \, y^3 \tag{80}$$

$$(\partial^2 \varkappa_4 / \partial x^2) = (\partial \, (yx^4 + 2 \, x^3y^2 + 2 \, x^2y^3 + xy^4 + 1/5 \, y^5) \, / \partial x) = 4yx^3 + 6 \, x^2y^2 + 4 \, xy^3 + y^4 \tag{81}$$

$$(\partial \varkappa_1 / \partial y) = (\partial \, (\tfrac{1}{2} \, yx^2 + \tfrac{1}{2} \, xy^2) \, / \partial y) = \tfrac{1}{2} \, x^2 + xy \tag{82}$$

$$(\partial \varkappa_2 / \partial y) = (\partial \, (1/3 \, yx^3 + \tfrac{1}{2} \, x^2y^2 + 1/3 \, xy^3) \, / \partial y) = 1/3 \, x^3 + x^2y + xy^2 \tag{83}$$

$$(\partial \varkappa_3 / \partial y) = (\partial \, (1/4 \, yx^4 + 1/6 \, x^3y^2 + 1/3 \, x^3y^2 + 1/3 \, x^2y^3 + 1/6 \, x^2y^3 + \tfrac{1}{4} \, xy^4) \, / \partial y) =$$

$$1/4 \, x^4 + 1/3 \, x^3y + 2/3 \, x^3y + x^2y^2 + 1/2 \, x^2y^2 + xy^3 \tag{84}$$

$$(\partial \varkappa_4 / \partial y) = (\partial \, (1/5 \, yx^5 + \tfrac{1}{2} \, x^4y^2 + 2/3 \, x^3y^3 + \tfrac{1}{2} \, x^2y^4 + 1/5 \, xy^5) \, / \partial y) = 1/5 \, x^5 + x^4y + 2$$

$$x^3y^2 + 2 \, x^2y^3 + xy^4 \tag{85}$$

$$(\partial^2 \varkappa_1 / \partial y \partial x) = (\partial \, (\tfrac{1}{2} \, x^2 + xy) \, / \partial x) = x + y \tag{86}$$

$$(\partial^2 \varkappa_2 / \partial y \partial x) = (\partial\,(\,1/3\,x^3 + x^2y + xy^2\,)\,/\partial x) = x^2 + 2\,xy + y^2 \tag{87}$$

$$(\partial^2 \varkappa_3 / \partial y \partial x) = (\partial\,(1/4\,x^4 + 1/3\,x^3y + 2/3\,x^3y + x^2y^2 + 1/2\,x^2y^2 + xy^3)\,/\partial x) = x^3 + x^2y + 2\,x^2y$$

$$+\,2\,xy^2 + xy^2 + y^3 \tag{88}$$

$$(\partial^2 \varkappa_4 / \partial y \partial x) = (\partial\,(1/5\,x^5 + x^4y + 2\,x^3y^2 + 2\,x^2y^3 + xy^4)\,/\partial x) =$$

$$x^4 + 4\,x^3y + 6\,x^2y^2 + 4\,xy^3 + y^4 \tag{89}$$

$$(\partial^2 \varkappa_1 / \partial y^2) = (\partial\,(\,\tfrac{1}{2}\,x^2 + xy\,)\,/\partial y) = x \tag{90}$$

$$(\partial^2 \varkappa_2 / \partial y^2) = (\partial\,(\,1/3\,x^3 + x^2y + xy^2)\,/\partial y) = x^2 + 2\,xy \tag{91}$$

$$(\partial^2 \varkappa_3 / \partial y^2) = (\partial\,(\,1/4\,x^4 + 1/3\,x^3y + 2/3\,x^3y + x^2y^2 + 1/2\,x^2y^2 + xy^3)\,/\partial y) = 1/3\,x^3 + 2/3\,x^3 +$$

$$2\,x^2y + x^2y + 3\,xy^2 \tag{92}$$

$$(\partial^2 \varkappa_4 / \partial y^2) = (\partial\,(\,1/5\,x^5 + x^4y + 2\,x^3y^2 + 2\,x^2y^3 + xy^4\,)\,/\partial y) = x^4 + 4\,x^3y + 6\,x^2y^2 + 4\,xy^3 \tag{93}$$

The starting equation becomes:

$$\varrho(x, y) = \{\,\varkappa_4\,\lambda_1 + \varkappa_3\,\lambda_2 + \varkappa_2\,\lambda_3 + \varkappa_1\,\lambda_4 + \lambda_5\,xy\,\}\,\}\,/\,\{\,\lambda_6 * xy - \{\,\alpha_2 * a\,[6\,(yx^2/2 + xy^2/2) +$$

$$xy\,] + 2\,\alpha_3 * a\,xy\,\}\,\}$$

$$(\partial\,(\,\varrho(x, y)\,)\,/\partial x) = \{\,\{\,(\partial \varkappa_4 /\partial x)\,\lambda_1 + (\partial \varkappa_3 /\partial x)\,\lambda_2 + (\partial \varkappa_2 /\partial x)\,\lambda_3 + (\partial \varkappa_1 /\partial x)\,\lambda_4 + \lambda_5\,y\,\} * \{$$

$$\lambda_6 * xy - \{\,\alpha_2 * a\,[6\,(yx^2/2 + xy^2/2) + xy\,] + 2\,\alpha_3 * a\,xy\,\}\,\} - \{\,\varkappa_4\,\lambda_1 + \varkappa_3\,\lambda_2 + \varkappa_2\,\lambda_3 + \varkappa_1\,\lambda_4$$

$$+\,\lambda_5\,xy\,\} * \{\,\lambda_6 * y - \{\,\alpha_2 * a\,[6\,(yx + y^2/2) + y\,] + 2\,\alpha_3 * a\,y\,\}\,\}\,\}\,/\,\{\,\lambda_6 * xy - \{\,\alpha_2 * a\,[6$$

$$(yx^2/2 + xy^2/2) + xy\,] + 2\,\alpha_3 * a\,xy\,\}\,\}^2 \tag{94}$$

$$(\partial^2\,(\,\varrho(x, y)\,)\,/\partial x^2) = (\partial\,\{\,\{\,(\partial \varkappa_4 /\partial x)\,\lambda_1 + (\partial \varkappa_3 /\partial x)\,\lambda_2 + (\partial \varkappa_2 /\partial x)\,\lambda_3 + (\partial \varkappa_1 /\partial x)\,\lambda_4 + \lambda_5\,y$$

$$\} * \{\,\lambda_6 * xy - \{\,\alpha_2 * a\,[6\,(yx^2/2 + xy^2/2) + xy\,] + 2\,\alpha_3 * a\,xy\,\}\,\} - \{\,\varkappa_4\,\lambda_1 + \varkappa_3\,\lambda_2 + \varkappa_2\,\lambda_3 +$$

$$\varkappa_1\,\lambda_4 + \lambda_5\,xy\,\} * \{\,\lambda_6 * y - \{\,\alpha_2 * a\,[6\,(yx + y^2/2) + y\,] + 2\,\alpha_3 * a\,y\,\}\,\}\,\}\,/\,\{\,\lambda_6 * xy - \{\,\alpha_2$$

$$* a\,[6\,(yx^2/2 + xy^2/2) + xy\,] + 2\,\alpha_3 * a\,xy\,\}\,\}^2 /\partial x\,) =$$

$$\{\,\{\,\{\,\{\,(\partial^2 \varkappa_4 /\partial x^2)\,\lambda_1 + (\partial^2 \varkappa_3 /\partial x^2)\,\lambda_2 + (\partial^2 \varkappa_2 /\partial x^2)\,\lambda_3 + (\partial^2 \varkappa_1 /\partial x^2)\,\lambda_4\,\} * \{\,\lambda_6 * xy - \{\,\alpha_2$$

$$* a\,[6\,(yx^2/2 + xy^2/2) + xy\,] + 2\,\alpha_3 * a\,xy\,\}\,\} + \{\,(\partial \varkappa_4 /\partial x)\,\lambda_1 + (\partial \varkappa_3 /\partial x)\,\lambda_2 + (\partial \varkappa_2 /\partial x)$$

$\lambda_3 + (\partial \varkappa_1 / \partial x) \lambda_4 + \lambda_5 y \} * \{ \lambda_6 * y - \{ \alpha_2 * a [6 (yx + y^2 / 2) + y] + 2 \alpha_3 * a y \} \} - \{ (\partial \varkappa_4$

$/ \partial x) \lambda_1 + (\partial \varkappa_3 / \partial x) \lambda_2 + (\partial \varkappa_2 / \partial x) \lambda_3 + (\partial \varkappa_1 / \partial x) \lambda_4 + \lambda_5 y \} * \{ \lambda_6 * y - \{ \alpha_2 * a [6 (yx + y^2$

$/ 2) + y] + 2 \alpha_3 * a y \} \} - \{ \varkappa_4 \lambda_1 + \varkappa_3 \lambda_2 + \varkappa_2 \lambda_3 + \varkappa_1 \lambda_4 + \lambda_5 xy \} * \{ 6y \alpha_2 * a \} \} \} * \{ \lambda_6 *$

$xy - \{ \alpha_2 * a [6 (yx^2 / 2 + xy^2 / 2) + xy] + 2 \alpha_3 * a xy \} \}^2 - \{ \{ (\partial \varkappa_4 / \partial x) \lambda_1 + (\partial \varkappa_3 / \partial x) \lambda_2 +$

$(\partial \varkappa_2 / \partial x) \lambda_3 + (\partial \varkappa_1 / \partial x) \lambda_4 + \lambda_5 y \} * \{ \lambda_6 * xy - \{ \alpha_2 * a [6 (yx^2 / 2 + xy^2 / 2) + xy] + 2 \alpha_3 * a$

$xy \} \} - \{ \varkappa_4 \lambda_1 + \varkappa_3 \lambda_2 + \varkappa_2 \lambda_3 + \varkappa_1 \lambda_- + \lambda_5 xy \} * \{ \lambda_6 * y - \{ \alpha_2 \times a [6 (yx + y^2 / 2) + y] +$

$2 \alpha_3 * a y \} \} \} * \{ 2 * \{ \lambda_6 * y - \{ \alpha_2 * a [6 (yx + y^2 / 2) + y] + 2 \alpha_3 * a y \} \} * \{ \lambda_6 * xy - \{ \alpha_2 * a [5 (yx^2$

$/ 2 + xy^2 / 2) + xy] + 2 \alpha_3 * a xy \} \} \} / \{ \lambda_6 * xy - \{ \alpha_2 * a [6 (yx^2 / 2 + xy^2 / 2) + xy] + 2 \alpha_3 * a xy \} \}^4$

(94.a)

$(\partial (\varrho(x, y)) / \partial y) = \{ \{ (\partial \varkappa_4 / \partial y) \lambda_1 + (\partial \varkappa_3 / \partial y) \lambda_2 + (\partial \varkappa_2 / \partial y) \lambda_3 + (\partial \varkappa_1 / \partial y) \lambda_4 + \lambda_5 x \} * \{$

$\lambda_6 * xy - \{ \alpha_2 * a [6 (yx^2 / 2 + xy^2 / 2) + xy] + 2 \alpha_3 * a xy \} \} - \{ \varkappa_4 \lambda_1 + \varkappa_3 \lambda_2 + \varkappa_2 \lambda_3 + \varkappa_1 \lambda_4$

$+ \lambda_5 xy \} * \{ \lambda_6 * x - \{ \alpha_2 * a [6 (x^2 / 2 + xy) + x] + 2 \alpha_3 * a x \} \} \} / \{ \lambda_6 * xy - \{ \alpha_2 * a [6$

$(yx^2 / 2 + xy^2 / 2) + xy] + 2 \alpha_3 * a xy \} \}^2$

(95)

$(\partial^2 (\varrho(x, y)) / \partial y^2) = (\partial \{ \{ (\partial \varkappa_4 / \partial y) \lambda_1 + (\partial \varkappa_3 / \partial y) \lambda_2 + (\partial \varkappa_2 / \partial y) \lambda_3 + (\partial \varkappa_1 / \partial y) \lambda_4 + \lambda_5 x$

$\} * \{ \lambda_6 * xy - \{ \alpha_2 * a [6 (yx^2 / 2 + xy^2 / 2) + xy] + 2 \alpha_3 * a xy \} \} - \{ \varkappa_4 \lambda_1 + \varkappa_3 \lambda_2 + \varkappa_2 \lambda_3 +$

$\varkappa_1 \lambda_4 + \lambda_5 xy \} * \{ \lambda_6 * x - \{ \alpha_2 * a [6 (x^2 / 2 + xy) + x] + 2 \alpha_3 * a x \} \} \} / \{ \lambda_6 * xy - \{ \alpha_2 *$

$a [6 (yx^2 / 2 + xy^2 / 2) + xy] + 2 \alpha_3 * a xy \} \}^2 / \partial y) =$

$\{ \{ \{ \{ (\partial^2 \varkappa_4 / \partial y^2) \lambda_1 + (\partial^2 \varkappa_3 / \partial y^2) \lambda_2 + (\partial^2 \varkappa_2 / \partial y^2) \lambda_3 + (\partial^2 \varkappa_1 / \partial y^2) \lambda_4 \} * \{ \lambda_6 * xy - \{ \alpha_2 *$

$a [6 (yx^2 / 2 + xy^2 / 2) + xy] + 2 \alpha_3 * a xy \} \} + \{ (\partial \varkappa_4 / \partial y) \lambda_1 + (\partial \varkappa_3 / \partial y) \lambda_2 + (\partial \varkappa_2 / \partial y) \lambda_3 +$

$(\partial \varkappa_1 / \partial y) \lambda_4 + \lambda_5 x \} * \{ \lambda_6 * x - \{ \alpha_2 * a [6 (x^2 / 2 + xy) + x] + 2 \alpha_3 * a x \} \} - \{ (\partial \varkappa_4 / \partial y)$

$\lambda_1 + (\partial \varkappa_3 / \partial y) \lambda_2 + (\partial \varkappa_2 / \partial y) \lambda_3 + (\partial \varkappa_1 / \partial y) \lambda_4 + \lambda_5 x \} * \{ \lambda_6 * x - \{ \alpha_2 * a [6 (x^2 / 2 + xy) +$

$x] + 2 \alpha_3 * a x \} \} - \{ \varkappa_4 \lambda_1 + \varkappa_3 \lambda_2 + \varkappa_2 \lambda_3 + \varkappa_1 \lambda_4 + \lambda_5 xy \} * \{ 6x \alpha_2 * a \} \} \} * \{ \lambda_6 * xy$

$- \{ \alpha_2 * a [6 (yx^2 / 2 + xy^2 / 2) + xy] + 2 \alpha_3 * a xy \} \}^2 - \{ \{ (\partial \varkappa_4 / \partial y) \lambda_1 + (\partial \varkappa_3 / \partial y) \lambda_2 + (\partial$

$\varkappa_2/\partial y)\,\lambda_3 + (\partial\,\varkappa_1/\partial y)\,\lambda_4 + \lambda_5\,x\,\} * \{\lambda_6 * xy - \{\alpha_2 * a\,[6\,(yx^2/2 + xy^2/2) + xy\,] + 2\,\alpha_3 * a$

$xy\,\}\,\} - \{\varkappa_4\,\lambda_1 + \varkappa_3\,\lambda_2 + \varkappa_2\,\lambda_3 + \varkappa_1\,\lambda_4 + \lambda_5\,xy\,\} * \{\lambda_6 * x - \{\alpha_2 * a\,[6\,(x^2/2 + xy) + x\,] + 2$

$\alpha_3 * a\,x\,\}\,\}\,\} * \{2 * \{\lambda_6 * x - \{\alpha_2 * a\,[6\,(x^2/2 + xy) + x\,] + 2\,\alpha_3 * a\,x\,\}\,\} * \{\lambda_6 * xy - \{\alpha_2 * a\,[6\,(yx^2$

$/2 + xy^2/2) + xy\,] + 2\,\alpha_3 * a\,xy\,\}\,\}\,\} / \{\lambda_6 * xy - \{\alpha_2 * a\,[6\,(yx^2/2 + xy^2/2) + xy\,] + 2\,\alpha_3 * a\,xy\,\}\,\}^4$

$$(95.\text{a})$$

$(\partial^2\,(\varrho(x, y))\,/\partial x\partial y) = (\partial\,\{\{(\partial\,\varkappa_4/\partial x)\,\lambda_1 + (\partial\,\varkappa_3/\partial x)\,\lambda_2 + (\partial\,\varkappa_2/\partial x)\,\lambda_3 + (\partial\,\varkappa_1/\partial x)\,\lambda_4 + \lambda_5$

$y\,\} * \{\lambda_6 * xy - \{\alpha_2 * a\,[6\,(yx^2/2 + xy^2/2) + xy\,] + 2\,\alpha_3 * a\,xy\,\}\,\} - \{\varkappa_4\,\lambda_1 + \varkappa_3\,\lambda_2 + \varkappa_2\,\lambda_3 +$

$\varkappa_1\,\lambda_4 + \lambda_5\,xy\,\} * \{\lambda_6 * y - \{\alpha_2 * a\,[6\,(yx + y^2/2) + y\,] + 2\,\alpha_3 * a\,y\,\}\,\}\,\} / \{\lambda_6 * xy - \{\alpha_2$

$* a\,[6\,(yx^2/2 + xy^2/2) + xy\,] + 2\,\alpha_3 * a\,xy\,\}^2/\partial y) =$

$\{\{\{(\partial^2\,\varkappa_4/\partial x\partial y)\,\lambda_1 + (\partial^2\,\varkappa_3/\partial x\partial y)\,\lambda_2 + (\partial^2\,\varkappa_2/\partial x\partial y)\,\lambda_3 + (\partial^2\,\varkappa_1/\partial x\partial y)\,\lambda_4 + \lambda_5\} * \{\lambda_6 *$

$xy - \{\alpha_2 * a\,[6\,(yx^2/2 + xy^2/2) + xy\,] + 2\,\alpha_3 * a\,xy\,\}\,\} + \{(\partial\,\varkappa_4/\partial x)\,\lambda_1 + (\partial\,\varkappa_3/\partial x)\,\lambda_2 + (\partial$

$\varkappa_2/\partial x)\,\lambda_3 + (\partial\,\varkappa_1/\partial x)\,\lambda_4 + \lambda_5\,y\,\} * \{\lambda_6 * x - \{\alpha_2 * a\,[6\,(x^2/2 + xy) + x\,] + 2\,\alpha_3 * a\,x\,\}\,\} - \{$

$(\partial\,\varkappa_4/\partial y)\,\lambda_1 + (\partial\,\varkappa_3/\partial y)\,\lambda_2 + (\partial\,\varkappa_2/\partial y)\,\lambda_3 + (\partial\,\varkappa_1/\partial y)\,\lambda_4 + \lambda_5\,x\,\} * \{\lambda_6 * y - \{\alpha_2 * a\,[6\,(yx$

$+ y^2/2) + y\,] + 2\,\alpha_3 * a\,y\,\}\,\} - \{\varkappa_4\,\lambda_1 + \varkappa_3\,\lambda_2 + \varkappa_2\,\lambda_3 + \varkappa_1\,\lambda_4 + \lambda_5\,xy\,\} * \{\lambda_6 - \{\alpha_2 * a\,[6\,(x$

$+ y) + 1\,] + 2\,\alpha_3 * a\,\}\,\}\,\} * \{\lambda_6 * xy - \{\alpha_2 * a\,[6\,(yx^2/2 + xy^2/2) + xy\,] + 2\,\alpha_3 * a\,xy\,\}^2$

$- \{\{(\partial\,\varkappa_4/\partial x)\,\lambda_1 + (\partial\,\varkappa_3/\partial x)\,\lambda_2 + (\partial\,\varkappa_2/\partial x)\,\lambda_3 + (\partial\,\varkappa_1/\partial x)\,\lambda_4 + \lambda_5\,y\,\} * \{\lambda_6 * xy - \{\alpha_2 * a$

$[6\,(yx^2/2 + xy^2/2) + xy\,] + 2\,\alpha_3 * a\,xy\,\}\,\} - \{\varkappa_4\,\lambda_1 + \varkappa_3\,\lambda_2 + \varkappa_2\,\lambda_3 + \varkappa_1\,\lambda_4 + \lambda_5\,xy\,\} * \{\lambda_6$

$* y - \{\alpha_2 * a\,[6\,(yx + y^2/2) + y\,] + 2\,\alpha_3 * a\,y\,\}\,\}\,\} * \{2 * \{\lambda_6 * x - \{\alpha_2 * a\,[6\,(x^2/2 + xy)$

$+ x\,] + 2\,\alpha_3 * a\,x\,\}\,\} * \{\lambda_6 * xy - \{\alpha_2 * a\,[6\,(yx^2/2 + xy^2/2) + xy\,] + 2\,\alpha_3 * a\,xy\,\}\,\}\,\} /$

$\{\lambda_6 * xy - \{\alpha_2 * a\,[6\,(yx^2/2 + xy^2/2) + xy\,] + 2\,\alpha_3 * a\,xy\,\}^4$

$$(96)$$

$(\partial^2\,(\varrho(x, y))\,/\partial y\partial x) = (\partial\,\{\{(\partial\,\varkappa_4/\partial y)\,\lambda_1 + (\partial\,\varkappa_3/\partial y)\,\lambda_2 + (\partial\,\varkappa_2/\partial y)\,\lambda_3 + (\partial\,\varkappa_1/\partial y)\,\lambda_4 + \lambda_5$

$x\,\} * \{\lambda_6 * xy - \{\alpha_2 * a\,[6\,(yx^2/2 + xy^2/2) + xy\,] + 2\,\alpha_3 * a\,xy\,\}\,\} - \{\varkappa_4\,\lambda_1 + \varkappa_3\,\lambda_2 + \varkappa_2$

$\lambda_3 + \varkappa_1\,\lambda_4 + \lambda_5\,xy\,\} * \{\lambda_6 * x - \{\alpha_2 * a\,[6\,(x^2/2 + xy) + x\,] + 2\,\alpha_3 * a\,x\,\}\,\}\,\} / \{\lambda_6 * xy - \{$

$\alpha_2 * a [6 (yx^2/2 + xy^2/2) + xy] + 2 \alpha_3* a xy \} \}^2 / \partial x) =$

$\{ \{ \{ (\partial^2 \varkappa_4 / \partial y \partial x) \lambda_1 + (\partial^2 \varkappa_3 / \partial y \partial x) \lambda_2 + (\partial^2 \varkappa_2 / \partial y \partial x) \lambda_3 + (\partial^2 \varkappa_1 / \partial y \partial x) \lambda_4 + \lambda_5 \} * \{ \lambda_6 *$

$xy - \{ \alpha_2 * a [6 (yx^2/2 + xy^2/2) + xy] + 2 \alpha_3* a xy \} \} + \{ (\partial \varkappa_4 / \partial y) \lambda_1 + (\partial \varkappa_3 / \partial y) \lambda_2 + (\partial$

$\varkappa_2 / \partial y) \lambda_3 + (\partial \varkappa_1 / \partial y) \lambda_4 + \lambda_5 x \} * \{ \lambda_6 * y - \{ \alpha_2 * a [6 (yx + y^2/2) + y] + 2 \alpha_3* a y \} \} - \{$

$(\partial \varkappa_4 / \partial x) \lambda_1 + (\partial \varkappa_3 / \partial x) \lambda_2 + (\partial \varkappa_2 / \partial x) \lambda_3 + (\partial \varkappa_1 / \partial x) \lambda_4 + \lambda_5 y \} * \{ \lambda_6 * x - \{ \alpha_2 * a [6$

$(x^2/2 + xy) + x] + 2 \alpha_3* a x \} \} - \{ \varkappa_4 \lambda_1 + \varkappa_3 \lambda_2 + \varkappa_2 \lambda_3 + \varkappa_1 \lambda_4 + \lambda_5 xy \} * \{ \lambda_6 - \{ \alpha_2 * a [6$

$(x + y) + 1] + 2 \alpha_3* a \} \} \} * \{ \lambda_6 * xy - \{ \alpha_2 * a [6 (yx^2/2 + xy^2/2) + xy] + 2 \alpha_3* a xy \}$

$\}^2 - \{ \{ (\partial \varkappa_4 / \partial y) \lambda_1 + (\partial \varkappa_3 / \partial y) \lambda_2 + (\partial \varkappa_2 / \partial y) \lambda_3 + (\partial \varkappa_1 / \partial y) \lambda_4 + \lambda_5 x \} * \{ \lambda_6 * xy - \{ \alpha_2$

$* a [6 (yx^2/2 + xy^2/2) + xy] + 2 \alpha_3* a xy \} \} - \{ \varkappa_4 \lambda_1 + \varkappa_3 \lambda_2 + \varkappa_2 \lambda_3 + \varkappa_1 \lambda_4 + \lambda_5 xy \} * \{$

$\lambda_6 * x - \{ \alpha_2 * a [6 (x^2/2 + xy) + x] + 2 \alpha_3* a x \} \} \} * \{ 2 * \{ \lambda_6 * y - \{ \alpha_2 * a [6 (yx + y^2$

$/2) + y] + 2 \alpha_3* a y \} \} * \{ \lambda_6 * xy - \{ \alpha_2 * a [6 (yx^2/2 + xy^2/2) + xy] + 2 \alpha_3* a xy \} \} \} /$

$\{ \lambda_6 * xy - \{ \alpha_2 * a [6 (yx^2/2 + xy^2/2) + xy] + 2 \alpha_3* a xy \} \}^4$ (97)

Equation (98) is the total curvature of the bivariate resilient signal in the variables x and y and it is also the total curvature of the bivariate cubic resilient-curvature signal reconstruction formula.

$C (R((x, y), f)) = (\partial^2 (\varrho(x, y)) / \partial x^2) + (\partial^2 (\varrho(x, y)) / \partial y^2) + (\partial^2 (\varrho(x, y)) / \partial x \partial y) +$

$(\partial^2 (\varrho(x, y)) / \partial y \partial x)$ (98)

Trivariate Cubic SRI

The starting equation is (56) and it requires the position:

$f (0, 0, 0) = \{ 6 (\omega_1 a)^2 \Pi^{(4)} - [4 (\omega_1 a)^2 + 2 \omega_2 \omega_1 * a^2 + 6 \omega_2 \omega_1 * a^2] \Pi^{(3)} + [6/4 (\omega_1 a)^2 +$

$1/2 (\omega_1 a)^2 + \omega_2 \omega_1 * a^2 + 12 \omega_2 \omega_1 * a^2 + \omega_1 \omega_2 a^2 + 2 (\omega_2 a)^2] \Pi^{(2)} + [6 (\omega_1 a)^2 + 1/4 (\omega_1$

$a)^2 + 1/2 \omega_2 \omega_1 * a^2 + 6 \omega_2 \omega_1 * a^2 + 2 \omega_1 \omega_2 a^2 + 4 (\omega_2 a)^2] \Pi^{(1)} + [(\omega_1 a)^2 + 2 \omega_2 \omega_1 * a^2$

$+ \omega_1 \omega_2 a^2 + 2 (\omega_2 a)^2] x y z \} / \{ \{ \omega_1 * a + 2 \omega_2 * a \} * xyz - \{ \omega_1 * a [6 (1/2 yzx^2 + 1/2$

$xzy^2 + 1/2 xyz^2) + xyz] + 2 \omega_2 * a xyz \} \} = \varrho (x, y, z)$ (99)

Let us posit:

SIGNAL RESILIENT TO INTERPOLATION

$$\Phi_1 = 6 \, (\omega_1 \, a)^2 \tag{100}$$

$$\Phi_2 = [\, 4 \, (\omega_1 \, a)^2 + 2 \, \omega_2 \, \omega_1 {}^* \, a^2 + \, 6 \, \omega_2 \, \omega_1 {}^* \, a^2 \,] \tag{101}$$

$$\Phi_3 = [6/4 \, (\omega_1 \, a)^2 + 1/2 \, (\omega_1 \, a)^2 \, + \omega_2 \, \omega_1 {}^* \, a^2 + 12 \, \omega_2 \, \omega_1 {}^* \, a^2 + \omega_1 \, \omega_2 \, a^2 + 2 \, (\omega_2 \, a)^2] \tag{102}$$

$$\Phi_4 = [\, 6 \, (\omega_1 \, a)^2 + 1/4 \, (\omega_1 \, a)^2 + 1/2 \, \omega_2 \, \omega_1 {}^* \, a^2 + 6 \, \omega_2 \, \omega_1 {}^* \, a^2 + 2 \, \omega_1 \, \omega_2 \, a^2 + 4 \, (\omega_2 \, \, a)^2] \tag{103}$$

$$\Phi_5 = [\, (\omega_1 \, a)^2 + \, 2 \, \omega_2 \, \omega_1 {}^* \, a^2 \, + \omega_1 \, \omega_2 \, a^2 + 2 \, (\omega_2 \, a)^2] \tag{104}$$

Equations (105), (106), (107), (108) will make use of the positions made in equations (1) through (50) of chapter 9 and such positions will be useful to calculate the first and second order partial derivatives of $\Lambda_{(1)}$, $\Lambda_{(2)}$, $\Lambda_{(3)}$, $\Lambda_{(4)}$ (which are reported in chapter 9), and also the total curvature in the subsequent equations (127) through (162) of this chapter.

$$\varphi_1 = [\Phi_1 \, \Pi^{(4)}] \quad = [\Phi_1 \, \Lambda_{(4)}] \, xyz \tag{105}$$

$$\varphi_2 = [\Phi_2 \, \Pi^{(3)}] = [\Phi_2 \, \Lambda_{(3)}] \, xyz \tag{106}$$

$$\varphi_3 = [\Phi_3 \, \Pi^{(2)}] = [\Phi_3 \, \Lambda_{(2)}] \, xyz \tag{107}$$

$$\varphi_4 = [\Phi_4 \, \Pi^{(1)}] \quad = [\Phi_4 \, \Lambda_{(1)}] \, xyz \tag{108}$$

$$\varphi_5 = [\Phi_5 \, x \, y \, z \,] \tag{109}$$

$$\Phi_6 = [\, \omega_1 {}^* \, a + 2 \, \omega_2 {}^* \, a \,] \tag{110}$$

$$\varphi_6 = \{ \, \Phi_6 {}^* \, xyz - \{ \, \omega_1 {}^* \, a \, [\, 6 \, (\, \tfrac{1}{2} \, yzx^2 + \tfrac{1}{2} \, xzy^2 + \tfrac{1}{2} \, xyz^2) + xyz \,] \, + 2 \, \omega_2 {}^* \, a \, xyz \, \}\} =$$

$$\{ \, \Phi_6 - \{ \, \omega_1 {}^* \, a \, [\, 6 \, (\, \tfrac{1}{2} \, x + \tfrac{1}{2} \, y + \tfrac{1}{2} \, z) + 1 \,] \, + 2 \, \omega_2 {}^* \, a \, \}\} \, xyz = \Lambda_6 \, xyz \tag{111}$$

$$\Lambda_6 = \{ \, \Phi_6 - \{ \, \omega_1 {}^* \, a \, [\, 6 \, (\, \tfrac{1}{2} \, x + \tfrac{1}{2} \, y + \tfrac{1}{2} \, z) + 1 \,] \, + 2 \, \omega_2 {}^* \, a \, \}\} \tag{111.a}$$

It follows that:
$$(\partial \, (\Lambda_6) \, / \partial x) = - \, \{ \, 3 \, \omega_1 {}^* \, a \, \} \tag{112}$$

$$(\partial \, (\Lambda_6) \, / \partial y) = - \, \{ \, 3 \, \omega_1 {}^* \, a \, \} \tag{113}$$

$$(\partial \, (\Lambda_6) \, / \partial z) = - \, \{ \, 3 \, \omega_1 {}^* \, a \, \} \tag{114}$$

$$(\partial^2 \, (\Lambda_6) \, / \partial x^2) = (\partial^2 \, (\Lambda_6) \, / \partial y^2) = (\partial^2 \, (\Lambda_6) \, / \partial z^2) = (\partial^2 \, (\Lambda_6) \, / \partial x \partial y) = (\partial^2 \, (\Lambda_6) \, / \partial y \partial x) = \, (\partial^2 \, (\Lambda_6)$$

$$/ \partial x \partial z) = (\partial^2 \, (\Lambda_6) \, / \partial z \partial x) = (\partial^2 \, (\Lambda_6) \, / \partial y \partial z) = (\partial^2 \, (\Lambda_6) \, / \partial z \partial y) = 0 \tag{115}$$

It is immediate to determine:
$$(\partial \, (\varphi_5) \, / \partial x) = (\partial \, [\Phi_5 \, x \, y \, z \,] \, / \partial x) = [\Phi_5 \, y \, z \,] \tag{116}$$

$$(\partial \, (\varphi_5) \, / \partial y) = (\partial \, [\Phi_5 \, x \, y \, z \,] \, / \partial y) = [\Phi_5 \, x \, z \,] \tag{117}$$

$$(\partial\, (\varphi_5)\, /\partial z) = (\partial\, [\Phi_5\, x\, y\, z\,]\, /\partial z) = [\Phi_5\, x\, y\,] \tag{118}$$

$$(\partial^2\, (\varphi_5)\, /\partial x^2) = (\partial\, [\Phi_5\, y\, z\,]\, /\partial x) = 0 \tag{119}$$

$$(\partial^2\, (\varphi_5)\, /\partial y^2) = (\partial\, [\Phi_5\, x\, z\,]\, /\partial y) = 0 \tag{120}$$

$$(\partial^2\, (\varphi_5)\, /\partial z^2) = (\partial\, [\Phi_5\, x\, y\,]\, /\partial z) = 0 \tag{121}$$

$$(\partial^2\, (\varphi_5)\, /\partial x \partial y) = (\partial^2\, (\varphi_5)\, /\partial y \partial x) = [\Phi_5\, z\,] \tag{122}$$

$$(\partial^2\, (\varphi_5)\, /\partial y \partial z) = (\partial^2\, (\varphi_5)\, /\partial z \partial y) = [\Phi_5\, x\,] \tag{123}$$

$$(\partial^2\, (\varphi_5)\, /\partial x \partial z) = (\partial^2\, (\varphi_5)\, /\partial z \partial x) = [\Phi_5\, y\,] \tag{124}$$

Equations (105), (106), (107), (108), (109), (110), (111) and (111.a) permit us to rewrite equation (99), for which the value of $\varrho\,(x, y, z)$ pertains, in the following simplified form:

$$\varrho\,(x, y, z) = \{\, 6\, (\omega_1\, a)^2\, \Pi^{(4)} + [\, 4\, (\omega_1\, a)^2 + 2\, \omega_2\, \omega_1 * a^2 + 6\, \omega_2\, \omega_1 * a^2\,]\, \Pi^{(3)} + [6/4\, (\omega_1\, a)^2 +$$

$$1/2\, (\omega_1\, a)^2 + \omega_2\, \omega_1 * a^2 + 12\, \omega_2\, \omega_1 \times a^2 + \omega_1\, \omega_2\, a^2 + 2\, (\omega_2\, a)^2\,]\, \Pi^{(2)} + [\, 6\, (\omega_1\, a)^2 + 1/4\, (\omega_1$$

$$a)^2 + 1/2\, \omega_2\, \omega_1 * a^2 + 6\, \omega_2\, \omega_1 * a^2 + 2\, \omega_1\, \omega_2\, a^2 + 4\, (\omega_2\ a)^2\,]\, \Pi^{(1)} + [\, (\omega_1\, a)^2 + 2\, \omega_2\, \omega_1 * a^2$$

$$+ \omega_1\, \omega_2\, a^2 + 2\, (\omega_2\, a)^2\,]\, x\, y\, z\,\}\, /\, \{\{\alpha_1 * a + 2\, \omega_2 * a\,\} * xyz - \{\, \omega_1 * a\, [\, 6\, (\, \tfrac{1}{2}\, yzx^2 + \tfrac{1}{2}$$

$$xzy^2 + \tfrac{1}{2}\, xyz^2) + xyz\,] + 2\, \omega_2 * a\, xyz\,\}\}$$

$$= \{\, [\, \Phi_1\, \Lambda_{(4)}\,]\, xyz + [\, \Phi_2\, \Lambda_{(3)}\,]\, xyz - [\, \Phi_3\, \Lambda_{(2)}\,]\, xyz + [\, \Phi_4\, \Lambda_{(1)}\,]\, xyz + [\, \Phi_5\, x\, y\, z\,]\,\}\, /\, \{\, \varphi_6$$

$$\} = \{\, [\, \Phi_1\, \Lambda_{(4)}\,]\, xyz + [\, \Phi_2\, \Lambda_{(3)}\,]\, xyz + [\, \Phi_3\, \Lambda_{(2)}\,]\, xyz + [\, \Phi_4\, \Lambda_{(1)}\,]\, xyz + [\, \Phi_5\, x\, y\, z\,]\,\}\, /\, \{\, \Lambda_5\, xyz\,\} \tag{125}$$

$$\varrho\,(x, y, z) = \{\, [\, \Phi_1\, \Lambda_{(4)}\,] + [\, \Phi_2\, \Lambda_{(3)}\,] + [\, \Phi_3\, \Lambda_{(2)}\,] + [\, \Phi_4\, \Lambda_{(1)}\,] + [\, \Phi_5\,]\,\}\, /\, \{\, \Lambda_6\,\} \tag{126}$$

Calculation of First and Second Order Partial Derivatives of $\varrho\,(x, y, z)$

Keeping in mind that from equation (104) that:

$$\Phi_5 = [\, (\omega_1\, a)^2 + 2\, \omega_2\, \omega_1 * a^2 + \omega_1\, \omega_2\, a^2 + 2\, (\omega_2\, a)^2\,]$$

is constant with respect to the x, y, and z variates, its derivative is always equal to zero.

$$(\partial\, (\varrho\,(x, y, z))\, /\partial x) = \{\, \{\, \Phi_1 * (\partial\, (\Lambda_{(4)})\, /\partial x) + \Phi_2 * (\partial\, (\Lambda_{(3)})\, /\partial x) + \Phi_3 * (\partial\, (\Lambda_{(2)})\, /\partial x) + \Phi_4$$

$$* (\partial\, (\Lambda_{(1)})\, /\partial x)\,\} * \{\, \Lambda_6\,\} - \{\, \Phi_1 * \Lambda_{(4)} + \Phi_2 * \Lambda_{(3)} + \Phi_3 * \Lambda_{(2)} + \Phi_4 * \Lambda_{(1)} + \Phi_5\,\} * (\partial\, (\Lambda_6)\, /\partial x)\,\}\, /$$

$$[\, \Lambda_6\,]^2 \tag{127}$$

SIGNAL RESILIENT TO INTERPOLATION

Let us posit:

$$N_L = \{ \Phi_1 * \Lambda_{(4)} + \Phi_2 * \Lambda_{(3)} + \Phi_3 * \Lambda_{(2)} + \Phi_4 * \Lambda_{(1)} + \Phi_5 \} \tag{128}$$

$$(\partial (N_L) / \partial x) = \{\Phi_1 * (\partial (\Lambda_{(4)}) / \partial x) + \Phi_2 * (\partial (\Lambda_{(3)}) / \partial x) + \Phi_3 * (\partial (\Lambda_{(2)}) / \partial x) + \Phi_4 * (\partial (\Lambda_{(1)}) / \partial x)\} \tag{129}$$

$$(\partial^2 (N_L) / \partial x^2) = \{ \Phi_1 * (\partial^2 (\Lambda_{(4)}) / \partial x^2) + \Phi_2 * (\partial^2 (\Lambda_{(3)}) / \partial x^2) + \Phi_3 * (\partial^2 (\Lambda_{(2)}) / \partial x^2) + \Phi_4 * (\partial^2 (\Lambda_{(1)})$$

$$/ \partial x^2) \} \tag{130}$$

It follows that:

$$(\partial (\varrho (x, y, z)) / \partial x) = \{ (\partial (N_L) / \partial x) * \Lambda_6 - N_L * (\partial (\Lambda_6) / \partial x) \} / [\Lambda_6]^2 \tag{131}$$

$$(\partial^2 (\varrho (x, y, z)) / \partial x^2) = \{ \{ (\partial^2 (N_L) / \partial x^2) * [\Lambda_6] + (\partial (N_L) / \partial x) * (\partial (\Lambda_6) / \partial x) - (\partial (N_L)$$

$$/ \partial x) * (\partial (\Lambda_6) / \partial x) - N_L * (\partial^2 (\Lambda_6) / \partial x^2) \} * [\Lambda_6]^2 - \{ (\partial (N_L) / \partial x) * \Lambda_6 - N_L * (\partial (\Lambda_6) / \partial x)$$

$$\} * \{ 2 * \Lambda_6 * (\partial (\Lambda_6) / \partial x) \} \} / [\Lambda_6]^4 \tag{132}$$

Let us posit:

$$(\partial (N_L) / \partial y) = \{\Phi_1 * (\partial (\Lambda_{(4)}) / \partial y) + \Phi_2 * (\partial (\Lambda_{(3)}) / \partial y) + \Phi_3 * (\partial (\Lambda_{(2)}) / \partial y) + \Phi_4 * (\partial (\Lambda_{(1)}) / \partial y)\} \tag{133}$$

$$(\partial^2 (N_L) / \partial y^2) = \{ \Phi_1 * (\partial^2 (\Lambda_{(4)}) / \partial y^2) + \Phi_2 * (\partial^2 (\Lambda_{(3)}) / \partial y^2) + \Phi_3 * (\partial^2 (\Lambda_{(2)}) / \partial y^2) + \Phi_4 *$$

$$(\partial^2 (\Lambda_{(1)}) / \partial y^2) \} \tag{134}$$

$$(\partial (N_L) / \partial z) = \{\Phi_1 * (\partial (\Lambda_{(4)}) / \partial z) + \Phi_2 * (\partial (\Lambda_{(3)}) / \partial z) + \Phi_3 * (\partial (\Lambda_{(2)}) / \partial z) + \Phi_4 * (\partial (\Lambda_{(1)}) / \partial z)\} \tag{135}$$

$$(\partial^2 (N_L) / \partial z^2) = \{ \Phi_1 * (\partial^2 (\Lambda_{(4)}) / \partial z^2) + \Phi_2 * (\partial^2 (\Lambda_{(3)}) / \partial z^2) + \Phi_3 * (\partial^2 (\Lambda_{(2)}) / \partial z^2) + \Phi_4 *$$

$$(\partial^2 (\Lambda_{(1)}) / \partial z^2) \} \tag{136}$$

$$(\partial^2 (N_L) / \partial x \partial y) = \{\Phi_1 * (\partial^2 (\Lambda_{(4)}) / \partial x \partial y) + \Phi_2 * (\partial^2 (\Lambda_{(3)}) / \partial x \partial y) + \Phi_3 * (\partial^2 (\Lambda_{(2)}) / \partial x \partial y) +$$

$$\Phi_4 * (\partial^2 (\Lambda_{(1)}) / \partial x \partial y) \} \tag{137}$$

$$(\partial^2 (N_L) / \partial y \partial x) = \{\Phi_1 * (\partial^2 (\Lambda_{(4)}) / \partial y \partial x) + \Phi_2 * (\partial^2 (\Lambda_{(3)}) / \partial y \partial x) + \Phi_3 * (\partial^2 (\Lambda_{(2)}) / \partial y \partial x) +$$

$$\Phi_4 * (\partial^2 (\Lambda_{(1)}) / \partial y \partial x) \} \tag{138}$$

$(\partial^2 (N_L) / \partial x \partial z) = \{ \Phi_1 * (\partial^2 (\Lambda_{(4)}) / \partial x \partial z) + \Phi_2 * (\partial^2 (\Lambda_{(3)}) / \partial x \partial z) + \Phi_3 * (\partial^2 (\Lambda_{(2)}) / \partial x \partial z) +$

$\Phi_4 * (\partial^2 (\Lambda_{(1)}) / \partial x \partial z) \}$ (139)

$(\partial^2 (N_L) / \partial z \partial x) = \{ \Phi_1 * (\partial^2 (\Lambda_{(4)}) / \partial z \partial x) + \Phi_2 * (\partial^2 (\Lambda_{(3)}) / \partial z \partial x) + \Phi_3 * (\partial^2 (\Lambda_{(2)}) / \partial z \partial x) +$

$\Phi_4 * (\partial^2 (\Lambda_{(1)}) / \partial z \partial x) \}$ (140)

$(\partial^2 (N_L) / \partial y \partial z) = \{ \Phi_1 * (\partial^2 (\Lambda_{(4)}) / \partial y \partial z) + \Phi_2 * (\partial^2 (\Lambda_{(3)}) / \partial y \partial z) + \Phi_3 * (\partial^2 (\Lambda_{(2)}) / \partial y \partial z) +$

$\Phi_4 * (\partial^2 (\Lambda_{(1)}) / \partial y \partial z) \}$ (141)

$(\partial^2 (N_L) / \partial z \partial y) = \{ \Phi_1 * (\partial^2 (\Lambda_{(4)}) / \partial z \partial y) + \Phi_2 * (\partial^2 (\Lambda_{(3)}) / \partial z \partial y) + \Phi_3 * (\partial^2 (\Lambda_{(2)}) / \partial z \partial y) +$

$\Phi_4 * (\partial^2 (\Lambda_{(1)}) / \partial z \partial y) \}$ (142)

It follows that:

$(\partial (\varrho (x, y, z)) / \partial y) = \{ (\partial (N_L) / \partial y) * \Lambda_6 - N_L * (\partial (\Lambda_6) / \partial y) \} / [\Lambda_6]^2$ (143)

$(\partial^2 (\varrho (x, y, z)) / \partial y^2) = \{ \{ (\partial^2 (N_L) / \partial y^2) * [\Lambda_6] + (\partial (N_L) / \partial y) * (\partial (\Lambda_6) / \partial y) - (\partial (N_L)$

$/ \partial y) * (\partial (\Lambda_6) / \partial y) - N_L * (\partial^2 (\Lambda_6) / \partial y^2) \} * [\Lambda_6]^2 - \{ (\partial (N_L) / \partial y) * \Lambda_6 - N_L * (\partial (\Lambda_6) / \partial y)$

$\} * \{ 2 * \Lambda_6 * (\partial (\Lambda_6) / \partial y) \} \} / [\Lambda_6]^4$ (144)

$(\partial (\varrho (x, y, z)) / \partial z) = \{ (\partial (N_L) / \partial z) * \Lambda_6 - N_L * (\partial (\Lambda_6) / \partial z) \} / [\Lambda_6]^2$ (145)

$(\partial^2 (\varrho (x, y, z)) / \partial z^2) = \{ \{ (\partial^2 (N_L) / \partial z^2) * [\Lambda_6] + (\partial (N_L) / \partial z) * (\partial (\Lambda_6) / \partial z) - (\partial (N_L)$

$/ \partial z) * (\partial (\Lambda_6) / \partial z) - N_L * (\partial^2 (\Lambda_6) / \partial z^2) \} * [\Lambda_6]^2 - \{ (\partial (N_L) / \partial z) * \Lambda_6 - N_L * (\partial (\Lambda_6) / \partial z)$

$\} * \{ 2 * \Lambda_6 * (\partial (\Lambda_6) / \partial z) \} \} / [\Lambda_6]^4$ (146)

$(\partial^2 (\varrho (x, y, z)) / \partial x \partial y) = \{ \{ (\partial (\partial (N_L) / \partial x) / \partial y) * [\Lambda_6] - (\partial (N_L) / \partial y) * (\partial (\Lambda_6) / \partial x) + (\partial$

$(N_L) / \partial x) * (\partial (\Lambda_6) / \partial y) - N_L * (\partial^2 (\Lambda_6) / \partial x \partial y) \} * [\Lambda_6]^2 - \{ (\partial (N_L) / \partial x) * \Lambda_6 - N_L * (\partial$

$(\Lambda_6) / \partial x) \} * \{ 2 * \Lambda_6 * (\partial (\Lambda_6) / \partial y) \} \} / [\Lambda_6]^4$ (147)

$(\partial^2 (\varrho (x, y, z)) / \partial y \partial x) = \{ \{ (\partial (\partial (N_L) / \partial y) / \partial x) * [\Lambda_6] - (\partial (N_L) / \partial x) * (\partial (\Lambda_6) / \partial y) + (\partial$

(N_L) $/\partial y$) * $(\partial (\Lambda_6) /\partial x)$ - N_L * $(\partial^2 (\Lambda_6) /\partial y \partial x)$ } * $[\Lambda_6]^2$ - { $(\partial (N_L) /\partial y)$ * Λ_6 - N_L * $(\partial$

$(\Lambda_6) /\partial y)$ } * { 2 * Λ_6 * $(\partial (\Lambda_6) /\partial x)$ } } / $[\Lambda_6]^4$ (148)

$(\partial^2 (\varrho (x, y, z)) /\partial z \partial x) = $ { { $(\partial (\partial (N_L) /\partial z) /\partial x)$ * $[\Lambda_6]$ - $(\partial (N_L) /\partial x)$ * $(\partial (\Lambda_6) /\partial z)$ + $(\partial$

$(N_L) /\partial z)$ * $(\partial (\Lambda_6) /\partial x)$ - N_L * $(\partial^2 (\Lambda_6) /\partial z \partial x)$ } * $[\Lambda_6]^2$ - { $(\partial (N_L) /\partial z)$ * Λ_6 - N_L * $(\partial$

$(\Lambda_6) /\partial z)$ } * { 2 * Λ_6 * $(\partial (\Lambda_6) /\partial x)$ } } / $[\Lambda_6]^4$ (149)

$(\partial^2 (\varrho (x, y, z)) /\partial x \partial z) = $ { { $(\partial (\partial (N_L) /\partial x) /\partial z)$ * $[\Lambda_6]$ - $(\partial (N_L) /\partial z)$ * $(\partial (\Lambda_6) /\partial x)$ + $(\partial$

$(N_L) /\partial x)$ * $(\partial (\Lambda_6) /\partial z)$ - N_L * $(\partial^2 (\Lambda_6) /\partial x \partial z)$ } * $[\Lambda_6]^2$ - { $(\partial (N_L) /\partial x)$ * Λ_6 - N_L * $(\partial$

$(\Lambda_6) /\partial x)$ } * { 2 * Λ_6 * $(\partial (\Lambda_6) /\partial z)$ } } / $[\Lambda_6]^4$ (150)

$(\partial^2 (\varrho (x, y, z)) /\partial y \partial z) = $ { { $(\partial (\partial (N_L) /\partial y) /\partial z)$ * $[\Lambda_6]$ - $(\partial (N_L) /\partial z)$ * $(\partial (\Lambda_6) /\partial y)$ + $(\partial$

$(N_L) /\partial y)$ * $(\partial (\Lambda_6) /\partial z)$ - N_L * $(\partial^2 (\Lambda_6) /\partial y \partial z)$ } * $[\Lambda_6]^2$ - { $(\partial (N_L) /\partial y)$ * Λ_6 - N_L * $(\partial$

$(\Lambda_6) /\partial y)$ } * { 2 * Λ_6 * $(\partial (\Lambda_6) /\partial z)$ } } / $[\Lambda_6]^4$ (151)

$(\partial^2 (\varrho (x, y, z)) /\partial z \partial y) = $ { { $(\partial (\partial (N_L) /\partial z) /\partial y)$ * $[\Lambda_6]$ - $(\partial (N_L) /\partial y)$ * $(\partial (\Lambda_6) /\partial z)$ + $(\partial$

$(N_L) /\partial z)$ * $(\partial (\Lambda_6) /\partial y)$ - N_L * $(\partial^2 (\Lambda_6) /\partial z \partial y)$ } * $[\Lambda_6]^2$ - { $(\partial (N_L) /\partial z)$ * Λ_6 - N_L * $(\partial$

$(\Lambda_6) /\partial z)$ } * { 2 * Λ_6 * $(\partial (\Lambda_6) /\partial y)$ } } / $[\Lambda_6]^4$ (152)

Calculation of the Total Curvature of the Resilient Signal

$(\partial^2 (R((x, y, z), f)) /\partial x^2) = (\partial^2 (\varrho (x, y, z)) /\partial x^2)$ (153)
$(\partial^2 (R((x, y, z), f)) /\partial y^2) = (\partial^2 (\varrho (x, y, z)) /\partial y^2)$ (154)
$(\partial^2 (R((x, y, z), f)) /\partial z^2) = (\partial^2 (\varrho (x, y, z)) /\partial z^2)$ (155)

$(\partial (\partial (R((x, y, z), f)) /\partial y)\partial x) = (\partial^2 (\varrho (x, y, z)) /\partial y \partial x)$ (156)
$(\partial (\partial (R((x, y, z), f)) /\partial x)\partial y) = (\partial^2 (\varrho (x, y, z)) /\partial x \partial y)$ (157)

$(\partial (\partial (R((x, y, z), f)) /\partial y)\partial z) = (\partial^2 (\varrho (x, y, z)) /\partial y \partial z)$ (158)
$(\partial (\partial (R((x, y, z), f)) /\partial z)\partial y) = (\partial^2 (\varrho (x, y, z)) /\partial z \partial y)$ (159)

$(\partial (\partial (R((x, y, z), f)) /\partial x)\partial z) = (\partial^2 (\varrho (x, y, z)) /\partial x \partial z)$ (160)
$(\partial (\partial (R((x, y, z), f)) /\partial z)\partial x) = (\partial^2 (\varrho (x, y, z)) /\partial z \partial x)$ (161)

The total curvature C(R((x, y, z), f)) is:

$C(R((x, y, z), f)) = (\partial^2 (R((x, y, z), f)) /\partial x^2) + (\partial^2 (R((x, y, z), f)) /\partial y^2) + (\partial^2 (R((x, y, z), f)) /\partial z^2) + (\partial (\partial (R((x, y, z), f)) /\partial y)\partial x) + (\partial (\partial (R((x, y, z), f)) /\partial x)\partial y) + (\partial (\partial (R((x, y, z), f)) /\partial y)\partial z) + (\partial (\partial (R((x, y, z), f)) /\partial z)\partial y) + (\partial (\partial (R((x, y, z), f)) /\partial z)\partial x) + (\partial (\partial (R((x, y, z), f)) /\partial x)\partial z)$

$$(162)$$

Analogously to the mono-variate and the bivariate formulas, equation (162) is the total curvature of the trivariate resilient signal in the variables x, y and z and it is also the total curvature of the trivariate cubic resilient-curvature signal reconstruction formula.

SUMMARY

For each of the three interpolators called with the guffix g_4, specifically: g_4 (x) in 1D, g_4 (x, y) in 2D, and g_4 (x, y, z) in 3D, it is possible to produce three new interpolation functions called classic-curvature, resilient-curvature, and classic-resilient-curvature hybrid formulae. They are derived from the simplified scheme: I $(\Omega) = f (\Omega_0) + PIC (\Omega)$, classic-curvature formula; R $(\Omega) = f (\Omega_0) + PIC (\Omega)$, resilient-curvature formula, and HYB $(\Omega) = I (\Omega)$ if I $(\Omega) < R (\Omega)$ or HYB $(\Omega) = R (\Omega)$ if I $(\Omega) \geq R (\Omega)$, classic-resilient-curvature hybrid formula. Where Ω_0 is f(0) in 1D; is f(0, 0) in 2D; and is f(0, 0, 0) in 3D. Likewise Ω is x in 1D; is (x, y) in 2D; and is (x, y, z) in 3D. Results of these new classes of interpolators will be presented in chapter 12.

REFERENCES

Ciulla, C. (2009). *Improved Signal and Image Interpolation in Biomedical Applications: The Case of Magnetic Resonance Imaging (MRI)* – Medical Information Science Reference - IGI Global Publisher, Hershey, PA, U.S.A.

CHAPTER 12

RESILIENT INTERPOLATION: THE RESULTS OF THE INVESTIGATION ON THE DIMENSIONALITY OF THE POLYNOMIALS

INTRODUCTION

This chapter presents the results of the application of the mathematics developed in chapter 11. It also serves the purpose of giving to the reader a summary of the procedures employed and the main findings of exploratory research carried out here.

Given an interpolation function, two issues have been the object of study regarding the interpolation error. One is the arbitrary use of the transfer function used to process the convolution of pixel intensity and misplacements, and the other one is the minimization of the error as a result of introducing an improved paradigm. Along these lines of research, another relevant finding is that an improved interpolation paradigm does not necessarily guarantee its superiority in all of the pixels of the image (volume in 3D), thus it is a good idea to derive an hybrid form of interpolation formula which makes use of both classic-curvature and resilient-curvature interpolation formulae on a pixel-by-pixel basis, depending on which one gives the lower Mean Absolute Error (MAE).

Three functions were studied in chapter 11 and the research query which is relevant to their math form is that which aims to clarify how they work with the same data set, given that the only difference between them is the dimensionality (mono-dimensional, bivariate in x and y, and trivariate in x, y and z). Also, a study on the effect on the interpolation error of the constant parameter of the three polynomials of chapter 11 is presented. The chapter presents the theoretical concepts accompanied by the results and the formulae used at each step of the exploratory research; images showing the importance and the meaningfulness of the curvature either in the classic or in the resilient form are also an object of discussion.

HOW TO CALCULATE THE RESILIENT SIGNAL FROM AN INTERPOLATION FORMULA

Given an interpolation formula such as the $g_4(x)$ in equation (1), let us calculate the second order derivatives under the assumption that they are not null and continuous in their interval of definition, which equates to the function $g_4(x)$ being second order differentiable. The second step is that of calculating the intensity-curvature terms (Ciulla, 2009) before and after interpolation, which are called $E_o = E_o(x)$ and $E_{IN} = E_{IN}(x)$ respectively. The signal resilient to interpolation is derived from the solution of the equation $E_{IN}(x) = E_o(x)$, which implies the assumption that the resulting signal is the same when interpolated at the location x and when not interpolated. In chapter 11 we have seen that given:

$$g_4(x) = f(0) + \alpha_2 * a\ [x^3 + 1/2\ x^2 + \frac{1}{4}\ x + 1] + \alpha_3 * a\ [x^2 + 2\ x + 1] \tag{1}$$

We obtained equation (52) of chapter 11 given below:

$$f(0) = \{ (\alpha_2\ a)^2\ [6/5\ x^5 + \frac{3}{4}\ x^4 + 1/2\ x^3 + 3x^2 + x^4/4 + 1/6\ x^3 + 1/8\ x^2 + x] + \alpha_2 * a\ [x^4/4 +$$

$$1/6\ x^3 + 1/8\ x^2 + x] * 2\ \alpha_3\ a + \alpha_3\ \alpha_2\ a^2 * [3/2\ x^4 + 4\ x^3 + 3x^2 + x^3/3 + x^2 + x] + \alpha_3\ a\ [x^3/3$$

$$+ x^2 + x] * 2\ \alpha_3\ a\ \} / \{ \{ \alpha_2 * a + 2\ \alpha_3 * a \} * x - \{ \alpha_2 * a\ [3x^2 + x] + 2\ \alpha_3 * a\ x \} \} \tag{2}$$

The arbitration in the use of the transfer function employed to process the convolution of pixel intensity and misplacements was the main concern of the author and a relevant finding of this exploratory research is that the signal resilient to interpolation, as much as the classic signal, also offers the chance to derive the transfer function from the math form of the interpolation formula, thus avoiding the eventuality that the interpolation error is subject to the bias introduced with the transfer function of the choice.

HOW TO OBTAIN THE TRANSFER FUNCTION FROM THE MATH FORM OF THE INTERPOLATION FORMULA

The first step is that of calculating the second order derivative (curvature) of the signal (signal resilient to interpolation or classic signal). It is thus possible to elicit the idea that this major step of calculation of the second order derivatives is sufficient to derive two classes of interpolators, which in this book have been called classic-curvature and resilient-curvature signal reconstruction formulae.

Why do we need to use the total curvature? Simply because the set of second order derivatives ensures comprehensive calculation of the declination (angle) subtended between the horizontal and the tangent to the first order derivative, so that the angle is calculated locally at the intra-node (1D), intra-pixel (2D) or intra-voxel (3D) locations which are being reconstructed.

What is the meaning of comprehensive? It means complete, in considering all the covariates such as xy and yx in the bivariate case or xy, xz, yx, yz, zx, zy in the trivariate case, while calculating the total curvature, which is representative of the angle.

But why then would we need such an angle? Chapter 4 introduces the concept of Pixel Intensity Correction (PIC) and shows the geometrical meaning the PIC in two dimensions. The angle is necessary for us to arrive at the following formula: PIC = Shift * tan (*total curvature*) in order to calculate the fraction of the pixel intensity so as to reconstruct the signal at the re-sampling location, where the shift is calculated with the Pythagorean Theorem.

Why should there be two classes of interpolators? Because the total curvature can and is calculated from the interpolation function in its original form, as in, for example, equation (1), or from the Signal Resilient to Interpolation (SRI) deriving from the interpolation function and seen in equation (2). When the total curvature is calculated, the formula PIC = Shift * tan (*total curvature*) provides us with the customized transfer function, which eliminates the arbitration of the choice and the bias in the calculation of the interpolation error. The two classes of interpolators are called classic-curvature formula and resilient-curvature formula.

At this point another issue of relevance must be raised. What happens if one or the other of the two interpolators outperforms the other in a specific pixel re-sampling task. As we have said, it is a good idea to form a hybrid formula out of the two and this hybrid formula enables us to reconstruct the signal on a pixel-by-pixel basis depending on which of the two fomulae, classic-curvature and resilient-curvature, furnishes the lowest error. Both of the two formulae make use of the customized transfer function derived mathematically through an unbiased process that starts from the math form of the initial interpolation formulae (for example see equations (1) and (2)). At this point we are ready to see the results of the investigation.

RESULTS

Figures 1 through 3 show in each of them, the ratio of improvement of the hybrid formula, herein called classic-resilient-curvature interpolation (or signal reconstruction) formula, over the classic-curvature (C.CR) and the resilient curvature (R.CR) formulae, and the three graphs relevant to the Mean Absolute Error (MAE) of the three curvature-based classes of interpolation functions.

Classic-curvature With E	Resilient-curvature With E
G4	
G41DRCCurvature	
G42DRCCurvature	
G43DRCCurvature	

Table I: The G4 interpolation formulae employed in the research relating to differences in results when the dimensionality of the formula changes from 1D to 2D to 3D and differences in the results when the constant parameter 'a' changes its value (see equation (1) in 1D or equation (9) in 3D). All formulae were calculated embedding ('With E') the pixel (node or voxel) to be reconstructed.

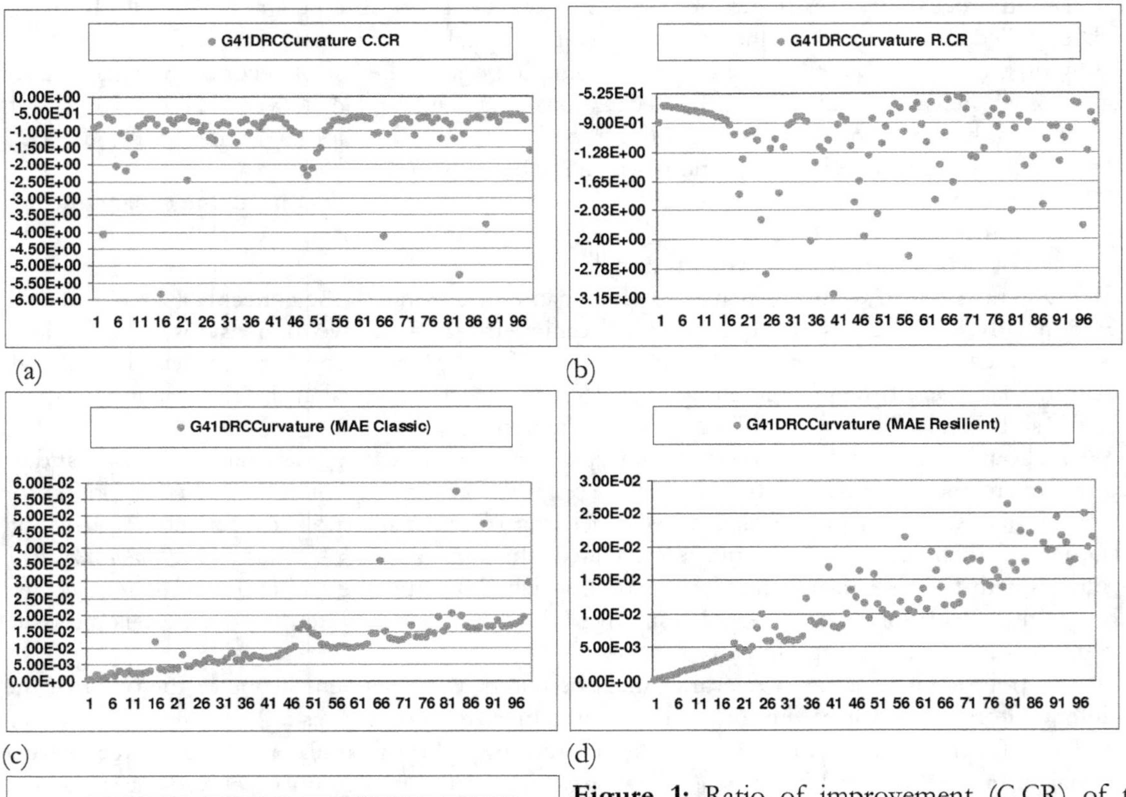

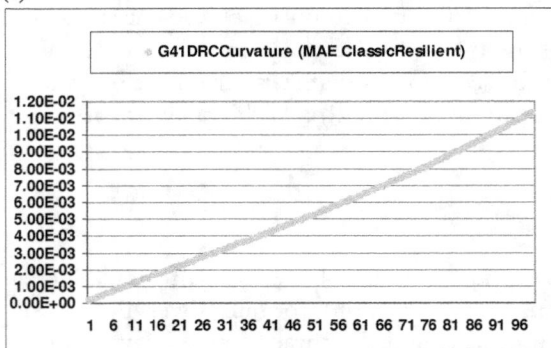

(a)

(b)

(c)

(d)

(e)

Figure 1: Ratio of improvement (C.CR) of the hybrid formula over the classic-curvature formula in (a), ratio of improvement (R.CR) of the hybrid formula over the resilient-curvature formula in (b), MAE of the classic-curvature formula in (c), MAE of the resilient-curvature formula in (d), MAE of the classic-resilient-curvature hybrid formula in (e). Experiments are relevant to the mono-dimensional case named G41DRCCurvature.

As far as regards the nomenclature of the graphs, Table I gives the names of the headings of the graphs for each dimension for which the experiments were conducted. In all of the three dimensions the interpolation formula is named with the suffix G4 in accordance with the math developments in chapter 11. Naturally the suffix 1D, 2D and 3D in the name of the formulae (c.f. entries in the tables) signifies one-dimensional in (x), bivariate in (x, y) and trivariate in (x, y, z).

In figure 1, the graphs are relevant to the mono-dimensional case study named G41DRCCurvature. Misplacements varied from 0.01 to 0.99 and the abscissa of each graph shows this dimension times 100.

In what follows the ratio of improvement of the classic-resilient-curvature hybrid formula over the classic-curvature formula will be referred to as C.CR, whereas the ratio of improvement of the hybrid formula over the resilient-curvature formula will be referred to as R.CR. In (a) of figure 1, the C.CR presents scattered behavior between approximately 50% and approximately 250%, in (b) the R.CR is scattered between approximately 52.5% and approximately 278%.

It can be inferred that both of the formulae, classic-curvature and resilient-curvature, contribute almost equally to the improvement obtained through the hybrid signal reconstruction formula. The resilient-curvature formula performs in a similar way to the classic-curvature formula and this is discernible in the graphs of the MAE in (c) and (d) where the behavior is generally linear, reaching a maximum error of 0.02 and 0.025 respectively. In (e) the value of the MAE of the hybrid formula linearly increases with the increase of the misplacement and reaches approximately 0.011. In the experimental data reported in figure 1 the value of the parameter *'the_A_const'* was set equal to -8.4554.

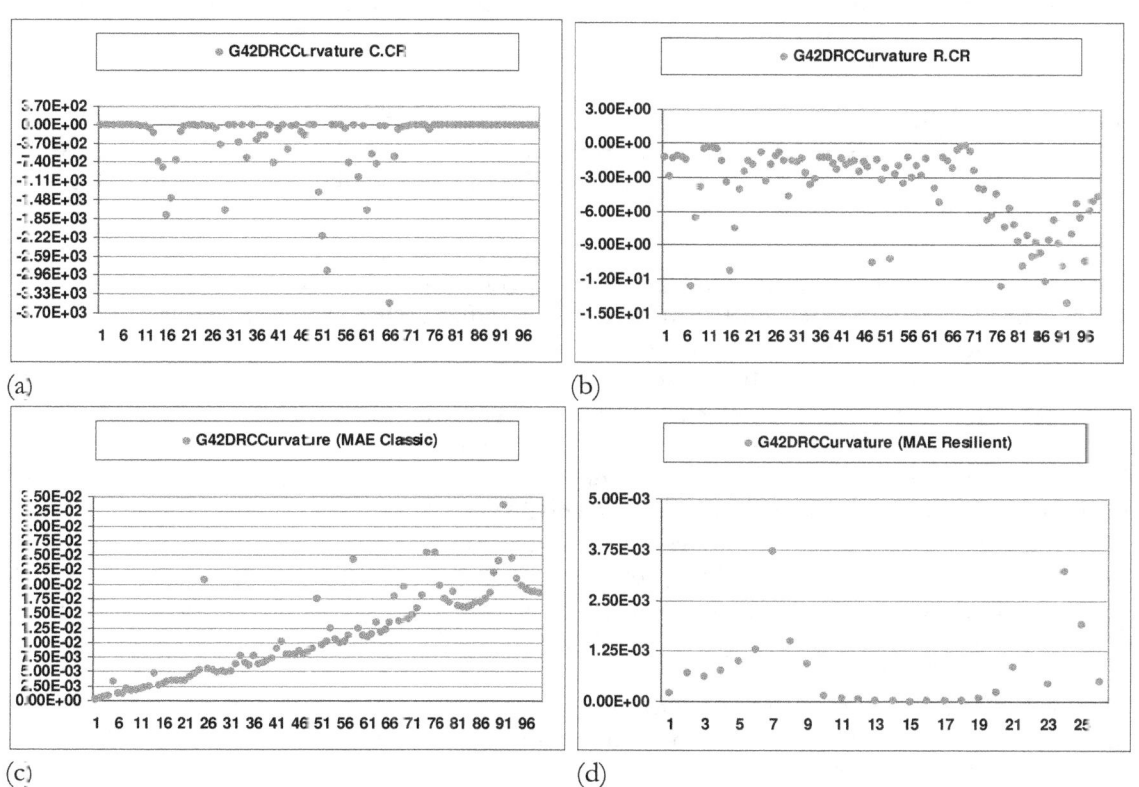

(a)   (b)

(c)   (d)

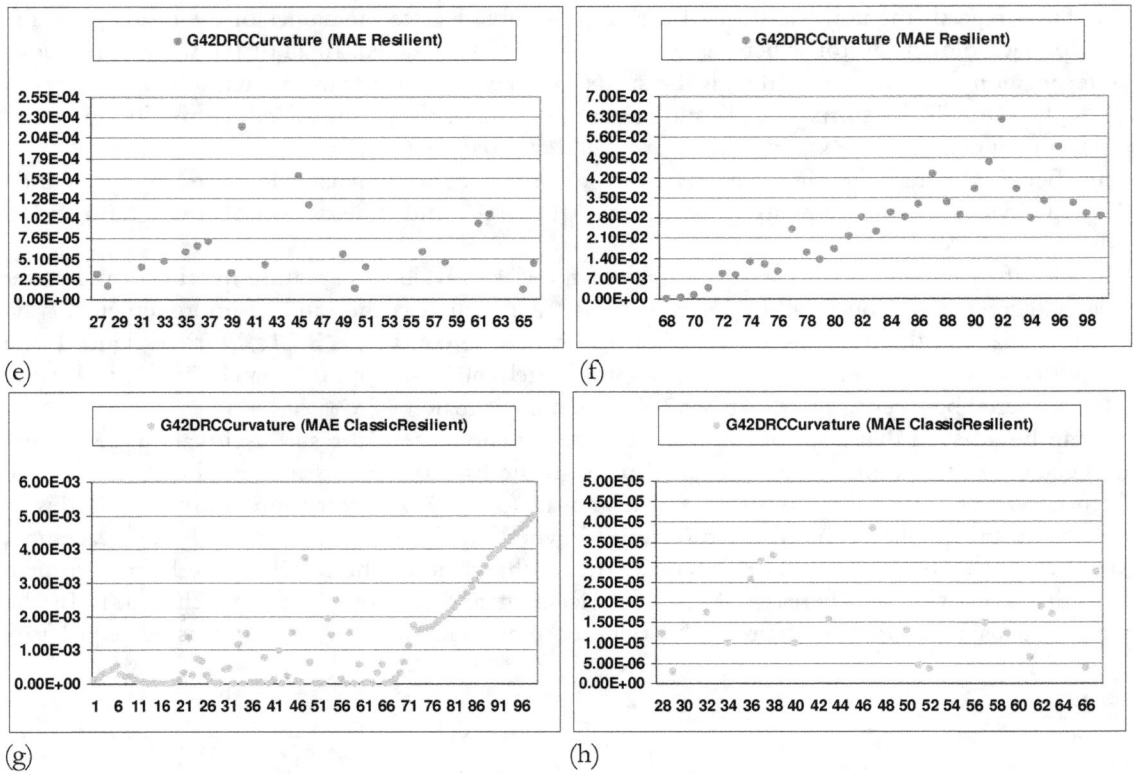

Figure 2: Ratio C.CR of the hybrid formula in (a), ratio R.CR of the hybrid formula in (b), MAE of the classic-curvature formula in (c), MAE of the resilient-curvature formula in (d), (e) and (f), MAE of the classic-resilient-curvature hybrid formula in (g) and (h). Experiments are relevant to the bivariate case named G42DRCCurvature.

In figure 2, the graphs are relevant to the bivariate case study named G42DRCCurvature. Misplacements varied from $(x, y) = (0.01, 0.01)$ to $(x, y) = (0.99, 0.99)$ with a progressive increase of $(0.01, 0.01)$ at each step (along the diagonal of the pixel) and the abscissa of each graph displays one of the two dimensions x or y times 100.

This case study shows behavior which is largely different from that seen in figure 1 in the case of the mono-dimensional shift and we may observe this as soon as we look at the graphs of the ratios C.CR and R.CR of improvement of the hybrid formula versus the classic-curvature and the resilient-curvature formulae, which are scattered between 0 and 333,000 % (C.CR) (see (a)) and between 0 and 1,500% (R.CR) (see (b)), respectively. We can see a large number of points in the graph in (a) between $(x, y) = (0.29, 0.29)$ and $(x, y) = (0.68, 0.68)$, compared to those seen in the same interval in (b), which show a remarkable superiority of the resilient-curvature formula over the classic-curvature formula (the values of the C.CR ratio between $(x, y) = (0.29, 0.29)$ and $(x, y) = (0.68, 0.68)$ are higher in (a) than the values of the R.CR in (b)).

This means that the gain of the hybrid formula is greater for the classic-curvature formula than it is for the resilient-curvature formula. When looking at the portion of the graphs in (c) and (e) between $(x, y) = (0.29, 0.29)$ and $(x, y) = (0.68, 0.68)$ we see that the MAE is between approximately 5.00 E-03 and approximately 1.25 E-02 (see (c)) and below approximately 2.30 E-04 (see (e)), showing the superiority of the resilient-curvature formula.

Again, between (x, y) = (0.29, 0.29) and (x, y) = (0.68, 0.68), the graph in figure 2(c) shows MAE values below 5.00 E-03 and 1.25 E-02 respectively, and in (e) the majority of the points show the MAE to be below 1.53 E-04, once again underlining the superiority of the resilient-curvature formula over the classic-curvature formula.

When looking at the interval from (x, y) = (0.68, 0.68) to (x, y) = (0.99, 0.99), in figure 2(c) (classic-curvature) and figure 2(f) (resilient-curvature), the behavior of the MAE is generally similar, with a slight superiority in the case of the classic-curvature formula, showing an approximate linear increase with the increase of the misplacement up to values approximately below 2.00 E-02 (see (c)) and approximately below 5.50 E-02 (see (f)).

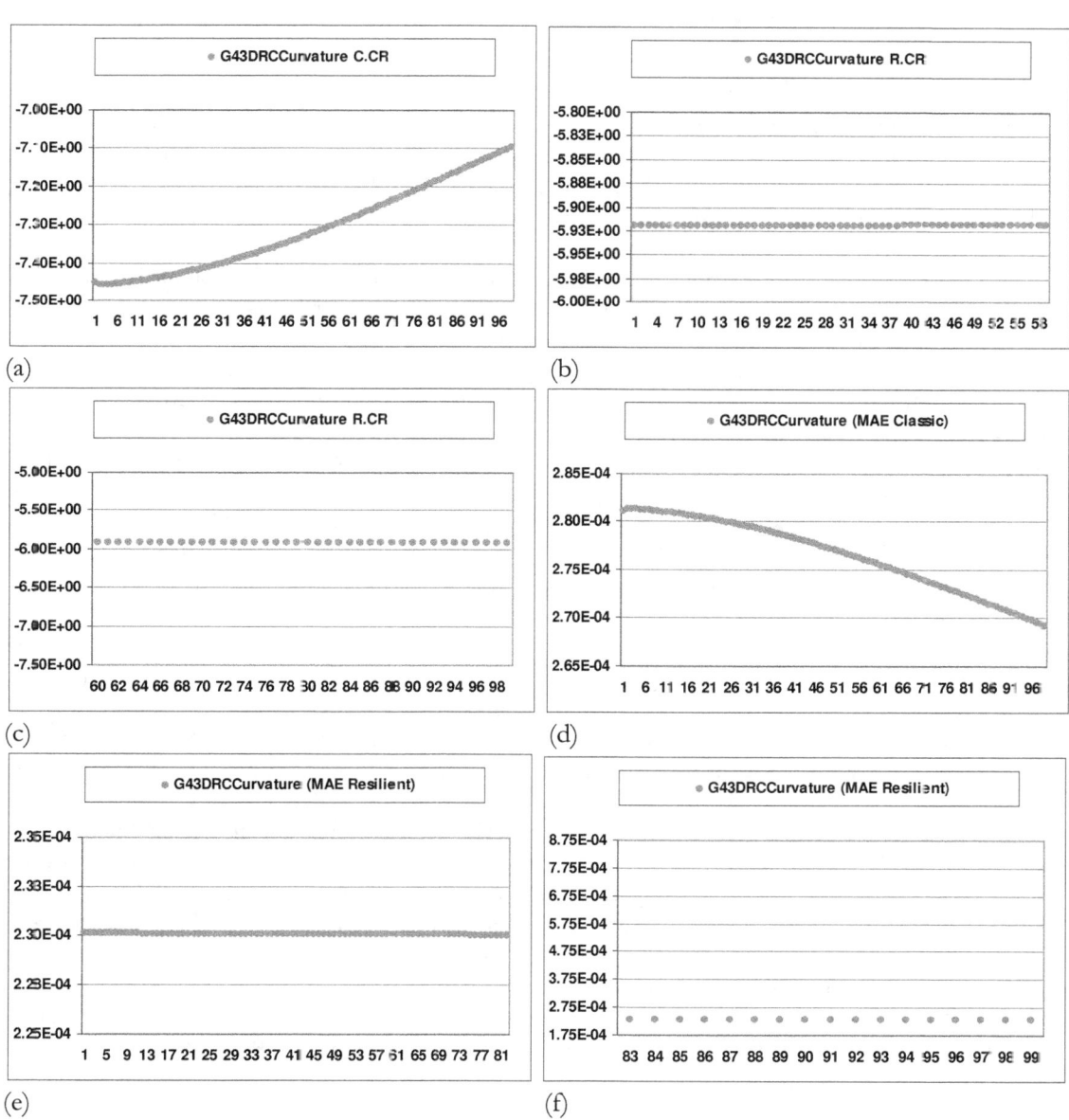

(a)

(b)

(c)

(d)

(e)

(f)

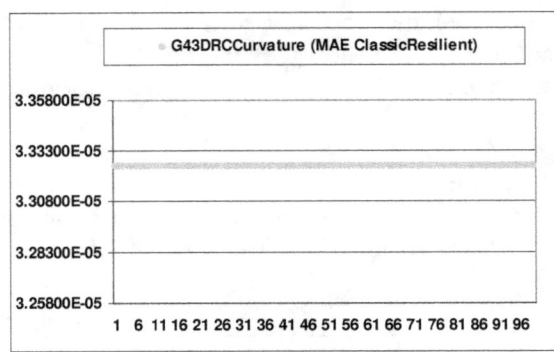

(g)

Figure 3: Ratio C.CR of the hybrid formula in (a), ratio R.CR of the hybrid formula in (b) and (c), MAE of the classic-curvature formula in (d), MAE of the resilient-curvature formula in (e) and (f), MAE of the classic-resilient-curvature hybrid formula in (g). Experiments are relevant to the trivariate case named G43DRCCurvature.

Interestingly, in figure 2(g), between (x, y) = (0.29, 0.29) and (x, y) = (0.68, 0.68), the MAE of the hybrid formula is quite low, descending to approximately 4.00 E-05 and below (not visible in (g), this portion of the graph is expanded to improve visibility in figure 2(h)) because of the gain derived from the resilient-curvature formula; and for values of the misplacement beyond the location (x, y) = (0.68, 0.68), it increases dramatically in view of the increase of the MAE of both classic-curvature and resilient-curvature formulae.

When comparing the mono-dimensional to the bivariate set of experimentations it is interesting to grasp the difference between figure 1(e) and figure 2(g) where a large discrepancy between the values of MAE of the hybrid formulae can clearly be seen.

In particular, in figure 1(e) the MAE of the hybrid formula linearly increases with increasing misplacement and reaches approximately 0.011, whereas in figure 2(g) the MAE is below approximately 5.00 E-03 and in figure 2(h) mainly below approximately 4.00 E-05.

In figure 3, the graphs are relevant to the trivariate case study named G43DRCCurvature, and the abscissa of each graph reports the 99 sessions without specific reference to the value of the three coordinates. Misplacements varied 99 times from (x, y, z) = (9.94 E-03, 1.00 E-02, 1.006 E-02) to (x, y, z) = (9.59 E-03, 9.99 E-03, 1.04 E-02) and were obtained through rigid transformation. The three rotational angles about X, Y and Z axis incremented themselves to the tune of 0.001 deg at each step, starting from the value of 0.01 at the first step, and the misplacement was kept constant to 0.01 for x, y and z coordinates. The numerical values in the abscissa of the graphs in figure 3 correspond to the progressive steps of the misplacements ((from (x, y, z) = (9.94 E-03, 1.00 E-02, 1.006 E-02) to (x, y, z) = (9.59 E-03, 9.99 E-03, 1.04 E-02)) obtained through the rigid transformation.

Figures 3(a) and 3(b) and 3(e) are partially in favor of the resilient-curvature formula since the ratios C.CR and R.CR are such as to be between approximately 710 % and approximately 750 % (see (a)), and approximately below 590 % and approximately above 600 % (see (b) and (c) respectively), indicating the improvement of the hybrid over the classic-curvature and the resilient-curvature formulae respectively. Consistently the graphs of the MAE in (d) and (e) show values between approximately 2.7 E-04 and approximately 2.8 E-04 (see (d)) and approximately 2.3 E-04 (see (e)). Both graphs behave well and show a non linear decrease (d) and constancy (e). Finally figure 3(g) shows the MAE of the hybrid formula with values between approximately 3.308 E-05 and approximately 3.333 E-05 with a well-defined behavior of constancy.

It is possible to derive a brief summary of the results present in figures 1 through 3 considering that the data is the same as in the T2 Magnetic Resonance Imaging (MRI) seen in chapter 10 (Courtesy of: R. S. Swenson, www.Dartmouth.edu/~rswenson/Atlas), and that the misplacement ranges in 1D from x = 0.01 to x = 0.99, in 2D, from (x, y) = (0.01, 0.01) to (x, y) = (0.99, 0.99), and in 3D from (x, y, z) = (9.94 E-03, 1.00 E-02, 1.006 E-02) to (x, y, z) = (9.59 E-03, 9.99 E-03, 1.04 E-02). The MAE in the mono-dimensional experiment reaches approximately 0.011 (see figure 1(e)), whereas in the bivariate experiment MAE is approximately below 5.00 E-03 and mainly approximately below 4.00 E-05 (see figure 2(g) and figure 2(h) respectively). In the trivariate experiment, the MAE of the hybrid formula presents values between approximately 3.308 E-05 and approximately 3.333 E-05 (see figure 3(g)). In general, we may make a provisional summary by saying that the performances of the bivariate and trivariate polynomials are better than those of the mono-variate polynomials, at least to the extent that the data seen up to this point in this chapter.

TOTAL CURVATURE MAPS AND RESIDUAL IMAGES

This section discusses the meaning of the total curvature and the corresponding maps derived in each of the three experiments: mono-dimensional, bivariate and trivariate. This paragraph brings to our attention that the meaning of the curvature is relevant to the calculation of the Pixel Intensity Correction (PIC): PIC = Shift * tan(*total curvature*). Thus the relevance of the total curvature is two-fold. One aspect is that it allows us to calculate the customized transfer function removing the arbitration of the choice and the bias in the interpolation error, and the other aspect is that it allows us to finalize the signal reconstruction process since the PIC is the fraction of the pixel to be added to f() (in 1D), f(0, 0) (in 2D), f(0, 0, 0) (in 3D) to obtain the interpolated value.

For the commodity of the reader, we give below the six formulae that were used in chapter 11 to calculate the total curvature in 1D, 2D and 3D and they are seen here in equations (3), (4) and (5) for the classic signals I((x), f), I((x, y), f) and I((x, y, z), f) (interpolation formulae) respectively, and in equations (6), (7) and (8) for the resilient signal R((x), f), R((x, y), f) and R((x, y, z), f) respectively (where ϱ is defined differently depending on the interpolation formula which we have chosen to study and corresponds to the resilient signal). These formulae act upon the classic or resilient signal producing maps which resemble the original features of the original image, as seen already at the end of chapter 10.

$$C(I((x), f)) = (\partial^2 (g_4(x)) / \partial x^2) \tag{3}$$

$$C(I((x, y), f)) = (\partial^2 (g_4(x, y)) / \partial x^2) + (\partial^2 (g_4(x, y)) / \partial y^2) + (\partial^2 (g_4(x, y)) / \partial x / \partial y) + (\partial^2 (g_4(x, y)) / \partial y / \partial x) \tag{4}$$

$$C(I((x, y, z), f)) = (\partial^2 (g_4(x, y, z)) / \partial x^2) + (\partial^2 (g_4(x, y, z)) / \partial y^2) + (\partial^2 (g_4(x, y, z)) / \partial z^2) + (\partial^2 (g_4(x, y, z)) / \partial x \partial y) + (\partial^2 (g_4(x, y, z)) / \partial y \partial x) + (\partial^2 (g_4(x, y, z)) / \partial x \partial z) + (\partial^2 (g_4(x, y, z)) / \partial y \partial z) + (\partial^2 (g_4(x, y, z)) / \partial z \partial x) + (\partial^2 (g_4(x, y, z)) / \partial z \partial y) \tag{5}$$

$$C (R ((x), f)) = (\partial^2 (\varrho(x)) / \partial x^2) \tag{6}$$

$$C (R((x, y), f)) = (\partial^2 (\varrho(x, y)) / \partial x^2) + (\partial^2 (\varrho(x, y)) / \partial y^2) + (\partial^2 (\varrho(x, y)) / \partial x \partial y) + (\partial^2 (\varrho(x, y)) / \partial y \partial x) \tag{7}$$

$$C(R((x, y, z), f)) = (\partial^2 (\varrho(x, y, z), f)) / \partial x^2) + (\partial^2 (\varrho ((x, y, z), f)) / \partial y^2) + (\partial^2 (\varrho ((x, y, z), f)) / \partial z^2) + (\partial (\partial (\varrho ((x, y, z), f)) / \partial y)\partial x) + (\partial (\partial (\varrho ((x, y, z), f)) / \partial x)\partial y) + (\partial (\partial (\varrho ((x, y, z), f)) / \partial y)\partial z) + (\partial (\partial (\varrho ((x, y, z), f)) / \partial z)\partial y) + (\partial (\partial (\varrho ((x, y, z), f)) / \partial z)\partial x) + (\partial (\partial (\varrho ((x, y, z), f)) / \partial x)\partial z) \tag{8}$$

Hybrid Signal Reconstruction

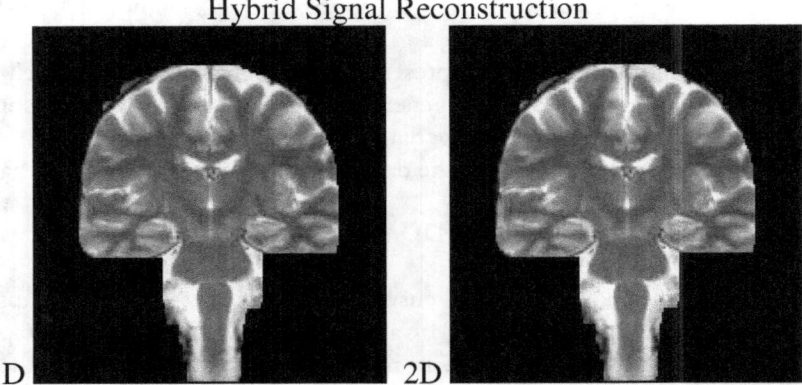

Figure 4: Hybrid signal reconstruction shown after a mono-dimensional interpolation task with 0.77 as the misplacement (left), and a bivariate interpolation task with (x, y) = (0.77, 0.77) as the misplacement. The polynomials employed were cubic in 1D and 2D. The T2 MRI is courtesy of: R. S. Swenson, www.Dartmouth.edu/~rswenson/Atlas.

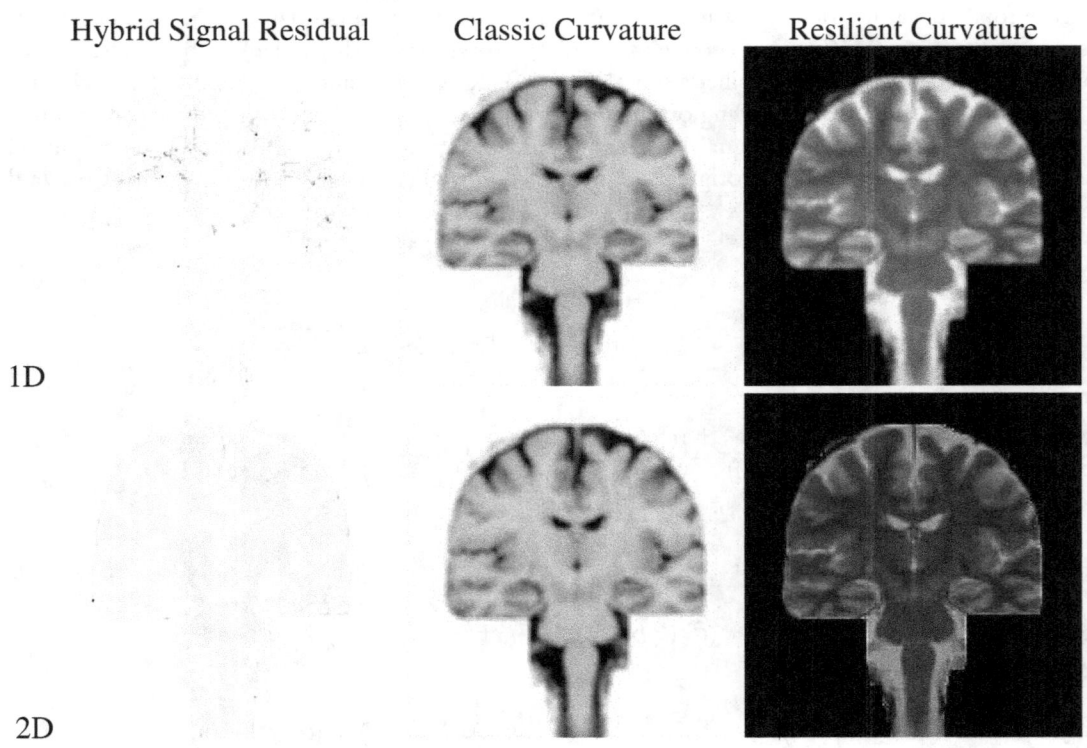

Figure 5: Hybrid signal residuals in 1D and 2D interpolation of misplacement x = 0.77 (upper left) and (x, y) = (0.77, 0.77), along the diagonal of the two dimensional pixel (bottom left). Classic curvature images (center column), and resilient curvature images (right column).

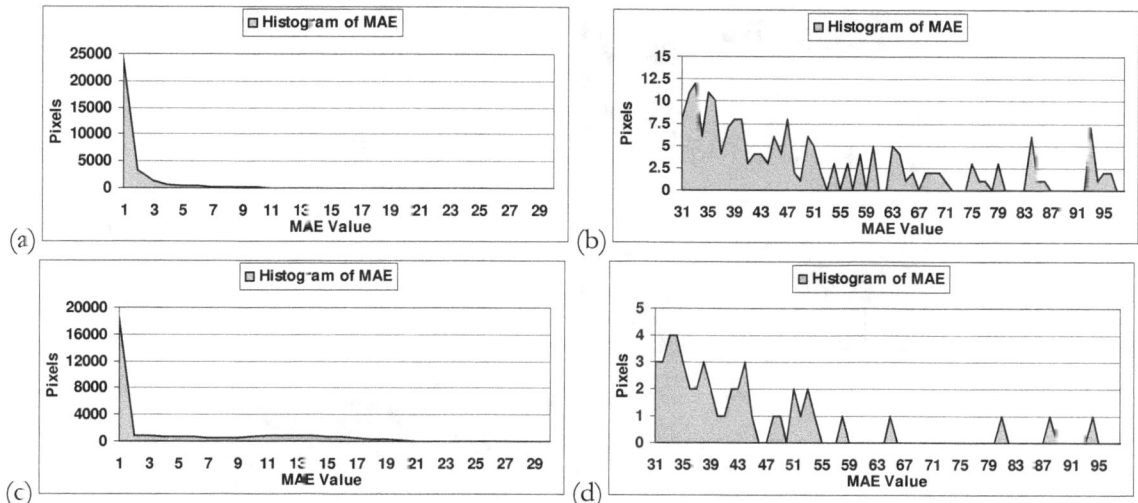

Figure 6: Histograms of the residual images in the range [0, 255] where the residual is not zero, mono-dimensional interpolation (a) and (b), and bivariate interpolation (c) and (d).

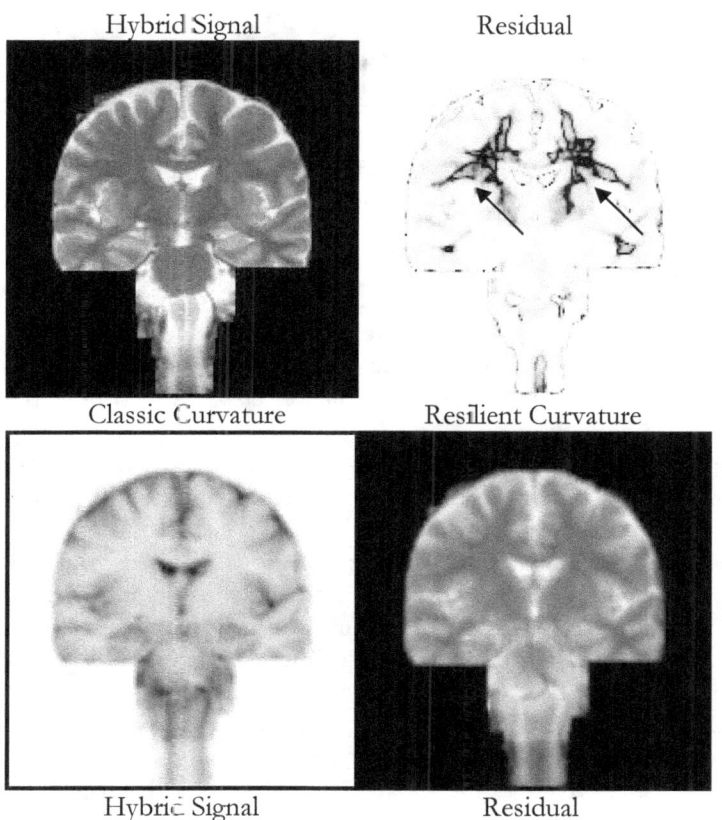

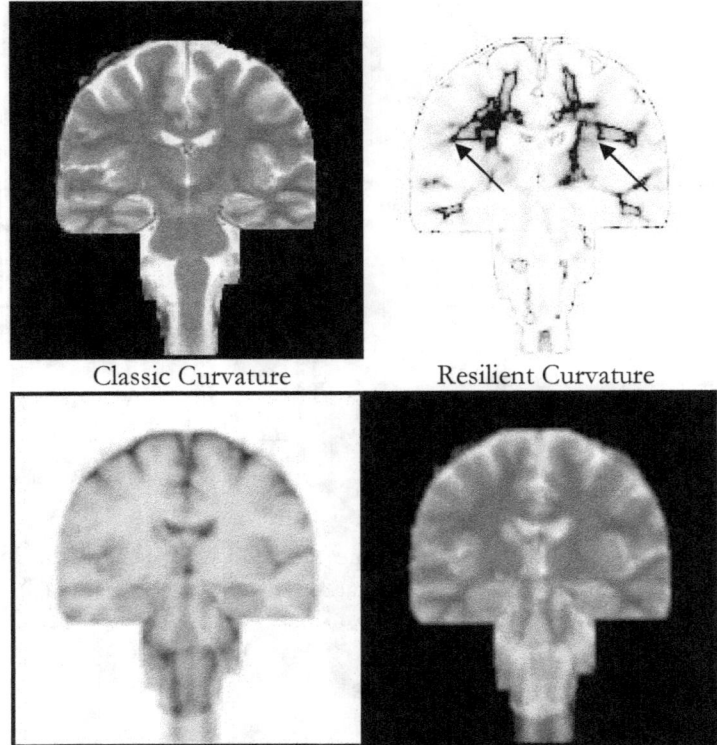

Classic Curvature   Resilient Curvature

Figure 7: In the upper and middle rows the hybrid reconstruction image and the residuals are shown, in the second and fourth row from the top, the curvature images corresponding to upper and middle rows' hybrid images are shown.

In figure 4, the MAE of the signal reconstruction with the hybrid formula is 8.27 E-03 and 1.77 E-03 for the mono-dimensional and the bivariate interpolation tasks respectively. Visual inspection reveals a high level of similarity between the two images, suggesting that, regardless of the misplacement and the dimensionality of the interpolation task, the two polynomials reconstruct the signal efficiently and quite similarly.

Nevertheless, the qualitative evaluation also requires analyzing the difference images and this is seen in figure 5, along with the curvature maps. Figure 5 shows the residuals of the interpolation task in 1D and 2D. Residual images were inverted and had the contrast and brightness changed so as to reveal details hidden in the small error values shown in the MAE. This processing applies to all the residual images seen in figure 5 and figures 7 through 9. The histograms of the residual images (see figure 6, where the MAE is clearly the residual seen in the residual images of figure 5) were built in the nonzero regions so as to have a more focused set of information regarding the performance of the signal reconstruction done with the classic-resilient-curvature hybrid formula.

Regarding the curvature images, it is important at this stage to confirm that formulae (3) and (4) for the classic curvature images in 1D and 2D respectively and (6) and (7) for the resilient curvature images in 1D and 2D respectively had the capability of making such images, which in turn had the capability of finalizing the pixel intensity correction so as to obtain the hybrid signal reconstruction shown in figure 4. *This explains the capability of the total curvature within the context of signal reconstruction, though we must still remember that the total curvature is also capable of defining the transfer function from the math form of the interpolation function without any bias or arbitration.*

The histograms shown in figure 6 are of the nonzero areas of the two residual images seen in figure 5 in the left-most column. Each of the pixel intensities of the residual images was scaled into the range [0, 255] and the histogram built on the basis of such a range. The totals of the histogram values were calculated and they are 3.128 E+04 in (a), 1.950 E+02 in (b), 3.145 E+04 in (c) and 4.90 E+01 in (d). What, then, produces the difference in the results? The MAE is favorable towards the bivariate interpolation task, its value being 1.77 E-03 versus 8.27 E-03 for the mono-dimensional formula, and the visual inspection is favorable towards the mono-dimensional formula.

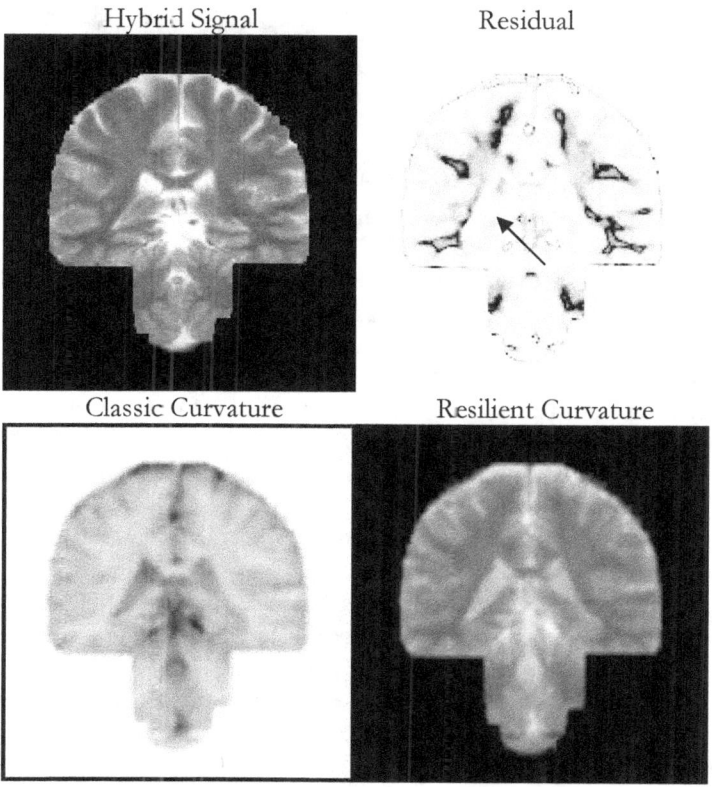

Figure 8: The hybrid reconstruction image and its residual are located in the upper row, the curvature images are located in the second row from the top.

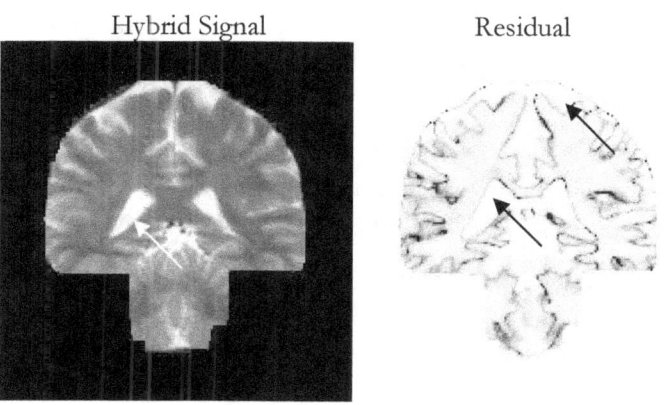

Classic Curvature   Resilient Curvature

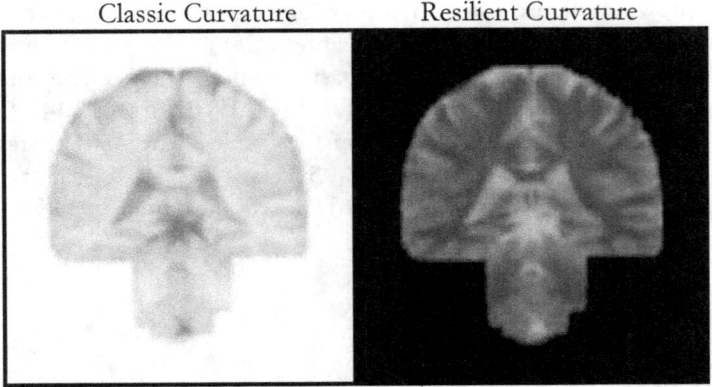

Figure 9: The hybrid reconstruction image and its residual are located in the top, whereas the curvature images are located in the second row from the top.

Changing the contrast and brightness in the residual image was done solely for the purpose of highlighting the features of the images, thus visual inspection cannot be the appropriate tool for judging which one of the two interpolation tasks out of the 1D one and the 2D one was more successful. It is plausible to infer that most of the difference in the MAE is explained through the different values of the histogram totals, 3.128 E+04 in (a) and 3.145 E+04 in (c), acting to the benefit of the mono-dimensional interpolation task. The different values 1.950 E+02 in (b) and 4.90 E+01 in (d) are explicative of the larger distribution of pixels with nonzero value making more features visible in the 2D residual image than in the 1D residual image. In figures 7 through 9 the interpolation task is solved with the hybrid signal reconstruction formula in 3D while the misplacement is (x, y, z) = (9.94 E-03, 1.00 E-02, 1.006 E-02) and the MAE is 3.33 E-05.

Looking at the ventricles and the grey and white matter of the T2 Magnetic Resonance Imaging of the human brain it is possible to distinguish a clear difference between the classic and the resilient curvature formulae and we can see that the two curvature images have values which are almost the opposite of each other. This means that in figures 7 through 9, the arc tangent of the angle subtended by the horizontal and the tangent to the first order derivative (which is the curvature) is in the classic curvature image the opposite of that one in the resilient curvature image. Also, the level of details reproduced is quite distinct and resembles the reconstructed hybrid image. To go into any further detail as regards this aspect of the curvature images goes beyond our scope here since this issue was already addressed in chapter 10 with the same T2 MRI data set.

Something else can be observed in the residual images in figure 7, namely that in the first one at the top right we find the grey matter having the higher residual (see dark arrows), and also in the third one from the top right we find the grey matter of the brain having the higher residual (see dark area in the residual image and dark arrows). Similarly, in figures 8 and 9 it is the ventricles and the most of the white matter which have the lower residual (see bright regions), whereas it is the grey matter which has the higher residual. This is observable while looking at the dark arrows pointing to the extensive brain regions in the bright area in the residual images.

Something which can be observed on the basis of the residual images of figures 8 and 9 is the gradient of the MAE, which can be seen at the borderline of the ventricles, which means that the value of the MAE tends to shift from one extreme to the other at the border of the brain structure, which includes a given region of quasi constant pixel intensity (see the white region, which represents the ventricles, pointed out by the white arrow in the hybrid image at the top left of figure 9).

STUDY OF THE BEHAVIOR OF THE MAE WHILE CHANGING THE VALUE OF THE PARAMETER OF THE POLYNOMIAL FUNCTIONS

The three functions shown and studied in chapter 11, whose results are presented in this chapter, are all parametric in the value of the parameter called *'the a_const'*. It is reasonable to infer that some changes to the signal reconstruction performance of the three formulae: classic-curvature, resilient curvature and classic-resilient-curvature hybrid, happen and are manifest in terms of the interpolation error, when the value of *'the a_const'* changes. It follows that to study the behavior of the MAE while keeping constant the misplacement and changing the value of *'the a_const'* is useful and informative.

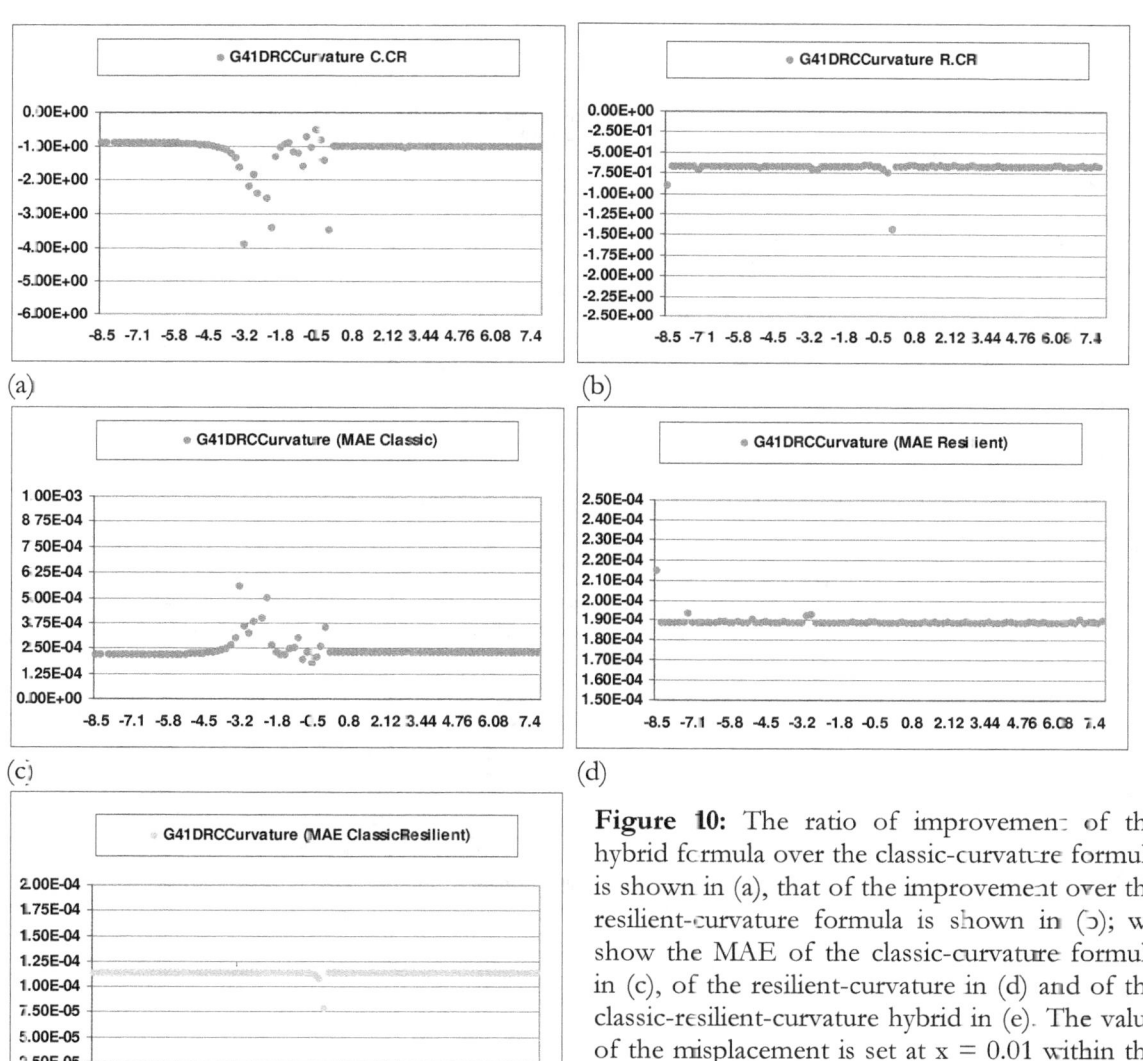

(a)

(b)

(c)

(d)

Figure 10: The ratio of improvement of the hybrid formula over the classic-curvature formula is shown in (a), that of the improvement over the resilient-curvature formula is shown in (b); we show the MAE of the classic-curvature formula in (c), of the resilient-curvature in (d) and of the classic-resilient-curvature hybrid in (e). The value of the misplacement is set at x = 0.01 within the entire mono-dimensional block of experiments.

(e)

In this section we present the block of experiments in 1D, 2D and 3D. In figures 10, 11 and 12 (mono-dimensional block of experiments) the value of the misplacement was x = 0.01, x = 0. 45 and x =0.99 respectively. In figures 13, 14 and 15 (bivariate block of experiments) the misplacement was (x, y) = (0.01, 0.01), (x, y) = (0.45, 0.45) and (x, y) = (0.99, 0.99) respectively. In figures 16, 17 and 18 (trivariate block of experiments) the misplacement was (x, y) = (0.01, 0.01, 0.01), (x, y) = (0.45, 0.45, 0.45) and (x, y) = (0.99, 0.99, 0.99) respectively.

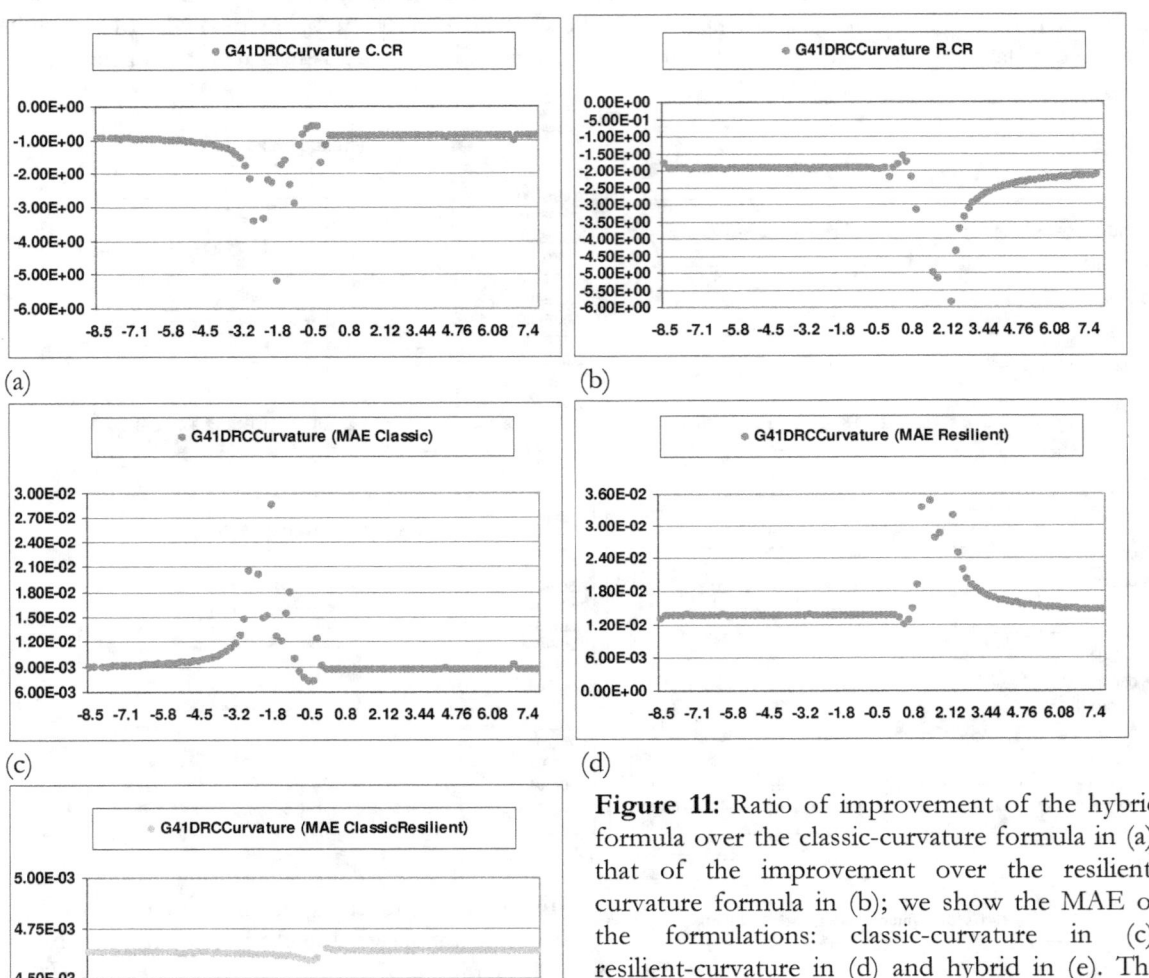

(a)

(b)

(c)

(d)

(e)

Figure 11: Ratio of improvement of the hybrid formula over the classic-curvature formula in (a), that of the improvement over the resilient-curvature formula in (b); we show the MAE of the formulations: classic-curvature in (c), resilient-curvature in (d) and hybrid in (e). The value of the misplacement is set at x = 0.45 within the entire mono-dimensional block of experiments.

In figure 10, 'the a_const' is the parameter that changes while the misplacement in 1D experimentations is kept constant and equal to the value of 0.01 along the x direction. The figure gives the graphs of the ratio of improvement of the hybrid formula over classic-curvature and resilient-curvature formulae, and the MAE of the three types of signal reconstruction. The parameter 'the a_const' changed its value inside the range [-8.4554 E+00, 7.7342 E+00] at each experiment while increasing by steps of 0.1652 from -8.4554 E+00 to 7.7342 E+00. On the other hand this setting was kept the same while changing instead the value of the misplacement in each block of experiments. An overall view of figures 10(a) and 10(b) shows that the ratio of improvement is quasi constant (see figure 10(b)), or confined within a small range of fluctuation except for the middle part of the graph where there is a consistent betterment of the hybrid formula (see figure 10(a) where the values of the C.CR ratio are smaller).

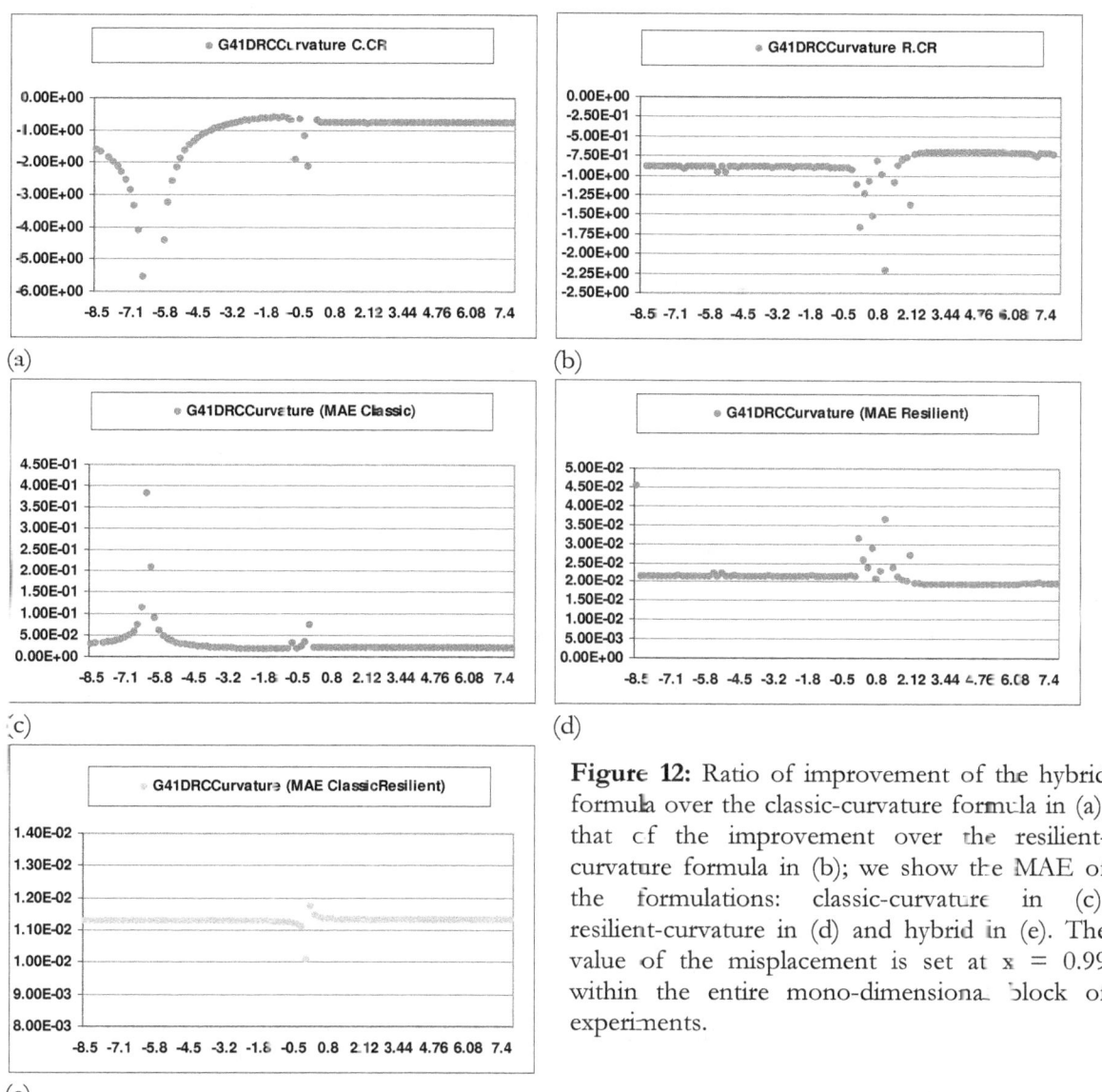

(a)

(b)

(c)

(d)

(e)

Figure 12: Ratio of improvement of the hybrid formula over the classic-curvature formula in (a), that of the improvement over the resilient-curvature formula in (b); we show the MAE of the formulations: classic-curvature in (c), resilient-curvature in (d) and hybrid in (e). The value of the misplacement is set at x = 0.99 within the entire mono-dimensional block of experiments.

This tells us that it is important to choose the parameter *'the a_const'* carefully because of its effect on the interpolation error, given the polynomials studied herein. This is visible in figures 10(c), and 10(e) where the three formulations show a sensitive fluctuation of the MAE while changing *'the a_const'*, although this happens to a relative extent and without an apparent and defined trend and/or behavior pattern, except that there is likely to be a sub-range of [-8.4554 E+00, 7.7342 E+00] inside which the MAE changes to a relatively greater extent than in the rest of the range.

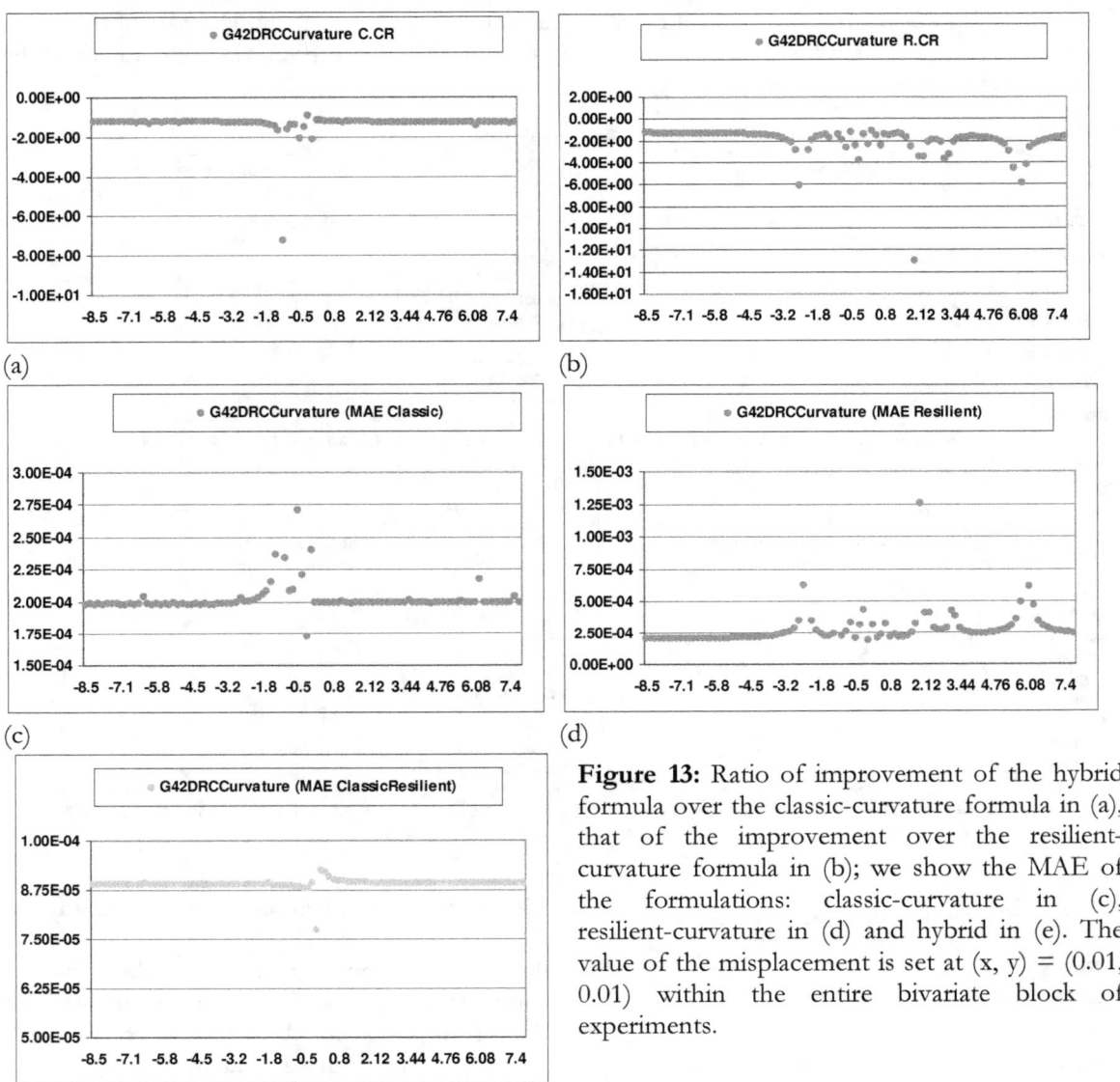

Figure 13: Ratio of improvement of the hybrid formula over the classic-curvature formula in (a), that of the improvement over the resilient-curvature formula in (b); we show the MAE of the formulations: classic-curvature in (c), resilient-curvature in (d) and hybrid in (e). The value of the misplacement is set at (x, y) = (0.01, 0.01) within the entire bivariate block of experiments.

When looking at figure 10(c), compared with figure 11(c), we may observe that the fluctuation of the ratio of improvement is higher in the case of the classic-curvature formula than it is in the case of the resilient-curvature formula, thus '*the_A_const*' makes the former more sensitive than the latter.

The behavior of the ratio of improvement of the hybrid formula over classic-curvature and resilient-curvature formulae observed in figure 10(c), 11(c), is not the same of the one behavior observable in figures 11(c) and 11(d), because the fluctuations of the ratio of improvement are in: approximately between the range [6.00 E-03, 3.00 E-02] (figure 11(c)) and approximately between the range [1.20 E-02, 3.60 E-02] (figure 11(d)). Since (3.00 E-02 - 6.00 E-03) = (3.60 E-02 - 1.20 E-02) = 0.024

the range of fluctuation of the MAE is the same, thus the sensitivity of both of classic-curvature and resilient-curvature formulae can be deemed the same.

The behavior of the ratio of improvement of the hybrid formula over classic-curvature and resilient-curvature formulae observed in figures 10(c) and 10(d), is similar to that one behavior observable in figures 12(c) and 12(d) respectively, where we can see that the MAE of the resilient formula is quasi constant (similarly to what seen in figure 10(d)) whereas the MAE of the classic formula has a larger fluctuation (similarly to what seen in figures 10(c)).

What seen in figures 10(c), 10(d) and 12(c), 12(d) simply means that the resilient-curvature formula is more robust than the classic-curvature formula (see the smaller fluctuations of the MAE the resilient-curvature formula when compared to the fluctuations of the MAE of the classic-curvature formula, with varying numerical values of 'the_A_const').

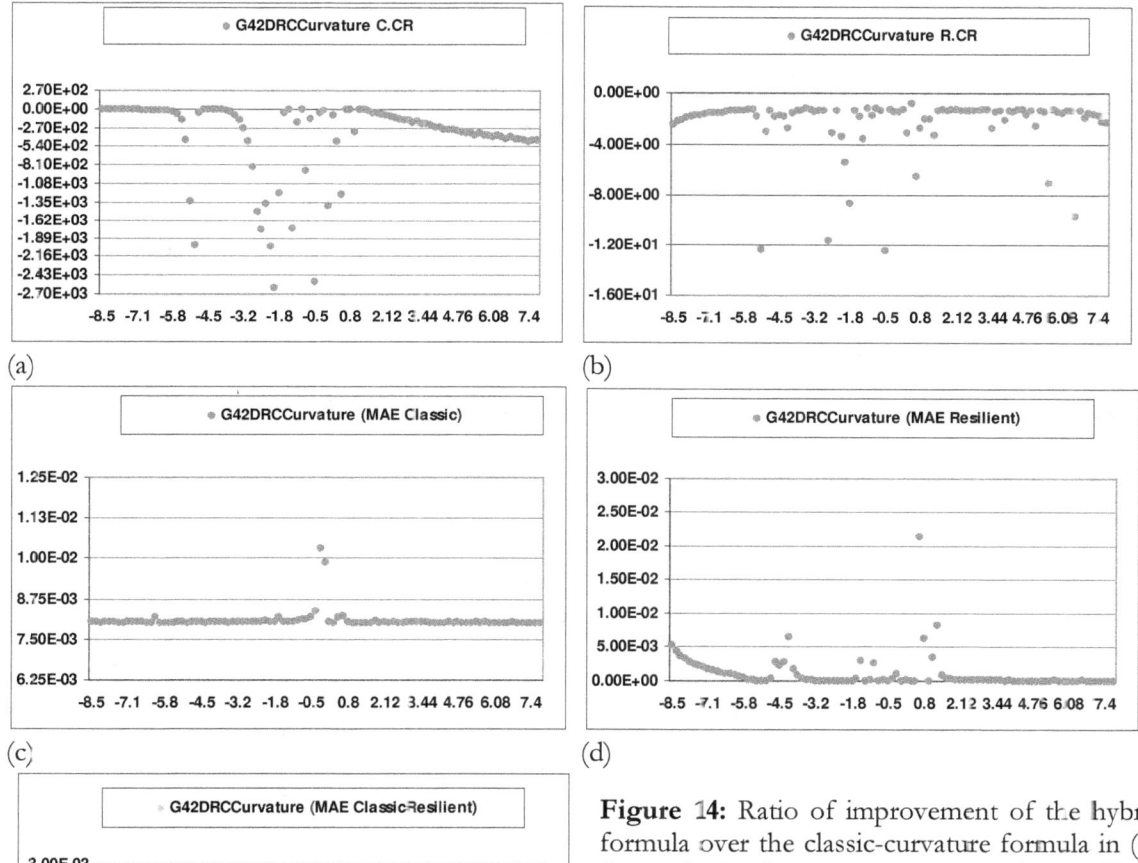

(a)

(b)

(c)

(d)

(e)

Figure 14: Ratio of improvement of the hybrid formula over the classic-curvature formula in (a), that of the improvement over the resilient-curvature formula in (b); we show the MAE of the formulations: classic-curvature in (c), resilient-curvature in (d) and hybrid in (e). The value of the misplacement is set at (x, y) = (0.45, 0.45) within the entire bivariate block of experiments.

SIGNAL RESILIENT TO INTERPOLATION

Looking at figures 10 through 14, the most robust resilient-curvature and classic-curvature signal reconstructions block of experiments is seen in figure 10(d) (range [1.90 E-04, 2.10 E-04]) and figure 13(c) (range [1.75 E-04, 2.75 E-04]) respectively.

This statement is to be treated with a dose of caution, however; therefore the inference to be drawn only regards the robustness of the resilient-curvature formula being more accentuated than the robustness of the classic-curvature formula, in relation to the block of experiments shown in figures 10 through 14. We may further conclude that there is likely to be, for each experimental session, a value or a set of values of *'the_A_const'*, which make the MAE particularly sensitive, regardless of the misplacement.

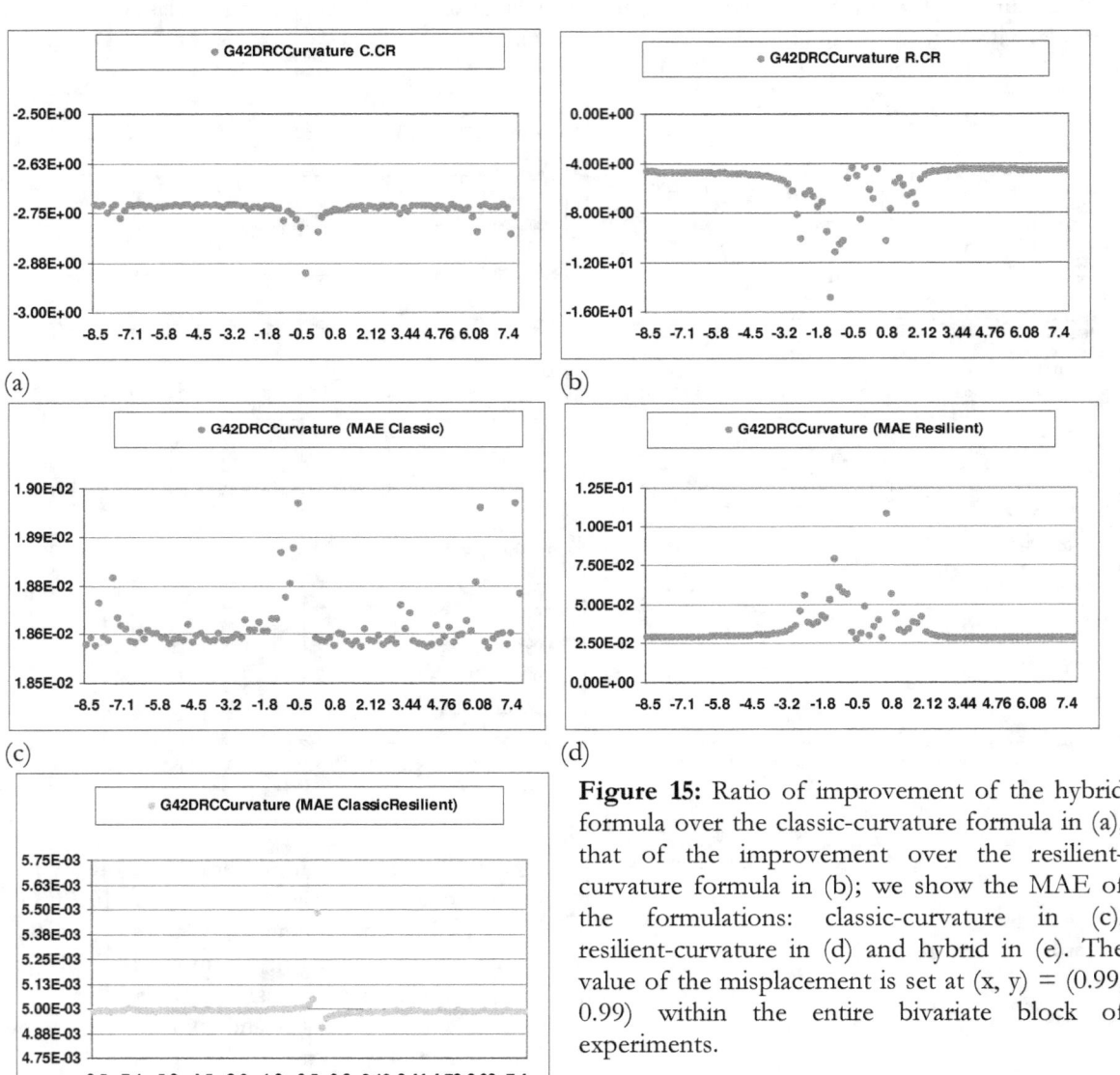

(a)

(b)

(c)

(d)

(e)

Figure 15: Ratio of improvement of the hybrid formula over the classic-curvature formula in (a), that of the improvement over the resilient-curvature formula in (b); we show the MAE of the formulations: classic-curvature in (c), resilient-curvature in (d) and hybrid in (e). The value of the misplacement is set at (x, y) = (0.99, 0.99) within the entire bivariate block of experiments.

Another observation which it is possible to make, on the basis of figure 13, regards the larger fluctuations in the value of the ratio of improvement compared to the value of the MAE (see (a), (b), (c) and (d)). This can be summarized by saying that at a given value of the parameter '*the_A_const*' values of the MAE in classic-curvature and resilient-curvature formulae differ substantially, and the overall behavior of the MAE shows fluctuations that are less pronounced than the fluctuations of the ratio of improvement. The aforementioned observation is generally supported by the values of the ratios of improvement C.CR and R.CR of the hybrid formula over classic-curvature and resilient-curvature formulae respectively.

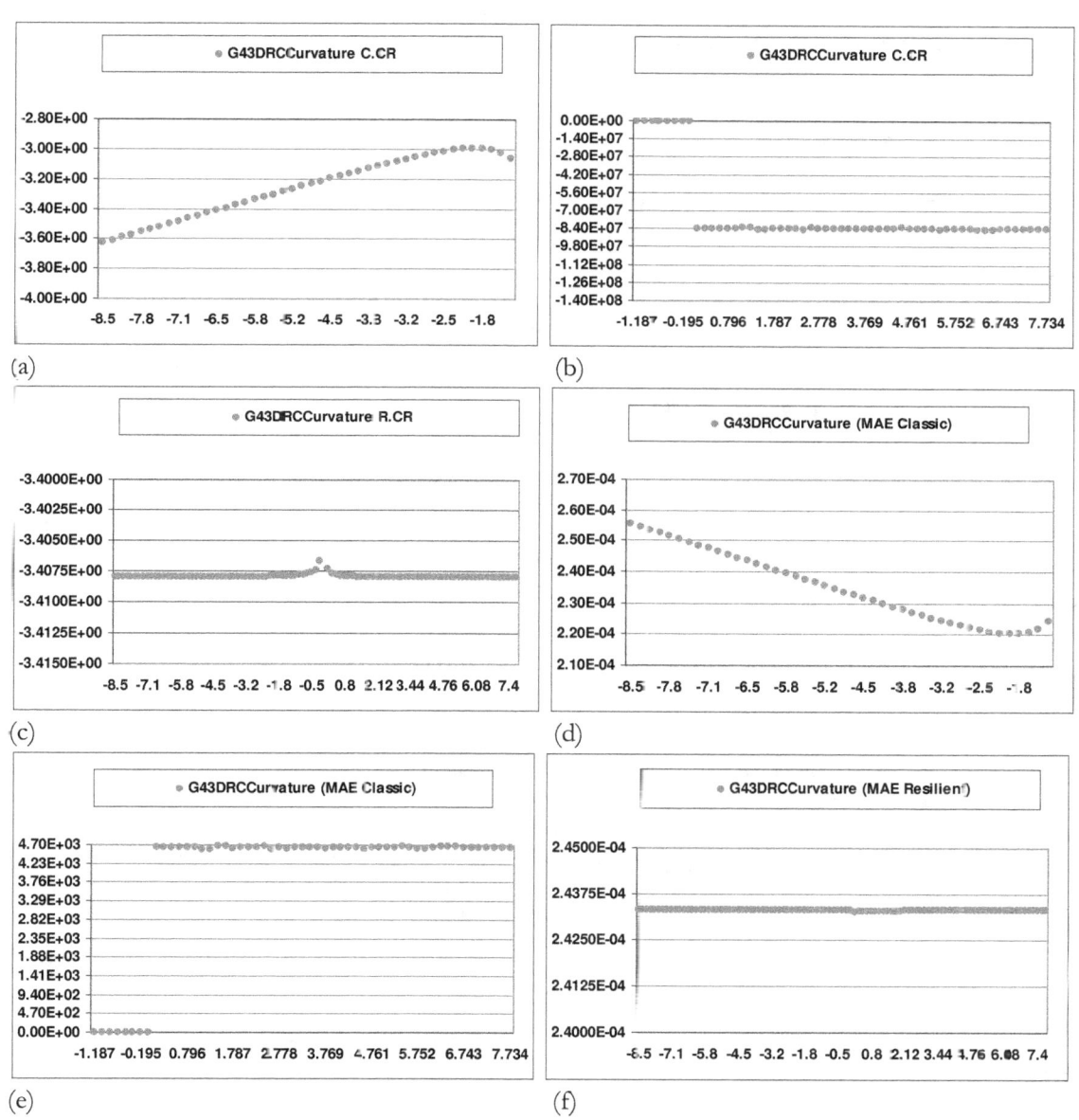

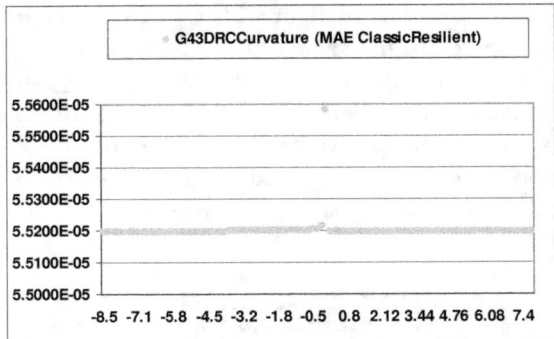

(g)

Figure 16: Ratio of improvement of the hybrid formula over the classic-curvature formula in (a) and (b), that of the improvement over the resilient-curvature formula in (c); we show the MAE of the formulations: classic-curvature in (d) and (e), resilient-curvature in (f) and hybrid in (g). The value of the misplacement is set at (x, y, z) = (0.01, 0.01, 0.01) within the entire trivariate block of experiments.

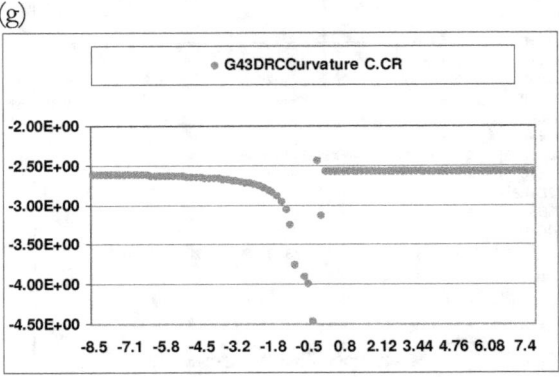

(a)

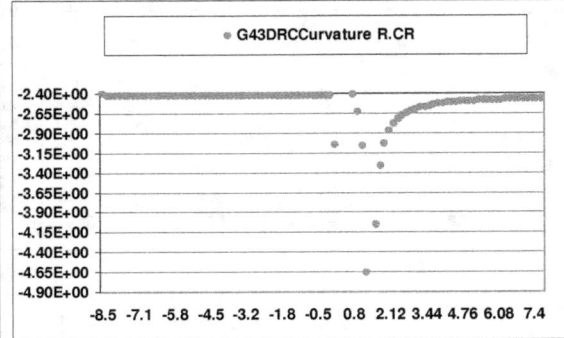

(b)

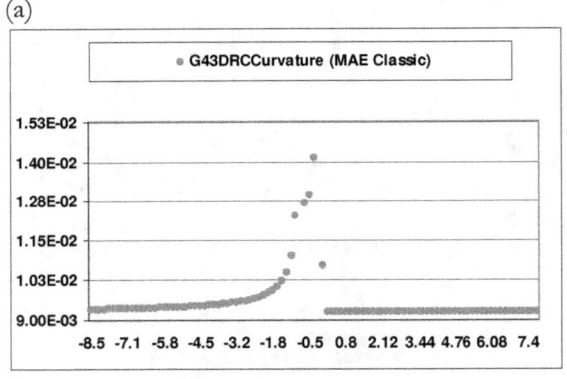

(c)

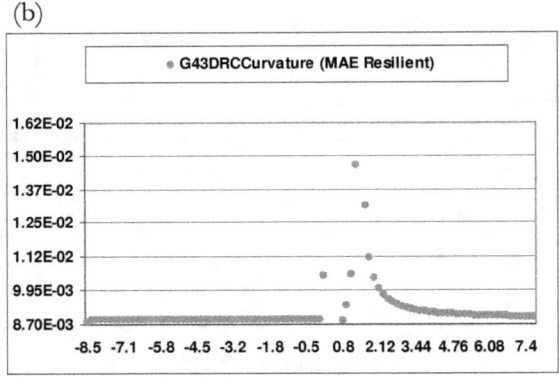

(d)

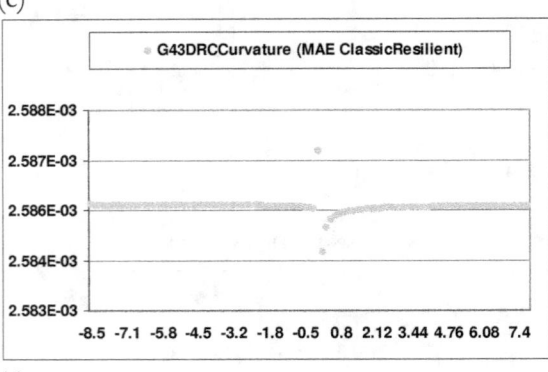

(e)

Figure 17: Ratio of improvement of the hybrid formula over the classic-curvature formula in (a), that of the improvement over the resilient-curvature formula in (b); we show the MAE of the formulations: classic-curvature in (c), resilient-curvature in (d) and hybrid in (e). The value of the misplacement is set at (x, y, z) = (0.45, 0.45, 0.45) within the entire trivariate block of experiments.

The case of figure 14 highlights, for the specific block of experiments with (x, y) = (0.45, 0.45) the superiority of the resilient-curvature formula over the classic-curvature formula and this is readily observable in (a), and (b), which show the improvement made by the hybrid formula compared to the classic-curvature formula, which is higher than that of the hybrid over the resilient-curvature formula; and in (c) and (d) where the MAE is clearly to the benefit of the resilient-curvature signal reconstruction. The aforementioned observation is not disproved, moreover, by the values of the ratios C.CR and R.CR seen in the overwhelming majority of the points in figures 14(a), 14(b), 14(c) and 14(d), which are in favor of the resilient-curvature formula.

The opposite can be said in figure 15 while comparing the MAE values of classic-curvature versus resilient-curvature formulae. We can see that while the former keeps MAE values in the range of approximately [1.85 E-02, 1.90 E-02] (see figure 15(c)), the latter shows oscillations between the value of approximately 2.50 E-02 and the value of approximately 5.00 E-02 with some scattered points above 5.00 E-02 (see figure 15(d)).

Furthermore, it is interesting to note the correlation existing between the profiles of figure 15(a) and figure 15(e) (see oscillation between the value of *the_A_const* of approximately -0.5258 and *the_A_const* of approximately 1.262, and see the peak of MAE value of 5.48 E-03 in figure 15(e) corresponding to the value of the *the_A_const* of -0.0302). Such correlation is within the context of the improvement that we may obtain to the benefit of the hybrid formula from the classic-curvature formula. To summarize, care should be taken in choosing the value of any constants in the parametric interpolation functions, as such values indirectly relate to the interpolation error.

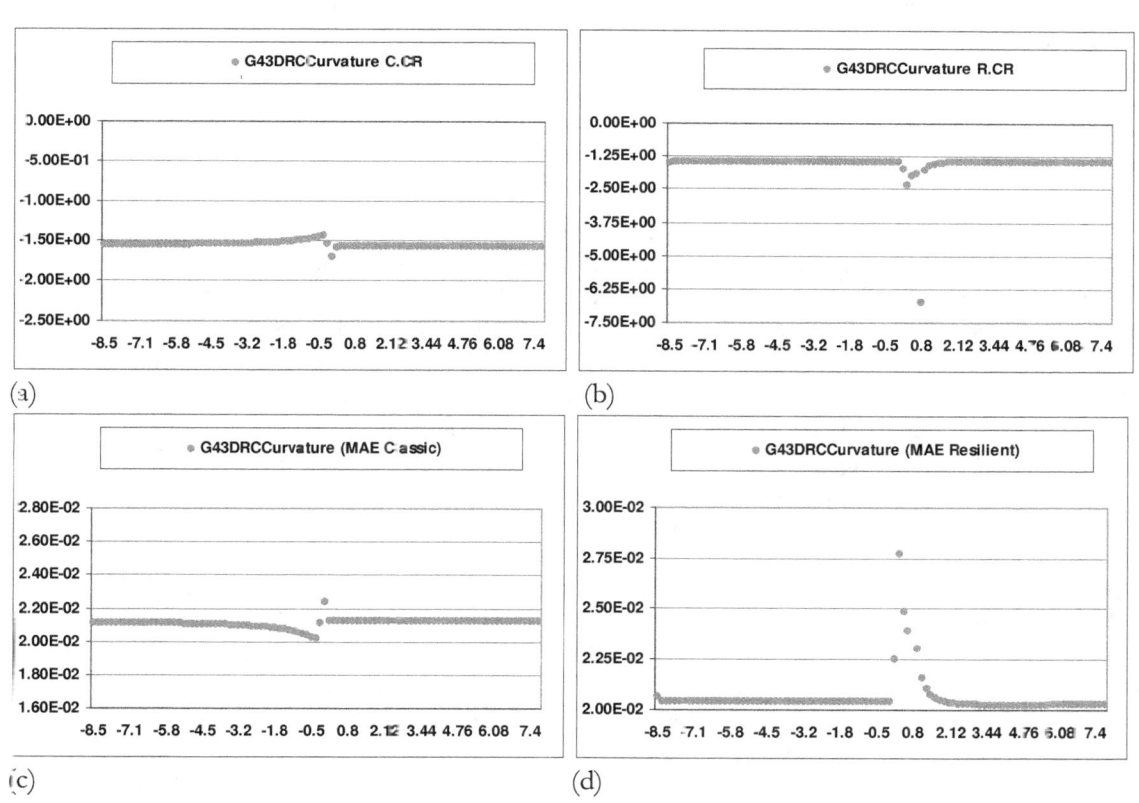

(a)

(b)

(c)

(d)

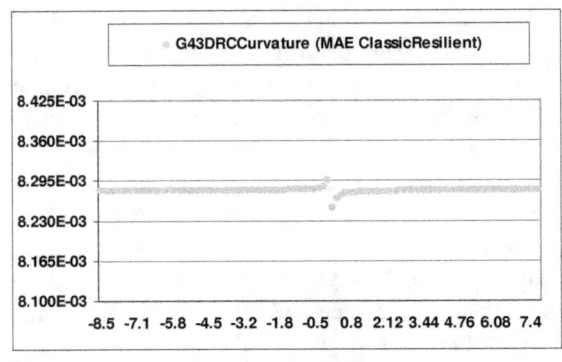

Figure 18: Ratio of improvement of the hybrid formula over the classic-curvature formula in (a), that of the improvement over the resilient-curvature formula in (b); we show the MAE of the formulations: classic-curvature in (c), resilient-curvature in (d) and hybrid in (e). The value of the misplacement is set at (x, y, z) = (0.99, 0.99, 0.99) within the entire trivariate block of experiments.

(e)

Figure 16 presents a case which deserves our immediate attention, which is drawn to the results in the graphs of the ratio of improvement of the hybrid over the classic-resilient formula in (b) and the MAE of the classic-resilient formula in (e). There is a specific value of the parameter *'the_A_const'* which shifts the trend drastically and this was found to be 0.135, which brings the MAE of the classic-resilient formula to high values (see figure 16(e)), making the signal reconstruction practically impossible. Emphasis must be placed on the care that should be taken in choosing the value to assign to any constants of parametric interpolation functions. In this specific case, the interpolation function is in equation (9), where ω_1 and ω_2 are sums of the pixels' intensity values in the chosen neighborhood, and the parameter 'a' is the parameter called *'the_A_const'*.

$$g_4(x, y, z) = f(0, 0, 0) + \omega_1 * a\,[(x + y + z)^3 + 1/2\,(x + y + z)^2 + \tfrac{1}{4}\,(x + y + z) + 1] + \omega_2 *$$

$$a\,[(x + y + z)^2 + 2\,(x + y + z) + 1] \tag{9}$$

When looking at the graphs in figure 16(a) and 16(c) we can see that the C.CR is between approximately 280 % and approximately 380 % and the R.CR is approximately 340 %, demonstrating that for this particular value of misplacement (x, y, z) = (0.01, 0.01, 0.01) the resilient-curvature formula is superior to the classic-curvature formula inside the range of the *'the_A_const'* of [-8.4554, -6.473], and this is also seen in the MAE of figure 16(d) and 16(f) respectively, where the values are between approximately 2.60 E-04 and approximately 2.20 E-04 (16(d)) and are around the quasi constant value of 2.437 E-04 (16(f)). Still, even with the imperfection caused because of the values of *'the_A_const'* above 0.135, the hybrid formula is functional (see quasi constant value 5.52 E-05 of the MAE of the hybrid formula in figure 16(g)) in signal reconstruction and this occurs because of the effectiveness of the resilient-curvature formula across the full range of values of the constant parameter.

As can be seen in figures 17(a) and 17(b) in this block of experiments with the value of the misplacement kept constant to (x, y, z) = (0.45, 0.45, 0.45), the resilient-curvature formula outperforms the classic-curvature formula, with both having a fluctuation inside the range of the parameter *'the_A_const'*, while the overwhelming majority of the points of classic-curvature formula have their C.CR between approximately above 250 % and the overwhelming majority of the points of the resilient-curvature formula have their R.CR between approximately above 240 % (the reader is recalled that negative values of the ratios C.RC and R.CR indicate improvement of the hybrid formula over classic-curvature and resilient-curvature formulae respectively (see section titled 'DEFINITION OF TECHNICAL TERMS')). Because of its capability of taking advantage of both classic-curvature and resilient curvature formulae, the MAE of the hybrid is quasi constant to the value of approximately 2.586 E-03 (see (e)) notwithstanding the fluctuations seen in (c) and (d).

In figure 18(c), which shows the graph of the MAE of the classic-curvature formula, there is an analogous situation, already seen in figure 16(e) in the MAE of the classic-curvature formula, with the difference being that signal reconstruction is still effective in figure 18, where the misplacement is (x, y, z) = (0.99, 0.99, 0.99), whereas it is not in figure 16(e), where the misplacement is (x, y, z) = (0.01, 0 01, 0.01). The value of 'the_A_const' which determines the drastic change in the profile of the MAE in 18(c) is -0.0302. In (a) and (b) we find the ratios C.CR and R.CR, which show that the resilient-curvature formula is almost totally superior to the classic-curvature formula (see quasi constant value of R.CR of approximately 125 % in (b)). The hybrid formula has a quasi constant value of the MAE of approximately 8.295 E-03 (see 18(e)). The fluctuation seen in figures 18(a) and 18(c) (classic-curvature formula), and figures 18(b) and 18(d) (resilient-curvature formula) where an oscillation in the behavior of the graphs of C.RC and R.CR (figures 18(a) and 18(b) respectively), and an oscillation in the behavior of the graphs of the MAE (figures 18(c) and 18(d) respectively), is determined, suggests that, overall, there are sub-ranges of the range 'the_A_const' inside the one employed for the experimentations ([-8.4554 E+00, 7.7342 E+00]) which have to be carefully considered when choosing the value of the constant parameter called 'the_A_const'.

SUMMARY

To derive the math form of the transfer function, the curvature and more specifically, the total curvature is a necessary tool and it is just as effective as it is important in producing the pixel intensity correction which finalizes the signal reconstruction. To this extent the curvature images are well-defined and useful for the purpose. The bivariate and trivariate polynomials studied here handle re-sampling quite well at the given misplacement, which is a little better than the mono-dimensional polynomial. Generally, care must be taken in choosing the value of any constant parameter in the math form of the interpolation function. There might be intervals of the value of the constant parameter which produce not quite well-defined behavior or worse, poor signal reconstruction; on the other hand, there might be values of the constant parameter which exalt the performance in signal reconstruction.

REFERENCES

Ciulla, C. (2009). *Improved Signal and Image Interpolation in Biomedical Applications: The Case of Magnetic Resonance Imaging (MRI)* – Medical Information Science Reference - IGI Global Publisher, Hershey, PA, U.S.A.

CONCLUSION

When addressing students in academic disciplines of applied mathematics, signal-image interpolation, signal-image processing, biomedical imaging - biomedical engineering, in Colleges and Universities, Carlo Ciulla writes of a unified theory for the improvement of the interpolation error as a continuing effort from the unifying theory reported in earlier works.

This research monograph has the following main theme. Given an interpolation function, which is supposed to determine an estimate of the unknown signal value, the reader can use the traditional approach (traditional interpolation function) to estimate the numerical value of the signal. Alternatively, the reader can follow the theoretical developments offered in the book and so design, on the basis of the theory described in the book, three new interpolation functions with improved approximation capabilities. That means, that under the unified theory, the book offers three new classes of interpolation functions with improved approximation capabilities of the true and unknown signal to estimate.

There are likely to be three main types of readership of this research monograph. The primary readership is composed of users of libraries which may adopt the book as reference. The secondary readership is composed of the population of instructors/professors of a course in the aforementioned academic disciplines. In such case, the book can be used as an additional educational resource to be available both to undergraduate and to graduate students, in order to assign homework and/or projects to be included in the coursework. The tertiary readership is composed of apprentices and/or passionate of math. In such case, the book would be used to employ time while following the desire of intellectual enrichment. The apprentices and/or passionate of math would study the methodology of the unified theory, would apply the unified theory such to design new interpolation functions, and should there be the desire of furthering the interest, the apprentices and/or passionate would proceed further to the analysis of the results, and possibly further into the discussion and the dissemination of the knowledge made out of this book.

More generally, the reader of this book, while provided with an interpolation function, through the math developments outlined in the book, can practice both of algebra and calculus. Algebraic equations are solved in the book along with the calculation of first and second order derivatives, in one, two and three dimensional variables.

Consistently, the research in this book explores the approximation properties of the mathematical functions, while providing readers with basic concepts of applied mathematics through the illustration of algebra and calculus. The book offers innovative methods aimed: to the generation of signal estimation, to the increase of students' potentiality in creativity, and to the demonstration of the methodology employed to validate the proposed math exploration. While offering the aforementioned, the book proposes the concept related to education in thinking, which is that of making creativity as a viable option which can be used to develop both of students' skills and students' performance in disciplines related to computing. The students and more generally the readers are also given the possibility to expand on the basis of the book contents and so to make new knowledge which goes beyond what is outlined in the book, and along the process of expansion, the students can realize how to make research.

The wording Signal Resilient to Interpolation (SRI) conveys the meaning 'bounce back' or 'to go up again'. The SRI is thus obtained from the signal which provides with the basis of its calculation. Thus, the wording Signal Resilient to Interpolation (SRI) is just another way to write Interpolation.

The book includes illustrations of the math processes, the logical reasoning, and the results (both of quantitative and qualitative nature), which show how to generate and obtain the Signal Resilient to Interpolation. Within the book, the SRI is calculated from the math of quadratic and cubic polynomials in one, two and three dimensions, in both of the cases of embedding and of not embedding the pixel (node in one dimension, voxel in three dimensions) to be re-sampled.

The following premise shall be given. The term convolution in this book is used to indicate the fraction of the estimate of the signal to be added to the numerical value of the signal such to obtain the interpolated value of the signal. The term transfer function is any math function employed to scale the fraction of the estimate of the signal.

The two classes of new interpolation functions called in the book classic-curvature and resilient-curvature interpolation functions are both composed of the sum of the numerical value of the signal and the intensity correction (in 1D), or the Pixel-Intensity-Correction (in 2D and 3D). Through the aforementioned sum, the interpolated value of the signal can be calculated in the form of the sum of the signal intensity value with the output of the transfer function (which is scaling the fraction of the estimate of the signal, at any intra-node/intra-pixel/intra-voxel sampling locations). In all of the three cases of 1D, 2D and 3D, the intensity-correction (so called IC in 1D), and the Pixel-Intensity-Correction (abbreviated as PIC, in both of 2D and 3D) assumes the role of customized transfer function (customized to the math form of the traditional interpolation function).

The customization entails that both of IC and PIC are calculated as the multiplication between the misplacement and the tangent (tan) of the total curvature of the new interpolation function (which is either the classic-curvature interpolation function or the resilient-curvature interpolation function). Therefore, the arbitration of the choice of the transfer function is conceptually and practically removed because of the customized transfer function. The customized transfer function is customized to the math form of the traditional interpolation function. The third class of new interpolation functions is the classic-resilient-curvature (hybrid) interpolation function, which is defined on a pixel-to-pixel basis as the one interpolation function between the classic-curvature and the resilient-curvature interpolation functions which result with the lowest Mean Absolute Error (MAE).

The scholarly value of this book mainly consists in the contribution of a special theory for the improvement of the interpolation error. The special theory for the improvement of the interpolation error, customizes the transfer function directly from the math form of the traditional interpolation function. Along such process, the special theory achieves one main objective. The objective is to avoid the bias introduced in the approximation properties of the interpolation function through an arbitrary choice of the transfer function, so to replace the transfer function with the customized transfer function obtained from the math form of the traditional interpolation function. Consequentially to the achievement of the objective is the removal of the arbitration of the math form of the transfer function.

The field of signal-image interpolation is in need of originality as much as is in need of innovation. Among the many methods proposed in literature to improve the interpolation error of the math functions, there is not a unified special theory which groups the math functions under the same methodological approach, like this book actualize.

There is not direct competition to the book that Carlo Ciulla has written. The reason is that, in classroom settings, the book might be chosen either as complement reading or as the textbook depending on the instructor/professor choice in relationship to the course requirements.

DEFINITION OF TECHNICAL TERMS

Terms Introduced Throughout the Book

Classic curvature: Sum of all of the second order derivatives of the interpolation function. The classic curvature is employed to calculate the *Pixel-Intensity Correction (PIC)* using the interpolation function.

Classic-curvature interpolation formula: The interpolation formula made of the sum of: the numerical value of the pixel to re-sample and the *Pixel-Intensity Correction (PIC)*. The Classic-curvature interpolation formula is calculated using the *classic-curvature* of the interpolation function.

C.RC ratio of improvement: Is the ratio indicating the relativity of performance between the *classic-resilient-curvature (hybrid) formula* and the *classic-curvature interpolation formula*. It is calculated as percentage value of the ratio between the MAE of the *classic-curvature interpolation formula* and the MAE of the *hybrid formula*, with the math expression: (1.0 - (MAE of the *classic-curvature interpolation formula* / MAE of the *hybrid formula*)).

Classic-resilient-curvature interpolation formula (hybrid formula): The formula which combines the two formulae (*classic-curvature* and *resilient-curvature*) depending on which one of the two gives lowest *Mean Absolute Error (MAE)*. The concept applies on a node-to-node basis (in 1D), and on a pixel-to-pixel basis (in 2D), and on a voxel-to-voxel basis (in 3D). The *hybrid formula* is employed to improve the performance of both *classic-curvature formula* and *resilient-curvature formula*.

Classic Signal: The signal in its original form as modeled through an interpolation function without any processing. The interpolation function is any of those treated in this book (e.g. bivariate quadratic B-Spline interpolation function, trivariate cubic B-Spline, etc.).

Intensity-curvature multiplication: In 1D is the term comprising of the multiplication between the node intensity ($f(0)$) and the second order derivative. In 2D and 3D, the intensity-curvature multiplication is the multiplication between the pixel intensity ($f(0, 0)$) and the sum of second order derivatives calculated respect to the dimensional variables. The *intensity-curvature multiplication* is employed to measure the combined information content of both pixel intensity and curvature of the interpolation function. This concept was initially introduced in: Ciulla & Deek (2005).

Intensity-curvature term before interpolation (E_0): Is the integral of the *intensity-curvature multiplication* calculated: inside the closed interval $[0, x]$ in 1D, inside the closed interval $[(0, 0), (x, y)]$ in 2D, inside the closed interval $[(0, 0, 0), (x, y, z)]$ in 3D. In 1D the *intensity-curvature multiplication* is calculated through the multiplication between the sampled intensity value and the second order derivative. In 2D and 3D, the *intensity-curvature multiplication* is calculated through the multiplication between the sampled intensity value and the sum of second order derivatives calculated respect to the dimensional variables. The second order derivatives are calculated at the node (in 1D) or grid point (in 2D or 3D). The intensity-curvature term before interpolation (Ciulla, (2009)) is employed to measure the energy level of the non-interpolated signal (for given interpolation function).

1D Example: $\int_{0}^{x} f(0) * (\partial (f(x)) / \partial x^2) (0) \, dx$.

2D Example: $\int\limits_{0}^{x}\int\limits_{0}^{y} f(0, 0) * [(\partial^2 (f(x, y)) /\partial x^2) (0, 0) + (\partial^2 (f(x, y)) /\partial y^2) (0, 0) +$
$(\partial^2 (f(x, y)) /\partial x\partial y) (0, 0) + (\partial^2 (f(x, y)) /\partial y\partial x) (0, 0)]$ dx dy.

3D Example: $\int\limits_{0}^{x}\int\limits_{0}^{y}\int\limits_{0}^{z} f(0, 0, 0) * [(\partial^2 (f(x, y, z)) /\partial x^2) (0, 0, 0) + (\partial^2 (f(x, y, z)) /\partial y^2) (0, 0, 0) +$
$(\partial^2 (f(x, y, z)) /\partial z^2) (0, 0, 0) + (\partial^2 (f(x, y, z)) /\partial x\partial y) (0, 0, 0) +$
$(\partial^2 (f(x, y, z)) /\partial y\partial x) (0, 0, 0) + (\partial^2 (f(x, y, z)) /\partial x\partial z) (0, 0, 0) +$
$(\partial^2 (f(x, y, z)) /\partial z\partial x) (0, 0, 0) + (\partial^2 (f(x, y, z)) /\partial y\partial z) (0, 0, 0) +$
$(\partial^2 (f(x, y, z)) /\partial z\partial y) (0, 0, 0)]$ dx dy dz.

A note to the reader is due. There is a difference between the intensity-curvature term before interpolation calculated in this book and the intensity-curvature term before interpolation reported in Ciulla, (2009). The difference is that in: Ciulla, (2009), the second order derivatives employed to calculate (E_o) are 3 in the covariates in 3D, are 2 in the covariates in 2D, and 1 in the univariate in 1D.

Intensity-curvature term after interpolation (E_{IN}): Is the integral of the *intensity-curvature multiplication* calculated: inside the closed interval [0, x] in 1D, inside the closed interval [(0, 0), (x, y)] in 2D, inside the closed interval [(0, 0, 0), (x, y, z)] in 3D. In 1D the *intensity-curvature multiplication* is calculated at the intra-nodal point x through the multiplication between the intensity value given by the interpolation function and the second order derivative. In 2D and 3D the *intensity-curvature multiplication* is calculated at the grid location (x, y) through the multiplication between the intensity value given by the interpolation function and the sum of second order derivatives respect to the dimensional variables. The second order derivatives are calculated at the intra-nodal (1D) or intra-pixel (2D), or intra-voxel (3D) point identified with x in 1D, identified with (x, y) in 2D, identified with (x, y z) in 3D. The intensity-curvature functional after interpolation (Ciulla, (2009)) is employed to measure the energy level of the interpolated signal (for given interpolation function).

1D Example: $\int\limits_{0}^{x} f(x) * (\partial (f(x)) /\partial x^2) (x)$ dx.

2D Example: $\int\limits_{0}^{x}\int\limits_{0}^{y} f(x, y) * [(\partial^2 (f(x, y)) /\partial x^2) (x, y) + (\partial^2 (f(x, y)) /\partial y^2) (x, y) +$
$(\partial^2 (f(x, y)) /\partial x\partial y) (x, y) + (\partial^2 (f(x, y)) /\partial y\partial x) (x, y)]$ dx dy.

3D Example: $\int\limits_{0}^{x}\int\limits_{0}^{y}\int\limits_{0}^{z} f(x, y, z) * [(\partial^2 (f(x, y, z)) /\partial x^2) (x, y, z) + (\partial^2 (f(x, y, z)) /\partial y^2) (x, y, z) +$
$(\partial^2 (f(x, y, z)) /\partial z^2) (x, y, z) + (\partial^2 (f(x, y, z)) /\partial x\partial y) (x, y, z) +$
$(\partial^2 (f(x, y, z)) /\partial y\partial x) (x, y, z) + (\partial^2 (f(x, y, z)) /\partial x\partial z) (x, y, z) +$
$(\partial^2 (f(x, y, z)) /\partial z\partial x) (x, y, z) + (\partial^2 (f(x, y, z)) /\partial y\partial z) (x, y, z) +$
$(\partial^2 (f(x, y, z)) /\partial z\partial y) (x, y, z)]$ dx dy dz.

A note to the reader is due. There is a difference between the intensity-curvature term after interpolation calculated in this book and the intensity-curvature term before interpolation reported in Ciulla, (2009). The difference is that in: Ciulla, (2009), the second order derivatives employed to calculate (E_o) are 3 in the covariates in 3D, are 2 in the covariates in 2D, and 1 in the univariate in 1D.

Intensity-Curvature Functional: The ratio between two *intensity-curvature* terms before (E_o) and after (E_{IN}) interpolation, which is: E_o/E_{IN}. The *Intensity-Curvature Functional* (Ciulla, 2009) is employed to measure the energy level change determined through the interpolation function.

DEFINITION OF TECHNICAL TERMS

Mean Absolute Error (MAE): The absolute value of the difference between the value of the signal and the interpolated signal. The *MAE* can be calculated at a given node (in 1D), or pixel (in 2D), or voxel (in 3D). The *MAE* can be calculated for the *classic-curvature* interpolation formula, the *resilient-curvature* interpolation formula, and the *classic-resilient-curvature interpolation formula (hybrid formula)*. The MAE is employed to seek for energy preservation (quest for minimal energy state).

Pixel-Intensity Correction (PIC): The fraction of pixel intensity employed to calculate the signal at a given location not sampled through the sampling device. The *PIC* is computed through the product between the misplacement and the trigonometric tangent function of the total curvature (which is defined as the sum of all of second order derivatives) of the interpolation function. The *PIC* can be calculated using either *classic curvature* or *resilient curvature*. The Pixel-Intensity-Correction (PIC) acts as the transfer function obtained from the math form of the interpolation function.

Ratio of improvement: Generally, *(i)* the more negative the *ratio of improvement* is, the better is the improvement of one interpolation formula over the another interpolation formula; or *(ii)* the more negative the ratio of improvement is, the more effective the gain in merging one interpolation formula (for instance the *classic-curvature interpolation formula*) with another interpolation formula (for instance the *resilient-curvature interpolation formula*) on a pixel-by-pixel basis is, and the higher the gain of the resulting interpolation formula which transforms itself into the resulting interpolation formula (for instance the *classic-resilient-curvature interpolation formula* (or so called *hybrid formula*)).

(i) For instance negative values of the *ratio of improvement* are indicative of superiority in signal reconstruction of *resilient interpolation* over classic interpolation (Newton & Huygens, trans. 1934), whereas positive values of the ratio of improvement are indicative of superiority in signal reconstruction of classic interpolation over *resilient interpolation*. The *ratio of improvement* is calculated with the math expression: (1.0 - (MAE of classic interpolation / MAE of *resilient interpolation*)) when measuring the relativity of performance between classic interpolation and *resilient interpolation*.

(ii) For instance merging the *classic-curvature interpolation formula* and the *resilient-curvature interpolation formula* on a pixel-by-pixel basis entails a gain visible through the *classic-resilient-curvature interpolation formula* (or so called *hybrid formula*). Such gain can be measured through the *C.RC ratio of improvement* with the math expression: (1.0 - (MAE of the *classic-curvature interpolation formula* / MAE of the *hybrid formula*)), or through the *R.CR ratio of improvement* with the math expression: (1.0 - (MAE of the *resilient-curvature interpolation formula* / MAE of the *hybrid formula*)).

R.CR ratio of improvement: Is the ratio indicating the relativity of performance between the *classic-resilient-curvature (hybrid) formula* and the *resilient-curvature interpolation formula*. It is calculated as percentage value of the ratio between the MAE of the *resilient-curvature interpolation formula* and the MAE of the *hybrid formula*, with the math expression: (1.0 - (MAE of the *resilient-curvature interpolation formula* / MAE of the *hybrid formula*)).

Resilient curvature: Sum of all of the second order derivatives of the *Signal Resilient to Interpolation (SRI)* derived from the interpolation function. The resilient curvature is employed to calculate the *Pixel-Intensity Correction (PIC)* using the *Signal Resilient to Interpolation (SRI)* derived from the interpolation function.

Resilient-curvature interpolation formula: The interpolation formula made of the sum of: the numerical value of the pixel to re-sample and the *Pixel-Intensity Correction (PIC)*. The Resilient-curvature interpolation formula is calculated using the *resilient-curvature* of the interpolation function.

Resilient interpolation: The concept *resilient interpolation* is closely linked to that of the equation between the intensity curvature terms before (no interpolation) and after interpolation (see chapter 1). The term *resilient interpolation* is also used in the book as acronym of *Resilient-curvature interpolation formula*. The term *resilient interpolation* is used in chapter 4, whereas the term *Resilient-curvature interpolation formula* is used in chapters 8 and 12. Both of the terms are related to the formula used to re-sample the signal, which is: the sum of the numerical value of the pixel to re-sample and the *Pixel-Intensity Correction (PIC)*. Whereas in chapter 4 and 8 the aforementioned formula is used in 2D, in chapter 12 the formula is used in 1D, 2D and 3D.

Signal Resilient to Interpolation (SRI): The signal obtained through the solution of the equation of the two intensity-curvature terms before *(E$_o$)* and after *(E$_{IN}$)* interpolation. The *SRI* is derived univocally on the basis of a given interpolation function. The *SRI* is employed to obtain the *Resilient-curvature interpolation formula*.

Total curvature of an interpolation function: Is defined to be the sum of second order derivatives of the math form of the interpolation function respect to the dimensional variables.
1D Example: $(\partial (f(x)) / \partial x^2) (x)$.
2D Example: $[(\partial^2 (f(x, y)) / \partial x^2) (x, y) + (\partial^2 (f(x, y)) / \partial y^2) (x, y) + (\partial^2 (f(x, y)) / \partial x \partial y) (x, y) + (\partial^2 (f(x, y)) / \partial y \partial x) (x, y)]$.
3D Example: $[(\partial^2 (f(x, y, z)) / \partial x^2) (x, y, z) + (\partial^2 (f(x, y, z)) / \partial y^2) (x, y, z) + (\partial^2 (f(x, y, z)) / \partial z^2) (x, y, z) + (\partial^2 (f(x, y, z)) / \partial x \partial y) (x, y, z) + (\partial^2 (f(x, y, z)) / \partial y \partial x) (x, y, z) + (\partial^2 (f(x, y, z)) / \partial x \partial z) (x, y, z) + (\partial^2 (f(x, y, z)) / \partial z \partial x) (x, y, z) + (\partial^2 (f(x, y, z)) / \partial y \partial z) (x, y, z) + (\partial^2 (f(x, y, z)) / \partial z \partial y) (x, y, z)]$.

Total curvature of the Signal Resilient to Interpolation: Is defined to be the sum of second order derivatives of the math form of the Signal Resilient to Interpolation (SRI) respect to the dimensional variables.
1D Example: $(\partial (SRI(x)) / \partial x^2) (x)$.
2D Example: $[(\partial^2 (SRI(x, y)) / \partial x^2) (x, y) + (\partial^2 (SRI(x, y)) / \partial y^2) (x, y) + (\partial^2 (SRI(x, y)) / \partial x \partial y) (x, y) + (\partial^2 (SRI(x, y)) / \partial y \partial x) (x, y)]$.
3D Example: $[(\partial^2 (SRI(x, y, z)) / \partial x^2) (x, y, z) + (\partial^2 (SRI(x, y, z)) / \partial y^2) (x, y, z) + (\partial^2 (SRI(x, y, z)) / \partial z^2) (x, y, z) + (\partial^2 (SRI(x, y, z)) / \partial x \partial y) (x, y, z) + (\partial^2 (SRI(x, y, z)) / \partial y \partial x) (x, y, z) + (\partial^2 (SRI(x, y, z)) / \partial x \partial z) (x, y, z) + (\partial^2 (SRI(x, y, z)) / \partial z \partial x) (x, y, z) + (\partial^2 (SRI(x, y, z)) / \partial y \partial z) (x, y, z) + (\partial^2 (SRI(x, y, z)) / \partial z \partial y) (x, y, z)]$.

Virtual shift-rotation: In 1D, a signal is virtually shifted-rotated when the sampling location x is not moved and the signal is simply re-calculated at the location x. In 2D, a signal is virtually shift-rotated when the grid sampling location (x, y) is not moved and the signal is simply re-calculated at (x, y). Similarly, in 3D, a signal is virtually shift-rotated when the 3D grid sampling location (x, y, z) is not moved and the signal is simply re-calculated at (x, y, z). Re-calculation is made through an interpolation formula. An implication of the virtual shift-rotation is that since the grid (node in 1D) is not shifted-rotated, interpolation alone does not suffice to produce any real shift-rotations.

With embedding: In 1D is the act of embedding (including) the signal intensity value located at the center of the neighborhood (see common term section) into the math form of the function, which is usually the interpolation function. The concept is the same in 2D and 3D except that the signal intensity values are located in a 2D neighborhood grid or a 3D neighborhood grid.

Without embedding (WE): In 1D is the act of not embedding (excluding) the signal intensity value located at the center of the neighborhood (see common term section) from the math form of the function, which is usually the interpolation function. The concept is the same in 2D and 3D except that the signal intensity values in a 2D neighborhood grid or a 3D neighborhood grid.

Common terms

Analog: Discontinuous, discrete (Boyer, (1968)).

Bivariate: Math expression and/or formula in two independent variables (Amidror, 2002).

B-Spline: Math function made of convolutions of: signal intensity values with polynomial forms. The polynomial forms are usually curved accordingly with the exponents of the independent variables. The B-Splines are interpolation functions (De Boor, (1978); Unser, et al. 1993a).

Curvature of the interpolation function: Length of an arc. Conveys the meaning as to how curved is the shape of the interpolation function. The math expressing the curvature is the second order derivative of the interpolation function. Geometrically, the curvature of the signal is strictly related to the angle formed in between the horizontal and the tangent to the first order derivative curve of the signal (Stewart, (2009)).

Differential: The differential (Gauthier-Villars, (1882); Weber, (1876, 1953); Weierstrass, (1894/1927)) is a local property of the mathematical function (curve) defined as the linear increment of the mathematical function $y = f(x)$, versus the increment ($\Delta x = dx$) of the independent variable x. Example: $dy = (\partial\ f(x)\ /\ \partial x) * dx$.

Differentiability: The property of a math function which consists in that of admitting, in its interval of definition, continuity and existence of non null second order derivatives (Gauthier-Villars, (1882); Weber, (1876, 1953); Weierstrass, (1894/1927)).

Discrete sample: The discontinuous signal value sampled through the sampling device. The discrete sample (Newton & Huygens, trans. 1934) is not any continuous signal. A continuous signal can be any math functions defined into any intervals of the domain of the Real numbers, with the Real numbers acting as independent variable (Nyquist, 2002).

Down-sampling: Act upon the original signal aimed to decrease the number of sampled nodes (1D), or the sampled matrix size (in 2D and 3D) (Haacke et. al., (1999)).

First Order Derivative: Is a math function (f ') calculated through the operation known in calculus as derivation. The derivation is calculated on a math function (another one, for instance: f). The first order derivative conveys the information as to if the math function is increasing or decreasing when plotted against the independent variable. The geometrical meaning of the first order derivative is that one of the tangent to the math function. The first order derivative is a local property of the math functions (Gauthier-Villars, (1882); Weber, (1876, 1953); Weierstrass, (1894/1927)). Example: $f' = (\partial\,(\,f(x)\,)\,/\partial x)$.

First Order Partial Derivative. Is a math function (f_x') calculated through the derivation respect to one of the multiple independent variables. The first order partial derivative (Gauthier-Villars, (1882); Weber, (1876, 1953); Weierstrass, (1894/1927)) is a local property of the mathematical functions. Example: $f_x' = (\partial\,(\,f(x, y)\,)\,/\partial x)$.

Interpolation function. Mathematical function employed as model in order to fit the sampled pixel intensities and so to determine the estimation (interpolated value) of the intensity value at a spatial or temporal location where the signal has not been sampled (Whittaker, 1915; Whittaker, 1935).

Interpolator. Abbreviation of the term interpolation function (Lehmann, et. al. 1999).

Interpolation error. Because of the true and unknown value of the signal to approximate is not easy to define the interpolation error. The reader is referred to works of: Schoenberg, 1946a; Schoenberg, 1946b; Schoenberg, 1969; Strang & Fix, 1973; which relate within the context of the approximation theory, the true and unknown value of the signal to approximate, to the estimate of the signal.

Lagrange interpolation function. Math function made of convolutions of: signal intensity values with polynomial forms which include terms having all of the exponents (Whittaker, 1915; Whittaker, (1935)). For instance, if the maximum exponent of the monovariate Lagrange interpolation function is 3, the polynomial forms will be of type $ax^3 + bx^2 + cx + d$.

Misplacement (Shift): In 1D is the distance measured from the origin of the node to any intra-nodal point x (Newton & Huygens, trans. 1934; Blu et al. 2004). In 2D the misplacement is the distance measured from the origin of the node to any intra-pixel point (x, y). In 3D the misplacement is the distance measured from the origin of the node to any intra-voxel point (x, y, z).

Monovariate (Univariate): Math expression and/or formula in one single independent variable (Blu et al. 2004).

Neighborhood. Concept with applicability (Lehmann, et. al., (1999)) in 1D, 2D and 3D. In 1D is the set of intensity values with center into the node which is object of estimation through the interpolation function. In 2D and 3D the concept is the same except that the node is called pixel (2D) or voxel (3D).

Order (degree) of the polynomial function: The order (degree) of the interpolation function is the maximum exponent of the formula (Meijering, 2002).

Sampling location. Spatial or temporal location (Newton & Huygens, trans. 1934) where the signal is sampled (acquired through the sampling device).

Second Order Derivative. Is a math function (f'') calculated through two iterations of the calculus operation called derivation. The derivation is calculated on a math function (another one, for instance: f). The second order derivative conveys the information as to if the math function is concave or convex when plotted against the independent variable. The geometrical meaning of the second order derivative is that one of the tangent to the first order derivative of the math function. The second order derivative (Gauthier-Villars, (1882); Weber, (1876, 1953); Weierstrass, (1894/1927)) is a local property of the mathematical functions. Example: $f'' = (\partial^2\,(\,f(x)\,)\,/\partial x^2)$.

Second Order Partial Derivative: Is a math function (f_{xy}'') calculated through two iterations of the calculus operation called derivation. The derivation is calculated on a math function (another one). The derivation (Gauthier-Villars, (1882); Weber, (1876, 1953); Weierstrass, (1894/1927)) is calculated respect to more than one of the multiple independent variables, and it is a local property of the mathematical functions. Example: $f_{xy}'' = (\partial^2\,(\,f(x,y)\,)\,/\partial x \partial y)$.

Transfer function: Math function employed to process (scale) the convolution of: pixel intensities with the polynomial forms of the interpolation function. The transfer function determines the intensity correction to apply to the signal to obtain the signal estimate at locations not sampled through the sampling device. The transfer function can also be used to calculate the B-Spline (or more generally polynomials) coefficients (Unser et al. 1993a, 1993b; Ramani et al. 2010).

Trivariate: Math expression and/or formula in three independent variables (Grevera & Udupa, 1996, 1998).

Up-sampling: Act upon the original signal aimed to increasing the number of sampled nodes (1D), or the sampled matrix size (in 2D and 3D) (Haacke et. al. (1999)).

REFERENCES

Amidror, I. (2002). *Scattered Data Interpolation Methods for Electronic Imaging Systems: A Survey*. J Electr Imag, 11(2), 157-176.

Blu, T., Thevenaz, P. & Unser, M. (2004). *Linear interpolation revitalized*. IEEE Transactions on Image Processing, 13(5), 710-719.

Boyer, C. B. (1968). *A History of Mathematics*. New York, NY: John Wiley & Sons, U.S.A.

Gauthier-Villars, (1882). *Cauchy, Augustin-Louis, Oeuvres completes*, (reissued by Cambridge University Press, 2009).

Ciulla C. & Deek, F. P. (2005). *On the approximate nature of the bivariate linear interpolation function: A novel scheme based on intensity-curvature*. ICGST - International Journal on Graphics, Vision and Image Processing, 5(7), 9-19.

Ciulla, C. (2009). *Improved Signal and Image Interpolation in Biomedical Applications: The Case of Magnetic Resonance Imaging (MRI)* – Medical Information Science Reference - IGI Global Publisher, Hershey, PA, U.S.A.

De Boor, C. (1978). *A practical guide to splines. Applied mathematical sciences*. Springer-Verlag, U.S.A.

Grevera, G. J. & Udupa, J. K. (1996). *Shape-based interpolation of multidimensional grey-level images*. IEEE Transactions on Medical Imaging, 15(6), 881-892.

Grevera, G. J. & Udupa, J. K. (1998). *An objective comparison of 3-D image interpolation methods*. IEEE Transactions on Medical Imaging, 17(4), 642-652.

Haacke, E. M., R. W. Brown, M. R. Thompson & R. Venkatesan. (1999) *Magnetic Resonance Imaging: Physical Principles and Sequence Design*. New York, NY: John Wiley & Sons, Inc., U.S.A.

Lehmann, T. M., Gonner, C., & Spitzer, K. (1999). *Survey: Interpolation methods in medical image processing*. IEEE Transactions on Medical Imaging, 18(11), 1049-1075.

Meijering, E. (2002). *A chronology of interpolation: From ancient astronomy to modern signal and image processing.* Proceedings of the IEEE, 90(3), 319-342.

Newton, I. & Huygens, C (1934). *The motion of the moon's nodes.* In R. M. Aynard Hutchins (Ed.), Mathematical Principles of Natural Philosophy: Optics, Treatise on light (A. Motte, Trans.). (pp. 338-339). William Benton., U.S.A.

Nyquist, H. (2002). *Certain topics in telegraph transmission theory.* Proceedings of the IEEE, 90(2), 280-305.

Ramani, S., Thevenaz, P. & Unser, M. (2010). *Regularized Interpolation for Noisy Images.* IEEE Transactions on Medical Imaging, 29(2), 543-558.

Schoenberg, I. J. (1946a). *Contributions to the problem of approximation of equidistant data by analytic functions.* Part A. On the problem of smoothing or graduation. A first class of analytic approximation formulae. Quarterly of Applied Mathematics, 4, 45-99.

Schoenberg, I. J. (1946b). *Contributions to the problem of approximation of equidistant data by analytic functions.* Part A. On the problem of osculatory interpolation. A second class of analytic approximation formulae. Quarterly of Applied Mathematics, 4, 112–141.

Schoenberg, I. J. (1969). *Cardinal interpolation and spline functions.* Journal of Approximation Theory 2, 167-206.

Stewart, J. (2009). *Calculus Imperial.* Thomson Brooks/Cole.

Strang, G., & Fix, G. (1971). *A Fourier analysis of the finite element variational method.* In, Constructive Aspect of Functional Analysis (pp. 796-830). Rome, Italy: Edizioni Cremonese.

Unser, M., Aldroubi, A., & Eden, M. (1993a). *B-spline signal processing: Part I – theory.* IEEE Transactions on Signal Processing, 41(2), 821-833.

Unser, M., Aldroubi, A., & Eden, M. (1993b). *B-spline signal processing: Part II – efficient design and applications.* IEEE Transactions on Signal Processing, 41(2), 834-848.

Weber, H. (1876). *Gesammelte mathematische Werke und wissenschaftlicher Nachlass.* (Bernard Riemann Works), Leipzig, Germany.

Weber, H. (1953). *The Collected Works.* H. Weber, Ed., assisted by R. Dedekind, with supp. by M. Noether and Wirtinger and a new introduction by Hans Lewy. New York, NY, U.S.A.

Weierstrass, K. (1894/1927). *Mathematische Werke.* Vol. 7., Mayer & Muller, Berlin, Germany. English translation: On continuous functions of a real argument that do not possess a well-defined derivative for any value of their argument, in: G.A. Edgar, Classics on Fractals, Addison-Wesley Publishing Company, 1993, pp 3-9, U.S.A.

Whittaker, E. T. (1915). *On the functions which are represented by the expansions of interpolation theory.* Proceedings of the Royal Society Edinburgh, UK, Sec. A(35), 181-194.

Whittaker, J. M. (1935). *Interpolatory function theory* (Cambridge Tracts in Mathematics and Mathematical Physics No. 33). New York, NY: Cambridge University Press, U.S.A.

QUESTIONS AND ANSWERS

Q & A. 1; Why the Signal Resilient to Interpolation is called in such way?
The wording Signal Resilient to Interpolation (SRI) does not want to convey the meaning that there exists a signal which is resilient (resistant) to the math functions that determine the approximation through interpolation. The word resilient ought to be such to convey the meaning 'bounce back' or 'to go up again', and this signifies that the SRI is derived from the signal from which is calculated. Thus, the wording Signal Resilient to Interpolation (SRI) is just another way to say interpolation.

Q & A. 2; Is the Signal Resilient to Interpolation any signal?
No. The following description is provided in the one-dimensional case. The Signal Resilient to Interpolation (SRI) is derived at any intra-node location from the solution of the equation $E_0(x) = E_{IN}(x)$ (which poses the equality between the intensity-curvature term before interpolation and the intensity-curvature term after interpolation), such to explicit the value $f(0)$ from $f(x_i) = f(0)$ and $f(x_k) = f(0) + \xi(\Omega\, x_k)$. The value of the $f(0)$ is the value of the SRI (thus changing at each node) and $\xi(\Omega\, x)$ (with $x = x_i$ and $x = x_k$) is the convolution of the neighborhood pixel intensities with the polynomial form of the model function. The SRI is given by $\int [f(0)] * [(\partial^2 f(x)/\partial\, x^2)]_{(x\,=\,xi)}\, dx = \int [f(0) + \xi(\Omega\, x_k)] * [(\partial^2 f(x)/\partial\, x^2)]_{(x\,=\,xk)}\, dx$, which yields $f(0) = \{ \int [\xi(\Omega\, x_k)] * [(\partial^2 f(x)/\partial\, x^2)]_{(x\,=\,xk)}\, dx \} / \{ \int [(\partial^2 f(x)/\partial\, x^2)]_{(x\,=\,xi)}\, dx - \int [(\partial^2 f(x)/\partial\, x^2)]_{(x\,=\,xk)}\, dx \}$.

Q & A. 3; Why the Signal Resilient to Interpolation is derived?
As the empirics show, the SRI can be used to improve the approximation capabilities of the math functions.

Q & A. 4; Why is the Mean Absolute Error (MAE) used to quantify the interpolation error?
The answer is provided in the one-dimensional (1D) case. The search for the MAE is coherent with the search for flat curvature in between nodes. The following figures 1(a), 1(b) and 1(c) give the idea and the congruency of the idea with the search for flat curvature. The curve f ' is the first order derivative of the signal curve. The arc tangent of the angles θ_1 and θ_2 is the curvature of the signal curve. The search for the angles θ_1 and θ_2 correlates with minimal change of the MAE.

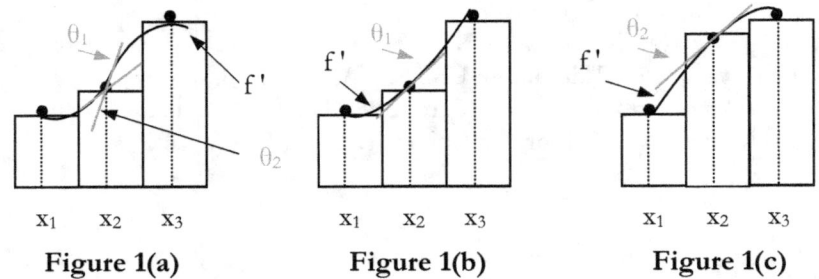

| Figure 1(a) | Figure 1(b) | Figure 1(c) |

Q & A. 5; How the unifying theory relates to the unified theory?
The unifying theory reported in the year 2009 has been focused in this book to visualize (to see) the unified theory. The tree of knowledge shown in figure 1(a) of chapter 1 renders the answer: *the unifying theory yields the unified theory*; in the following terms. The unifying theory has been derived from scattered knowledge through deduction. The two intensity-curvature terms (before *(E_o)* and after *(E_{IN})* interpolation) consists of the special theory, which is the set of formulae of the unifying theory resulting from focusing (i.e. assessing the specifics). It is possible through empirics to induce the unified theory.

ACKNOWLEDGEMENTS

This book is dedicated to my son Roman Alexander Ciulla. I am in debt with my parents Angela Parlavecchio, Pietro Ciulla and my aunt Mariella Parlavecchio for their constant support during the time spent in this exploratory research. The author is also grateful because of the invaluable suggestions provided from Dr. Toru Kumagai at the National Institute of Advanced Industrial Science and Technology (AIST), Tsukuba, Japan, in order to improve important aspects of the book which are relevant to the definition of the technical terms, the review of the literature and the placement of these works within the context of the current signal-image processing literature. I am also thankful to Charlie Smith, the reviewer whom contributed to the improvement of the language of the manuscript.

CARLO CIULLA had been undergraduate and graduate student at the University of Palermo, Italy, RUTGERS University, U.S.A. and the New Jersey Institute of Technology, U.S.A. from the year 1987 to the year 2002. He has earned the following graduate degrees: Laurea in Management Engineering (Italy); an M.S. in Information Systems and a Ph.D. in Computer and Information Science (U.S.A.). Carlo was pre-doctoral student at the National Institute of Bioscience and Human Technology (NIBH) in Tsukuba, Japan (1995-1997) and he worked with Magnetoencephalography (MEG) studying the spontaneous alpha rhythm of the human brain. Following the completion of the Doctoral degree, Carlo's former academic appointments were: Research Associate at Yale University (2002-2003); Postdoctoral Scholar at the University of Iowa (2004-2005); Postdoctoral Scholar at Wayne State University (2005-2007); Assistant Professor of Computer Science at Lane College (2007-2009). During the years 2009-2012 Carlo is a self employed scholar whom devoted his time to his research interest related to the development of innovative methods of signal interpolation and also to the development of educational software for students. During the course of his career the research interests remain in the domain of mathematics in computational engineering: Artificial Neural Networks, Image Registration in fMRI, Signal-Image Interpolation, and MEG Alpha Rhythm. He has authored and co-authored numerous papers in journals and conference proceedings, and is the author of the books: (i) *Improved Signal and Image Interpolation in Biomedical Applications: The Case of Magnetic Resonance Imaging (MRI)*; (ii) *AUTOALIGN: Methodology and Technology for the Alignment of Functional Magnetic Resonance Imaging Time Series: Image Registration: The Case of Functional MRI*; (iii) *SIGNAL RESILIENT TO INTERPOLATION: An Exploration on the Approximation Properties of the Mathematical Functions*; and (iv) *Computer Science Signal Processing Applications in Higher Learning*.